NEB 15 – Recent Developments in Gravity 2012

Journal of Physics: Conference Series Volume 453

Chania, Greece
20-23 June 2012

ISBN: 978-1-62993-091-6
ISSN: 1742-6588

Printed from e-media with permission by:

Curran Associates, Inc.
57 Morehouse Lane
Red Hook, NY 12571

Some format issues inherent in the e-media version may also appear in this print version.

Printed by Curran Associates, Inc. (2012)

Institute of Physics
Dirac House, Temple Back
Bristol BS1 6BE UK

Phone: 44 1 17 929 7481
Fax: 44 1 17 920 0979

techtracking@iop.org

Additional copies of this publication are available from:

Curran Associates, Inc.
57 Morehouse Lane
Red Hook, NY 12571 USA
Phone: 845-758-0400
Fax: 845-758-2634
Email: curran@proceedings.com
Web: www.proceedings.com

TABLE OF CONTENTS

012001 A Newtonian Problem As an Insightful Tool for the Behavior of Gravitational-wave Sources 1
T A Apostolatos, G Pappas, K Chatziioannou
012002 Canonical Quantization of Static Spherically Symmetric Geometries 6
T Christodoulakis, N Dimakis, P A Terzis, G Doulis, Th Grammenos, E Melas, A Spanou
012003 The Cosmic Battery and the Inner Edge of the Accretion Disk 15
I Contopoulos, D B Papadopoulos
012004 The IWM–CFC Approximation in Differentially Rotating Relativistic Stars 25
P Iosif, N Stergioulas
012005 Mind the Resonances: Final Stages of Accretion Into Bumpy Black Holes 30
G Lukes-Gerakopoulos, G Contopoulos
012006 Acceleration in Weyl Geometry 40
J Miritzis
012007 Short-range Gravity Experiment Using Digital Image Analysis 47
K Ninomiya, R Kishi, H Murakami, H Nishio, N Ogawa, A Taketani, J Murata
012008 Future Stability for Reflection Symmetric Solutions of the Einstein-Vlasov System of Bianchi Type VI_0 57
E Nungesser
012009 Using Noether Symmetries to Specify F(R) Gravity 68
A Paliathanasis
012010 Local Interior Estimates for the Euler-Einstein System 84
T Papakostas, D Pliakis
012011 The Evolution of the F-mode Instability and Gravitational Wave Detection Prospectives 102
A Passamonti, E Gaertig, K D Kokkotas
012012 Testing General Relativity Using the Growth Rate of Structure 110
A Pouri, S Basilakos
012013 Asymptotic Quasi-normal Frequencies, Horizon Area Spectra and Multi-horizon Spacetimes 118
J Skákala
012014 Induced Gravity from Gauge Theories 124
R F Sobreiro, A A Tomaz, V J V Otoya
012015 Cosmological Constant and Vacuum Energy: Old and New Ideas 141
J Solà
012016 Constraints on the Nuclear Symmetry Energy Via Asteroseismology 189
H Sotani, K Nakazato, K Iida, K Oyamatsu
012017 Thermodynamic Phase Structure of Charged Anti-de Sitter Scalar-tensor Black Holes 195
I Z Stefanov, D D Doneva, S S Yazadjiev, K D Kokkotas, M D Todorov
012018 MTV-G Experiment : Probing a Non-standard Strong Gravitational Field at Nuclear Scale Using Geodetic Precession 203
S Tanaka, Y Nakaya, K Ninomiya, H Nishio, J Onishi, R Openshaw, M Pearson, Y Totsuka, J Murata
012019 Experimental Determination of Gravitomagnetic Effects by Means of Ring Lasers 210
A Tartaglia
012020 Geometrization of Lie and Noether Symmetries with Applications in Cosmology 219
M Tsamparlis
012021 Decoupling of Field Equations in Einstein and Modified Gravity 234
S I Vacaru
012022 Horava-Lifshitz Gravity with Detailed Balance 263
D Vernieri, T P Sotiriou
012023 Causal Sets Dynamics: Review & Outlook 267
P Wallden
012024 Radiation from Horizons, Chirality and the Principle of Effective Theory 279
A Zampeli
012025 Quasinormal Modes and Thermodynamical Aspects of the 3D Lifshitz Black Hole 286
B Cuadros-Melgar, J De Oliveira, C E Pellicer
Author Index

Preface

Taxiarchis Papakostas[1] and Demetrios Pliakis[2]
[1]Department of Natural Resources and Environment, Technological Educational Institute of Crete, Chania GR 73103
[2]Department of Electronics, Technological Educational Institute of Crete, Chania GR 73103

In June 20–23 2012 the Technological Institute of Crete in Chania hosted the organization of the conference 'Recent Developments in Gravity, NEB15', arranged biennially by the Greek Society for Relativity and Cosmology (http://www.hsrgc.gr). The talks ranged over Classical and Quantum Gravity and Cosmology, Holography, Black Hole Theory, Relativistic Astrophysics, Mathematical and Numerical Relativity. These proceedings include contributions from speakers of the conference.

T Papakostas and D Pliakis, as members of the local organizing committee, wish to express their gratitude to all the participants of the conference for the warm atmosphere they created.

Peer review statement

All papers published in this volume of *Journal of Physics: Conference Series* have been peer reviewed through processes administered by the proceedings Editors. Reviews were conducted by expert referees to the professional and scientific standards expected of a proceedings journal published by IOP Publishing.

NEB 15 – Recent Developments in Gravity IOP Publishing
Journal of Physics: Conference Series **453** (2013) 012001 doi:10.1088/1742-6596/453/1/012001

A Newtonian problem as an insightful tool for the behavior of gravitational-wave sources

Theocharis A. Apostolatos[1]†, George Pappas[1,2] and Katerina Chatziioannou[3]

[1] Section of Astrophysics, Astronomy, and Mechanics, Department of Physics, University of Athens, Panepistimiopolis Zografos GR15783, Athens, Greece
[2] Theoretical Astrophysics, IAAT, Eberhard Karls University of Tübingen, Tübingen 72076, Germany
[3] Department of Physics, Montana State University, Bozeman, MT 59717, USA.

E-mail: † thapostol@phys.uoa.gr

Abstract. We present the intriguing mathematical and physical similarities arising from the study of the Kerr metric and the Euler's problem of two fixed gravitational centers. We show how one could extend these similarities to physical problems that emerge from the above integrable problems, when both are slightly perturbed.

1. Introduction

There is an old problem of classical mechanics which is characterized by a lot of properties that are similar to those of an important astrophysical problem; that of a Kerr black hole. That problem is the so called Euler's problem of two fixed gravitational centers. Although the two problems do not seem to describe analogous physical configurations (moreover the two problems arise in a completely different theoretical framework), the similarities between the two problems are so deep that not only one can identify a number of quantitative and qualitative correspondences between the two, but one can use one of them to get insight about the other. Especially the newtonian problem could be used as a simple Newtonian analogue of a Kerr black hole in order to investigate the behavior of the latter one when this is somehow perturbed.

In Section 2 we give a short description of the Euler's problem. In Section 3 we present a list of all similarities between the two physical problems. In Section 4 we show how one could exploit this intriguing similarity between the two problems to get insight about one of them by performing numerical simulations with the other one.

2. The Euler's problem

The Newtonian gravitational field of two stationary gravitating point masses m_1, m_2 located at a fixed distance $2a$ from each other, one at $a\hat{\mathbf{z}}$ and the other at $-a\hat{\mathbf{z}}$ is given by

$$V(\mathbf{r}) = -\frac{Gm_1}{r_1} - \frac{Gm_2}{r_2} = -\frac{Gm_1}{|\mathbf{r} - a\hat{\mathbf{z}}|} - \frac{Gm_2}{|\mathbf{r} + a\hat{\mathbf{z}}|}, \tag{1}$$

where r_1, r_2 are the distances between the two masses and the point -located at $\mathbf{r}$- where the potential is measured. This problem is better analyzed in spheroidal coordinates:

$$\xi \equiv \frac{r_1 + r_2}{2a}, \eta \equiv \frac{r_1 - r_2}{2a}. \tag{2}$$

By construction this problem actually describes a prolate (along the z-axis) configuration, while the gravitational field of a Kerr black hole is oblate due to the black-hole's spin. Moreover if the two masses in the Euler's problem are not equal, the gravitational field is not reflection symmetric in contrary to what happens with a Kerr black hole. The above discrepancies between the two problems can be easily fixed by replacing a by an imaginary distance ia and setting $m_1 = m_2 = M/2$ in the Euler's problem. Although the replacement of a distance by an imaginary one cannot be actually realized by a physical nonsingular distribution of matter, the corresponding potential is perfectly real-valued (when $m_1 = m_2$) and can be easily studied.

The actual form of the potential in the latter case -which is the case that we are going to study as the Newtonian analogue of Kerr metric- assumes the following form in spherical and spheroidal coordinates:

$$V = -\frac{GM}{\sqrt{2}R^2}\sqrt{R^2 + r^2 - a^2} = -\frac{GM\xi}{a(\xi^2 + \eta^2)}, \tag{3}$$

where

$$R^2 = \sqrt{(r^2 - a^2)^2 + (2a\mathbf{r}\cdot\hat{\mathbf{z}})^2}. \tag{4}$$

The complicated form in spherical coordinates arises from the summation of two complex congugate quantities (cf. Eq. 1). It should be noted that the quantity under the square root is always positive as one could easily verify by a simple inspection of the form of R. It only becomes zero on the equatorial $\mathbf{r}\cdot\hat{\mathbf{z}} = 0$ circle $r = a$.

Henceforth, we will use geometrized units $G = c = 1$ for the Newtonian problem so that it is easier to compare with the relativistic one.

3. Similarities

(i) Both the Euler's problem (henceforth the Euler's problem will refer to the gravitational field of two equal masses at an imaginary distance $2ia$ apart, as it is described in Eq. (3)) and the Kerr black hole are gravitational fileds in which the equations of motion (either Newtonian or relativistic) of an orrbiting test body are integrable. This is a very important similarity that renders both problems seperable and thus easily solvable by the Hamilton-Jacobi technique.

(ii) Both problems describe oblate fields in which the geodesic orbits are precessing around the equatorial plane. Thus the spherical azimuthal angle θ of a point test particle performs oscillations around $\pi/2$.

(iii) Apart from the obvious symmetries (stationarity and axisymmetry) of the two problems that lead to the conservation of the energy E and the z-angular momentum L_z, there is an extra conserved quantity in both cases: The third integral b in the Euler's problem (see [1]) and the Carter constant Q for the Kerr metric. Both integrals of motion are quadratic with respect to momenta. However the analogy between the two quantities is not only restricted to general qualitative characteristics. The Carter constant can be written as a function of θ and θ-momentum (p_θ) as

$$Q = p_\theta^2 + \cos^2\theta\left[a^2(1 - E^2) + L_z^2/\sin^2\theta\right]. \tag{5}$$

If we make the following replacement of the spheroidal coordinate η of the Euler's problem $\eta \to \cos\theta$, then $p_\eta^2(1-\eta^2) \to p_\theta^2$ (the latter is a direct consequence of the former one), and the Newtonian approximation of the relativistic energy $E^2 \to 1+2E_N$ where E_N is the conserved Newtonian energy of the Euler's problem then the third integral b assumes exactly the same form as Q (apart from some additive terms related to the rest conserved quantities):

$$b = -Q - L_z^2 - 2a^2 E_N. \tag{6}$$

The same expression connecting b and Q, could be shown to hold true also when the two conserved quantities are written as functions of r and p_r (for Q), and $r = \xi a$ and $p_r = p_\xi / a$ (for b respectively). The only difference here is that besides the correspondence between E_N and E mentioned above suitable approximations of various functions of r should be made.

(iv) Will showed [2] that the only axisymmetric and reflection symmetric Newtonian potential that leads to a conserved quantity which is bilinear with respect to momenta is the original Euler's problem (the one with a real distance a) that has exactly the same spectrum of mass multipole moments as the Kerr one apart from a sign; namely two consecutive non-vanishing mass moments of the initial Euler's problem satisfy the equation

$$M_{2l} = a^2 M_{2l-2}, \tag{7}$$

with $M_0 = M$. (Due to reflection symmetry only the even multipoles are non-zero.) If a is replaced by ia in the original Euler's problem one gets back the exact multipolar structure of a Kerr black hole; that is

$$M_{2l} = -a^2 M_{2l-2}{}^{1} \tag{8}$$

with $M_0 = M$. This particular similarity manifests more clearly the physical similarity between the two fields.

(v) Usually a Newtonian gravitational potential (like the monopole one) does not have something analogous to an innermost stable circular orbit (ISCO). However the Euler's problem actually possesses an ISCO. The radius of ISCO for the Euler's problem is

$$\xi_{\rm ISCO} = \sqrt{3}. \tag{9}$$

It is rather clumsy to compare the corresponding radius $r_{\rm ISCO} = a\xi_{\rm ISCO}$ with the a dependence of the radius of ISCO in Kerr, since there are two such radii; one for retrograde orbits and one for prograde ones. Here we simply note the existence of ISCO in the Newtonian problem as an analogy with orbits in Kerr.

(vi) Since both problems (the Newtonian and the relativistic) are integrable all bound orbits are characterized by 3 fundamental frequencies related to the r-, θ-oscillations and ϕ-libration. Actually these are the frequencies that one could measure through gravitational waves emitted from small objects orbiting around massive Kerr black holes. The behavior of these frequencies, though not analytically equal between the two problems, they show some analogies. For example for both problems $\omega_r/\omega_\theta \to 1$ for large radii. Also for a large range of parameters (a, and inclination angle, ι, of the orbit) the ratios of frequencies have qualitatively similar behavior as a function of orbital semi-latus rectum. Finally the resonance condition $\omega_r/\omega_\theta = 2$ was never met in all cases we examined for both cases. The avoidance of such a resonance is quite essential to keep orbits that start almost circular ($r =$ const) to remain circular after an adiabatic evolution of the orbit (see [3]).

[1] Here the comparison is made only between the mass moments of the two gravitational fields. The analogue of the current-mass moments in relativity does not affect the gravitational field in Newtonian gravity.

(vii) In a recent paper [4] it has been shown that the wave equation in a Newtonian gravitational potential becomes separable only for the monopole potential and the Euler problem (either the prolate or the oblate one). It is quite interesting that the oblate Euler' problem leads to a set of differential equations that resembles a lot the corresponding wave equations that describe propagation of scalar waves in a Kerr background. To be more specific the azimuthal part is exactly the same for the two problems, while the radial part, though not analytically equal, they both have similar properties especially for some particular values of the parameters of the Newtonian wave equation.

4. Exploiting the analogies

The most interesting similarity though between the two problems is that when their integrability is slightly destroyed (for example by constructing exact solutions of vaccuum Einstein equations that are parametrized by a single parameter which measures deviations from a Kerr metric, and adding a small point mass at the origin in the Euler's problem) both problems demonstrate very similar behavior. This is a generic outcome of the KAM theorem for all slightly non-integrable problems. Thus according to that theorem in the Poincaré sections of such problems chains of Birkhoff islands form near resonances (when the ratio of fundamental frequencies ω_r/ω_θ is rational). The idea then is to exploit such similarities to check how a perturbed Kerr metric is expected to behave under gravitational radiation evolution by studying the adiabatic evolution of an orbit in the perturbed Euler's problem. In the Newtonian case it is easy to assume a suitably small dissipative force acting on the particle and study the behavior of the orbit when it passes a resonance. Is the corresponding crossing slower or faster than what one yields when the orbit is evolved by means of the average energy and angular momentum loss due to the force, where the average losses are computed outside the resonance? This is a very difficult problem to address in relativity since the radiation self-force is quite difficult to compute even in a pure Kerr background. In a generic perturbed Kerr background the computation of self-force seems rather intractable.

The first numerical runs show that the process of crossing a resonance island is actually rather slower than what one obtains from averaging the losses. If this is a general outcome, then the plateau effect suggested in [5, 6] might be further enhanced in perturbed Kerr metrics. Therefore by seeking such plateaus in gravitational-wave-data analysis, one could pursue exotic non-Kerr objects. More details will be presented in a future paper [7]

Acknowledgments

We would like to thank Kostas Kokkotas for the hospitality at the University of Tübingen. This work has been supported by the I.K.Y. (IKYDA 2010) and G.P. would also like to acknowledge DAAD scholarship number A/12/71258.

References

[1] Landau L. D and Lifshitz E. M., 1976, Mechanics (Third Edition), Butterworth-Heinemann.
[2] Will C., 2010, *Phys. Rev. Lett.* **102**, 061101
[3] Kennefick D. and Ori A., 1996, *Phys. Rev. D* **53**, 4319
[4] Glampedakis K. and Apostolatos T. A., submitted in *Class. & Quant. Gravity*
[5] Apostolatos T. A., Lukes-Gerakopoulos G. and Contopoulos G., 2009, *Phys. Rev. Lett.* **103**, 111101
[6] Lukes-Gerakopoulos G., Apostolatos T. A. and Contopoulos G., 2010, *Phys. Rev. D* **81**, 124005
[7] T.A. Apostolatos, G. Pappas and K. Chatziioannou, paper in preparation

Journal of Physics: Conference Series **453** (2013) 012001 doi:10.1088/1742-6596/453/1/012001

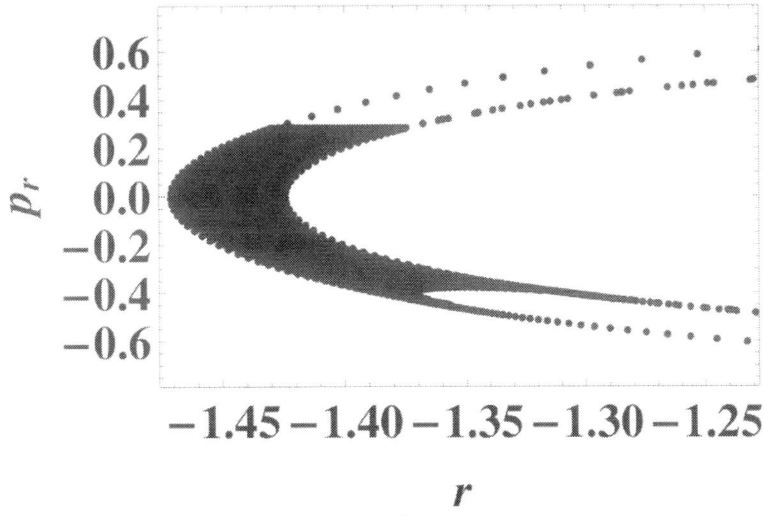

Figure 1. The Euler's problem with an added small mass at the origin was transformed to a slightly perturbed integrable problem that could be used to mimic a perturbed Kerr black hole. Here is plotted the Poincaré section $\theta = \pi/2$ of a phase orbit that hits the resonance of $2:3$, due to adiabatic inspiral caused by an artificial dissipative force. The orbit spends a much longer time inside the resonance island, compared to the time one would expect based on the average loss of energy and angular momentum when the orbit was still outside the resonance region.

Canonical quantization of static spherically symmetric geometries

T Christodoulakis[1], N Dimakis[2], P A Terzis[3], G Doulis[4], Th Grammenos[5], E Melas[6] and A Spanou[7]

[1,2,3] Nuclear and Particle Physics Section, Physics Department, University of Athens, GR 157–71 Athens
[4] Department of Mathematics and Statistics, University of Otago, P.O. Box 56, Dunedin 9010, New Zealand
[5] Department of Civil Engineering, University of Thessaly, GR 383–34 Volos
[6] Logistics Department, GR 32-200, Thiva, Technological Educational Institution of Chalkida
[7] School of Applied Mathematics and Physical Sciences, National Technical University of Athens, GR 157–80, Athens

E-mail: [1]tchris@phys.uoa.gr, [2]nsdimakis@gmail.com, [3]pterzis@phys.uoa.gr, [4]gdoulis@phys.uoa.gr, [5]thgramme@civ.uth.gr, [6]evangelosmelas@yahoo.co.uk, [7]aspanou@central.ntua.gr

Abstract. The conditional symmetries of the reduced Einstein–Hilbert action emerging from a static, spherically symmetric geometry are used as supplementary conditions on the wave function. Based on their integrability conditions, only one of the three existing symmetries can be consistently imposed, while the unique Casimir invariant, being the product of the remaining two symmetries, is calculated as the only possible second condition on the wave function. This quadratic integral of motion is identified with the reparametrization generator, as an implication of the uniqueness of the dynamical evolution, by fixing a suitable parametrization of the r-lapse function. In this parametrization, the determinant of the supermetric plays the role of the mesure. The combined Wheeler - DeWitt and linear conditional symmetry equations are analytically solved. The solutions obtained depend on the product of the two "scale factors".

1. Introduction

A vast amount of work on the use of symmetries in the quantization of minisuperspace cosmological models fills the relevant literature of the last fifteen years. In fact, the application of the variational symmetries approach to classical Bianchi cosmologies has started in [1]. For the case of classical and quantum cosmology, in either Bianchi or higher derivative models, the first relevant works are [2], [3], [4], [5] and [6] (where conditional symmetries are used), while a revival of recent work on the subject can be found in [7], [8], [9] and [10]. Essentially, the method consists in the application of the standard theory of variational symmetries [11], [12] to the Lagrangian of some minisuperspace model. The steps of the usual procedure are (i) gauge fixing the lapse function to some convenient value, (ii) applying the first prolongation in the velocity phase space of a vector in the configuration space to this gauge fixed Lagrangian,

Journal of Physics: Conference Series **453** (2013) 012002 doi:10.1088/1742-6596/453/1/012002

and (iii) demanding its action to be zero. A slightly different approach exploiting the notions of the special projective and/or the homothetic group is adopted in [13], [14] and [15].

In the present work, we do not gauge fix the lapse function. This is justified by the fact that the full content of the corresponding reparametrization generator, i.e., the quadratic constraint, is kept intact by the presence of the lapse. This fact is important not only for the symmetries but also in the solution of the classical equations of motion, a task that is greatly facilitated by inserting the algebraic solution for N from the quadratic constraint into the other equations of motion [16], [17], [18] and [19]. In [20], a phase space point of view is adopted and a conditional symmetry is defined [21] as a conformal field of both the supermetric and the potential. Here, it is proven that this definition leads to the uncovering of all variational symmetries.

An application of the method to the case of static, spherically symmetric geometries is presented in this work. A minisuperspace Lagrangian for the aforesaid set of metrics has been given in [22] and [23], where a $3+1$ decomposition is introduced along the radial coordinate r which is taken to be the dynamical variable. In [24] and [25], the symmetries of the reduced Schwarzschild action have been used in the quantization procedure. Here, the quantum analogues of the linear integrals of motion are also adopted as supplementary conditions imposed on the wave function. A careful examination of their role and integrability conditions leads to the unique Casimir invariant of their algebra. Through a particular redefinition of the lapse function, the Casimir invariant is identified to the corresponding r - reparametrization generator.

The paper is organized as follows: In section 2 the reduced valid Lagrangian reproducing Einstein's field equations for static, spherically symmetric configurations is given, and passing to its Hamiltonian formulation [26], the three conditional symmetries are revealed. In section 3, the dynamical system is quantized by use of Dirac's canonical quantization procedure for constrained systems [27]. Finally, some concluding remarks are presented in the discussion.

2. Hamiltonian formulation of static, spherically symmetric geometries

Our starting point is the static, spherically symmetric line element

$$ds^2 = -a(r)^2 dt^2 + n(r)^2 dr^2 + b(r)^2 (d\theta^2 + \sin^2\theta\, d\phi^2). \tag{2.1}$$

In the usual ADM $3+1$ decomposition, one foliates space-time in t-hypersurfaces and the coefficient of dt^2 is the lapse function. Here, we adopt a $3+1$ foliation in the r coordinate and, therefore, the role of the lapse is attributed to the coefficient of dr^2 in the line element. Thus, we consider $n(r)$ in (2.1) to be the r-lapse function while $a(r)$ and $b(r)$ are the dependent "dynamical" variables on the r-hypersurface. Previous authors use also a "shift" term $2B(r)\,dr\,dt$. However, this is not relevant for the following discussion, as $B(r)$ does not enter the Einstein tensor while it can be absorbed by a time redefinition of the form $t = \tilde{t} + \int \frac{B(r)}{a(r)^2} dr$.

The Einstein - Hilbert action $A_{E-H} = \int \sqrt{-g}\, R\, d^4x$ for the geometries (2.1) leads to the reduced action $\mathcal{A} = \int L dr$ with the following Lagrange function $L(a, b, a', b', n)$:

$$L = 2\,a\,n + \frac{4\,b\,a'\,b'}{n} + \frac{2\,a\,b'^2}{n}, \tag{2.2}$$

where $'$ denotes differentiation with respect to the spatial coordinate r. It is easy to verify that the Euler - Lagrange equations obtained from (2.2) are identical to Einstein's equations $\mathcal{G}_{\mu\nu} = 0$ for the line element (2.1).

In order to proceed with the Hamiltonian formalism we calculate the conjugate momenta,

$$\begin{aligned}
\pi_n &= \frac{\partial \mathcal{L}}{\partial n'} = 0, \\
\pi_a &= \frac{\partial \mathcal{L}}{\partial a'} = \frac{4b\, b'}{n}, \\
\pi_b &= \frac{\partial \mathcal{L}}{\partial b'} = \frac{4b\, a'}{n} + \frac{4\, a\, b'}{n}.
\end{aligned} \tag{2.3}$$

Obviously, π_n is a primary constraint. The Legendre transformation leads to the Hamiltonian

$$H = n\,\mathcal{H}_c,$$

where

$$\mathcal{H}_c = -2\,a - \frac{a\,\pi_a^2}{8\,b^2} + \frac{\pi_a\,\pi_b}{4\,b}. \tag{2.4}$$

The preservation of the primary constraint π_n in the r−evolution, i.e.

$$\pi_n' = \{\pi_n, H\} \approx 0,$$

leads to the secondary constraint

$$\mathcal{H}_c \approx 0. \tag{2.5}$$

The minisuperspace metric inferred from (2.4) is

$$G^{\alpha\beta} = \begin{pmatrix} -\frac{a}{4b^2} & \frac{1}{4b} \\ \frac{1}{4b} & 0 \end{pmatrix}. \tag{2.6}$$

According to our definition of a conditional symmetry [20], as a vector ξ on the configuration space, the following relations must hold

$$\mathcal{L}_\xi G^{\alpha\beta} = \phi(q) G^{\alpha\beta} \quad , \quad \mathcal{L}_\xi V(q) = \phi(q) V(q) \tag{2.7}$$

where q's are the configuration space variables. Thus, we are led to the following three conformal Killing fields of both $G_{\alpha\beta}$ and the potential $V = -2a$

$$\xi_1 = (-a, b), \quad \xi_2 = \left(\frac{1}{a\,b}, 0\right), \quad \xi_3 = \left(-\frac{a}{2\,b}, 1\right) \tag{2.8}$$

which, contracted with (π_a, π_b), provide us with the three integrals of motion

$$Q_1 = -a\,\pi_a + b\,\pi_b, \quad Q_2 = \frac{\pi_a}{a\,b}, \quad Q_3 = -\frac{a\,\pi_a}{2\,b} + \pi_b. \tag{2.9}$$

We calculate the Poisson brackets of these conserved quantities with the canonical Hamiltonian H and the Poisson algebra that they satisfy

$$\{Q_1, H\} = n\mathcal{H}_c, \quad \{Q_2, H\} = -\frac{n}{a^2\,b}\mathcal{H}_c, \quad \{Q_3, H\} = \frac{n}{2\,b}\mathcal{H}_c, \tag{2.10a}$$

$$\{Q_1, Q_3\} = Q_3, \quad \{Q_2, Q_1\} = Q_2, \quad \{Q_3, Q_2\} = 0. \tag{2.10b}$$

As expected, the Poisson brackets (2.10a) are weakly vanishing on the constraint surface $\mathcal{H}_c \approx 0$ and, therefore, the three Q_I's are constants of motion.

At this point, it is interesting and useful for what follows in the quantization, to adopt a new parametrization of the lapse $n(r) = \frac{\bar{n}(r)}{2a(r)}$ which makes the potential constant (see [20]). The Lagrangian and the corresponding Hamiltonian are now given by

$$\bar{L} = \bar{n} + \frac{8\,a\,b\,a'\,b'}{\bar{n}} + \frac{4\,a^2\,b'^2}{\bar{n}} \quad , \qquad \bar{H} = \bar{n}\bar{\mathcal{H}}_c = \bar{n}\frac{1}{2a}\mathcal{H}_c. \tag{2.11}$$

If the value of $\bar{n}$ specified by the constrained equation is substituted into the Euler - Lagrange equations for $a(r)$ and $b(r)$ the system can be solved for only one acceleration, say $a''(r)$, and the general solution of the entire system is

$$\bar{n}(r) = 2\,c\,b'(r) \quad , \quad a(r) = c\sqrt{1 - \frac{2M}{b(r)}}. \tag{2.12}$$

The constants of integration have been rearranged so that the ensuing line element

$$ds^2 = -c^2\left(1 - \frac{2M}{b(r)}\right)dt^2 + \left(1 - \frac{2M}{b(r)}\right)^{-1} b'(r)^2 dr^2 + b(r)^2(d\theta^2 + \sin^2\theta\, d\phi^2) \tag{2.13}$$

bears the closest possible resemblance to the standard form of the Schwarzschild metric, while the presence of the arbitrary function $b(r)$ reflects the r-reparametrisation covariance of the system.

The new supermetric is now given by

$$\bar{G}^{\alpha\beta} = \begin{pmatrix} -\frac{1}{8b^2} & \frac{1}{8a\,b} \\ \frac{1}{8a\,b} & 0 \end{pmatrix}, \tag{2.14}$$

and the fields (2.8) are now turned into Killing fields of $\bar{G}^{\alpha\beta}$; they are also, trivially, symmetries of the constant potential $\bar{V} = 1$ as well. The algebra satisfied by the new quantities $\bar{Q}_I = \xi_I^\alpha \bar{\pi}_\alpha$ and $\bar{\mathcal{H}}_c$ can be easily seen to be

$$\{\bar{Q}_1, \bar{\mathcal{H}}_c\} = 0, \quad \{\bar{Q}_2, \bar{\mathcal{H}}_c\} = 0, \quad \{\bar{Q}_3, \bar{\mathcal{H}}_c\} = 0. \tag{2.15}$$

Therefore, the quantities $\bar{Q}_I$ still remain constants of motion. It is in this conformal gauge that one can explicitly verify that the prolonged ξ_I's and X_I, are satisfying a form of the standard condition of variational symmetries $pr^{(1)}X_I(\bar{L}) = 0$, thus ensuring that they are generators of Noether symmetries. It is important to note that if one had inserted the X_I's into the full relation $pr^{(1)}X(L) + L\mathrm{Div}\Xi = \mathrm{Div}F$ with L given by (2.2), the result would be negative.

If we invert the definitions for $\bar{\pi}_\alpha$ and use (2.12) in $\bar{Q}_I$ we calculate the integrals' values on the solution space

$$Q_1 = \bar{Q}_1 = 4\,c\,M \quad , \quad Q_2 = \bar{Q}_2 = \frac{4}{c} \quad , \quad Q_3 = \bar{Q}_3 = 2\,c. \tag{2.16}$$

As it is known, the Schwarzschild solution involves only one essential constant, the mass M. The second constant appearing in the solution (2.12), (2.13) can be seen to be absorbable by a scaling of the time coordinate $t \to \frac{t}{c}$, allowing to set $c = 1$ but not $c = 0$. So, on the solution

Journal of Physics: Conference Series **453** (2013) 012002 doi:10.1088/1742-6596/453/1/012002

space, we can set $\bar{Q}_1 = 4M$, $\bar{Q}_2 = 4$, $\bar{Q}_3 = 2$. It is noteworthy that the values of the last two integrals of motion can be changed at will and that they do not involve the essential parameter M which characterizes the geometry. The above argument relies on the underlying geometry.

But what if we were deprived of the line element and we were just given the dynamical system (2.11)? How could we differentiate between the constants κ_1 and κ_2, κ_3? Interestingly enough, there is an argument that leads to a distinction between them: The crucial observation is that $\bar{Q}_1$ has a vanishing Poisson bracket with each kinetic term of $\bar{H}$, while $\bar{Q}_2$ and $\bar{Q}_3$ need the entire kinetic term in order to produce vanishing Poisson brackets. This fact is reflected in the following property concerning the one-parameter family of canonical transformations generated by the charge $\bar{Q}_1$ (see [29] and [30] for the generalization of Noether symmetries for constrained systems and for Noether's theorem in phase space),

$$a \to e^{\lambda} a,\ \bar{\pi}_a \to e^{-\lambda} \bar{\pi}_a,\ b \to e^{-\lambda} b,\ \bar{\pi}_b \to e^{\lambda} \bar{\pi}_b. \tag{2.17}$$

Under such transformations, $\bar{Q}_1$ remains, of course, unchanged, while $\bar{Q}_2$ and $\bar{Q}_3$ are scaled by $e^{-\lambda}$ and e^{λ}, respectively. One can, thus, use the freedom of λ to arbitrarily change the values κ_2 and κ_3, but not κ_1. Furthermore, the Hamiltonian $\bar{\mathcal{H}}_c$ remains, due to the particular scaling of $\bar{Q}_2$ and $\bar{Q}_3$, unchanged.

If we return to the phase space, we can write

$$\bar{Q}_{23} = \bar{Q}_2\,\bar{Q}_3 = -\frac{1}{2\,b^2}\bar{\pi}_\alpha^2 + \frac{1}{a\,b}\,\bar{\pi}_\alpha\,\bar{\pi}_\beta \tag{2.18}$$

with the relevant Poisson bracket algebra now becoming

$$\{\bar{Q}_{23}, \bar{\mathcal{H}}_c\} = 0, \tag{2.19a}$$

$$\{\bar{Q}_I, \bar{Q}_{23}\} = 0, \qquad I = 1, 2, 3. \tag{2.19b}$$

As expected $\bar{Q}_{23}$ is a quadratic integral of motion. From a group theoretical view, $\bar{Q}_{23}$ is an element of the centre of the universal enveloping algebra (uea) generated by $\bar{Q}_I$'s, i.e., it is the Casimir invariant. The Hamiltonian $\bar{\mathcal{H}}_c$ belongs also to the centre of uea (it commutes with all $\bar{Q}_I$'s), thus, it can only differ from $\bar{Q}_{23}$ by an additive and/or a multiplicative constant. It is an easy matter for one to check that indeed

$$\bar{Q}_{23} = 8\,(\bar{\mathcal{H}}_c + 1). \tag{2.20}$$

To sum up, we have constructed a gauge independent, quadratic in the momenta, integral of motion which commutes with the only integral of motion that entangles the sole essential constant of the Schwarzschild solution. In the next section, and in order to proceed with the quantization, we will rely on these two quantities.

3. Quantization

In order to quantize our system, we must turn into operators $\bar{\mathcal{H}}_c = \frac{1}{8}\,(\bar{Q}_{23} - 8)$, $\bar{Q}_1$, $\bar{Q}_2$ and $\bar{Q}_3$ (hereafter, for the shake of simplicity, we will omit the *bars* from the symbols of the corresponding operators). The corresponding quantum operators can be inferred from the general definitions

$$\widehat{\mathcal{H}}_c := -\frac{1}{2\mu}\partial_\alpha(\mu\,G^{\alpha\beta}\partial_\beta) + V(q), \tag{3.1}$$

$$\widehat{Q}_I := -\frac{\mathrm{i}}{2\mu}(\mu\,\xi_I^\alpha\,\partial_\alpha + \partial_\alpha\mu\,\xi_I^\alpha), \tag{3.2}$$

where μ is a suitable measure which transforms as a scalar density. Definition (3.2) supplies us with a Hermitian operator [28] for the previously mentioned conditional symmetries (2.8). It can be seen [20] that, the measure entering the quantum operators ought to be taken as $\mu(a,b) = \lambda\sqrt{det|\bar{G}_{\alpha\beta}|} \propto a\,b$. The constancy of λ is forced by the combined requirement that the Q_I's must be realized as Hermitian operators and, at the same time, retain their classical geometrical character by acting as derivatives. Thus, the extra term $\xi_I^\alpha \partial_\alpha \lambda$ must vanish for all $I = 1,2,3$, which leads to a constant λ. Two further arguments in favor of this choice of measure are:

- The fact that the quantum analogue of the algebra (2.19b) is made isomorphic to the classical, i.e.,
$$[\widehat{Q}_I, \widehat{Q}_{23}]F(a,b) = 0, \quad I = 1,2,3 \quad \text{for any } F(a,b), \tag{3.3}$$
a fact that is highly non trivial, since it depends on the choice of both the factor ordering and the measure.
- At the classical level, the only linear integral of motion involving the essential constant is Q_1. If we seek the functions on the configuration space which are invariant under the point transformations generated by Q_1 we find $\{Q_1, f(a,b)\} = 0 \Rightarrow f(a,b) = f(a\,b)$.

The above arguments lead to the following linear operators corresponding to the elements of the classical algebra, the Casimir invariant and the Hamiltonian

$$\widehat{Q}_1 = -\mathbb{i}\,(b\,\partial_b - a\,\partial_a)\,, \tag{3.4}$$
$$\widehat{Q}_2 = -\frac{\mathbb{i}}{a\,b}\partial_a, \tag{3.5}$$
$$\widehat{Q}_3 = -\mathbb{i}\left(\partial_b - \frac{a}{2\,b}\,\partial_a\right), \tag{3.6}$$
$$\widehat{Q}_{23} = \frac{2}{b^2}\,\partial_a\partial_a - \frac{1}{a\,b}\,\partial_a\partial_b + \frac{1}{2a\,b^2}\,\partial_b, \tag{3.7}$$
$$\widehat{\mathcal{H}}_c = \frac{1}{8}\left(\widehat{Q}_{23} - 8\right). \tag{3.8}$$

It is an easy task to check that these operators satisfy not only the relations

$$[\widehat{Q}_I, \widehat{Q}_J]F(a,b) = \mathbb{i}\,C^K{}_{IJ}\widehat{Q}_K F(a,b) \tag{3.9}$$

for any test function $F(a,b)$, but also (3.3) as well.

The integrability condition for the equations emerging from the enforcement of the $\widehat{Q_I}$'s upon the wave function

$$\widehat{Q}_I\Psi = \kappa_I\Psi \tag{3.10}$$

leads to the selection rule

$$C^M_{IJ}\kappa_M = 0. \tag{3.11}$$

Due to the constraint condition (3.11), applied to the specific structure constants inferred from (2.10b), we conclude that only the eigenvalue κ_1 is free while κ_2 and κ_3 must necessarily be zero. However, the latter is impossible since on the classical solution space (2.16) hold. This means that not all three operators can be applied simultaneously. The latter results in the need to consider the two- and/or one- dimensional subalgebras together with the Wheeler - DeWitt equation $\widehat{\mathcal{H}}_c\Psi = 0$. The investigation of these cases can be easily carried out.

As far as the $2d$ subalgebras are concerned, the results obtained are briefly the following:

(a) For the two non Abelian subgroups either κ_2 or κ_3 are forced to be zero, something that is inconsistent with their classical values.

(b) For the Abelian subgroup, the two linear equations lead to the solutions $\Psi(a,b) = A\exp\left(\frac{\mathbb{i}}{2}(\kappa_2\, a^2\, b + 2\kappa_3\, b)\right)$, where A is constant and the quadratic constraint enforces the restriction $\kappa_2\,\kappa_3 = 8$. It is of course doubtful, if one can accept such a wave function to represent the geometry, knowing that it does not contain the essential constant M. However, one could interpret it as plane waves representing the limiting flat space-time $M = 0$.

The one-dimensional subalgebras spanned by $\widehat{Q}_2$ and $\widehat{Q}_3$ give solutions which are special cases of the solution described in (b), as expected since they commute.

Consequently, the only possibility is to adopt $\widehat{\mathcal{H}}_c$ and $\widehat{Q}_1$ as conditions on the wave function. This is indeed possible, since they commute with each other and, therefore, can be considered as physical quantities on the phase space that can be measured "simultaneously" (our dynamical parameter is the distance r). The ensuing eigenvalue equations are

$$\widehat{\widehat{\mathcal{H}}}_c\Psi = 0 \Rightarrow a\,\partial_a\partial_a\,\Psi - 2\,b\,\partial_a\partial_b\,\Psi + \partial_a\,\Psi - 16\,a\,b^2\,\Psi = 0, \tag{3.12a}$$

$$\widehat{Q}_1\Psi = \kappa_1\,\Psi \Rightarrow \mathbb{i}(-a\,\partial_a\,\Psi + b\,\partial_b\,\Psi) = \kappa_1\,\Psi. \tag{3.12b}$$

The solution of the linear partial differential equation (3.12b) is

$$\Psi(a,b) = a^{\mathbb{i}\kappa_1}\,S(a\,b). \tag{3.13}$$

If we insert the above solution into the Hamiltonian constraint (3.12a), we arrive at the following ordinary differential equation for $S(u)$ $(u = a\,b)$

$$u^2\,S''(u) + u\,S'(u) + \left(\kappa_1^2 + 16\,u^2\right)\,S(u) = 0 \tag{3.14}$$

which has the general solution

$$S(u) = c_1\,J_{\mathbb{i}\kappa_1}(4\,u) + c_2\,Y_{\mathbb{i}\kappa_1}(4\,u), \tag{3.15}$$

in terms of the Bessel functions of imaginary order.

In order to gain some insight on the normalizability of the formal probability, instead of these Bessel functions and because of their imaginary order, we can use the functions $F_{\mathbb{i}\kappa_1}(4\,u)$ and $G_{\mathbb{i}\kappa_1}(4\,u)$ defined in [31] through the Hankel functions $H^{(1)}_\mu(u) = J_\mu(u) + \mathbb{i}\,Y_\mu(u)$ and $H^{(2)}_\mu(u) = J_\mu(u) - \mathbb{i}\,Y_\mu(u)$, $\mu \in \mathbb{C}$. Thus, the solution can be written as

$$S(u) = c_1\,F_{\mathbb{i}\kappa_1}(4\,u) + c_2\,G_{\mathbb{i}\kappa_1}(4\,u) \tag{3.16}$$

with

$$F_{\mathbb{i}\kappa_1}(4\,u) = \frac{1}{2}\left(e^{-\kappa_1\pi/2}\,H^{(1)}_{\mathbb{i}\kappa_1}(4u) + e^{\kappa_1\pi/2}\,H^{(2)}_{\mathbb{i}\kappa_1}(4u)\right), \tag{3.17}$$

$$G_{\mathbb{i}\kappa_1}(4\,u) = \frac{1}{2\mathbb{i}}\left(e^{-\kappa_1\pi/2}\,H^{(1)}_{\mathbb{i}\kappa_1}(4u) - e^{\kappa_1\pi/2}\,H^{(2)}_{\mathbb{i}\kappa_1}(4u)\right). \tag{3.18}$$

These functions are linearly independent solutions of (3.14) and have the following properties: a) when $u \in (0,+\infty)$ they are real, b) they are oscillatory with a phase difference of $\frac{\pi}{2}$ and c) when both u, and/or κ_1 tend to zero, $F_{\frac{1}{2}\kappa_1}(4u)$ tends to 1, while $G_{\frac{1}{2}\kappa_1}(4u)$ becomes infinite. The final form of the wave function $\Psi(a,b)$ is

$$\Psi(a,b) = a^{\frac{1}{2}\kappa_1}\, S(a\,b), \tag{3.19}$$

so we can define a probability density of the form

$$\mu(a\,b)\,\Psi^*(a,b)\,\Psi(a,b) \propto u\,S^*(u)\,S(u). \tag{3.20}$$

4. Discussion

In this paper we used a method for the quantizations of *minisuperspace* actions which can be summarized in the following steps:

(i) Go over to the Hamiltonian $\bar{H}$.

(ii) Calculate the Noether symmetries as Killing fields of the metric $\bar{G}^{\alpha\beta}(q)$ and, trivially, symmetries of the constant potential.

(iii) Identify the essential constants of the metric. Promote the allowed Q_I's to operators according to (3.2) with $\mu = |\bar{G}|^{1/2}$.

(iv) Promote the Hamiltonian constrain $\bar{\mathcal{H}}_c$ (which is a linear function of the Casimir invariant) via (3.1) to a Hermitian operator acting on the wave function.

We applied the above method for the case of static, spherically symmetric geometries. First, we begin from the Lagrangian (2.2) emanating from the line element (2.1). We find the simultaneous conformal Killing fields (2.8) of the supermetric and the potential, which define the three conserved charges (2.9). The unique Casimir invariant of their algebra is Q_{23}. In order to make it numerically proportional to the kinetic part of the Hamiltonian we are led to (2.11).

Our specific example (described in sections 2 and 3) has also been the subject of [24]. In that work, Vakili finds two of the Q_I's but then he uses a linear combination of them in order to reproduce the essential constant M of the Schwarzchild metric. He thus reaches to the unique acceptable linear quantum operator equivalent to our $\widehat{Q}_1$. The clever choice of the lapse function, his equation (4), along with the somewhat unorthodox choice of factor ordering for the operators (see below his equation (55)) leads essentially to a constant potential (his equation (53)) and to the Laplacian operator (his equation (56)). As a result, the solution spaces found both by us and Vakili essentially coincide. Of course, our general theory given in [20] constitutes a systematic explanation of the various choices of his work.

5. References

[1] Capozziello S, Marmo G, Rubano C and Scudellaro P 1997 *Int. J. Mod. Phys.* D **6**, 4 491-503.
[2] Capozziello S and Lambiase G 2000 *Gen. Rel. Grav.* **32** 295-311
[3] Capozziello S and Lambiase G 2000 *Gen. Rel. Grav.* **32** 673-96
[4] Cotsakis S, Leach P G L and Pantazi H 1998 *Grav. Cosmol.* **4** 314-25
[5] Leach P G L, Cotsakis S and Miritzis J 2001 *Grav. Cosmol.* **7** 311-20
[6] Christodoulakis T, Gakis T and Papadopoulos G O 2002 *Class. Quantum Grav.* **19** 101325
[7] Vakili B and Khazaie F, 2012 *Class. Quantum Grav.* **29** 035015
[8] Falciano F T, Pereira R, Pinto-Neto N and Santini E S 2012 *Phys. Rev.* D **86** 063504
[9] Capozziello S, De Laurentis M and Odintsov S D 2012 *Preprint* gr-qc/1206.4842v1
[10] Sarkar K, Nayem Sk, Debnath S, and Sanyal A K 2012 *Preprint* astro-ph.CO/1207.3219v1
[11] Olver P J 2000, *Applications of Lie Groups to Differential Equations* 2nd ed. (Berlin, Heidelberg, New York: Springer - Verlag)
[12] Olver P J 1995 *Equivalence, Invariants and Symmetry* (Cambridge: Cambridge University Press) p 237

[13] Tsamparlis M and Paliathanasis A 2010 *Gen. Rel. Grav.* **42** 2957-80
[14] Tsamparlis M and Paliathanasis A 2011 *J. Phys.* A **A44** 175202
[15] Tsamparlis M and Paliathanasis A 2011 *Gen. Rel. Grav.* **43** 1861-81
[16] Christodoulakis T and Terzis P A 2006 *J. Math. Phys.* **47** 102502
[17] Christodoulakis T and Terzis P A 2007 *Class. Quantum Grav.* **24** 87587
[18] Terzis P A and Christodoulakis T 2009 *Gen. Rel. Grav.* **41** 46995
[19] Terzis P A and Christodoulakis T 2010 *Preprint* gr-qc/1007.1561
[20] Christodoulakis T, Dimakis N, Terzis P A, Doulis G, Grammenos Th, Melas E and Spanou A 2012 *Preprint* gr-qc/1208.0462
[21] Kŭchar K V 1982 *J. Math. Phys.* **23** 1647-61
[22] Cavaglià M, de Alfaro V, and Filippov A T 1995 *Int. J. Mod. Phys.* D **4** 661-72
[23] Cavaglià M, de Alfaro V and Filippov A T 1996 *Int. J. Mod. Phys.* D **5** 227-50
[24] Vakili B 2012 *Int. J. Theor. Phys.* **51** 1 133-45
[25] Jalalzadeh S, Vakili B 2012 *Int. J. Theor. Phys.* **51** 1 263-75
[26] Sundermeyer K 1982 *Constrained Dynamics* (Berlin, Heidelberg, New York: Springer - Verlag)
[27] Dirac P A M 1964 *Lectures on Quantum Mechanics* (New York: Yeshiva University, Academic Press)
[28] Christodoulakis T and Korfiatis E 1991 *Phys. Let.* B **256** 484-90
[29] Garcia J A and Pons J M 2000 *Int. J. Mod. Phys.* A **15** 4681
[30] Struckmeier J 2012 *Preprint* math-ph/1206.4449v1
[31] Dunster T M 1990 *SIAM J. Math. Anal.* **21** 4 995-1018

Journal of Physics: Conference Series **453** (2013) 012003 doi:10.1088/1742-6596/453/1/012003

The Cosmic Battery and the Inner Edge of the Accretion Disk

Ioannis Contopoulos[1] **and Demetrios B. Papadopoulos**[2]

[1]Research Center for Astronomy, Academy of Athens, Athens 11527, Greece
[2]Department of Physics, Aristotle University of Thessaloniki, Thessaloniki 54124, Greece

E-mail: `icontop@academyofathens.gr`

Abstract. The Cosmic Battery (CB) proposes that the innermost part of the accretion disk around an astrophysical black hole is threaded by a large scale dipolar magnetic field generated in situ due to the Poynting-Robertson radiation abberation effect on the plasma electrons. In that scenario, the return part of the field diffuses outward through the accretion disk. This is different from the scenario that the field originates at large distances and is carried inward by the accretion flow. In view of the importance of large scale magnetic fields in regulating the processes of accretion and outflows, we study the stability of the inner edge of a magnetized disk in general relativity when the distribution of the magnetic field is the one predicted by the Cosmic Battery. We found that as the field grows, the inner edge of the disk gradually moves outward. In a fast spinning black hole with $a \gtrsim 0.8M$ the inner edge moves back in towards the black hole horizon as the field grows beyond some threshold value. In all cases, the inner part of the disk undergoes a dramatic structural change as the field approaches equipartition.

1. Introduction

Accretion of matter onto astrophysical black holes is a complex physical process involving several physical phenomena (general relativity, small and large scale magnetic fields, turbulence, plasma microphysics, etc.) Accretion proceeds in the form of a disk in which material moves slowly inward until it reaches an inner edge, beyond which it plunges in on dynamical timescales. Understanding where the accretion disk ends is of paramount importance in the two methods for measuring black hole spin, namely fitting the thermal X-ray continuum (Zhang, Cui & Chen 1997), and fitting the profile of the relativistically broadened iron Kα line (Fabian *et al.* 1989; see also McClintock et al. 2011 and references therein).

The inner edge of the disk is expected to be close to the radius of the Innermost Stable Circular Orbit (hereafter ISCO). It is well known that, for test particles, the position of the ISCO depends on the mass and angular momentum of the black hole. The situation is more complicated when, instead of test particles, one considers plasma in orbit around the black hole. It has been shown numerically that any turbulent magnetic field that may thread the plasma modifies its dynamics with respect to the idealized case of test particles, and the inner edge of the disk moves inward (Penna *et al.* 2010). It has even been suggested that since the bulk of the disk is turbulent, there is no feature in the flow related to the ISCO, and the inner edge of the disk moves inward very close to the event horizon (Balbus 2012). What is still not very clear is what is the role of any large scale dipolar magnetic field present in the immediate vicinity of the central black hole. Igumenshchev (2008), Tchekhovskoy, Narayan & McKinney (2011) and others have shown numerically that a dipolar magnetic field advected inward from large distances can grow to equipartition, and can disrupt the process of accretion forming a 'magnetically arrested disk' (Narayan, Igumenshchev & Abramowicz 2003) interspersed with inward spiraling bundles of low-density plasma.

An independent line of thought, the Cosmic Battery (hereafter CB; Contopoulos & Kazanas 1998; Kylafis, Contopoulos, Kazanas & Christodoulou 2011), proposes that the central dipolar magnetic field is generated in situ through the Poynting-Robertson radiation abberation effect on the electrons of the plasma at the inner edge of the disk. In that scenario, the return part of the field diffuses outward through the turbulent accretion disk. In other words, the origin of the large scale magnetic field is situated in the vicinity of the inner edge of the accretion disk, and not at large distances as in the former case. In view of the importance of large scale magnetic fields in regulating the processes of accretion and outflows, we believe that there is still a lot to be learned from the theoretical study of axisymmetric accretion within the framework of the CB. We, therefore, decided to follow on the footsteps of Thorne & Macdonald (1982) and Mobarry & Lovelace (1986) and determine the inner edge of a magnetized disk in general relativity when the distribution of the large scale magnetic field is the one predicted by the CB. For simplicity, we ignore here the potentially important role of turbulent magnetic stresses (Balbus 2012). We begin in the next section with a presentation of our equations and physical simplifications of the problem. In § 3 we obtain our first results for a non-rotating (Schwarzschild) black hole, and in § 4 we extend our study in the most general case of a rotating (Kerr) black hole. Our conclusions are drawn in § 5.

2. The Relativistic MHD Equations

In order to derive the fundamental equation that governs the dynamics of the inner edge of the accretion disk around an astrophysical black hole we follow closely the 3+1 formulation of Thorne & Macdonald (1982) used by most researchers in the astrophysical community. We restrict our analysis to steady-state and axisymmetric space times where $(\ldots)_{,t} = (\ldots)_{,\phi} = 0$. In that case, the general 4-dimensional space-time geometry may be written in Boyer-Lindquist spherical coordinates $x^\mu \equiv (t, r, \theta, \phi)$ as

$$\mathrm{d}s^2 = g_{\mu\nu}\mathrm{d}x^\mu \mathrm{d}x^\nu = -\alpha^2\mathrm{d}t^2 + \frac{A\sin^2\theta}{\Sigma}(\mathrm{d}\phi - \omega\mathrm{d}t)^2 + \frac{\Sigma}{\Delta}\mathrm{d}r^2 + \Sigma\mathrm{d}\theta^2 \ . \tag{1}$$

Here, $\alpha \equiv (\Delta\Sigma/A)^{1/2}$ and $\omega \equiv 2aMr/A$ are the lapse function and angular velocity of '*zero-angular momentum*' fiducial observers (ZAMOs) respectively,

$$\Delta \equiv r^2 - 2Mr + a^2 \ \ , \ \ \Sigma \equiv r^2 + a^2\cos^2\theta \ \ , \ \ A \equiv (r^2 + a^2)^2 - a^2\Delta\sin^2\theta \ , \tag{2}$$

M is the black hole mass, and a its angular momentum ($0 \le a \le M$). We adopt geometric units where $G = c = 1$. Semicolon stands for covariant derivative, comma for partial derivative. Latin indices denote spatial components $(1-3)$, Greek indices denote space-time components $(0-3)$, and '$\sim$' denotes the spatial part of vectors. ZAMOs move with 4-velocity $U^\mu = (1/\alpha\ , 0\ , 0\ , \omega/\alpha)$ orthogonal to hypersurfaces of constant t, and with acceleration

$$\mathrm{a}^\mu \equiv U^\mu_{;\nu}U^\nu \ . \tag{3}$$

Our black hole - accretion disk - magnetosphere hydromagnetic system is characterized by the following choice for the energy-momentum tensor

$$T^{\mu\nu} = \rho u^\mu u^\nu + \frac{1}{4\pi}(F^\mu_\alpha F^{\nu\alpha} - \frac{1}{4}F_{\alpha\beta}F^{\alpha\beta}g^{\mu\nu}) \ , \tag{4}$$

and the condition

$$T^{\mu\nu}_{;\nu} = 0 \ . \tag{5}$$

Here, $u^\mu \equiv dx^\mu/d\tau$ is the fluid 4-velocity as measured by ZAMOs, and ρ is the density of the fluid. We have set the pressure equal to zero, i.e. we have assumed a cold plasma. $F^{\mu\nu}$ is the electromagnetic field tensor related to the electric and magnetic fields E^μ, B^μ measured by ZAMOs through

$$F^{\mu\nu} = U^\mu E^\nu - U^\nu E^\mu + \epsilon^{\mu\nu\lambda\rho} B_\lambda U_\rho \ . \tag{6}$$

Here, $\epsilon_{\mu\nu\lambda\rho} \equiv \sqrt{|\det(g_{\mu\nu})|}[\mu\nu\lambda\rho]$ is the 4-dimensional Levi-Civita tensor. The GRMHD equations of motion are obtained from eq.(5) supplemented by Maxwell's equations of electrodynamics

$$F^{\mu\nu}_{;\nu} = 4\pi J^\mu, \quad F_{[\mu\nu;\alpha]} = 0 \ , \tag{7}$$

where J^μ is the electric current density that satisfies the charge conservation condition $J^\mu_{;\mu} = 0$. J^μ is related to E^μ through a generalized Ohm's law. Energy flow conservation now reads

$$D_\tau \tilde{S} + \tilde{\sigma} \cdot \tilde{S} + \varepsilon \tilde{a} + \frac{1}{\alpha} \tilde{\nabla} \cdot (\alpha \tilde{W}) = 0 \ , \tag{8}$$

where,

$$\tilde{S} = \Gamma^2 \rho \tilde{v} + \frac{1}{4\pi} \tilde{E} \times \tilde{B} \ , \quad \varepsilon = \Gamma^2 \rho + \frac{1}{8\pi}(\tilde{E}^2 + \tilde{B}^2) \ ,$$

$$\tilde{W} = \Gamma^2 \rho \tilde{v} \otimes \tilde{v} + \frac{1}{4\pi}\left[-(\tilde{E}\otimes\tilde{E} + \tilde{B}\otimes\tilde{B}) + \frac{1}{2}(\tilde{E}^2 + \tilde{B}^2)\tilde{\gamma}\right] \ ,$$

$$\sigma^{ij} = \frac{1}{2}\gamma^{i\mu}\gamma^{j\nu}(U_{\mu;\nu} + U_{\nu;\mu}) \ , \quad \Gamma \equiv (1 - \tilde{v}^2)^{-1/2} \equiv (1 + \tilde{u}^2)^{1/2} \ .$$

Here, ϵ is the internal energy, $\sigma^{\mu\nu}$ is the shear tensor of our system of reference, $\tilde{v}^i \equiv \alpha^{-1} dx^i/dt$ is the fluid 3-velocity as measured by ZAMOs ($\tilde{v} \equiv \Gamma^{-1}\tilde{u}$), and D_τ is the Fermi derivative (see Appendix). We simplify the problem even further by assuming a dipolar magnetic field geometry, by restricting our analysis to the equatorial plane $\theta = \pi/2$, and by considering only the simplest case with no azimuthal magnetic torques. In (r, θ, ϕ) coordinates, symmetry dictates that

$$\begin{aligned} \tilde{v} &= (0, 0, v^\phi) \\ \tilde{B} &= (0, B^\theta, 0) \quad \text{with} \quad B^r_{,\theta} \neq 0 \quad \text{in general} \ , \\ \tilde{E} &= (E^r, 0, 0) \quad \text{with} \quad E^\theta_{,\theta} \neq 0 \quad \text{in general} \ . \end{aligned} \tag{9}$$

For several applications in astrophysics, perfect (infinite) conductivity is a valid approximation. In that limit, Ohm's law yields the ideal MHD condition

$$u_\mu F^{\mu\nu} = 0 \tag{10}$$

which is equivalent to the statement that the electric field vanishes in a frame comoving with the fluid.

3. The Inner Disk around a Schwarzschild Black Hole

Before addressing the general case of an accreting spinning black hole, we will first consider the Schwarzschild metric on the equatorial plane where

$$ds^2 = -(1 - \frac{2M}{r})dt^2 + (1 - \frac{2M}{r})^{-1}dr^2 + r^2 d\theta^2 + r^2 d\phi^2 \ . \tag{11}$$

Here, M is the mass of the black hole. The lapse function is equal to $\alpha = (1 - 2M/r)^{1/2}$, $\omega = 0$ and $\sigma^{\mu\nu} = 0$. Furthermore,

$$U^\mu = \frac{1}{\alpha}(1\ ,0\ ,0\ ,0)\ ,\quad \mathrm{a}^i = \frac{M}{r^2}(1\ ,0\ ,0)\ . \tag{12}$$

On the equatorial plane, $B^r = E^\theta = 0$ due to symmetry, and according to eqs. (6) & (10),

$$E^r = \alpha r^2 \mathrm{v}^\phi B^\theta\ ,\quad E^\theta = -\frac{1}{\alpha}\mathrm{v}^\phi B^r\ , \tag{13}$$

with $B^r_{,\theta} \neq 0$ and $E^\theta_{,\theta} \neq 0$. The r-component of eq. (8) now yields

$$\left(1 - \frac{2M}{r}\right)\frac{(r\mathrm{v}^\phi)^2}{r} - \frac{M}{r^2} - \frac{(rB^\theta)}{4\pi\rho r}\left\{\alpha^2\left[r(rB^\theta)\right]_{,r} - B^r_{,\theta}\right\}\left[1 - (r\mathrm{v}^\phi)^2\right]^2$$
$$-\frac{(rB^\theta)^2}{4\pi\rho r}\left\{\frac{M}{r} - \alpha^2(r\mathrm{v}^\phi)^2\left[1 + \frac{r(r\mathrm{v}^\phi)_{,r}}{(r\mathrm{v}^\phi)}\right]\right\}\left[1 - (r\mathrm{v}^\phi)^2\right] = 0\ . \tag{14}$$

This is the generalization of the Newtonian radial force-balance in Schwarzschild geometry. It yields the distribution of the disk azimuthal angular velocity v^ϕ given the distributions of $\vec{B}$ and ρ with r in the disk. The latter can only be obtained through complex general relativistic magneto-hydrodynamic simulations. In this present work, however, we are only interested in determining the position of the innermost accretion disk layer where the CB is mostly active. We will thus address this problem analytically.

The vertical thickness h of the inner disk is a major uncertain parameter since it depends on several physical parameters (thermal pressure, radiation pressure from scattered photons, magnetic pressure), and is the subject of ongoing investigation. According to Fig. 1, we can approximate the value of the radial component of the magnetic field on the upper surface of the inner disk, $B^r|_{+h}$, with the value of the vertical component of the magnetic field on the disk midplane. Note that $B^r = 0$ on the disk midplane. In other words,

$$B^r|_{+h} = -B^r|_{-h} \approx -rB^\theta \text{ and } B^r_{,\theta} \approx \frac{r^2 B^\theta}{h} \equiv \lambda r B^\theta\ . \tag{15}$$

We have introduced here the parameter $\lambda \equiv r/h$. Furthermore, according to Contopoulos & Kazanas (1998), the dipole magnetic field originates in the innermost optically thin disk layer of radial extent δ such that

$$\delta\sigma_T \frac{\rho}{m_p} \approx 1\ , \tag{16}$$

in which photons coming from the central accretion disk region penetrate. This is where the azimuthal electric current due to the Poynting-Robertson effect on the plasma electrons at the inner edge of the disk is generated. Beyond that layer, the magnetic field reverses polarity (see Fig. 1). Here σ_T is the Thompson cross section of photons scattered by the inner disk plasma orbiting electrons, and m_p is the proton mass. Contopoulos & Kazanas (1998) showed that, in order for the CB battery to operate secularly (i.e. continuously), thus generating a large scale dipolar magnetic field interior to the disk inner edge, the reverse polarity magnetic field needs to diffuse outward through the accreting plasma. This occurs naturally in a turbulent accretion disk with magnetic Prandtl number $P_m \leq 1$. We will here assume that

$$\delta \sim h\ , \tag{17}$$

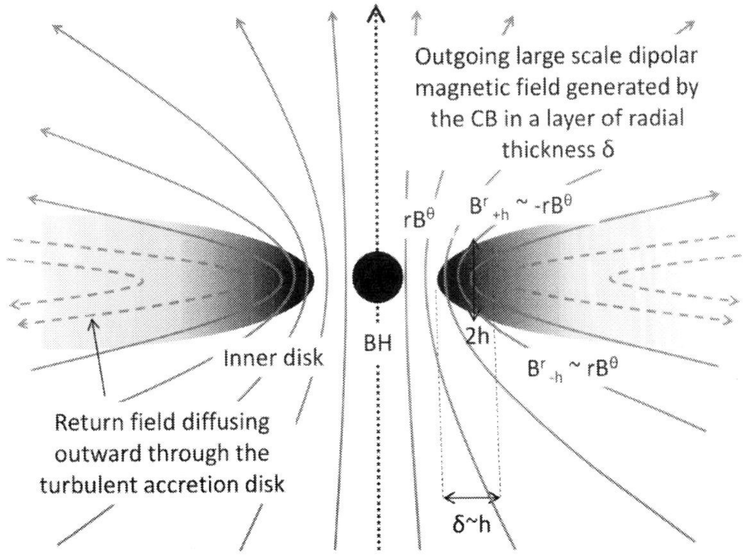

Figure 1. Sketch of innermost disk region where the large scale dipolar magnetic field is generated by the CB. Solid/dashed lines: outgoing/return field respectively. Dotted line: z-axis. The CB layer is the innermost disk region of radial extent δ. Adapted from Contopoulos & Papadopoulos (2012).

although this may be generalized. Therefore, in the CB inner disk layer,

$$(rB^\theta)_{,r} \approx -\frac{rB^\theta}{h} \equiv -\lambda B^\theta \ . \tag{18}$$

Finally, in order to get rid of the radial derivative of v^ϕ, we will assume a Keplerian velocity profile, namely

$$(r\mathrm{v}^\phi)_{,r} \approx -\frac{\mathrm{v}^\phi}{2} \ . \tag{19}$$

Under the above assumptions, eq. (14) becomes

$$\left(1-\frac{2M}{r}\right)\frac{(r\mathrm{v}^\phi)^2}{r} - \frac{M}{r^2} + \frac{\mathrm{v}_A^2(r)}{r}\left\{(2\lambda-1)\left(1-\frac{M}{r}\right)\right.$$
$$\left.-(r\mathrm{v}^\phi)^2\left(2\lambda-\frac{3}{2}-\frac{2M}{r}(\lambda-\frac{3}{2})\right)\right\}\left[1-(r\mathrm{v}^\phi)^2\right]=0\ , \tag{20}$$

where, $\mathrm{v}_A^2(r) \equiv (rB^\theta)^2/4\pi\rho$.

In order to determine the position of the disk inner edge, we will consider a virtual inward/outward displacement of the innermost layer of material of radial and vertical thickness $\delta \approx h$ and h respectively, and study its stability. During that displacement, mass conservation implies that

$$2\pi r\delta h\rho = \text{constant}. \tag{21}$$

In order to proceed, we need to make certain further assumptions about how the various physical quantities in our problem vary during the above displacement. Our first assumption is that the angular momentum parallel to the symmetry axis per unit energy l is conserved, i.e. that

$$\alpha v^{\phi} \equiv \frac{\mathrm{d}\phi}{\mathrm{d}t} = l\frac{\alpha^2}{r^2} . \tag{22}$$

Our second assumption is that of flux freezing during the displacement. This implies that magnetic flux per unit mass is also conserved, i.e.

$$\frac{rB^{\theta}}{\rho h} = \text{constant}. \tag{23}$$

Our final assumption is that

$$\lambda \equiv \frac{r}{h} = \text{constant} \tag{24}$$

during the infinitesimal displacement. We see that under these assumptions, v_A^2 varies inversely proportionally to r during the displacement away from the position of equilibrium, namely

$$\mathrm{v}_A^2(r) \approx \mathrm{v}_A^2 \frac{r_{\mathrm{ISCO}}}{r} , \tag{25}$$

where $\mathrm{v}_A^2 \equiv \mathrm{v}_A^2(r_{\mathrm{ISCO}})$.[1]

As we noted before, eq. (14) is the generalization of the Newtonian radial force-balance, which can also be seen as the zeroing of the first radial derivative of an effective potential V_{eff}. Therefore, in analogy to the Newtonian case, the displacement stability of the innermost disk layer is determined by the sign of the second radial derivative of V_{eff}. Marginal stability corresponds to zero second derivative of V_{eff}, i.e. to zero first derivative of eq. (14). When we substitute eqs. (22) & (25) in eq. (20), the condition for marginal stability, becomes

$$\mathcal{F}(x,\tilde{l};\lambda,\mathrm{v}_A^2 x_{\mathrm{ISCO}}) = 0 , \quad \text{and}$$
$$\frac{\partial}{\partial x}\mathcal{F}(x,\tilde{l};\lambda,\mathrm{v}_A^2 x_{\mathrm{ISCO}}) = 0 , \tag{26}$$

where,

$$\mathcal{F}(x,\tilde{l};\lambda,\mathrm{v}_A^2 x_{\mathrm{ISCO}}) \equiv \left(1-\frac{2}{x}\right)^2\frac{\tilde{l}^2}{x^3} - \frac{1}{x^2} + \frac{\mathrm{v}_A^2 x_{\mathrm{ISCO}}}{x^2}\left\{(2\lambda-1)\left(1-\frac{1}{x}\right)\right.$$
$$\left. -\frac{\tilde{l}^2}{x^2}(1-\frac{2}{x})\left(2\lambda-\frac{3}{2}-\frac{2\lambda-3}{x}\right)\right\}\left[1-\frac{\tilde{l}^2}{x^2}(1-\frac{2}{x})\right] , \tag{27}$$

and $x \equiv r/M$, $\tilde{l} \equiv l/M$. Notice that in this approach we do not need to derive an expression for $V_{\mathrm{eff}}(r)$.

The position x_{ISCO} of the innermost marginally stable circular orbit as a function of our parameters λ and v_A^2 is obtained numerically by eliminating $\tilde{l}$ in the above system of equations (see Fig. 2 for $a=0$). For $\mathrm{v}_A \ll 1$, the ISCO gradually moves outward from its unmagnetized position $6M$. In other words, *the magnetic field acts to destabilize the disk inner edge*, and the disk inner radius must be larger than $6M$ for stability, in accordance with Lovelace *et al.* (1986). However, for higher values of v_A an ISCO ceases to exist. We conclude that, *in the presence of a large scale dipolar magnetic field, the ISCO moves outward, and eventually disappears as the field grows beyond a threshold value.* In the case of a thin disk ($\lambda = 10$), that threshold value corresponds to $\mathrm{v}_A \sim 0.06$. We will return to this important point in the Discussion section.

[1] Eq. (25) may be generalized as $\mathrm{v}_A^2(r) \approx \mathrm{v}_A^2(r_{\mathrm{ISCO}}/r)^{\kappa}$.

4. The Inner Disk around a Spinning Black Hole

The above can be directly generalized on the equatorial plane ($\theta = \pi/2$) of a spinning black hole. Here,

$$\alpha = \left(1 - \frac{2M}{r} + \frac{a^2}{r^2}\right)^{1/2} \frac{r^2}{\sqrt{A}} \,, \quad U^\mu = \frac{1}{\alpha}\left(1\,,0\,,0\,,\frac{2aMr}{A}\right) \,, \tag{28}$$

$$\mathrm{a}^i = \frac{Mr^2}{A}\left(1 + 2\frac{a^2}{r^2} - 4\frac{Ma^2}{r^3} + \frac{a^4}{r^4}\,,0\,,0\right) \,. \tag{29}$$

As before, $B^r = E^\theta = 0$ due to symmetry,

$$E^r = \frac{\Delta}{\alpha}\mathrm{v}^\phi B^\theta \,, \quad E^\theta = -\frac{1}{\alpha}\mathrm{v}^\phi B^r \,, \tag{30}$$

with $B^r_{,\theta} \neq 0$ and $E^\theta_{,\theta} \neq 0$. In order to estimate the position of the ISCO when the CB is active, we will make the same simplifying assumptions as in the previous section. The r-component of eq. (8) simplifies then considerably and becomes

$$\Gamma^r_{\phi\phi}(\mathrm{v}^\phi)^2 + \left[\frac{1}{\alpha}(\Gamma^r_{t\phi} + \omega\Gamma^r_{\phi\phi}) + g_{\phi\phi}\sigma^{r\phi}\right]\mathrm{v}^\phi + \mathrm{a}^r$$

$$-\frac{\mathrm{v}_A^2 r_{\rm ISCO}}{r^2\Gamma^4}\{\frac{\Delta}{r^2}(\lambda+1)+\lambda\} - \frac{\mathrm{v}_A^2 r_{\rm ISCO}}{2r^3\Gamma^2}\{X(\mathrm{v}^\phi)^2 + Y\mathrm{v}^\phi + Z\} = 0 \,, \tag{31}$$

where the detailed expressions for $\sigma^{r\phi}$, X, Y, Z, the Christoffel symbols $\Gamma^r_{t\phi}$, $\Gamma^r_{\phi\phi}$ and the flow Lorentz factor Γ can be found in the Appendix. As before, $\mathrm{v}_A^2 \equiv \mathrm{v}_A^2(r_{\rm ISCO})$.

The assumption that angular momentum parallel to the symmetry axis per unit energy l is conserved during a virtual infinitesimal displacement of the innermost layer of the disk is now generalized as

$$\alpha\mathrm{v}^\phi \equiv \frac{\mathrm{d}\phi}{\mathrm{d}t} - \omega = l\frac{\left(1-\frac{2M}{r}\right)\frac{r^2}{A} + \omega^2}{1-\omega l} \tag{32}$$

(Bardeen, Press & Teukolsky 1972; hereafter BPT72). When we substitute the above in eq. (31), we obtain a complicated expression of the form

$$\mathcal{G}(x, \tilde{l}; \lambda, \mathrm{v}_A^2 x_{\rm ISCO}) = 0 \,, \tag{33}$$

and the condition for marginal stability becomes that

$$\frac{\partial}{\partial x}\mathcal{G}(x, \tilde{l}; \lambda, \mathrm{v}_A^2 x_{\rm ISCO}) = 0 \,. \tag{34}$$

As before, $x \equiv r/M$, and $\tilde{l} \equiv l/M$. The position of the ISCO as a function of our two free parameters λ and v_A^2 is obtained as a solution of the above system of equations (33) & (34) for x and $\tilde{l}$. The solution for various values of the black hole spin parameter a is obtained numerically in Fig. 2 for two values of $\lambda \equiv r/h$ that correspond to a thin and thick disk respectively. Similarly to the case of a Schwarzschild black hole, for small values of v_A, the ISCO always moves outward. However, for values of $a \gtrsim 0.8M$, an interesting transition takes place: as v_A grows beyond a threshold value, *the ISCO moves back inward.* As we will now see, this result complicates the observational determination of the black hole spin.

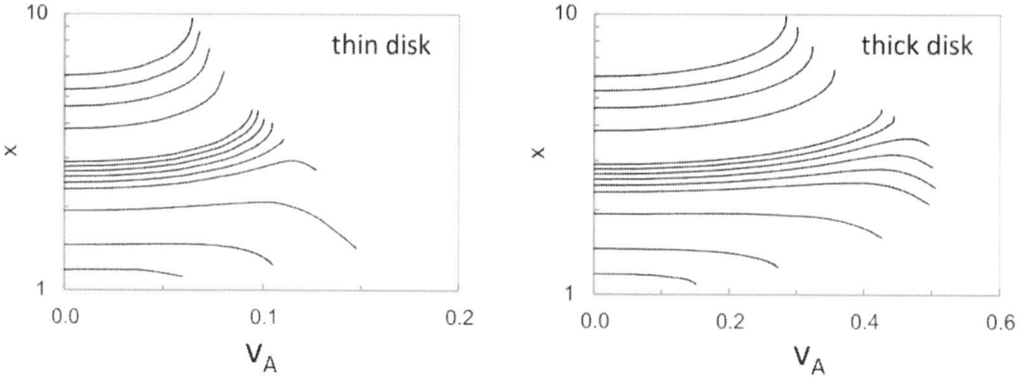

Figure 2. Dependence of $x_{\rm ISCO} \equiv r/M$ on the inner magnetic field strength for various black hole spin values a. At the points the curves end they begin to turn up and backwards. Left plot: thin disk ($\lambda \equiv r/h = 10$). Right plot: thick disk ($\lambda = 1$). The various curves from top to bottom correspond to $a = 0, 0.2, 0.4, 0.6, 0.8, 0.82, 0.84, 0.86, 0.88, 0.9, 0.95, 0.99, 0.999M$ respectively. Adapted from Contopoulos & Papadopoulos (2012).

5. Discussion

The displacement of the ISCO in the presence of a large scale magnetic field generated in situ has direct implications in the currently most promising method for measuring black hole spin, namely fitting the thermal X-ray continuum (Zhang et al. 1997). This method requires knowledge of both the Schwarzschild radius and the radius of the inner edge of the disk, i.e. it requires knowledge of $x_{\rm ISCO}$. As is shown in BPT72, in an unmagnetized disk, $x_{\rm ISCO}$ is one-to-one related to the spin of the black hole. In the present work, we showed that $x_{\rm ISCO}$ depends also on the value of the magnetic field accumulated in the innermost region of the accretion disk. This result introduces one extra complication.

In particular, the disappearance of the ISCO as the field grows towards equipartition (defined in this context as $v_A \rightarrow 1$) is an indication for a dramatic change in the structure of the accretion disk in the presence of a strong enough magnetic field generated by the CB. It is natural to speculate that, when the conditions for marginal stability of the inner edge cannot be satisfied, the accumulated magnetic field will escape outward through magnetic Rayleigh-Taylor-type instabilities, as proposed in İgumenshchev (2008), Tchekhovskoy, Narayan & McKinney (2011), and Kylafis *et al.* (2012). As a result, the accumulated magnetic field will decrease, the ISCO will reappear close to its unmagnetized disk position, and the disk will undergo a new phase of field growth and evolution. For a stellar mass black hole, the time required for equipartition field growth is on the order of a few hours to a few days (Contopoulos & Kazanas 1998; Kylafis *et al.* 2012), whereas the intervals of dramatic disk structure change are expected to be much shorter (dynamic timescales).

For black holes with spin parameter $a < 0.8M$, the inner edge of the disk is expected to be outside its unmagnetized position, therefore the thermal X-ray continuum fitting method will on average yield a *systematically smaller* value for the black hole spin. On the other hand, for fast spinning black holes with $a \gtrsim 0.8M$, the inner edge of the disk moves to lower values, and in those cases the thermal X-ray continuum fitting method will on average yield *systematically larger* values for the black hole spin. This result introduces an *artificial segregation* in our observational method that may be related to the seeming absence of evidence for the black hole spin powering the jets in X-ray binaries (Fender, Gallo & Russell 2010; see however also Narayan & McClintock 2012). In other words, the introduction of one extra free parameter, namely the amount of magnetic flux accumulated around the inner edge of the disk, complicates the direct determination of the black hole spin through the knowledge of $x_{\rm ISCO}$ alone.

The above are summarized in Fig. 3 where we show the range of $x_{\rm ISCO}$ values as a function of a for a thick disk ($r/h = 1$ at its inner edge). The dashed curve corresponds to the unmagnetized

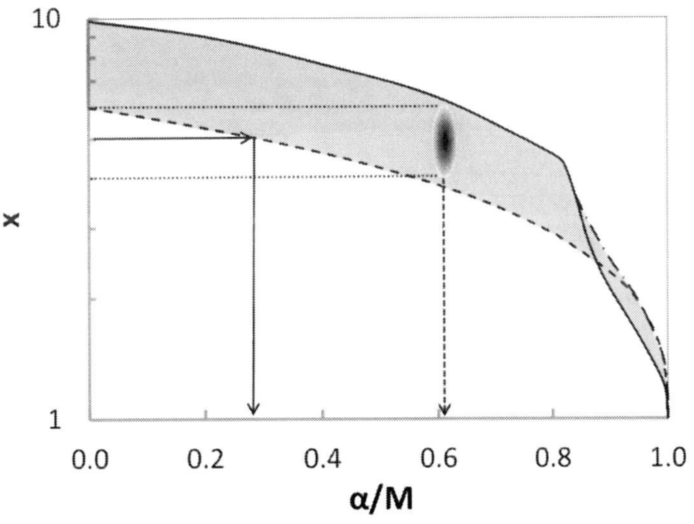

Figure 3. Range of $x_{\rm ISCO}$ values as a function of the black hole spin parameter a for a thick disk $r/h = 1$. Dashed curve: unmagnetized disk. Solid curve: maximally magnetized inner disk that supports an ISCO. Shadded region: range of $x_{\rm ISCO}$ values when $0 \leq {\rm v}_A \leq {\rm v}_A|_{\rm max}$. Diffuse cloud: range of $x_{\rm ISCO}$ values obtained for the X-ray black hole binary LMC X-3 over a period of 25 years. Left vertical arrow: average spin parameter of unmagnetized disk. Right vertical arrow: spin parameter that yields the observed range of $x_{\rm ISCO}$ values in the context of the CB. Adapted from Contopoulos & Papadopoulos (2012).

disk (BPT72), whereas the solid curve to a maximally magnetized disk that supports an ISCO. The dashed-dotted curve corresponds to the maximum values of $x_{\rm ISCO}$ reached as the inner field (${\rm v}_A$) grows in a spinning black hole with $a \gtrsim 0.8M$. In other words, according to our simplified model of the inner disk, observations of $x_{\rm ISCO}$ must lie in the shaded region. As an example, in the most studied object, LMC X-3, estimates of $x_{\rm ISCO}$ range from about 4 to about 6 (Steiner *et al.* 2010, assuming a canonical value for the black hole mass equal to $7.5M_\odot$). If the CB did not operate and the disk were unmagnetized, these values would yield an average black hole spin value $a \sim 0.3M$ (McClintock *et al.* 2011). It is natural, though, to expect that the CB does operate as described above, and that the various estimates of $x_{\rm ISCO}$ simply correspond to the various stages of magnetic field growth that the source is found to be in. As shown in Fig. 3, in this context, a more natural value of the black hole spin in LMC X-3 would then be $a \sim 0.6M$.

Finally, the reader may wonder in what respect the CB manifests itself in the inner disks structure differently from flux advection from large distances. One way to differentiate between the two is the radial width δ of the CB layer over which B^θ changes significantly. In the CB and a thin disk, our parameter $r/\delta \sim \lambda$ is expected to be on the order of 10. On the other hand, in the CB and a thick disk, as well as in a disk where the magnetic field is advected inward from large distance, λ is expected to be on the order of unity. Another way to differentiate between the two possibilities has to do with the particular cyclic disk variability central to the CB, namely intervals of secular field growth (hours, days) separated by intervals of strong (msec) disk variability. Further investigation on the role of the CB in determining the structure of the inner disk is under way.

Acknowledgments

This work was partially supported by the General Secretariat for Research and Technology of Greece and the European Social Fund in the framework of Action 'Excellence'.

Appendix

Various expressions used throughout the paper are defined below according to Thorne & Macdonald (1982):

$$\tilde{\nabla}\cdot\tilde{A} \equiv A^j_{;j}\ ,\quad (\tilde{\nabla}\times\tilde{A})^i \equiv \epsilon^{ijk}A_{k;j}\ ,\quad D_\tau A^\beta \equiv A^\beta_{;\mu}U^\mu - U^\beta U_{\mu;\nu}A^\mu U^\nu$$

$$\tilde{A}\cdot\tilde{B} \equiv A^i B_i\ ,\quad (\tilde{A}\times\tilde{B})^i \equiv \epsilon^{ijk}A_j B_k\ ,\quad (\tilde{A}\otimes\tilde{B})^{ij} \equiv A^i B^j\ ,\quad \gamma^{\mu\nu} \equiv g^{\mu\nu} + U^\mu U^\nu\ .$$

Here, $\epsilon^{ijk} \equiv [ijk]/\sqrt{|\det(\gamma_{ij})|}$ is the spatial Levi-Civita tensor. On the equatorial plane,

$$\Gamma^r_{t\phi} = -\frac{Ma\Delta}{r^4}\ ,\quad \Gamma^r_{rr} = -\frac{Mr-a^2}{\Delta r}\ ,\quad \Gamma^r_{\theta\theta} = -\frac{\Delta}{r}\ ,\quad \Gamma^r_{\phi\phi} = -\frac{\Delta}{r}\left[1-\frac{Ma^2}{r^3}\right]$$

$$\Gamma^k_{k1} \equiv \Gamma^r_{rr} + \Gamma^\theta_{\theta r} + \Gamma^\phi_{\phi r} = \frac{r}{\Delta}\left(2 - \frac{5M}{r} + \frac{2a^2}{r^2} - \frac{Ma^2}{r^3}\right) \quad (.1)$$

$$\sigma^{r\phi} = \sigma^{\phi r} = -\frac{Ma\Delta(a^2+3r^2)}{rA\sqrt{A\Delta}}\ ,\quad \mathrm{a}^r = \frac{Mr^2}{A}\left(1 + 2\frac{a^2}{r^2} - 4\frac{Ma^2}{r^3} + \frac{a^4}{r^4}\right)$$

$$\omega = \frac{2Mar}{A},\quad \alpha = \sqrt{\frac{\Delta r^2}{A}}\ ,\quad Q = \frac{\alpha_{,r}}{\alpha} + 2\Gamma^r_{rr} + \Gamma^\theta_{\theta r} + \Gamma^\phi_{\phi r}\ ,\quad \Gamma = \left[1 - \frac{A}{r^4}(r\mathrm{v}^\phi)^2\right]^{-1/2}$$

$$X \equiv \left(\frac{\Delta}{\alpha}\right)^2\{-\frac{\mathrm{a}^r r^2}{\Delta} + Q - \frac{1}{\Delta}\Gamma^r_{\theta\theta} - \frac{1}{\Delta}\Gamma^r_{\phi\phi}(\gamma^{\phi\phi}r^2)\} + 2\left(\frac{\Delta}{\alpha}\right)\left(\frac{\Delta}{\alpha}\right)_{,r} + 2\frac{\mathrm{v}^\phi_{,r}}{\mathrm{v}^\phi}\frac{\Delta^2}{\alpha^2} \quad (.2)$$

$$Y \equiv -2\frac{r^2}{\alpha}(\Gamma^r_{t\phi} + \omega\Gamma^r_{\phi\phi})]\ ,\quad Z \equiv -\mathrm{a}^r r^2 - \Delta Q - \Delta_{,r} + \Gamma^r_{\theta\theta} - \Gamma^r_{\phi\phi}(\gamma^{\phi\phi}r^2) \quad (.3)$$

References

[1] Balbus, S. A. 2012, MNRAS, 423L, 50
[2] Bardeen, J. M., Press, W. H. & Teukolsky, S. A. 1972, ApJ, **178**, 347 (BPT72)
[3] Contopoulos, I. & Kazanas, D. 1998, ApJ, **508**, 859
[4] Contopoulos, I., Kazanas, D., Christodoulou, D. M. & Gabuzda, D. C. 2009, ApJ, **702**, 148
[5] Contopoulos, I. & Papadopoulos, D. B. 2012, MNRAS, **425**, 147
[6] Fabian, A. C., Rees, M. J., Stella, L. & White, N. E. 1989, MNRAS, **238**, 729
[7] Fender, R. P., Gallo, E. & Russell, D. 2010, MNRAS, **406**, 1425
[8] Igumenshchev, I. V. 2008, ApJ, **677**, 317
[9] Kylafis, N. D., Contopoulos, I., Kazanas, D. & Christodoulou, D. M. 2012, A& A, **538**, 5
[10] Lovelace, R. V. E., Mehanian, C., Mobarry, C. M. & Sulkanen, M. E. 1986, ApJS, **62**, 1
[11] Macdonald, D. & Thorne, K., S. 1982, MNRAS, **198**, 345
[12] McClintock, J. E. *et al.* 2011, Class. & Quantum Grav., **28**, 114009
[13] Mobarry, C. M. & Lovelace, R. V. E. 1986, ApJ, **309**, 455
[14] Narayan, R. & McClintock, J. E. 2012, MNRAS, **419**, 69
[15] Narayan, R., Igumenshchev, I. V. & Abramowicz, M. A. 2003, PASJ, **55**, 69
[16] Penna, R. F., McKinney, J. C., Narayan, R., Tchekhovskoy, A., Shafee, R. & McClintock, J. E. 2010, MNRAS, **408**, 752
[17] Steiner, J. F., McClintock, J. E., Remillard, R. A., Gou, L., Yamada, S. & Narayan, R. 2010, ApJ, **718**, 117
[18] Tchekhovskoy, A., Narayan, R. & McKinnery, J. C. 2011, MNRAS, **418**, 79
[19] Thorne, K. S. & Macdonald, D. 1982, MNRAS, **198**, 339
[20] Zhang, S. N., Cui, W. & Chen, W. 1997, ApJ, **482**, L155

Journal of Physics: Conference Series **453** (2013) 012004 doi:10.1088/1742-6596/453/1/012004

The IWM–CFC approximation in differentially rotating relativistic stars

P Iosif and N Stergioulas

Department of Physics, Aristotle University of Thessaloniki, Thessaloniki 54124, Greece

E-mail: `piosif@auth.gr`, `niksterg@auth.gr`

Abstract. Many astrophysically relevant systems have been studied, in recent years, by means of the spatial conformal flatness condition (IWM–CFC), yielding very satisfactory results as far as the accuracy of IWM–CFC approximation is concerned. Here, we determine the accuracy of IWM–CFC for the case of single, but strongly differentially rotating stars, modeled by a polytropic equation of state. We find that for the fastest rotating models, the deviation from full general relativity is still below 2% for most integrated quantities and reaches up to 6% for sensitive quantities, such as the angular velocity at the equator. We construct a simple error indicator and demonstrate that it correlates well with the largest errors observed in physical quantities characterizing the model.

1. Introduction

The evaluation of the accuracy of the spatial conformal flatness condition (IWM–CFC) [1, 2] in various physical systems is crucial in order to better understand the limits within which it can be applied. Cook, Shapiro and Teukolsky (CST) [3] presented a first test of the CFC approximation for the case of single, uniformly rotating, relativistic stars (see also [4]). Miller, Gressman and Suen [5] used the Cotton–York tensor (which vanishes on spacelike slices that fulfill the IWM–CFC assumption) to evaluate the accuracy of initial data for binary compact objects. Here, we study the accuracy of the CFC approximation in stars that are highly deformed due to strong differential rotation.

The spacetime metric for a stationary, axisymmetric star in equilibrium is described by a line element of the form (see [6, 7] and references therein)

$$ds^2 = -e^{\gamma+\rho}dt^2 + e^{\gamma-\rho}r^2\sin^2\theta(d\phi - \omega dt)^2 + e^{2\mu}(dr^2 + r^2 d\theta^2) , \qquad (1)$$

where γ, ρ, ω and μ are metric potentials depending only on r and θ. We assume that the stellar matter behaves as a perfect fluid and that the equation of state obeys the polytropic relation

$$p = K\rho^{1+\frac{1}{N}} , \qquad (2)$$

where ρ is the rest mass density, K the polytropic constant and N the polytropic index. For the case of differential rotation we adopt the same rotation law as in Komatsu, Eriguchi and Hachisu (KEH) [8, 9]

$$u^t u_\phi := F(\Omega) = A^2(\Omega_c - \Omega) , \qquad (3)$$

Journal of Physics: Conference Series **453** (2013) 012004 doi:10.1088/1742-6596/453/1/012004

where A is a positive constant that determines the length scale over which the angular velocity Ω changes within the star and Ω_c is the angular velocity at the center of the configuration. Numerical models are obtained with the `rns` code by Stergioulas and Friedman [10, 11].

2. Method

In the 3+1 split, the line element is written as

$$ds^2 = -\alpha^2 dt^2 + \gamma_{ij}(dx^i + \beta^i dt)(dx^j + \beta^j dt), \tag{4}$$

where α is the lapse function, β^i is the shift vector and γ_{ij} is the spatial metric. In the IWM–CFC approximation, one assumes that the spatial part is conformally flat, that is

$$\gamma_{ij} = \psi^4 \eta_{ij}, \tag{5}$$

where ψ is a conformal factor and η_{ij} is the flat metric. The metric (1) takes the IWM–CFC form, if

$$\mu = \frac{\gamma - \rho}{2} . \tag{6}$$

Imposing the above condition between the three metric functions, instead of solving the actual equation for μ, allows for the construction of numerical models in the IWM–CFC approximation, without any other modification in the `rns` code.

For each model that is constructed, various physical quantities are calculated both in IWM–CFC and in full general relativity (GR). As one possible diagnostic of the accuracy of IWM–CFC we define the quantity

$$\Delta c = \frac{\mu - \frac{\gamma - \rho}{2}}{\mu} , \tag{7}$$

which vanishes in IWM–CFC, but is non-vanishing for models constructed in full GR with the metric (1). In addition, relative differences between IWM–CFC and full GR are calculated for every physical quantity.

We focus on the two sequences (A and B) of differentially rotating models with $N = 1$ and $K = 100$ that are presented in Table I of Stergioulas, Apostolatos and Font [12]. For comparison, we also consider their corresponding sequences of uniformly rotating models (AU and BU). Configurations in sequences A and AU have a fixed rest mass of $M_0 = 1.506 M_\odot$ and configurations in sequences B and BU have a fixed central mass density of $\rho_c = 1.28 \times 10^{-3}$ (a fixed central energy density of $\epsilon_c = 1.444 \times 10^{-3}$) in the non-dimensional units defined in [12]. We note that in addition to the models presented in [12], we constructed two models that have even higher rotation rate: model A12, with a polar to equatorial axis ratio of $r_p/r_e = 0.25$ and model B13 with $r_p/r_e = 0.34$.

3. Results

The following figures summarize our main result. Figure 1 shows the absolute value of the relative difference between CFC and full GR for three representative physical quantities (the gravitational mass M, the ratio $T/|W|$ of the rotational to gravitational binding energy and the Keplerian angular velocity Ω_K at the equator) as a function of $T/|W|$ along the four different sequences of equilibrium models. We observe that along sequences A and AU all relative differences saturate well below the 1% level, which is explained by the fact that higher rotation leads to smaller central densities, which counteracts the effect of larger flatness. Along sequence B, the central density is fixed and the error increases practically monotonously, reaching up to several percent.

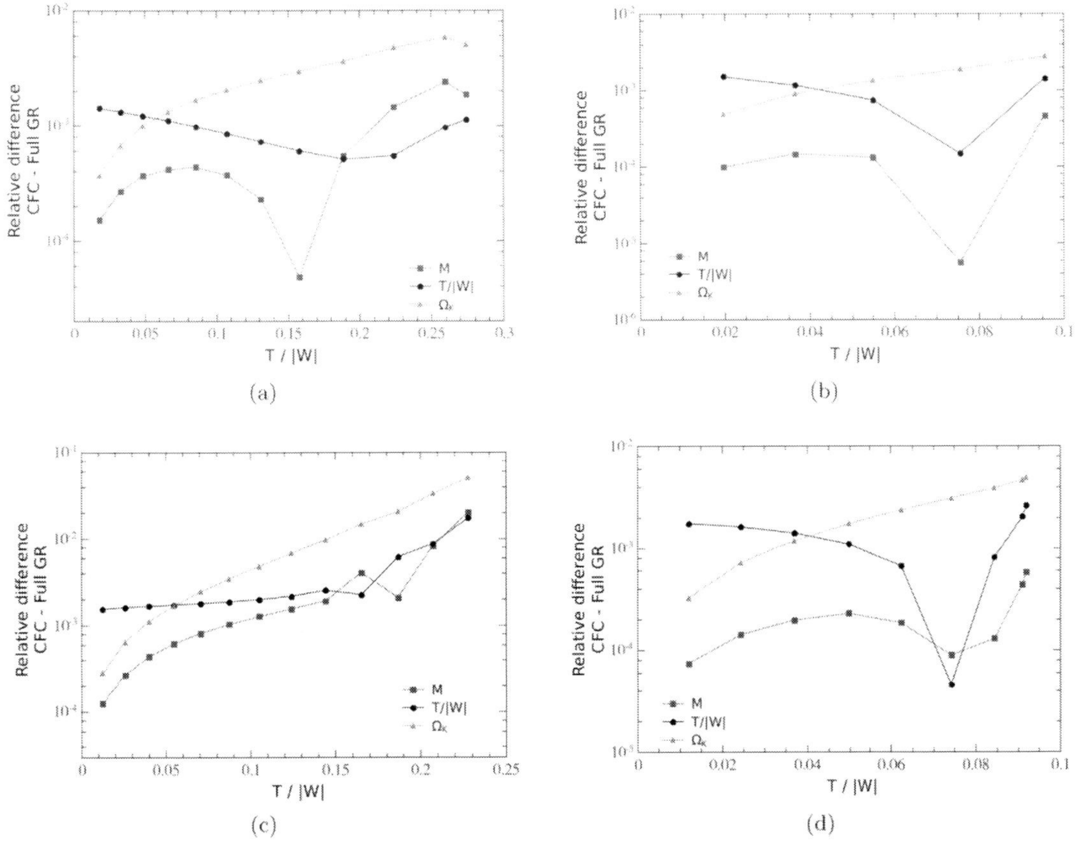

Figure 1: Absolute value of the relative differences between full GR and CFC approximation for the gravitational mass M, the rotational kinetic energy over gravitational binding energy ratio $T/|W|$ and the Keplerian angular velocity Ω_K. (a) sequence A, (b) sequence AU, (c) sequence B, (d) sequence BU.

Figure 2 displays the quantity Δc in a vertical plane for the fastest rotating models of each sequence. For model A12, Δc has a maximum value of roughly 2% inside the star. For model B13, the maximum value of Δc becomes 6%. Thus, the maximum values of Δc correlate well with the largest errors observed in physical quantities.

Figure 3 shows the profile of Δc for the four fastest rotating models in the equatorial plane, as a function of the compactified radial coordinate $s = r/(r + r_e)$. We notice that in all cases the relative error peaks at $s \simeq 0.4$, or $r \simeq 2/3\, r_e$.

4. Conclusions

The CFC approximation appears to be a robust method to study systems that exhibit differential rotation if the demands for accuracy are not particularly strict, i.e. if one can cope with a maximum error of several percent. For moderate rotation rates or small central densities the error can be less than the 1% level. We propose a simple error indicator, Δc, for the accuracy of the CFC approximation for isolated rotating stars, evaluated at 2/3 of the equatorial radius and demonstrate that it correlates well with the largest errors observed in physical quantities characterizing the model.

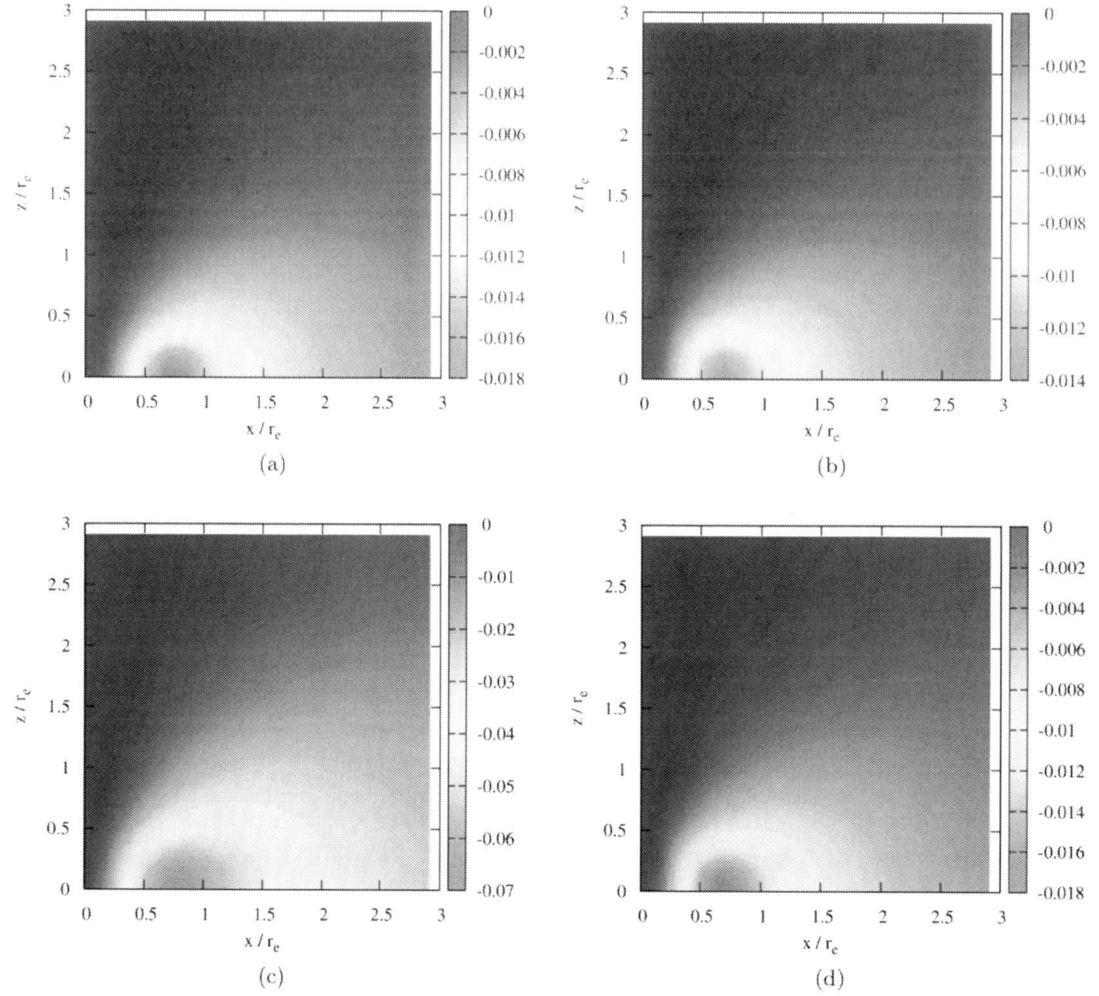

Figure 2: Relative difference Δc on the $x-z$ plane for the fastest rotating model of each sequence. (a) Model A12, $r_p/r_e = 0.25$, (b) Model AU5, $r_p/r_e = 0.575$, (c) Model B13, $r_p/r_e = 0.34$, (d) Model BU9, $r_p/r_e = 0.58$.

Acknowledgments

We are grateful to Andreas Bauswein for useful discussions during our stay at the MPA–Garching. This work was supported by an IKY–DAAD exchange grant (IKYDA 2012), an Excellence Grant for Basic Research (Research Committee of the Aristotle University of Thessaloniki – 87896), a MNiSW grant (N N203 511238) "Neutron stars and black holes as sources of GW" and a VESF fellowship for doctoral studies by the European Gravitational Observatory (EGO-DIR-126-2012).

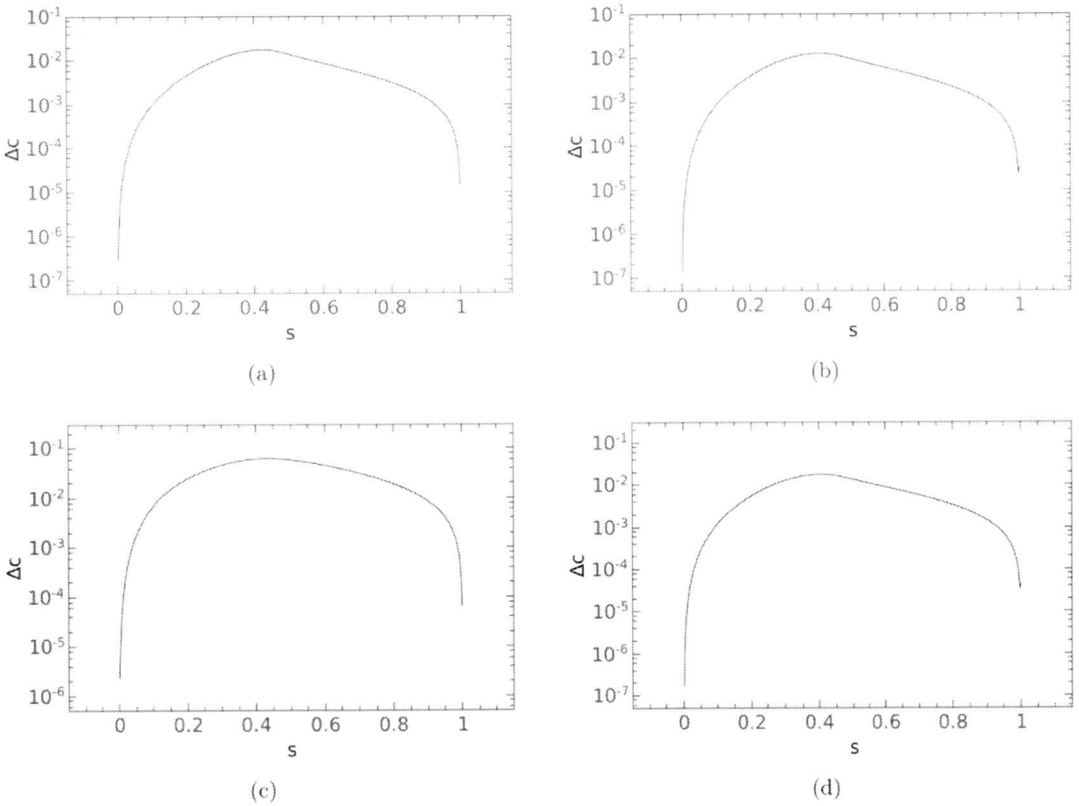

Figure 3: Relative difference Δc calculated on the equatorial plane for the fastest rotating model of each sequence. (a) Model A12, $r_p/r_e = 0.25$, (b) Model AU5, $r_p/r_e = 0.575$, (c) Model B13, $r_p/r_e = 0.34$, (d) Model BU9, $r_p/r_e = 0.58$.

References

[1] Isenberg J A 2008 *Int. J. Mod. Phys.* D **17** 265–273

[2] Wilson J R, Mathews G J and Marronetti P 1996 *Phys. Rev.* D **54** 1317–1331

[3] Cook G B, Shapiro S L and Teukolsky S A 1996 *Phys. Rev.* D **53** 5533–5540

[4] Bucciantini N and Del Zanna L 2011 *Astron. Astrophys.* **528** A101

[5] Miller M, Gressman P and Suen W 2004 *Phys. Rev.* D **69** 064026

[6] Stergioulas N 2003 *Living Rev. Relativity* **6** 3

[7] Friedman J L and Stergioulas N 2013 *Rotating Relativistic Stars* (Cambridge University Press)

[8] Komatsu H, Eriguchi Y and Hachisu I 1989 *Mon. Not. R. Astron. Soc.* **237** 355–379

[9] Komatsu H, Eriguchi Y and Hachisu I 1989 *Mon. Not. R. Astron. Soc.* **239** 153–171

[10] Stergioulas N and Friedman J L 1995 *Astrophys. J.* **444** 306–311

[11] Nozawa T, Stergioulas N, Gourgoulhon E and Eriguchi Y 1998 *Astron. Astrophys. Suppl. Ser.* **132** 431–454

[12] Stergioulas N, Apostolatos T A and Font J A 2004 *Mon. Not. R. Astron. Soc.* **352** 1089–1101

Journal of Physics: Conference Series **453** (2013) 012005
doi:10.1088/1742-6596/453/1/012005

Mind the Resonances: Final stages of accretion into bumpy black holes

Georgios Lukes-Gerakopoulos[1] **and George Contopoulos**[2]

[1] Theoretical Physics Institute, University of Jena, 07743 Jena, Germany

[2] Research Center for Astronomy and Applied Mathematics, Academy of Athens, Soranou Efesiou 4, GR-11527 Athens, Greece

E-mail: `gglukes@gmail.com`

Abstract. In this article we discuss a possible way of testing the Kerr black hole hypothesis by taking advantage of phenomena correlated with chaotic motion in the final stages of an accretion disk around a bumpy black hole. We anticipate that these phenomena should have an imprint in the electromagnetic spectrum coming from the accretion disk.

1. Introduction

The gravitational wave detectors are anticipated to provide observations which will help us to study massive compact objects and to test our theoretical expectations about their nature. Namely, the spacetime around black holes is expected to be described by the Kerr metric. One way to test the Kerr hypothesis is to search in the gravitational wave signals for any deviation from the corresponding signals we would expect to come from a Kerr spacetime background, see e.g., [1, 2, 3] and references therein. However, alternative tests for our expectation might be based on contemporary plausible astronomical observations. Namely, the Kerr hypothesis can be tested by studying the electromagnetic spectra coming from a black hole's surrounding environment (e.g., [4, 5, 6, 7, 8, 9, 10, 11]).

Both gravitational and electromagnetic Kerr hypothesis tests are usually applied on axially symmetric perturbations of the Kerr spacetime, such spacetimes are often called in the literature bumpy black holes (name given in [12]) or non-Kerr spacetimes. By perturbing the Kerr spacetime it seems that in general we destroy a generalized Noether symmetry (e.g., [13] and references therein) responsible for the existence of the Carter constant [14]. Even if there are attempts of finding higher order Killing tensors expressing this symmetry in specific bumpy black hole spacetimes ([15] and references therein), the numerical results show that chaos appears when we examine geodesic orbits in these spacetimes [16, 17, 18, 19, 20], which in turn implies that these spacetimes correspond to a non-integrable system and therefore no Carter-like constant can exist. On the other hand, if we demand the preservation of the Carter constant after the perturbation, it seems that we have to relax the requirement that the perturbed metric satisfies the Einstein equations [21]. Thus, it appears that after the perturbation of the Kerr metric we have to either lose the Carter constant or drop the requirement that the new metric has to satisfy the Einstein equations.

The non-integrability of the bumpy black holes is correlated with several interesting aspects of classical non-integrable systems. In particular, in [19] we have studied how the bifurcation of the periodic orbits at the resonances lead to escaping orbits in a classical system of two coupled harmonic oscillators and a relativistic system described by a subclass of the Manko-Novikov (MN) metric family [22]. In the present article we attempt to give a physical interpretation to the study done in [19]. Namely, we try to interpret our findings as possible imprints in the electromagnetic spectra indicating a method to distinguish bumpy black hole spacetimes from Kerr spacetimes by exploiting the non-linear phenomena studied in [19] at the final stages of accretion into a non-Kerr compact object, in a similar way as done by the authors of [23].

The article is organized as follows. We introduce the MN spacetime in section 2. Section 3 summarizes basic elements of geodesic motion in axisymmetric and stationary spacetimes. We discuss, by interpreting the numerical examples given in [19], the final stages of accreting matter into a MN bumpy black hole in section 4. Section 5 briefs our main conclusion.

2. The Manko-Novikov spacetime

The bumpy black hole spacetime we used in [19] is a spacetime which belongs to the so-called Manko-Novikov (MN) metric family [22]. Manko and Novikov found an exact vacuum solution of Einstein's equations which describes a stationary, axisymmetric, and asymptotically flat spacetime with arbitrary mass-multipole moments [22]. The MN metric subclass we used was introduced by [16] and deviates from the Kerr by the anomalous quadrupole moment q. The quantity q measures how much the MN quadrupole moment Q departs from the Kerr quadrupole moment $Q_{Kerr} = -S^2/M$ (i.e., $q = (Q_{Kerr} - Q)/M^3$), where M and S are the mass and spin of a Kerr black hole. The line element of the MN metric in the Weyl-Papapetrou cylindrical coordinates (ρ, φ, z) is

$$ds^2 = -f(dt - \omega d\varphi)^2 + f^{-1}[e^{2\gamma}(d\rho^2 + dz^2) + \rho^2 d\varphi^2], \tag{1}$$

where f, ω, γ are considered as functions of the prolate spheroidal coordinates v, w, while the coordinates ρ, z can be expressed as functions of v, w as well. Thus

$$\rho = k\sqrt{(v^2-1)(1-w^2)}, \quad z = kvw, \tag{2}$$

and

$$\begin{aligned}
f &= e^{2\psi}\frac{A}{B}, && \text{(3a)}\\
\omega &= 2ke^{-2\psi}\frac{C}{A} - 4k\frac{\alpha}{1-\alpha^2}, && \text{(3b)}\\
e^{2\gamma} &= e^{2\gamma'}\frac{A}{(v^2-1)(1-\alpha^2)^2}, && \text{(3c)}\\
A &= (v^2-1)(1+a\,b)^2 - (1-w^2)(b-a)^2, && \text{(3d)}\\
B &= [(v+1)+(v-1)a\,b]^2 + [(1+w)a + (1-w)b]^2, && \text{(3e)}\\
C &= (v^2-1)(1+a\,b)[(b-a) - w(a+b)]\\
&\quad +(1-w^2)(b-a)[(1+a\,b)+v(1-a\,b)], && \text{(3f)}\\
\psi &= \beta\frac{P_2}{R^3}, && \text{(3g)}\\
\gamma' &= \ln\sqrt{\frac{v^2-1}{v^2-w^2}} + \frac{3\beta^2}{2R^6}(P_3^2 - P_2^2)\\
&\quad + \beta\left(-2 + \sum_{\ell=0}^{2}\frac{v - w + (-1)^{2-\ell}(v+w)}{R^{\ell+1}}P_\ell\right), && \text{(3h)}
\end{aligned}$$

$$a = -\alpha \exp\left[-2\beta\left(-1+\sum_{\ell=0}^{2}\frac{(v-w)P_\ell}{R^{\ell+1}}\right)\right], \tag{4a}$$

$$b = \alpha \exp\left[2\beta\left(1+\sum_{\ell=0}^{2}\frac{(-1)^{3-\ell}(v+w)P_\ell}{R^{\ell+1}}\right)\right], \tag{4b}$$

$$R = \sqrt{v^2+w^2-1}, \tag{4c}$$

$$P_\ell = P_\ell(\frac{v\,w}{R}). \tag{4d}$$

Here $P_\ell(\zeta)$ is the Legendre polynomial of order l

$$P_\ell(\zeta) = \frac{1}{2^\ell \ell!}\left(\frac{d}{d\zeta}\right)^\ell (\zeta^2-1)^\ell, \tag{5}$$

while the parameters k, α, β are related to the mass M, the spin S, and the quadrupole deviation q through the expressions

$$\alpha = \frac{-1+\sqrt{1-\chi^2}}{\chi}, \quad k = M\frac{1-\alpha^2}{1+\alpha^2}, \quad \beta = q\left(\frac{1+\alpha^2}{1-\alpha^2}\right)^3. \tag{6}$$

while χ is the dimensionless spin parameter $\chi = S/M^2$. These formulae give the Kerr metric for $q=0$. We use the geometrical units $c = G = 1$ throughout the article, and without loss of generality we have set also $M = 1$.

3. Geodesic Motion in the Manko-Novikov spacetime

The geodesic orbits of a test particle of mass μ are described as equations of motion of the following Lagrangian

$$L = \frac{1}{2}\,\mu\, g_{\mu\nu}\, \dot{x}^\mu \dot{x}^\nu, \tag{7}$$

where the dots mean derivatives with respect to the proper time. Except from the Lagrangian itself $L = -\mu$, the MN metric has two more integrals of motion, namely the energy (per unit mass)

$$E = -\frac{\partial L}{\partial \dot{t}}/\mu = f(\dot{t} - \omega\,\dot{\varphi}), \tag{8}$$

and the z-component of the angular momentum (per unit mass)

$$L_z = \frac{\partial L}{\partial \dot{\varphi}}/\mu = f\omega(\dot{t}-\omega\,\dot{\varphi}) + f^{-1}\rho^2\dot{\varphi}, \tag{9}$$

The Kerr metric has one more integral of motion which is independent and in involution with the other integrals, the so-called Carter constant [14]. Thus the Kerr metric corresponds to an integrable system. However, the MN model misses such a constant, which means that MN corresponds to a non-integrable system, and therefore chaos may appear.

By simple algebraic manipulations of the metric (1) and the eqs. (8), (9) we find that the MN system satisfies the relation

$$\frac{1}{2}(\dot{\rho}^2+\dot{z}^2) + V_{\text{eff}}(\rho, z) = 0, \tag{10}$$

where the effective potential $V_{\text{eff}}(\rho, z)$ depends on q, E and L_z, and the geodesic motion can be restricted on a meridian plane (φ = const). Moreover, the motion takes place inside a region defined by $V_{\text{eff}}(\rho, z) \leq 0$, the border of which is the curve of zero velocity (CZV)

$$V_{\text{eff}} \equiv \frac{1}{2}e^{-2\gamma}\left[f - E^2 + \left(\frac{f}{\rho}(L_z - \omega E)\right)^2\right] = 0. \tag{11}$$

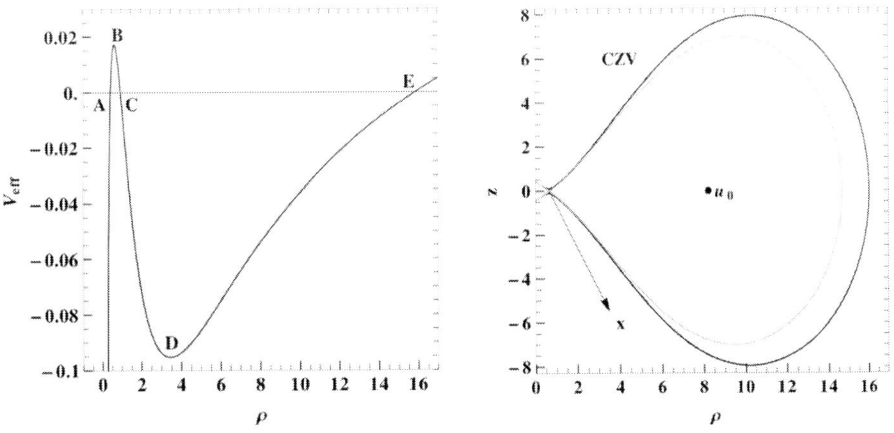

Figure 1. The left panel shows the portrait of the effective potential V_{eff} along the equatorial line ($z = 0$) for $E = 0.95$, $L_z = 2.5$, $\chi = 0.9$, and $q = 0$ (Kerr metric). The horizontal line $V_{\text{eff}} = 0$ indicates the upper limit reached by the non-escaping to infinity orbits, while the letters A, C, and E indicate the position of the roots of V_{eff}. Letters B, and D indicate the position of local extrema of V_{eff}. The right panel shows the CZV (black curve) on the meridian plane (ρ, z) for a slightly lower angular momentum than in the left panel ($L_z = 2.46$), the separatrix (gray curve) emanating from the unstable periodic orbit labeled by the letter "x" and the crossing of a stable periodic orbit through the $z = 0$ line (big dot) labeled by $\mathbf{u}_0$.

From the shape of the effective potential (11) we can discern the basic dynamics of the system like in a classical system. For instance in the integrable case of the Kerr metric the roots of the effective potential provide the boundaries of the non-escaping to infinity orbits (A, C, E in the left panel of Fig. 1). Between C and E lie the bounded non-plunging orbits, while from A leftwards lie the orbits that are plunging to the black hole (left panel of Fig. 1). Moreover, the local extrema of V_{eff} implies the existence of simple periodic orbits, for example the local minimum D indicates the existence of a simple stable periodic orbit indicated by $\mathbf{u}_0$ in the right panel of Fig. 1, while the local maximum B implies that a simple unstable periodic orbit will appear (x in the right panel of Fig. 1) for a specific range of V_{eff} parameters, namely if we lower the L_z in the case shown in the left panel of 1 to $L_z = 2.46$. When the local maximum B becomes negative, the separation between plunging and non-plunging orbits is defined by the separatrix (gray curve in the right panel) emanating from the unstable periodic orbit x. The orbits which lay between $\mathbf{u}_0$ and the part of the separatrix encircling $\mathbf{u}_0$ are bounded non-plunging orbits, all the other orbits inside the CZV plunge into the black hole. For a specific set of E and L_z values (lower than those in Fig. 1), when all the five points indicated by the letters shown in the left panel of Fig. 1 have merged into one point, the periodic orbits x and $\mathbf{u}_0$ merge into an indifferently stable periodic orbit, which is known in the bibliography as the Innermost Stable Circular Orbit (ISCO). Inside from an ISCO (leftwards in Fig. 1) only plunging orbits exist.

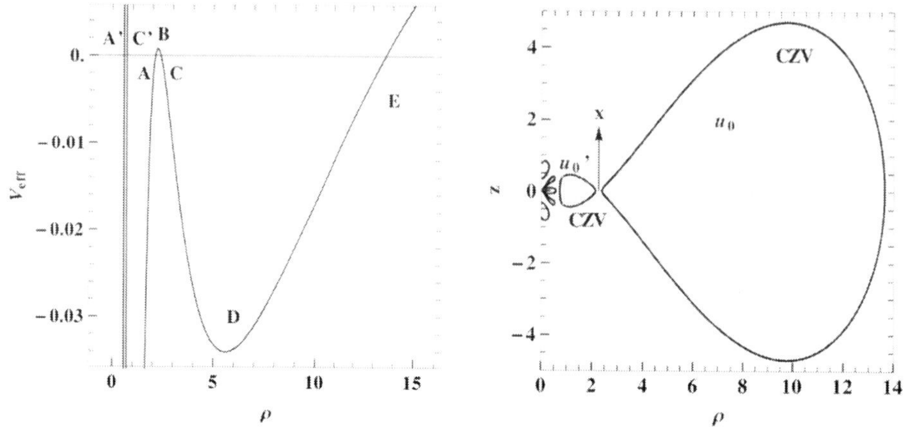

Figure 2. The left panel shows the portrait of the effective potential $V_{\rm eff}$ along the equatorial line ($z = 0$) for $E = 0.95$, $L_z = 3$, $\chi = 0.9$, and $q = 0.95$. The letters A′, C′, A, C and E indicate the position of the roots of $V_{\rm eff}$. Letters B, and D indicate the position of local extrema of $V_{\rm eff}$. The right panel shows the CZVs (black curves) on the meridian plane (ρ, z), and the projections on the plane of the simple stable periodic orbits $\mathbf{u}_0$, $\mathbf{u}_0'$ (gray curves). While x indicates the region where an unstable periodic orbit will appear for lower angular momentum L_z, or greater energy E.

In the case of the subclass of the MN family and for the parameters we have examined, the $V_{\rm eff}$ takes a little bit different shape (left panel of Fig. 2). Except from the roots we have seen in the Kerr case, two extra root appear the A′ and C′. Hence, an extra non-plunging region between C′ and A appear. The local minimum between C′ and A is extremely low (far out of the range shown in Fig. 2) and it indicates the existence of a new simple stable periodic orbit ($\mathbf{u}_0'$ in the right panel of Fig. 2). B and D again indicate a simple unstable periodic orbit (when B is below $V_{\rm eff}$) and a simple stable periodic orbit (x and $\mathbf{u}_0$ in the right panel of Fig. 2) respectively. Similarly as in the Kerr case, for a specific set of E and L_z values, when all the five points indicated by the not primed letters shown in the left panel of Fig. 2 have merged into one point, the periodic orbits x and $\mathbf{u}_0$ merge into an indifferently stable periodic orbit, which although it is not the innermost stable circular orbit, we are going to call it ISCO, like the authors of [23] did. Moreover, five leaf-like plunging regions appear on the left side of the right panel of Fig. 2.

If we cut the phase space along the along the equatorial plane ($z = 0$) plane and keep only the orbits crossing it with positive velocity, we produce a surface of section, also known as a Poincaré section. In Fig. 3 we show such a section and the periodic orbits x, $\mathbf{u}_0'$ and $\mathbf{u}_0$ on it. The scattered points seen in Fig. 3 belong to a chaotic region. Because the MN corresponds to a non-integrable system, this chaotic region is swept densely by the asymptotic manifolds emanating from x instead of the separatrix of the integrable case.

4. The final stages of an accretion disk in a Manko-Novikov spacetime

Bambi and Barausse in [23] found that during the accretion process in MN spacetime the gas crossing the ISCO can follow four different types of plunge:

(i) The gas plunges radially into the compact object like in the Kerr case.
(ii) The gas plunges radially, but gets trapped in a region near the compact object.
(iii) The gas plunges to the central object due to a vertical to the equator instability.

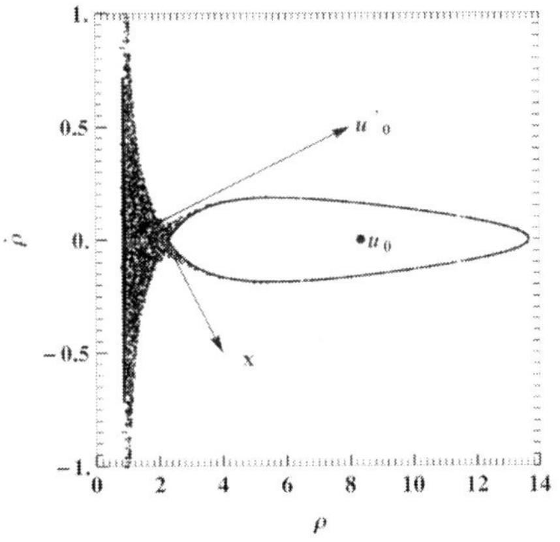

Figure 3. A Poincaré section $(\rho, \dot{\rho})$ along the $z = 0$ for $E = 0.95$, $L_z = 2.98$, $\chi = 0.9$, and $q = 0.95$. The position of the simple periodic orbit $\mathbf{u}_0$ is shown by a big dot, while the positions of the simple unstable periodic orbit x and of the simple periodic orbit $\mathbf{u}'_0$ are shown by arrows.

(iv) The gas plunges due to a vertical to the equator instability, but it gets trapped to two symmetrical to the equator regions.

Our study in a way follows up the second scenario, and explores what happens with the entrapped gas. The gas in our simple model consists of collisionless "test particle" fluid. We neglect the self-gravity of the gas, and the particles track the geodesic orbits of the MN spacetime background. The above approximation might seem a little irrelevant for a gas where the free path is short. However, when the gas particles collide, they just "exchange" their position in the phase space; in a great view they all still trace the background on which they are moving even if each of them individually does not follow geodesic orbit, but rather a great collection of small geodesic paths. In fact, we are not so interested on each particle's individual orbit, but rather we are concerned with the escape rate of a "swarm" of particles, thus, in our simple approximation we allow ourselves to ignore the collisions. Moreover, in our scenario the particles of the accretion matter share a constant angular momentum $L_z = 3$, but they radiate energy E away. This setup is like having a disk which radiates without losing its angular momentum.

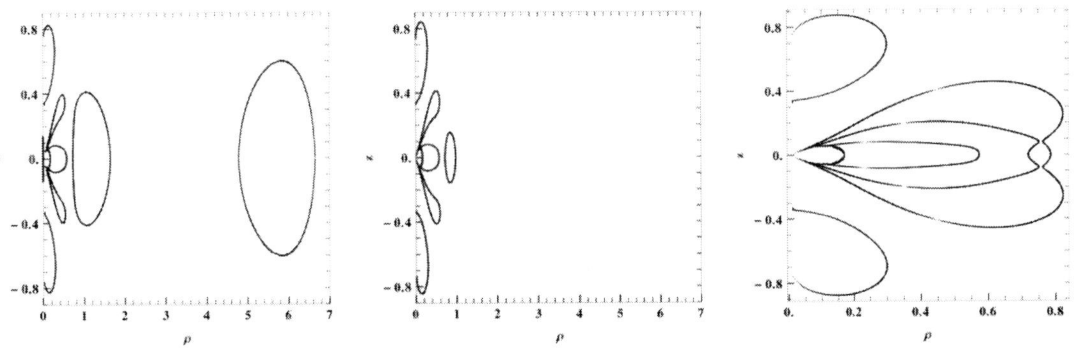

Figure 4. The CZVs as we reduce energy from the left panel to the right panel for constant $L_z = 3$ (beginning from left $E = 0.933$, $E = 0.7$, $E = 0.1899$).

We investigated the MN spacetime background for the values $q = 0.95$, and $\chi = 0.9$. By keeping the z-angular momentum $L_z = 3$ fixed and reducing the energy, the outer region around $\mathbf{u}_0$ (right panel of Fig. 2) gradually shrinks (left panel of Fig. 4), until it disappears and only the inner region around $\mathbf{u}'_0$ remains (central panel of Fig. 4). During the energy reduction the leaf like plunging region on either side of the inner region expand rightwards towards the inner region, until they reach it (right panel of Fig. 4). Our study in [19] focuses on the phase space transition from having bounded non-plunging orbits to getting plunging orbits. One must keep in mind that all the different energy states coexist, and the particles slowly drift with different rates of energy loss towards the inner regions.

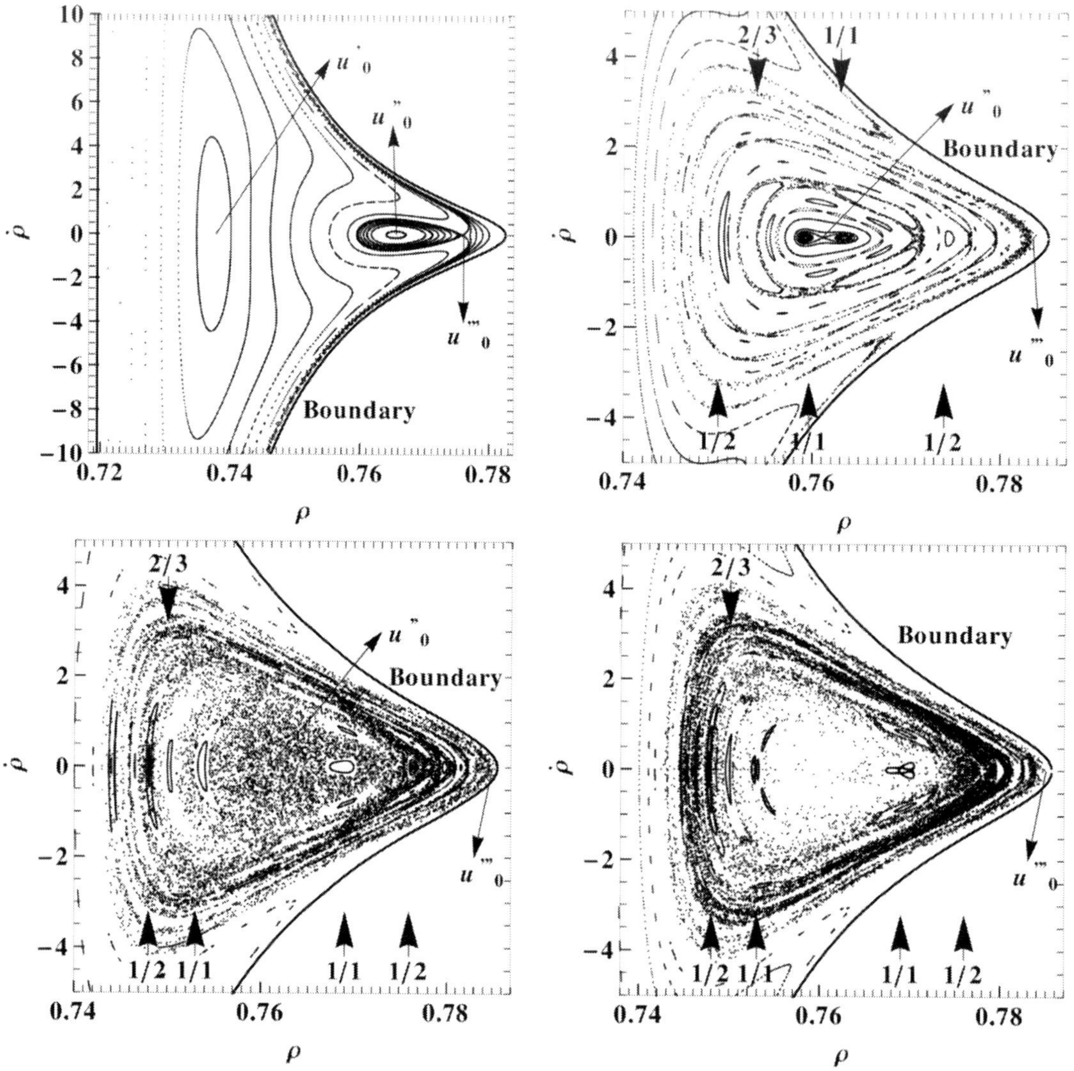

Figure 5. Surfaces of section $(\rho, \dot{\rho})$ for different energies: upper left panel $E = 0.192$, upper right panel $E = 0.19045$, lower left panel $E = 0.19015$ and lower right panel $E = 0.1901$. On the plots the simple periodic orbits $\mathbf{u}'_0$, $\mathbf{u}''_0$, $\mathbf{u}'''_0$ and the positions on the $\dot{\rho} = 0$ line of low order periodic orbits 1/1, 1/2, 2/3 are labeled.

It is quite interesting to monitor the phase space transition of the inner region from a region containing non-plunging orbits to a region with plunging orbits. Initially the region is occupied mostly by regular orbits, but about $E = 0.192$ a tangent bifurcation produces a couple of new simple periodic orbits $\mathbf{u}''_0$, $\mathbf{u}'''_0$ (upper left panel of Fig. 5), a stable and an unstable one respectively. The periodic orbit $\mathbf{u}_0''$ changes its stability to instability and vice versa an infinite number of times, while periodic orbits, like the low order ones seen in the upper right panel of Fig. 5, further bifurcate from it. This leads to a gradually increasing volume occupied by chaotic orbits (lower left panel of Fig. 5), and when the CZV around the inner region is reached by the leaf-like plunging region (right panel of Fig. 4), then the chaotic orbits lying around $\mathbf{u}_0''$ become plunging (lower right panel of Fig. 5). As the energy drops the period of $\mathbf{u}_0''$ increases and tends to infinity. In fact it becomes actually infinite when the CZV of the inner region opens up (see [19]). In a way the orbit of $\mathbf{u}_0''$ "traces" the path to the chaotic orbits in order for them to plunge.

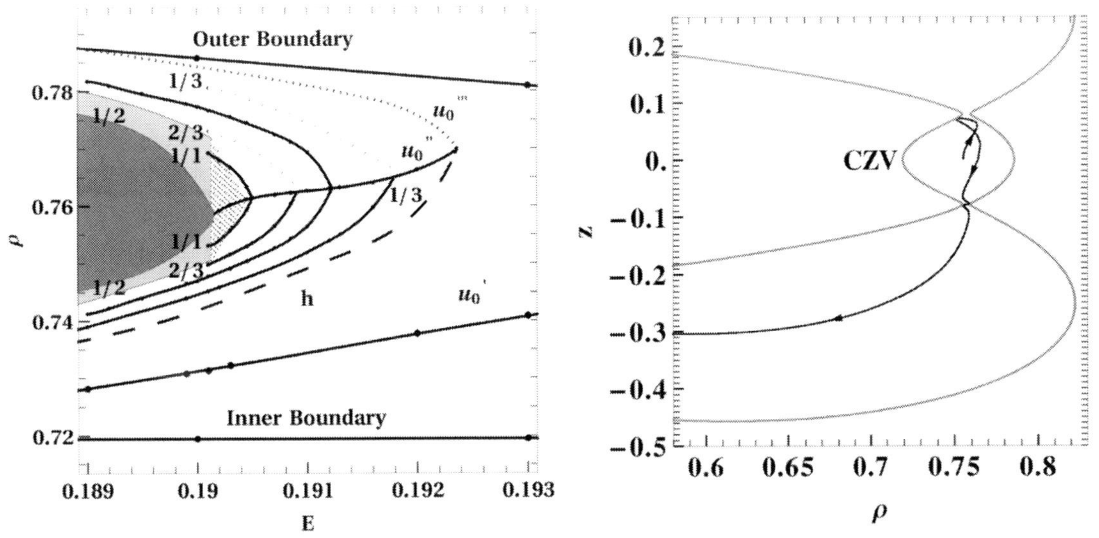

Figure 6. The left panel shows characteristics for the energy E of the families $\mathbf{u}_0'$, $\mathbf{u}_0''$, $\mathbf{u}_0'''$ and of some resonant bifurcations from the family $\mathbf{u}_0''$. The dotted region indicates regions where chaos is prominent. The gray region indicates regions with plunging orbits, while the red region gives the initial conditions (for $z = 0$) of orbits that plunge directly. The dashed line approximates the position of the first homoclinic section of the manifolds emanating from the orbit $\mathbf{u}_0'''$. The right panel shows a plunging orbit (black curve) and the CZV (red curve).

A summary of the procedure shown in Fig. 5 is given by the characteristic curves shown in the left panel of Fig. 6. This figure shows different bifurcations of the orbit $\mathbf{u}_0''$ and the creation of a prominent chaotic region (dotted region) around it as a function of the energy E, while E decreases. When the inner region joins the plunging region, the chaotic orbits become plunging (gray and red region). This mean that particles of our gas which move along a chaotic orbit will eventually plunge. Especially, particles moving along orbits corresponding to the red colored region (left panel of Fig. 6) will plunge directly, i.e., the particles will start from the equator but they will not cross it again with $\dot{z} > 0$ (e.g., in the right panel of Fig. 6 the orbit crosses the equator again but with $\dot{z} < 0$). The fact that the red region expands around the prolongation of the characteristic curve of the $\mathbf{u}_0''$ periodic orbit is in agreement with the aforementioned role of $\mathbf{u}_0''$ for creating a path for chaotic orbits to plunge, because it shows that the plunging paths lie around the $\mathbf{u}_0''$ periodic orbit. Thus, the left panel of Fig. 6 shows that the phase space undergoes an abrupt transition. Namely, the phase space ceases to be dominated by regular orbits and prominent chaotic regions appear just before the plunge begins.

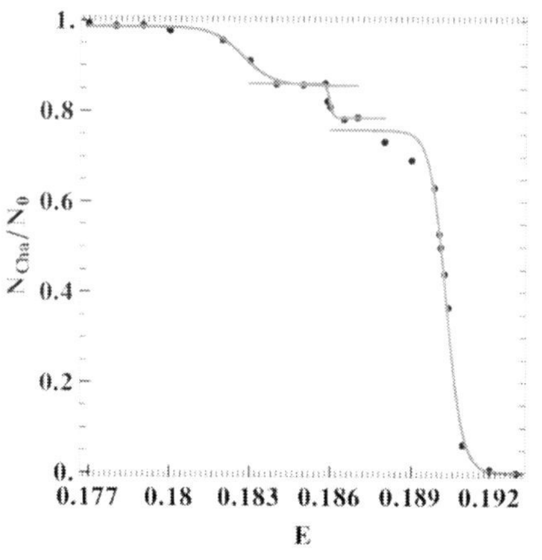

Figure 7. The proportion N_{cha}/N_0 of the chaotic-plunging orbits with initial conditions along the line $\dot{\rho} = 0$ as a function of the energy E.

This transition is shown better in Fig. 7, where we plot the proportion of chaotic-plunging orbits whose initial condition lie along the line $\dot{\rho} = 0$. The proportion changes abruptly at a narrow interval of decreasing energy values, and this implies that according to our simplistic model the majority of the gas particles will plunge along a non-equatorial trajectory, while their energy lies in this relatively narrow interval. Thus, in a electromagnetic spectrum coming from the vicinity of the MN central compact body we would expect a characteristic imprint of such a plunge, because the plunging particles are creating a non-radial cataclysmic inflow from the accretion disk. Moreover, in Fig. 7 we can see two more minor abrupt changes in the proportion N_{cha}/N_0, which means that we may have even some minor inflows as well, which will have their own characteristic imprints in the spectrum.

Since the particles move outside the equator our simplistic model should hold only when the enthalpy of the gas is small in comparison to the gravitational field of the central object. Otherwise, we have to take into account the gradient of the pressure of the gas and the geodesic approach has to be further investigated for its validity. On the other hand, we speculate that adding pressure would be just like putting a lid over a boiling pot, thus what we expect is that the energy interval of the cataclysmic inflow would just become narrower and the abrupt change in the fraction of chaotic plunging orbits (Fig. 7) more steep. Such a cataclysmic inflow of plunging matter should have an imprint on the radiation coming from the surrounding of the central compact object and could, in principle, be used as a method of detecting whether the black holes are described indeed by the Kerr metric or not, since such inflows are not expected to exist in the Kerr case.

5. Conclusions

By attempting to give an astrophysical interpretation to a dynamical study [19], we suggest a way that a cataclysmic inflow from an initially non-plunging region of the accreting matter towards the central compact object may lead to a method of testing the Kerr hypothesis by checking the electromagnetic spectra coming from the vicinity of the central compact object.

Acknowledgments

G. L-G is supported by the DFG grant SFB/Transregio 7.

References

[1] Amaro-Seoane P, Aoudia S, Babak S and et al 2012 *Classical Quant. Grav.* **29** 124016 (*Preprint* gr-qc/1202.0839)

[2] Bambi C 2011 *Mod. Phys. Lett. A* **26** 2453–2468 (*Preprint* gr-qc/1109.4256)
[3] Johannsen T 2012 *Adv. Astron.* **2012** 486750 (*Preprint* astro-ph.HE/1105.5645)
[4] Bambi C 2012 *Phys. Rev. D* **85** 043002 (*Preprint* gr-qc/1201.1638)
[5] Bambi C 2012 *J. Cosmol. Astropart. Phys.* **9** 14 (*Preprint* gr-qc/1205.6348)
[6] Johannsen T and Psaltis D 2010 *Astrophys. J.* **716** 187–197 (*Preprint* astro-ph.HE/1003.3415)
[7] Johannsen T and Psaltis D 2010 *Astrophys. J.* **718** 446–454 (*Preprint* astro-ph.HE/1005.1931)
[8] Johannsen T and Psaltis D 2011 *Astrophys. J.* **726** 11 (*Preprint* astro-ph.HE/1010.1000)
[9] Johannsen T and Psaltis D 2011 *Adv. Space Res.* **47** 528–532 (*Preprint* astro-ph.HE/1008.3902)
[10] Johannsen T and Psaltis D 2012 (*Preprint* astro-ph.HE/1202.6069)
[11] Pappas G 2012 *Mon. Not. R. Astron. Soc.* **422** 2581–2589 (*Preprint* astro-ph.HE/1201.6071)
[12] Collins N A and Hughes S A 2004 *Phys. Rev. D* **69** 124022 (*Preprint* gr-qc/0402063)
[13] Markakis C (*Preprint* astro-ph.SR/1202.5228)
[14] Carter B 1968 *Phys. Rev.* **174** 1559–1571
[15] Brink J 2011 *Phys. Rev. D* **84** 104015 (*Preprint* gr-qc/0911.4161)
[16] Gair J R, Li C and Mandel I 2008 *Phys. Rev. D* **77** 024035 (*Preprint* gr-qc/0708.0628)
[17] Lukes-Gerakopoulos G, Apostolatos T A and Contopoulos G 2010 *Phys. Rev. D* **81** 124005 (*Preprint* gr-qc/1003.3120)
[18] Contopoulos G, Lukes-Gerakopoulos G and Apostolatos T A 2011 *Int. J. Bifurcat. Chaos* **21** 2261 (*Preprint* gr-qc/1108.5057)
[19] Contopoulos G, Harsoula M and Lukes-Gerakopoulos G 2012 *Celest. Mech. Dyn. Astron.* **113** 255–278 (*Preprint* nlin.CD/1203.1010)
[20] Lukes-Gerakopoulos G 2012 *Phys. Rev. D* **86** 044013 (*Preprint* gr-qc/1206.0660)
[21] Vigeland S, Yunes N and Stein L C 2011 *Phys. Rev. D* **83** 104027 (*Preprint* gr-qc/1102.3706)
[22] Manko V S and Novikov I D 1992 *Classical Quant. Grav.* **9** 2477–2487
[23] Bambi C and Barausse E 2011 *Phys. Rev. D* **84** 084034 (*Preprint* gr-qc/1108.4740)

Journal of Physics: Conference Series **453** (2013) 012006 doi:10.1088/1742-6596/453/1/012006

Acceleration in Weyl geometry

J. Miritzis

Department of Marine Sciences, University of the Aegean, University Hill, Mytilene 81100, Greece

E-mail: imyr@aegean.gr

Abstract. We study homogeneous and isotropic cosmologies in a Weyl spacetime. We show that the field equations can be reduced to the Einstein equations with a two-fluid source and analyze the qualitative, asymptotic behavior of the models. Assuming an interaction of the two fluids we impose conditions so that the solutions of the corresponding dynamical system remain in the physically acceptable phase space. We show that in Weyl integrable spacetime, the corresponding scalar field acts as a phantom field and therefore, it may give rise to a late accelerated expansion of the Universe.

1. Introduction

The usual approaches for an explanation of the late-time acceleration of the universe are characterized by a departure from conventional cosmology. The proposed models, either assume the existence of dark energy [1, 2], or require a modification of general relativity at cosmological distance scales [3, 4], cf. [5, 6, 7, 8] for comprehensive reviews and references. Less explored is the idea that the geometry of spacetime is not the so far assumed Lorentz geometry (see for example [9]). Due to its simplicity Weyl geometry is considered as the most natural candidate for extending the Lorentzian structure.

We recall that a Weyl space is a manifold endowed with a metric $\mathbf{g}$ and a linear symmetric connection $\boldsymbol{\nabla}$ which are interrelated via

$$\nabla_\mu g_{\alpha\beta} = -Q_\mu g_{\alpha\beta}, \tag{1}$$

where the 1-form Q_μ is customarily called Weyl covariant vector field (see the Appendix in [10] for a detailed exposition of the techniques involved in Weyl geometry). We denote by D the Levi-Civita connection of the metric $g_{\alpha\beta}$.

A consistent way to incorporate an arbitrary connection into the dynamics of a gravity theory is the so-called constrained variational principle [11]. Applying this method in the context of Weyl geometry to the Lagrangian $L = R$, one obtains the field equations

$$G_{(\mu\nu)} = -\nabla_{(\mu} Q_{\nu)} + Q_\mu Q_\nu + g_{\mu\nu}\left(\nabla^\alpha Q_\alpha - Q^\alpha Q_\alpha\right) =: M_{\mu\nu},$$

where $G_{(\mu\nu)}$ is the symmetric part of the Einstein tensor (see equations (30) and (31) in [11]). If we express the tensors $G_{(\mu\nu)}$ and $M_{\mu\nu}$ in terms of the quantities formed with the Levi-Civita connection D, the field equations become

$$\overset{\circ}{G}_{\mu\nu} = \frac{3}{2}\left(Q_\mu Q_\nu - \frac{1}{2}Q^2 g_{\mu\nu}\right). \tag{2}$$

In the case of integrable Weyl geometry, i.e., when $Q_\mu = \partial_\mu \phi$, the source term is that of a massless scalar field. Taking the divergence of (2) and using the Bianchi identities we conclude that

$$D^\mu Q_\mu = 0.$$

In this paper we study Friedmann-Robertson-Walker (FRW) cosmologies in a Weyl framework. In Section 2 we explore the field equations derived from the Lagrangian $L = R + L_m$, where the matter Lagrangian, L_m, is chosen so that ordinary matter is described by a perfect fluid. It is shown that the presence of the Weyl vector field can be interpreted as a fluid and we analyze the asymptotic behavior of the models. Assuming an energy exchange between the two fluids we extend previous work [10]. In Section 3 we consider a modification of the Einstein-Hilbert Lagrangian, cf. (17), which may provide a mechanism of accelerating expansion.

2. Interacting fluids

In the following we assume an initially expanding FRW universe with expansion scale factor $a(t)$ and Hubble function $H = \dot{a}/a$. We adopt the metric and curvature conventions of [12]. An overdot denotes differentiation with respect to time t, and units have been chosen so that $c = 1 = 8\pi G$. Ordinary matter is described by a perfect fluid with energy-momentum tensor,

$$T_{\mu\nu} = (\rho_2 + p_2)\, u_\mu u_\nu + p_2 g_{\mu\nu}, \tag{3}$$

supplemented with an equation of state $p_2 = (\gamma_2 - 1)\,\rho_2$. Since for spatially homogeneous and isotropic spacetimes there is no preferred direction, Q^μ must be proportional to the fluid velocity u^μ, i.e.,

$$Q^\mu =: q u^\mu, \quad Q^2 = Q_\mu Q^\mu = -q^2.$$

In vacuum, we have to make the assumption that Q^μ is hypersurface orthogonal, i.e. it is proportional to the unit timelike vector field which is orthogonal to the homogeneous hypersurfaces. Formally the right-hand side of (2) can be rewritten as

$$\frac{3}{2}\left(Q_\mu Q_\nu - \frac{1}{2} Q^2 g_{\mu\nu}\right) = (\rho_1 + p_1)\, u_\mu u_\nu + p_1 g_{\mu\nu}, \tag{4}$$

with

$$\rho_1 = p_1 = \frac{3}{4} q^2, \tag{5}$$

i.e., the equation of state of the $q-$fluid corresponds to stiff matter. Therefore we are dealing with a two-fluid model with total energy density and pressure given by

$$\rho = \rho_1 + \rho_2, \quad p = p_1 + p_2, \tag{6}$$

respectively, where

$$p_1 = \rho_1, \quad p_2 = (\gamma_2 - 1)\,\rho_2, \tag{7}$$

i.e., $\gamma_1 = 2$ and $\gamma_2 < \gamma_1$.

The field equations are the Friedmann equation

$$H^2 + \frac{k}{a^2} = \frac{1}{3}\,(\rho_1 + \rho_2)\,, \tag{8}$$

and the Raychaudhuri equation

$$\dot{H} = -H^2 - \frac{1}{6}\left[(3\gamma_1 - 2)\,\rho_1 + (3\gamma_2 - 2)\,\rho_2\right]. \tag{9}$$

The Bianchi identities imply that the total energy-momentum tensor is conserved, so that an interaction between the two fluids is induced. It is necessary to make an assumption about the interaction between the two fluids cf [13], otherwise the field equations constitute an underdetermined system of differential equations. The simplest assumption is that the energy-momentum of each fluid is separately conserved, so that the two fluids do not interact and the densities decay independently,

$$\dot{\rho}_1 = -3\gamma_1 H\rho_1, \quad \dot{\rho}_2 = -3\gamma_2 H\rho_2. \tag{10}$$

The dynamical system (8)-(10) was analyzed in [10]. It was found that in expanding models the "real" fluid always dominates at late times and therefore the contribution of the Weyl fluid to the total energy-momentum tensor is important only at early times. The purpose of this section is to weaken the requirement of separate conservation of the two fluids.

In many cosmological situations the transfer of energy between two fluids is important, so one may assume that the two fluids exchange energy. The following simple model was proposed by Barrow and Clifton (see [14] for motivation and further examples),

$$\dot{\rho}_1 = -3\gamma_1 H\rho_1 - \beta H\rho_1 + \alpha H\rho_2, \tag{11}$$

$$\dot{\rho}_2 = -3\gamma_2 H\rho_2 + \beta H\rho_1 - \alpha H\rho_2, \tag{12}$$

where α and β are constants so that the total energy is conserved (see also [15] for a singularity analysis of the master equation derived in [14]).

Remark. In the case of separately conserved fluids, equations (10) imply that the sets $\rho_1 = 0$ and $\rho_2 = 0$, are invariant sets for the dynamical system and by standard arguments, if $\rho_i > 0$, $i = 1, 2$, for some initial time t_0, then $\rho_i(t) > 0$ throughout the solution. This fact can be made more transparent by the following argument. Assuming that $\gamma_1 > \gamma_2$, we define the transition variable $\chi \in [-1, 1]$

$$\chi = \frac{\rho_2 - \rho_1}{\rho_2 + \rho_1}, \tag{13}$$

which describes which fluid is dominant dynamically [13]. Applying the conservation equation to ρ_1 and ρ_2, one obtains the evolution equation of the variable χ,

$$\dot{\chi} = \frac{3}{2} H (\gamma_1 - \gamma_2)\left(1 - \chi^2\right), \tag{14}$$

which implies that the sets $\chi = \pm 1$ are invariant under the flow of the dynamical system. Furthermore, the transition variable χ is bounded, that is, if initially $\chi \in [-1, 1]$, it remains in that interval for all t. However, the choice (11) and (12), has the peculiarity that the sets $\rho_1 = 0$ and $\rho_2 = 0$, are no longer invariant sets for the dynamical system and therefore, the sign of the functions ρ_1 and ρ_2 is not conserved. This is also reflected to the fact that the transition variable χ no longer satisfies (14) and eventually escapes outside the interval $[-1, 1]$, thus exhibiting unphysical behavior.

In order to circumvent these difficulties, one has to impose further conditions on α and β. It turns out that the assumption $\beta = -\alpha$ in (11) and (12) is a sufficient condition for the boundness of the function χ. With this assumption we adopt the Coley and Wainwright formalism [13], for a general model with two fluids. The state of the system consists of the couple (χ, Ω), where $\Omega = \Omega_1 + \Omega_2$ is the total density parameter, $\Omega = \rho/3H^2$. In order to allow for closed models in our analysis, we define the compactified density parameter ω, (see [16])

$$\Omega = \frac{1}{\tan^2 \omega}, \tag{15}$$

or

$$\omega = \arctan\left(\frac{\sqrt{3}H}{\sqrt{\rho}}\right), \quad \text{with} \quad -\pi/2 \leq \omega \leq \pi/2.$$

We see that ω is bounded at the instant of maximum expansion ($H = 0$) and also as $\rho \to 0$, in ever-expanding models. Finally, defining a new time variable τ by

$$\frac{d\tau}{dt} = \frac{3(\gamma_1 - \gamma_2)}{2}\sqrt{\frac{\rho}{3}}\frac{1}{\cos\omega},$$

one obtains the following dynamical system

$$\begin{aligned}\frac{d\omega}{d\tau} &= -\frac{1}{2}(b - \chi)\cos 2\omega\cos\omega \\ \frac{d\chi}{d\tau} &= (\chi_\alpha^2 - \chi^2)\sin\omega,\end{aligned} \tag{16}$$

where the constant b is

$$b = \frac{3(\gamma_1 + \gamma_2) - 4}{3(\gamma_1 - \gamma_2)} > -1.$$

In our case, we always have $\gamma_1 = 2$. The parameter χ_α which determines the range of χ is given by

$$\chi_\alpha = \sqrt{1 - \frac{4\alpha}{3(\gamma_1 - \gamma_2)}}, \qquad 0 \leq \alpha \leq \frac{3(\gamma_1 - \gamma_2)}{4}.$$

The phase space of the two-dimensional system (16) is the closed rectangle

$$D = [-\pi/2, \pi/2] \times [-\chi_\alpha, \chi_\alpha]$$

in the $\omega - \chi$ plane (see Figure 1). Since $0 \leq \chi_\alpha \leq 1$, the rectangle D is shrinked compared to the phase space in [10] and [16].

The invariant sets of the system are denoted in the following table.

$\omega = -\pi/2$	contracting empty models	$\Omega = 0, H < 0$
$\chi = -\chi_\alpha$	scaling solution	$\frac{\Omega_2}{\Omega_1} = \frac{1-\chi_\alpha}{1+\chi_\alpha}$
$\chi = +\chi_\alpha$	scaling solution	$\frac{\Omega_1}{\Omega_2} = \frac{1-\chi_\alpha}{1+\chi_\alpha}$
$\omega = \pi/4$	expanding flat models	$\Omega = 1, H > 0$
$\omega = -\pi/4$	contracting flat models	$\Omega = 1, H < 0$
$\omega = \pi/2$	expanding empty models	$\Omega = 0, H > 0.$

It is easy to verify that the equilibrium points lie at the intersection of these sets and are denoted by $F_{1,2}^{\pm}$ (expanding or contracting flat model) and $O^{\pm}$ (expanding or contracting open model). The subscripts indicate which fluid dominates. Linearization around the equilibrium points is sufficient for the characterization of their stability and the result is the phase portrait is shown in Figure 2.

Regions III and IV correspond to expanding models. The F_1^+ is a past attractor of all models with $\Omega > 0$, i.e., the evolution near the big bang is approximated by the flat FRW model where the Weyl fluid dominates. Open models expand indefinitely and approach at late time a "scaling solution" where the Weyl fluid keeps a small fraction of the total energy density. Flat models expand indefinitely and the evolution is approximated by the flat FRW universe at late time. In both cases the "real" second fluid dominates at late times. On the other hand, any

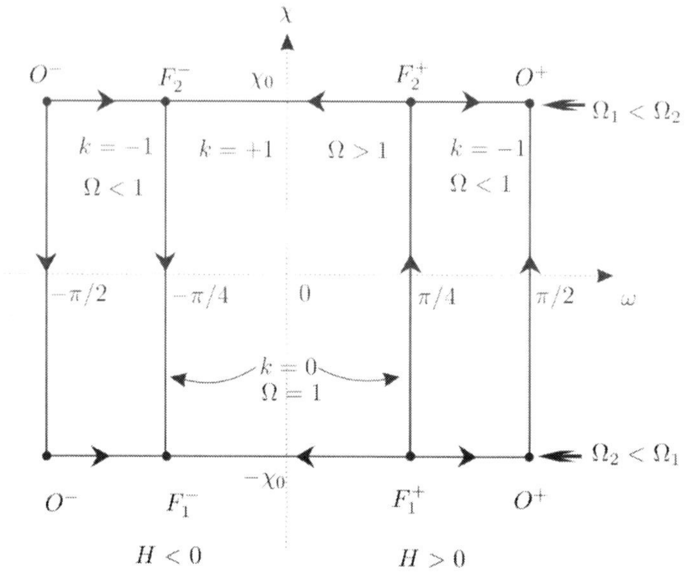

Figure 1. The invariant sets and equilibrium points of (16)

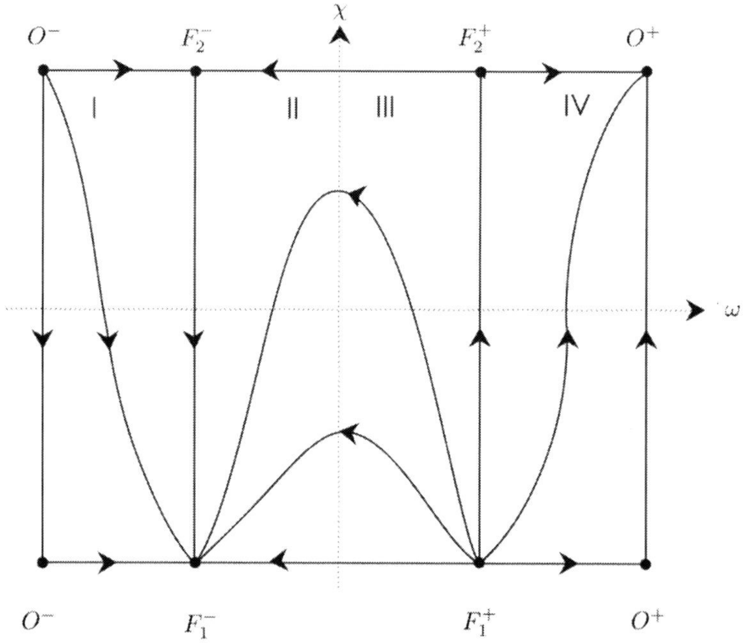

Figure 2. The phase portrait of (16) with $\gamma_2 = 1$.

initially expanding closed model in region III, however close to F_2^+, eventually recollapses and the evolution is approximated by the flat FRW model where the Weyl fluid dominates.

We therefore, conclude that the Weyl fluid has significant contribution only near the cosmological singularities. In expanding models, the "real" fluid always dominates at late times and therefore, the contribution of the Weyl fluid to the total energy-momentum tensor is

Journal of Physics: Conference Series **453** (2013) 012006 doi:10.1088/1742-6596/453/1/012006

important only at early times.

3. Phantom from pure geometry

The field equations (2) constitute the generalization of the Einstein equations in a Weyl spacetime in the sense that they come from the Lagrangian $L = R$. There is however an alternative view, namely that the pair $(Q, \mathbf{g})$ which defines the Weyl spacetime also enters into the gravitational theory and therefore, the field Q must be contained in the Lagrangian independently from $\mathbf{g}$. In the case of integrable Weyl geometry, i.e. when $Q_\mu = \partial_\mu \phi$ where ϕ is a scalar field, the pair $(\phi, g_{\mu\nu})$ constitutes the set of fundamental geometrical variables. We stress that the nature of the scalar field is purely geometric. A simple Lagrangian involving the set $(\phi, g_{\mu\nu})$ is given by

$$L = R + \xi \nabla^\mu Q_\mu + L_m, \tag{17}$$

where ξ is a constant and L_m corresponds to the Lagrangian yielding the energy-momentum tensor of a perfect fluid. Motivations for considering theory (17) can be found in [17, 18] (see also [19] for a multidimensional approach and [20] for an extension of (17) to include an exponential potential function of ϕ). By varying the action corresponding to (17) with respect to both $g_{\mu\nu}$ and ϕ one obtains

$$\overset{\circ}{G}_{\mu\nu} = \frac{3-4\xi}{2}\left(\partial_\mu\phi\partial_\nu\phi - \frac{1}{2}\left(\partial_\alpha\phi\partial^\alpha\phi\right) g_{\mu\nu}\right) + T_{\mu\nu}, \tag{18}$$

and

$$\overset{\circ}{\Box}\phi = \frac{1}{3-4\xi}\rho, \tag{19}$$

where $\overset{\circ}{\Box}$ is the D'Alembertian operator formed with the Levi-Civita connection D. As mentioned above, ordinary matter described by $T_{\mu\nu}$ is a perfect fluid with energy density ρ and pressure p. Setting

$$\lambda = \frac{4\xi - 3}{2},$$

we note that for $\lambda < 0$ the field equations are formally equivalent to general relativity with a massless scalar field coupled to a perfect fluid. In Weyl spacetime, the scalar field has a geometric nature and no restriction exists for the sign of the value of λ. For $\lambda > 0$, the Weyl field ϕ plays the role of a phantom scalar field and therefore, it may provide a mechanism of late time acceleration. Further investigation of this issue is the subject of future research.

References

[1] Sahni S and Starobinsky A 2000 *Int. J. Mod. Phys.* D**9** 373
[2] Peebles PJ and Ratra B 2003 *Rev. Mod. Phys.* **75** 559
[3] Carroll S, Duvvuri V, Trodden M and Turner M 2004 *Phys. Rev.* D**70** 043528
[4] Chiba T 2003 *Phys. Lett.* **B**575 1
[5] Copeland EJ , Sami M and Tsujikawa S 2006 *Int. J. Mod. Phys.* D**15** 1753
[6] Sotiriou T and Faraoni V 2010 *Rev. Mod. Phys.* **82** 451
[7] De Felice A and Tsujikawa S 2010 *Living Rev. Rel.* **13** 3
[8] Bamba K, Capozziello S, Nojiri S, Odintsov SD 2012 *Astrophysics and Space Science* **342**:155
[9] Capozziello S, Carloni S and Troisi A 2003 *Preprint* astro-ph/0303041; Capozziello S, Cianci R, Stornaiolo C and Vignolo S 2008 *Phys. Scripta* **78**:065010
[10] Miritzis J 2004 *Class. Quantum Grav.* **21** 3043
[11] Cotsakis S, Miritzis J and Querella L 1999 *J. Math. Phys.* **40** 3063
[12] Wainwright J and Ellis GFR 1997 *Dynamical Systems in Cosmology* (Cambridge: Cambridge University Press)
[13] Coley AA and Wainwright J 1992 *Class. Quantum Grav.* **9** 651
[14] Barrow JD and Clifton T 2006 *Phys. Rev.* D**73** 103520
[15] Cotsakis S and Kittou G 2012 *Phys. Lett.* B**712** 16

[16] Wainwright J 1996 *Relativistic Cosmology* In Proceedings of the 46th Scottish Universities Summer School in Physics Aberdeen pp 107-141 Eds GS Hall and JR Pulham (Institute of Physics Publishing)
[17] Novello M, Oliveira LAR, Salim JM and Elbas E 1993 *Int. J. Mod. Phys.* D**1** N 3-4 641
[18] Salim JM and Sautu SL 1996 *Class. Quantum Grav.* **13** 353
[19] Konstantinov MY and Melnikov VN 1995 *Int. J. Mod. Phys.* D**4** 339
[20] Oliveira HP, Salim JM and Sautu SL 1997 *Class. Quantum Grav.* **14** 2833

Journal of Physics: Conference Series **453** (2013) 012007 doi:10.1088/1742-6596/453/1/012007

Short-range Gravity experiment using digital image analysis

Kazufumi Ninomiya[1,2], **Reiko Kishi**[1], **Haruna Murakami**[1], **Hironori Nishio**[1], **Naruya Ogawa**[1], **Atsushi Taketani**[2] **and Jiro Murata**[1]

[1]Department of Physics, Rikkyo University, Nishi-Ikebukuro, Tokyo 171-8501, JAPAN
[2]RIKEN Nishina Center, 2-1 Hirosawa, Wako, Saitama 351-0198, JAPAN

E-mail: kazufumi@rikkyo.ac.jp

Abstract. According to a large extra dimension model, a deviation from Newton's inverse square law is expected at below sub-millimeter range. We have developed an experimental set up using a torsion balance pendulum and an online digital-image analysis system, aiming to test the Newtonian inverse square law at below millimeter scale. In addition, composition dependence of gravitational constant G is also tested at a millimeter scale, motivated to test the weak equivalence principle. In this paper, current status and results are described.

1. Introduction

1.1. Newtonian inverse square law

Extreme weakness of the gravitational force comparing to the other three gauge interactions is considered as one of the most severe problem in theoretical physics, which is known as the hierarchy problem. In order to build a unified theory including all four interactions, we cannot avoid resolving this problem. According to a large extra-dimension model which is known as the ADD model [1], the hierarchy problem can be naturally resolved in a geometrical understanding by introducing additional spatial dimensions outside of our four dimensional world in the brane world scenario. Astonishingly, the ADD model predicts that the size of the extra-dimension can be as large as a millimeter scale in order to set the high dimensional Planck energy to be around 1TeV. Existence of such additional spatial dimensions requires a modification of the gravitational inverse square law below the size of the extra dimension scale. For example, if we assume n large extra dimensions with their size of λ, the gravitational potential can be expressed in a modified form because of Gauss's law as followings;

$$\begin{aligned} V_{r\geq\lambda} &= -G\frac{Mm}{r} \\ V_{r\leq\lambda} &= -G'\frac{Mm}{r^{1+n}}, \end{aligned} \tag{1}$$

here, higher dimensional gravitational constant is defined as $G' = \lambda^n G$. To compare with experimental tests of various gravitational models, a modified gravitational potential is often expressed in a Yukawa interaction form.

$$V(r) = -G\frac{mM}{r^2}\left(1 + \alpha e^{-\frac{r}{\lambda}}\right), \tag{2}$$

here, α is coupling constant and λ is the range of the new interaction. Numbers of experiments have been tested the inverse square law, excluding the non-zero value of α in various length scales. However, as shown in Figure 1, high precision tests have been performed in a very limited region around astronomical scales. Little is known at below millimeter range, where deviation from the inverse square law is predicted by the ADD model.

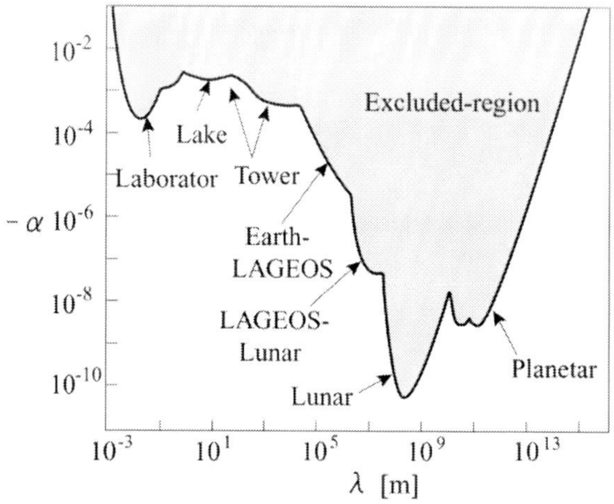

Figure 1. Summary of experimental search of the Yukawa term. Excluded region is illustrated as the shaded area [2]

1.2. Weak equivalence principle

We have developed a new type of experimental device using video imaging sensor to perform an experiment using a torsion balance pendulum to test the inverse square law at below millimeter scale. Our device can investigate not only the inverse square law, but also can test weak equivalence principle (WEP) using the same experimental technique. The weak equivalence principle is expressed as that, ratios between gravitational mass and inertia mass for every material are universal for any compositions. From the definition of the inertial mass $a = F/m_I$ and that of gravitational mass $F \propto m_g$, gravitational acceleration satisfies $a \propto m_g/m_I$. If the ratio between gravitational mass and inertia mass is constant for every material, the universality of free fall must be kept. Therefore, test of the universality of free fall can be regarded as a test of the weak equivalence principle.

Number of experimental tests confirmed that ratios between inertial mass for different compositions are same for the ratio between gravitational mass, using gravitational force from the Earth and the Sun. On the other hand, there are no experimental tests of this weak equivalence principle at a short range below cm scale. On this motivation, we have performed an experiment to test the weak equivalence principle by testing the universality of the gravitational constant G for different compositions [3]. There are several theoretical models which predict violation of the weak equivalence principle [4, 5, 6, 7]. Baryon number coupling force is one of such models. If a new interaction which couples to baryon number, the weak equivalence principle seems to be violated. Because nuclear mass is very similar to the baryon number, it is difficult to

Journal of Physics: Conference Series **453** (2013) 012007 doi:10.1088/1742-6596/453/1/012007

distinguish between them. However, the gravitational mass is slightly different from the baryon number time nucleon mass, because of the mass defect and proton/neutron mass difference. Therefore, existence of the new baryon number coupling force can be examined by measuring the gravity-like force between different substances.

2. Principle

2.1. Torsion pendulum

In our experiment, a gravitational force is measured using a torsion pendulum with an image analysis system to monitor the pendulum movement. Gravitational torque can be measured as an angular twisting displacement of the torsion balance bar, which can be associated with the torque in a form of Hooke's law with a spring constant κ.

$$\tau = -\kappa \cdot \Delta\theta, \tag{3}$$

here, τ is torque and κ is torsional spring constant. The torsional spring constant κ is estimated using the inertia moment and periodic oscillation of the torsion pendulum. The torsion pendulum is composed of two columns suspended by an aluminum bar. This column is called as "target". The torsion pendulum is hanged by $\phi 40\mu$m thin wire. A typical measuring procedure is shown in Figure 2. When the attractor position changes, balanced position between the spring force and gravitational force of the torsion pendulum is moved, resulting the angular displacement. The angular displacement of the pendulum is measured by a video camera viewing from the top of the apparatus. Figure 3 shows a typical time sequence data in such measurement, which clearly indicates the gravitational signal.

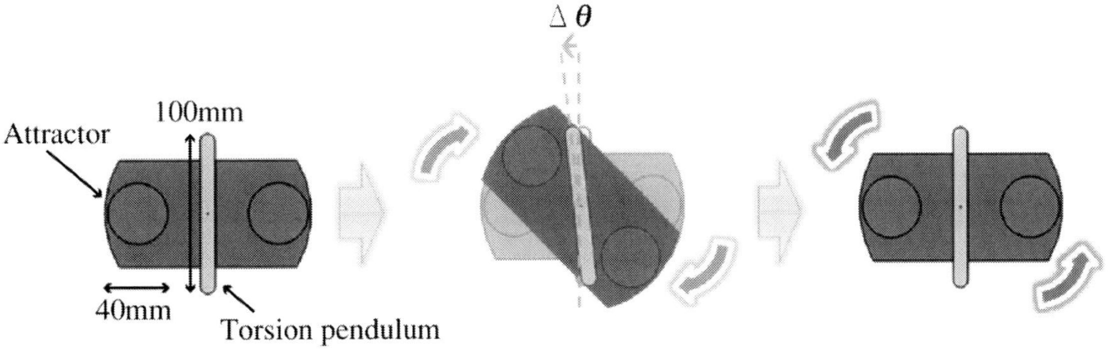

Figure 2. A typical measurement procedure, rotating the attractor position monitoring the angular displacement of the torsion pendulum.

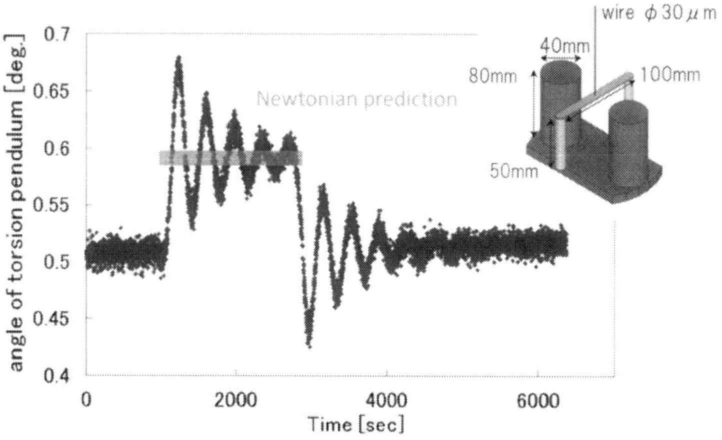

Figure 3. A typical time sequence data of the measurement like Figure 2.

2.2. Digital image analysis

We have developed a digital image analysis system for the position monitoring, which was originally developed for monitoring motion of the geometrical deformation of a large radiation detector which was used at a high energy collider experiment [8]. As shown in Figure 4, the video image is captured with a video camera, and stored as movie files recording the motion of the torsion pendulum. After then, image analysis is performed for all the static image frames extracted from the movie file. The angle information of the torsion pendulum can be extracted by performing a linear least square fitting on the pixel intensity data [9].

Angular resolution of around 1 micro degree is achieved using the image analysis system. This system can determine the angle and the offset position of the torsion pendulum at the same time, using the two dimensional information. It is a huge advantage comparing to a laser interferometer measurement which cannot distinguish the twisting motion from a parallel swinging motion without twisting, causing a big systematic error.

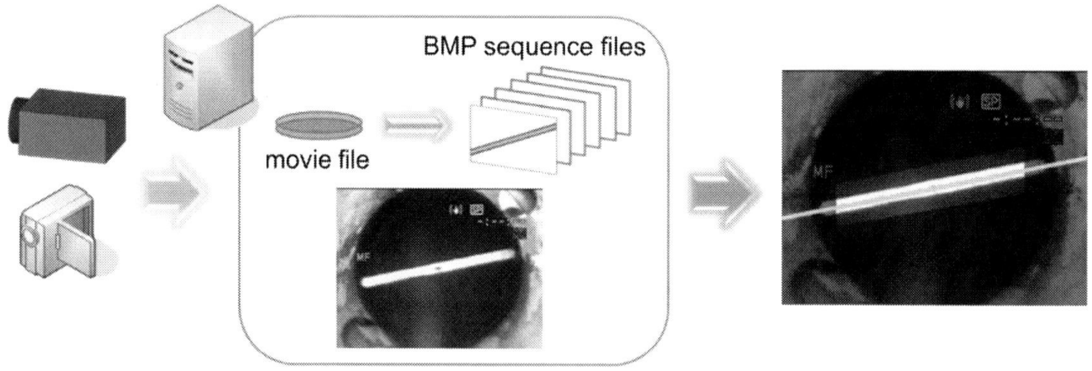

Figure 4. Digital image analysis technique determining the position and the angle of a line shaped object.

3. Experiment

3.1. Test of the Newtonian inverse square law

Figure 5 shows the experimental setup of Newton-SC experiment, which aims to test the inverse square law at centimeter scales. In this study, we are aiming to explore the existence of the extra dimension based on the ADD model. In this sense, we are aiming to perform the highest precision test of the inverse square law at below millimeter scale. Therefore, Newton-SC experiment is a performance test of our experimental devices towards sub-millimeter test. All the components of the device are electrically connected and made of non-magnetic metals. In addition, all the apparatus are set inside a vacuum chamber. In order to suppress mechanical vibration noises, the apparatus is set on a basement room at Rikkyo University.

Unlike the measurement shown in Figure 2, the attractors are continuously rotating around the target in a constant angular velocity in this measurement. Twisted angle of the torsion balance bar is measured as a function of the setting angle of the rotating attractor position relative to the target. Result from the Newton-SC experiment is shown in Figure 6. As shown in the Figure 6, the experimental data were obtained as continuous points, which are compared with theoretical predictions. Red line shows result of a numerical calculation supposing the Newtonian gravity. Other lines show examples of numerical calculations supposing modified gravitational potentials which include the strong Yukawa potential with different parameter settings. As a result, our data are consistent with Newtonian gravity in this measurement within experimental precision. We also set an excluded area on the $\alpha - \lambda$ plot in Figure 7. The yellow line is set using our experimental data. We succeeded to test the Newton's inverse square law at cm scale with 5% accuracy in this experiment.

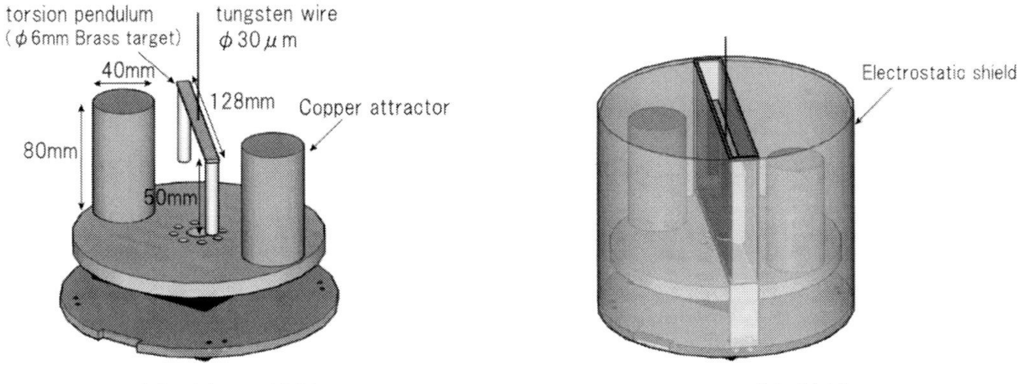

(a) without shield (b) shield

Figure 5. The Newton-SC experimental setup. Electrostatic shield is set between the attractors and the targets.

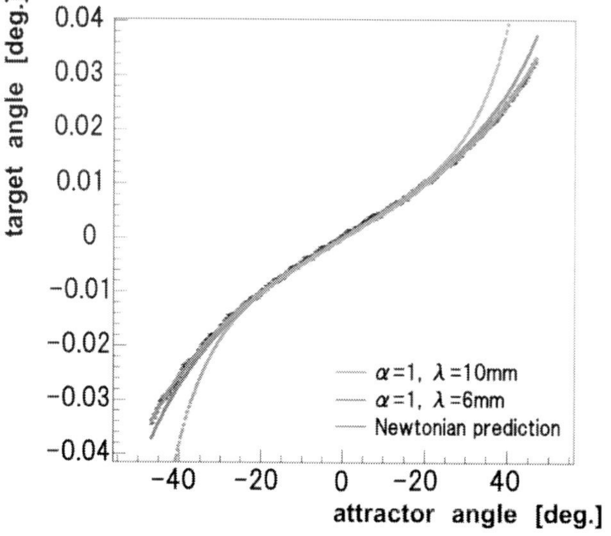

Figure 6. Experimental result of the Newton-SC experiment.

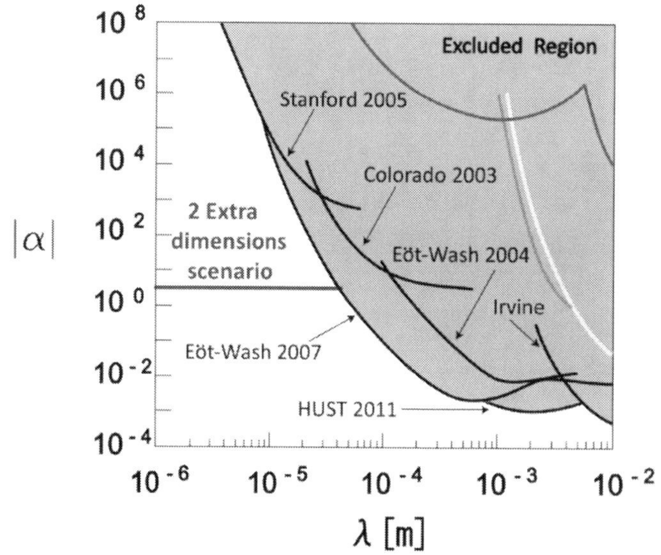

Figure 7. $\alpha - \lambda$ plot at below centimeter scale. Present result is also shown. [10, 11, 12, 13, 14, 15]

Unfortunately, our experimental sensitivity was not enough to set a new limit comparing to the past measurements. It is mainly because our apparatus is aiming to confirm its ability to observe gravitational signal using the simple video image analysis system, which was not designed to achieve the highest precision measurement. We are now performing a next generation experiment using a Null-type principle in order to maximize the sensitivity on the deviation from the inverse square law suppressing the Newtonian gravity.

3.2. Test of the weak equivalence principle

Test of the weak equivalence principle can be performed as a test of composition dependence of the gravitational constant, which requires measurements with different attractors. In order to suppress systematic errors caused by attractor exchanging, a next generation device, Newton-II, is developed. As shown in Figure 8, Newton-II is designed to be able to move the attractors rotating outside the torsion balance bar position, without changing the attractor materials by opening the vacuum chamber. The result is shown in Figure 9. Present configuration is designed to suppress the systematic error and maximize the sensitivity to the relative strength of gravitational force for different materials. For example, target zero angle position and attractor angular phase can be determined from the measured data without dominated by mechanical alignment precision.

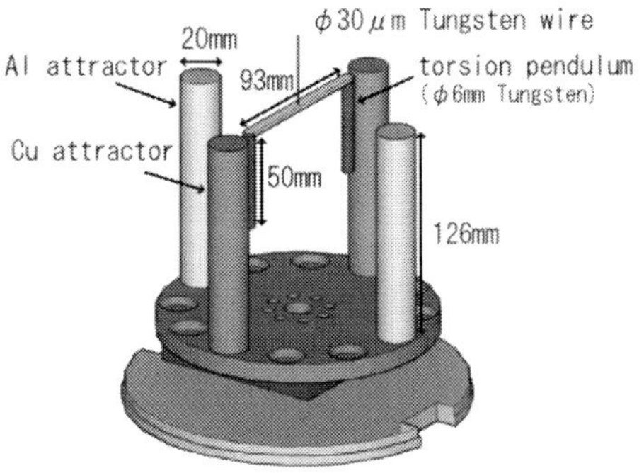

Figure 8. The Newton-II experimental setup

The Newton-II experiment is performed with continuously rotating attractors as same as for the Newton-SC experiment. Experimental results from the Newton-II experiment is shown in Figure 9. Contributions from the cupper attractors and aluminum attractors are clearly seen as different nodes, which can be compared with the theoretical calculation assuming known gravitational constant shown as a green line in the Figure 9 obtained as superimpose from the two compositions. A relative strength between different attractor materials provides precision information about the composition dependence of the gravitational constant. The obtained results are consistent with the universal gravitational constant which is independent of the compositions.

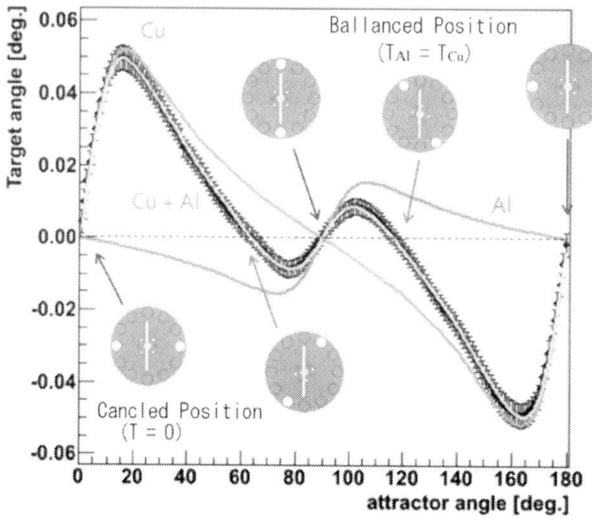

Figure 9. Experimental result from the Newton-II experiment.

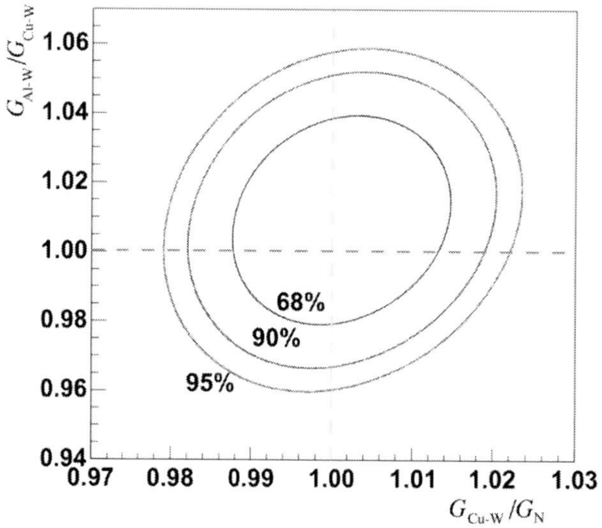

Figure 10. Allowed parameter region of the gravitational constant in the least square analysis.

In Figure 10, allowed region of the two parameters G_{Al-W} (G between aluminum and tungsten) and G_{Cu-W} (G between copper and tungsten) obtained in a least square analysis is shown. Best values from the analysis are obtained as;

$$\begin{aligned} G_{Al-W}/G_{Cu-W} &= 1.009 \pm 0.010_{stat} \pm 0.048_{syst} \\ G_{Al-W}/G_{Newton} &= 1.001 \pm 0.007_{stat} \pm 0.021_{syst}, \end{aligned}$$

here, G_{Newton} is the known gravitational constant obtained from the value on the particle data book [16]. This result is consistent with the composition independence of the gravitational constant within the present experimental precision. The obtained result also shows that the absolute value is consistent with the known gravitational constant. As same as for the new Yukawa interaction testing, we can set a similar constraint on a composition dependent new interaction. A new interaction coupling to the baryon number which violates the weak equivalence principle only at short range, can be parameterized in a new Yukawa term:

$$V(r) = -G\frac{mM}{r^2}\left(1 + \xi\frac{B_i}{\mu_i}\frac{B_j}{\mu_j}e^{-\frac{r}{\lambda}}\right), \tag{4}$$

here, B_A is baryon number for material A, and μ_A is gravitational mass in hydrogen mass unit. Existence of the baryon number coupling force can be constrained by the present study as shown in Figure 11 as a $\xi - \lambda$ plot. The upper limit of the new coupling constant ξ parameter is shown there. Figure 11 shows that we have succeeded to set the tightest upper limit of the ξ in a direct WEP test at the shortest range of below cm.

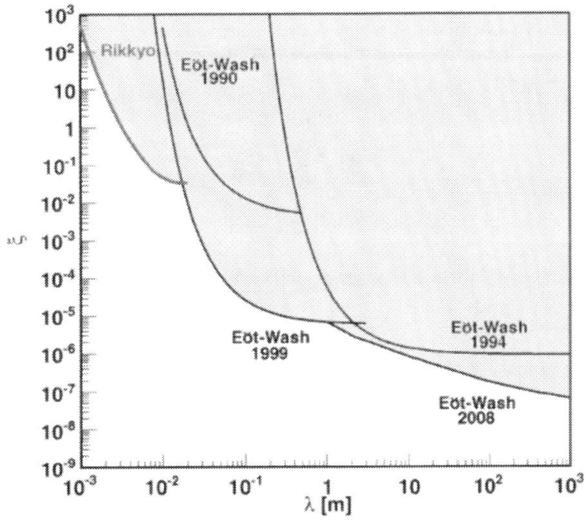

Figure 11. $\xi - \lambda$ plot testing composition dependent new Yukawa interaction. [17, 18, 19, 20]

4. Conclusions

In the Newton-SC experiment, we performed to test the Newtonian inverse square law at the centimeter scale. As a result, our data are consistent with Newtonian gravity in this measurement within experimental precision. In addition, we performed a direct measurement of the composition dependence of the gravitational constant G in Newton-II experiment. The obtaioned result consistent with the composition independence of the gravitational constant within the present experimental precision. This result can also be interpreted as a short-range test of WEP. We gives the most precise limit on the baryon number coupling ξ_B at the shortest scale. We are starting a new expeatriment named Newton-IV. In this experiment, we wilthe shortest scalel also be able to test the inverse square law at below millimeter scale with the highest precision by applying a null-type measurement.

Acknowledgments

Part of the present works are performed as undergraduate student experiments. The author thanks to Y. Miyano, M. Takahashi, T. Tsuneno, T. Amanuma, S. Danbara, T. Iino, S. Mizuno,

Y. Araki, T.Ohmori, Y. Sakurai, S. Yamaoka, R. Tsutsui, M. Hata, T. Akiyama, Y. Ikeda, Y. Sekiguchi and K. Watanabe for their vital eforts on this work.

References

[1] Arkani-Hamed N, Dimopoulos S and Dvali G 1998 *Physics Letters B* **429** 263–272
[2] Fischbach E and Talmadge C 1999 *The search for non-Newtonian gravity* (Springer Verlag) ISBN 0387984909
[3] Hata M *et al.* 2009 *Journal of Physics: Conference Series* vol 189 (IOP Publishing) p 012019
[4] Lee T and Yang C 1955 *Physical Review* **98** 1501–1501 ISSN 0031-899X
[5] Damour T 1996 *Classical and Quantum Gravity* **13** A33
[6] Zhang Y, Luo J and Nie Y 2000 *Arxiv preprint gr-qc/0006075*
[7] Damour T, Piazza F and Veneziano G 2002 *Phys. Rev. Lett.* **89** 81601 ISSN 1079-7114
[8] Murata J 2005 *IEEE Nuclear Science Symposium Conference Record* vol 675
[9] Ninomiya K *et al.* 2009 *Journal of Physics: Conference Series* vol 189 (IOP Publishing) p 012026
[10] Yang S, Zhan B, Wang Q, Shao C, Tu L, Tan W and Luo J 2012 *Physical Review Letters* **108** 81101
[11] Hoyle C, Kapner D, Heckel B, Adelberger E, Gundlach J, Schmidt U and Swanson H 2004 *Phys. Rev. D* **70** 42004 ISSN 1550-2368
[12] Kapner D, Cook T, Adelberger E, Gundlach J, Heckel B, Hoyle C and Swanson H 2007 *Physical review letters* **98** 21101
[13] Smullin S, Geraci A, Weld D, Chiaverini J, Holmes S and Kapitulnik A 2005 *Physical Review D* **72** 122001
[14] Long J, Chan H, Churnside A, Gulbis E, Varney M and Price J 2003 *Nature* **421** 922–925
[15] Hoskins J, Newman R, Spero R and Schultz J 1985 *Phys. Rev. D* **32** 3084–3095 ISSN 1550-2368
[16] PDG 2001 *Eur.Phys.J*
[17] Smith G, Hoyle C, Gundlach J, Adelberger E, Heckel B and Swanson H 1999 *Phys. Rev. D* **61** 22001 ISSN 1550-2368
[18] Su Y, Heckel B, Adelberger E, Gundlach J, Harris M, Smith G and Swanson H 1994 *Phys. Rev. D* **50** 3614–3636 ISSN 1550-2368
[19] Schlamminger S, Choi K, Wagner T, Gundlach J and Adelberger E 2008 *Phys. Rev. Lett.* **100** 41101 ISSN 1079-7114
[20] Adelberger E, Stubbs C, Heckel B, Su Y, Swanson H, Smith G, Gundlach J and Rogers W 1990 *Physical Review D* **42** 3267

Journal of Physics: Conference Series **453** (2013) 012008 doi:10.1088/1742-6596/453/1/012008

Future Stability for Reflection Symmetric Solutions of the Einstein-Vlasov System of Bianchi Type VI_0

Ernesto Nungesser
Department of Mathematics
Royal Institute of Technology
Lindstedtsvägen, 25
100 44 Stockholm, Sweden

Abstract. We present a sketch of the proof of future stability for reflection symmetric solutions of the Einstein-Vlasov system of Bianchi Type VI_0 with an extended introduction to motivate the interest for the Vlasov equation in the context of General Relativity.

1. Introduction

Cosmology describes the dynamics of the Universe as a whole. These dynamics can only be understood in terms of the theory of general relativity which was established by Albert Einstein in 1915. This theory is mathematically complicated enough to have given rise to a research field on its own, the area of Mathematical General Relativity.

A starting point to understand general models are the homogeneous models. In general, the focus has been on the fluid model since it appears (theoretically) relatively natural when dealing with isotropic universes and from observations we also know that the Universe is almost isotropic. However, in order to have a deeper understanding of the dynamics one should go beyond the study of isotropic universes. General statements may vary then depending on the choice of the matter model.

We will deal with the *future asymptotics* of some homogeneous cosmological models within the so called Bianchi class A while the matter is described via an ensemble of free falling particles also called collisionless matter. The late-time behaviour of Bianchi spacetimes with a non-tilted fluid is well understood. In particular, all non-tilted perfect fluid orthogonal Bianchi models except Type IX with a linear equation of state where $0 < \gamma < \frac{2}{3}$, are future asymptotic to the flat Friedmann-Lemître model. Note the restriction on γ here. One cannot expect isotropization for most of the Bianchi models. However, there are two important "conjectures" in the present work:

(i) The spacetimes considered tend to special (self-similar) solutions

(ii) For expanding models the dispersion of the velocities of the particles decays

This second "conjecture" means that asymptotically there is a dust-like behaviour for collisionless matter which is the matter model used here. The Einstein-Vlasov system remains a system of partial differential equations (PDE's) even if one assumes spatial homogeneity. The reason is that although the distribution function written in a suitable frame will not depend on the spatial point, the dependence with respect to the momenta remains. However, sometimes a reduction to a system of ordinary differential equations is possible due to additional symmetry assumptions. This is no longer possible if one drops some of these additional symmetries. Thus, if one wants to generalize the results obtained until now the theory of finite dimensional dynamical systems is not enough. Most of these results rely on the theory of dynamical systems. Thus, one might be tempted to use techniques from the theory of infinite-dimensional dynamical systems. The first important difficulty would be to choose the suitable (weighted) norm. Another difficulty is that important theorems which have been used for the finite-dimensional case cannot be applied here. All this may work, but this is not the approach taken here.

Here, the main tool used is a bootstrap argument which is often used in non-linear PDE's. For the reflection symmetric Bianchi Type VI_0 we have been able to show that the late-time behaviour remains the same if the LRS condition is dropped assuming small data. We will show that the spacetime will tend to solutions which are even more symmetric. In the case of Bianchi Type VI_0 there cannot be an LRS condition, however it is compatible with an additional discrete symmetry. The analysis of the asymptotics shows that the Bianchi Type VI_0 spacetimes tend to this special class. Note that for Bianchi Type VI_0 there is no corresponding LRS/previous result.

All the results show that the dust model usually assumed in observational cosmology in the 'matter-dominated' Era is robust. Another way of saying the same is that asymptotically collisionless matter is well approximated by the dust system.

2. Relativistic kinetic theory

We will consider collisionless matter as our matter model. Before introducing it more formally we want first to give the motivation for the restriction to the aforesaid kind of matter instead of considering the full Boltzmann equation.

First of all it is of course an enormous simplifying assumption. One assumes by modelling a galaxy as a particle that in a cosmological context the internal structure of the galaxy is irrelevant.

An important physical argument in favour of considering the collisionless model is that collisions between galaxies are not common and even if galaxies fly through each other not so many collisions between stars happen as one might expect. Further, in stellar dynamics collisionless matter is often used since collisions between stars are very unlikely. Actually this led Eddington to state:

> "The apparent analogy with the kinetic theory of gases is rejected altogether, and it is taken as a fundamental principle that *the stars describe paths under the general attraction of the stellar system without interfering with one another*"(p. 254 of [1]; italics from Eddington)

Two pages later he continues:

> "A regular progression may be traced through rigid dynamics, hydrodynamics, gas-dynamics to stellar dynamics. In the first all the particles move in a connected manner; in the second there is continuity between the motions of contiguous particles ; in the third the adjacent particles act on one another by collision, so that, although there is no mathematical continuity, a kind of physical continuity remains; in the last the adjacent particles are entirely independent."(p. 256 of [1])

Later, Jeans in the study of stellar dynamics referring to the collisionless Boltzmann equation writes:

> "This is the differential equation which must be satisfied by the distribution function f in every problem of stellar dynamics."(p. 230 of [2])

and on the same page as a footnote:

"The student of the Kinetic Theory will recognise that it is simply Boltzmann's well-known equation with the collisions left out."

However, this equation is usually named after Vlasov [3], in particular in the context of mathematical cosmology. Vlasov discovered in the context of plasma physics that pair collision terms do not describe correctly the plasma dynamics and also that these terms are not formally applicable since kinetic terms diverge. His point of view was that only the collective behaviour, i.e. the electromagnetic field created by the charged particles, explains the dynamics of the individual particles.

Maybe the emphasis on the difference of taking one or the other point of view has been the reason that the Vlasov equation is named after him not only in the context of plasma physics. Another reason maybe that although Boltzmann himself assumed that the particles interact only through:

- very long range forces which can be approximated by mean fields
- or very short range forces such as hard core interactions whose effect can be approximated by instantaneous collisions,

in practice the long range forces were often neglected since, for instance, the gravitational force is very weak.

Another fact which has attracted attention to the case of collisionless matter is the discovery that in analyzing the initial singularity of the Einstein-dust equations there arose singularities which are unphysical and related not to gravity but to the chosen matter model. One has seen that these problems do not occur when using collisionless matter as the matter model. Finally, the Vlasov equation is also used in astrophysics to model dark matter, where the particles are now elementary particles.

Of course, at some point it is of interest to know what happens if one includes collisions between the particles. A recent work on the Cauchy problem of the Einstein-Boltzmann system can be found in [4].

2.1. The Einstein-Vlasov system

A cosmological model represents a universe at a certain averaging scale. It is described via a Lorentzian metric $g_{\alpha\beta}$ (we will use the signature $-+++$) on a manifold M and a family of fundamental observers. The interaction between the geometry and the matter is described by the Einstein field equations (we use geometrized units, i.e. the gravitational constant G and the speed of light in vacuum c are set equal to one):

$$G_{\alpha\beta} = 8\pi T_{\alpha\beta},$$

where $G_{\alpha\beta}$ is the Einstein tensor and $T_{\alpha\beta}$ is the energy-momentum tensor. For the matter model we will take the point of view of kinetic theory. The Einstein summation convention to which repeated indices are to be summed over is used. Latin indices run from one to three and Greek ones from zero to three.

Consider a particle with non-zero unit rest mass moving under the influence of the gravitational field. The mean field will be described now by the metric and the components of the metric connection. The wordline x^α of a particle is a timelike curve in spacetime. The unit future-pointing tangent vector to this curve is the 4-velocity v^α and $p^\alpha = mv^\alpha$ is the 4-momentum of the particle. Let T_x be the tangent space at a point x^α in the spacetime M, then we define the *phase space P_1 for particles of unit mass*:

$$P_1 = \{(x^\alpha, p^\alpha) : \ x^\alpha \in M, \ p^\alpha \in T_x, \ p_\alpha p^\alpha = -1, \ p^0 > 0\},$$

which we will call the mass shell. The collection of particles (galaxies or clusters of galaxies) will be described (statistically) by a non-negative real valued distribution function $f(x^\alpha, p^\alpha)$ on P_1. This function represents the density of particles at a given spacetime point with given four-momentum. A free particle travels along a geodesic. Consider now a future-directed timelike geodesic parametrized by proper time s. The tangent vector is then at any time future-pointing unit timelike. Thus, the geodesic has a natural lift to a curve on P_1 by taking its position and tangent vector. The equations of motion thus define a flow on P_1 which is generated by a vector field L called geodesic spray or Liouville operator. Using the geodesic equations the restriction of the Liouville operator to the mass shell has the following form:

$$L = p^\alpha \frac{\partial}{\partial x^\alpha} - \Gamma^a_{\beta\gamma} p^\beta p^\gamma \frac{\partial}{\partial p^a}.$$

This operator is sometimes also called geodesic spray. Between collisions the particles follow geodesics. We will consider the collisionless case which is described via the Vlasov equation $L(f) = 0$.

2.2. *Energy momentum tensor and characteristics*

The unknowns of our system are a 4-manifold M, a Lorentz metric $g_{\alpha\beta}$ on this manifold and the distribution function f on the mass shell P_1 defined by the metric. We have the Vlasov equation defined by the metric for the distribution function and the Einstein field equations. It remains to define the energy-momentum tensor $T_{\alpha\beta}$ in terms of the distribution and the metric. Now considering p^0 as a dependent variable we denote the induced volume of the mass shell considered as a hypersurface in the tangent space at that point by ϖ. Now we define the energy momentum tensor as follows:

$$T_{\alpha\beta} = \int f(x^\alpha, p^a) p_\alpha p_\beta \varpi.$$

One can show that $T_{\alpha\beta}$ is divergence-free and thus it is compatible with the Einstein field equations. For collisionless matter all the energy conditions hold. The Vlasov equation in a fixed spacetime can be solved by the method of characteristics:

$$\frac{dX^a}{ds} = P^a; \quad \frac{dP^a}{ds} = -\Gamma^a_{\beta\gamma} P^\beta P^\gamma.$$

Let $X^a(s, x^\alpha, p^a)$, $P^a(s, x^\alpha, p^a)$ be the unique solution of that equation with initial conditions $X^a(t, x^\alpha, p^a) = x^a$ and $P^a(t, x^\alpha, p^a) = p^a$. Then the solution of the Vlasov equation can be written as:

$$f(x^\alpha, p^a) = f_0(X^a(0, x^\alpha, p^a), P^a(0, x^\alpha, p^a)),$$

where f_0 is the restriction of f to the hypersurface $t = 0$. It follows that if f_0 is bounded the same is true for f. We will assume that f has compact support in momentum space for each fixed t. This property holds if the initial datum f_0 has compact support and if each hypersurface $t = t_0$ is a Cauchy hypersurface. In particular, for the case we will deal with, the spacetime is future complete (theorem 2.1 of [5]).

3. Bianchi spacetimes

Bianchi spacetimes are defined as follows:

Definition 1. *A Bianchi spacetime is defined to be a spatially homogeneous spacetime whose isometry group possesses a three-dimensional subgroup G that acts simply transitively on the spacelike orbits.*

The only Bianchi spacetimes which admit a compact Cauchy hypersurface are Bianchi Types I and IX. In order to be not too restrictive we will consider locally spatially homogeneous spacetimes. Also, we will take the metric approach. If $\mathbf{W}^a$ denote the 1-forms dual to the frame vectors $\mathbf{E}_a$ the metric of a Bianchi spacetime takes the form:

$$^4g = -dt^2 + g_{ab}(t)\mathbf{W}^a\mathbf{W}^b, \tag{1}$$

where g_{ab} (and all other tensors) on the Lie group G will be described in terms of the frame components of the left invariant frame. A dot above a letter will denote a derivative with respect to the cosmological time t. There are different projections of the energy momentum tensor which are important: $\rho = T^{00}$, $j_a = T^0_a$ and $S_{ab} = T_{ab}$ where ρ is the energy density and j_a is the matter current. Let us also use the notation $S = g^{ab}S_{ab}$.

3.1. Time origin choice and new variables

With the 3+1 formulation our initial data are $(g_{ij}(t_0), k_{ij}(t_0), f(t_0))$, i.e. a Riemannian metric, a second fundamental form and the distribution function of the Vlasov equation, respectively, on a three-dimensional manifold $S(t_0)$. This is the initial data set at $t = t_0$ for the Einstein-Vlasov system. We assume that $k = g^{ab}k_{ab} < 0$ for all time which enables us to set without loss of generality $t_0 = -2/k(t_0)$. The reason for this choice will become clear later and is of a technical nature. We can decompose the second fundamental form introducing σ_{ab} as the trace-free part:

$$k_{ab} = \sigma_{ab} - Hg_{ab}. \tag{2}$$

Using the Hubble parameter $H = -\frac{1}{3}k$ we define:

$$\Sigma_a^b = \frac{\sigma_a^b}{H};\ \Sigma_+ = -\frac{1}{2}(\Sigma_2^2 + \Sigma_3^3);\ \Sigma_- = -\frac{1}{2\sqrt{3}}(\Sigma_2^2 - \Sigma_3^3)$$
$$\Omega = 8\pi\rho/3H^2;\ q = -1 - \frac{\dot{H}}{H^2};\ \frac{d\tau}{dt} = H.$$

The time variable τ is dimensionless. Since we use a left-invariant frame f we can express the Vlasov equation as follows

$$\frac{\partial f}{\partial t} + (p^0)^{-1}C^d_{ba}p^b p_d \frac{\partial f}{\partial p_a} = 0. \tag{3}$$

For the Vlasov equation we obtain by use of (7) for the Bianchi Type VI_0:

$$\frac{\partial f}{\partial t} + (p^0)^{-1}[p_2(p^3\frac{\partial f}{\partial p_1} - p^1\frac{\partial f}{\partial p_3}) + p_3(p^2\frac{\partial f}{\partial p_1} - p^1\frac{\partial f}{\partial p_2})] = 0.$$

From (3) it is also possible to define the characteristic curve V_a:

$$\frac{dV_a}{dt} = (V^0)^{-1}C^d_{ba}V^bV_d \tag{4}$$

for each $V_i(\bar{t}) = \bar{v}_i$ given $\bar{t}$. Note that if we define $V = g^{ij}V_iV_j$ due to the antisymmetry of the structure constants we have with (4):

$$\frac{dV}{dt} = \frac{d}{dt}(g^{ij})V_iV_j. \tag{5}$$

Journal of Physics: Conference Series **453** (2013) 012008 doi:10.1088/1742-6596/453/1/012008

4. Bianchi class A spacetimes

Definition 2. *A Bianchi class A spacetime is a Bianchi spacetime whose three-dimensional Lie algebra has traceless structure constants, i.e.* $C^a_{ba} = 0$.

In this case there is a unique symmetric matrix with components ν^{ij} such that the structure constants can be written as follows:

$$C^a_{bc} = \varepsilon_{bcd}\nu^{da}. \tag{6}$$

Bianchi Type VI_0 is of class A and $\nu = (0, 1, -1)$. We see that the structure constants in the case of Bianchi Type VI_0 are:

$$C^2_{31} = 1 = -C^2_{13}, \quad C^3_{21} = 1 = -C^3_{12}. \tag{7}$$

4.1. Reflection symmetry and the equations for diagonal Bianchi Type VI_0

We will restrict ourselves to the diagonal case assuming an additional symmetry, namely the reflection symmetry. In this case we have:

$$f(t, p_1, p_2, p_3) = f(t, -p_1, -p_2, p_3) = f(t, p_1, -p_2, -p_3).$$

One can see that the energy-momentum tensor is then diagonal. Thus, if the metric and the second fundamental form are diagonal initially, they will remain diagonal in the reflection symmetric case. This symmetry implies in particular that there is no matter current, which means that there is no "tilt". Let us define:

$$n_i = \nu_i \sqrt{\frac{g_{ii}}{g_{jj}g_{kk}}},$$

where (ijk) denotes a cyclic permutation of (123) and the Einstein summation convention is suspended. Define also

$$N_i = \frac{n_i}{H}.$$

The constraint equation is:

$$\Sigma_+^2 + \Sigma_-^2 = 1 - \Omega - \frac{1}{12}(N_2 - N_3)^2$$

and one obtains the evolution equation

$$\partial_t(H^{-1}) = \frac{3}{2} - \frac{1}{24}(N_2 - N_3)^2 + \frac{3}{2}(\Sigma_+^2 + \Sigma_-^2) + \frac{4\pi S}{3H^2}. \tag{8}$$

The evolution equations for the other variables are:

$$\dot{\Sigma}_+ = H[-\frac{1}{6}(N_2 - N_3)^2 - \Sigma_+(3 + \frac{\dot{H}}{H^2}) + \frac{4\pi}{3H^2}(S_2^2 + S_3^3 - 2S_1^1)] \tag{9}$$

$$\dot{\Sigma}_- = H[\frac{N_3^2 - N_2^2}{2\sqrt{3}} - (3 + \frac{\dot{H}}{H^2})\Sigma_- + \frac{4\pi(S_2^2 - S_3^3)}{\sqrt{3}H^2}] \tag{10}$$

$$\dot{N}_2 = -N_2 H(-2\Sigma_+ - 2\sqrt{3}\Sigma_- + 1 + \frac{\dot{H}}{H^2}) \tag{11}$$

$$\dot{N}_3 = -N_3 H(-2\Sigma_+ + 2\sqrt{3}\Sigma_- + 1 + \frac{\dot{H}}{H^2}). \tag{12}$$

In the case of Bianchi Type VI_0 there is a dust solution with diagonal metric discovered by Ellis and MacCallum:

$$g_{EM} = \mathrm{diag}(t^2, t^1, t^1).$$

Here, the values of the introduced variables are:

$$\Sigma_+ = -\frac{1}{4};\ \Sigma_- = 0;\ N_1 = 0;\ N_2 = -N_3 = \frac{3}{4};\ \Omega = \frac{3}{4}.$$

5. The bootstrap argument

Before coming to the bootstrap argument and in order to have a certain intuition for the bootstrap assumptions, let us consider the linearization in the dust case. For $S = 0$ the linearization gives us the expected estimates and, using the Ellis-MacCallum solution, we arrive at a plausible estimate of P. A first task is to find the suitable bootstrap assumptions. We choose a slightly slower decay for the anisotropy and the curvature variables than in the linearized cases with the hope that, using the central equations, we will be able to obtain the same decay as in the linearized case. For the estimate of P we start with a slower decay than the one obtained in the previous section as well. The assumption of small data here is in the sense that our solutions are not "far away" from our special solution. In general, in order to improve an estimate, the corresponding evolution equation will be integrated. The assumptions made exclude the vacuum case, since the values of Ω due to the constraint equation are near the corresponding values of Ω of the special solution, and thus they are far from being zero.

5.1. Bootstrap assumptions

$$\begin{aligned}
|\Sigma_+ + \frac{1}{4}| &\leq A_+(1+t)^{-\frac{3}{8}}, \\
|\Sigma_-| &\leq A_-(1+t)^{-\frac{3}{8}}, \\
|N_2 - \frac{3}{4}| &\leq A_{c1}(1+t)^{-\frac{3}{8}}, \\
|N_3 + \frac{3}{4}| &\leq A_{c2}(1+t)^{-\frac{3}{8}}, \\
P &\leq A_m(1+t)^{-\frac{1}{3}}.
\end{aligned}$$

5.2. Estimate of the mean curvature

The first variable we estimate is the trace of the second fundamental form or equivalently the Hubble variable. By integrating (8) and since $t_0 = \frac{2}{3}H^{-1}(t_0)$ (this choice was made in section (3.1)), one obtains from the different bootstrap assumptions for the Hubble variable :

$$\boxed{H = \frac{2}{3}t^{-1}(1 + O(\epsilon_{2/3}t^{-\frac{3}{8}}))}. \tag{13}$$

5.3. Estimate of the metric

Considering the components g^{22} and g^{33} one shows that:

$$\frac{d}{dt}(t^{-\gamma}\bar{g}^{ab}) = t^{-\gamma-1}\bar{g}^{ab}(-\gamma + \frac{p}{q} + \frac{\dot{g}^{ab}}{g^{ab}}t) \leq -\eta t^{-\gamma-1}\bar{g}^{ab},$$

with η positive by use of the bootstrap assumptions and choosing γ in a suitable way which implies that:

$$g^{ab}(t) \leq t_0^{-\gamma+\frac{p}{q}} g^{ab}(t_0) t^{-\frac{p}{q}+\gamma}. \tag{14}$$

We have with (13) for the components g^{22} and g^{33}:

$$\eta = \gamma + \frac{4}{3}(1 + O(\epsilon_{2/3} t^{-\frac{3}{8}}))(1 + \Sigma_+ \pm \sqrt{3}\Sigma_-) - \frac{p}{q},$$
$$\frac{4}{3}(1 + \Sigma_+) - \frac{p}{q} = O(A_+(1+t)^{-\frac{3}{8}}),$$

which enables us to choose γ in such a way that η is positive. Summarizing, this means that asymptotically up to a positive constant which depends only on t_0 the components (and their inverses) of the metrics g_{VI_0} for Bianchi Type VI$_0$ have the same decay up to an ϵ as the corresponding components of the Ellis-MacCallum solution respectively.

5.4. *Estimate of P*

It follows from (4) and using (5):

$$\dot{V} = \dot{g}^{bf} V_b V_f = 2H(\Sigma_a^b - \delta_a^b) g^{af} V_b V_f = 2H(\Sigma_1^1 g^{11} V_1^2 + \Sigma_2^2 g^{22} V_2^2 + \Sigma_3^3 g^{33} V_3^2) - 2HV.$$

The maximum of Σ_1^1, Σ_2^2 and Σ_3^3 is equal to $\frac{1}{4} + O(t^{-\frac{3}{8}})$. Thus, using now the estimate of H and integrating :

$$V \leq V(t_0)(t/t_0)^{-1+\epsilon}$$

from which it follows choosing $P(t_0) \leq A_m t_0^{\frac{1}{2}-\epsilon}$:

$$P \leq A_m t^{-\frac{1}{2}+\epsilon}$$

which is an improvement of the bootstrap assumption and which has a consequence that:

$$\boxed{\frac{S}{H^2} \leq C t^{-1+\epsilon}}. \tag{15}$$

5.5. *Closing the bootstrap argument*

What remains are the estimates for Σ_+, Σ_-, N_2 and N_3. In terms of the transformed linearization

$$\begin{pmatrix} \hat{\Sigma}_+ \\ \hat{\Sigma}_- \\ \hat{N}_2 \\ \hat{N}_3 \end{pmatrix} = \mathbf{M}_{\mathbf{VI0}}^{-1} \begin{pmatrix} \tilde{\Sigma}_+ \\ \tilde{\Sigma}_- \\ \tilde{N}_2 \\ \tilde{N}_3 \end{pmatrix}$$

and using the time variable τ we have:

$$\begin{pmatrix} \hat{\Sigma}_+ \\ \hat{\Sigma}_- \\ \hat{N}_2 \\ \hat{N}_3 \end{pmatrix}' = -\frac{3}{4} \begin{pmatrix} 1 & -\sqrt{3} & 0 & 0 \\ \sqrt{3} & 1 & 0 & 0 \\ 0 & 0 & 1 & -1 \\ 0 & 0 & 1 & 1 \end{pmatrix} \begin{pmatrix} \hat{\Sigma}_+ \\ \hat{\Sigma}_- \\ \hat{N}_2 \\ \hat{N}_3 \end{pmatrix} + O(A_m^2 e^{-\tau}) \begin{pmatrix} 1 \\ 1 \\ 1 \\ 1 \end{pmatrix}.$$

We arrive at

$$\frac{d}{d\tau}[\log(\hat{\Sigma}_+^2 + \hat{\Sigma}_-^2)] \leq -\frac{3}{2} + \epsilon e^{(-\frac{1}{16}-\xi)\tau}$$

and we obtain a similar expression for N_2 and N_3 such that in the end we obtain the desired estimates.

5.6. Results of the bootstrap argument
Let us summarize the results obtained in the following proposition:

Proposition 1. *Consider any C^∞ solution of the Einstein-Vlasov system with reflection Bianchi Type VI_0 symmetry and with C^∞ initial data. Assume that $|\Sigma_+(t_0) + \frac{1}{4}|$, $|\Sigma_-(t_0)|$, $|N_2(t_0) - \frac{3}{4}|$, $|N_3(t_0) + \frac{3}{4}|$ and $P(t_0)$ are sufficiently small. Then at late times the following estimates hold:*

$$\begin{aligned} H(t) &= \frac{2}{3}t^{-1}(1 + O(t^{-\frac{1}{2}+\epsilon})), \\ \Sigma_+ + \frac{1}{4} &= O(t^{-\frac{1}{2}+\epsilon}), \\ \Sigma_- &= O(t^{-\frac{1}{2}+\epsilon}), \\ N_2 - \frac{3}{4} &= O(t^{-\frac{1}{2}+\epsilon}), \\ N_3 + \frac{3}{4} &= O(t^{-\frac{1}{2}+\epsilon}), \\ P(t) &= O(t^{-\frac{1}{2}+\epsilon}). \end{aligned}$$

In the next section we will improve the estimates so that we can get rid of ϵ. However, the results stated in this section represent in fact the core of our results.

6. Main results
6.1. Arzela-Ascoli
We want to use the Arzela-Ascoli theorem. All relevant variables and their derivatives are uniformly bounded. The only variable which is not obviously bounded is the derivative of S. If we can bound the derivative of S then the second derivatives of Σ_-, Σ_+, N_2, N_3 and H are also bounded. The only term of the time derivative of S which could cause problems is the time derivative of the distribution function, but it can be handled with the Vlasov equation and through integration by parts . We obtain a term which can be bounded by S. Let $\{t_n\}$ be a sequence tending to infinity and let $(\Sigma_-)_n(t) = \Sigma_-(t+t_n)$, $(\Sigma_+)_n(t) = \Sigma_+(t+t_n)$, $(N_2)_n(t) = N_2(t+t_n)$, $(N_3)_n(t) = N_3(t+t_n)$, $H_n(t) = H(t+t_n)$ and $S_n(t) = S(t+t_n)$. Using the bounds, the Arzela-Ascoli theorem can be applied. This implies that, after passing to a subsequence, $(\Sigma_-)_n$, $(\Sigma_+)_n$, $(N_2)_n$, $(N_3)_n$, H_n and S_n converge uniformly on compact sets to a limit $(\Sigma_-)_\infty$, $(\Sigma_+)_\infty$, $(N_2)_\infty$, $(N_3)_\infty$, H_∞ and S_∞ respectively. The first derivative of these variables converges to the corresponding derivative of the limits since we have been able to bound the derivative of S in the last section. Going to this limit it is easy to see that the variable D_∞ is zero and consequently $H_\infty = \frac{2}{3}t^{-1}$. Thus, we obtain the optimal decay rates for the metric and for its derivative. This implies that we obtain the optimal decay rates for P. Since S/H^2 is zero asymptotically, we obtain the same estimates for Σ_-, Σ_+, N_2 and N_3 as in the Einstein-dust case. Introducing these estimates in (8), we also obtain the optimal estimate for H.

6.2. Optimal estimates

We can summarize the results:

Theorem 1. *Consider any C^∞ solution of the Einstein-Vlasov system with reflection Bianchi Type VI_0 symmetry and with C^∞ initial data. Assume that $|\Sigma_+(t_0)+\frac{1}{4}|$, $|\Sigma_-(t_0)|$, $|N_2(t_0)-\frac{3}{4}|$, $|N_3(t_0)+\frac{3}{4}|$ and $P(t_0)$ are sufficiently small. Then at late times the following estimates hold:*

$$\begin{aligned} H(t) &= \frac{2}{3}t^{-1}(1+O(t^{-\frac{1}{2}})), \\ \Sigma_+ + \frac{1}{4} &= O(t^{-\frac{1}{2}}), \\ \Sigma_- &= O(t^{-\frac{1}{2}}), \\ N_2 - \frac{3}{4} &= O(t^{-\frac{1}{2}}), \\ N_3 + \frac{3}{4} &= O(t^{-\frac{1}{2}}), \\ P(t) &= O(t^{-\frac{1}{2}}). \end{aligned}$$

From this theorem we can obtain some other results. Let λ_i be the eigenvalues of k_{ij} with respect to g_{ij}:

$$\det(k^i_j - \lambda\delta^i_j) = 0. \qquad (16)$$

We define $p_i = \frac{\lambda_i}{k}$ as the *generalized Kasner exponents*. From (2) we see that the eigenvalues (16) of the second fundamental form with respect to the induced metric are also the solutions of:

$$\det(\sigma^i_j - [\lambda - \frac{1}{3}k]\delta^i_j) = 0.$$

Let us define the eigenvalues of σ_{ij} with respect to g_{ij} by $\widehat{\lambda}_i$:

$$\widehat{\lambda}_i = \lambda_i - \frac{1}{3}k.$$

Using the optimal estimates for Σ_+, Σ_- and H and the fact that the sum of the generalized Kasner exponents is equal to one, we finally arrive at the generalized Kasner exponents for Bianchi Type VI_0 $(\frac{1}{2}, \frac{1}{4}, \frac{1}{4})$ up to an error of order $O(t^{-\frac{1}{2}})$. V_2 and V_3 become constant asymptotically. Now $f(t_0, p)$ has compact support on p and denoting by $\hat{p}$ the momenta in an orthonormal frame since $f(t, \hat{p})$ is constant along the characteristics, we have $|f(t,\hat{p})| \leq \|f_0\| = \sup\{|f(t_0,\hat{p})|\}$. Let us summarize this in the following corollary:

Corollary 1. *Consider the same assumptions as in Theorem 1. Then*

$$g_{VI_0} = t\,\mathrm{diag}(tK_1, K_2, K_3),$$

with $K_n = C_n + O(t^{-\frac{1}{2}})$ and where C_1-C_3 are independent of time. The corresponding result for the inverse metric also holds.

$$\begin{aligned} &p_{VI_0} = p_{EM} + O(t^{-\frac{1}{2}}) \\ &\rho = \rho_{EM}(1 + O(t^{-\frac{1}{2}})) \\ &S_{ij} \leq C|f_0|t^{-3} \end{aligned}$$

The fact that this quotient vanishes asymptotically means that matter behaves asymptotically as dust as it is expected.

7. Conclusions and Outlook

For Bianchi Type VI_0 even for the reflection symmetric case there is no analogous previous result. The reason is that it is not compatible with the LRS-symmetry. Thus, our result on Bianchi Type VI_0 clearly shows that the methods developed are powerful in the sense that one can obtain results which have been out of reach with the techniques developed until now. An important question is whether it is possible to remove the small data assumptions. We will work on this question using techniques developed in [6]. For details of the presented results we refer to [7].

Acknowledgement

The author thanks the Göran Gustafsson Foundation for Research in Natural Sciences and Medicine for their financial support.

References

[1] Eddington A S 1914 *Stellar movements and the structure of the Universe* (London, Macmillan and Co.)
[2] Jeans J H 1919 *Problems of Cosmogony and Stellar Dynamics* (Cambridge: Cambridge University Press)
[3] Vlasov A A 1968 *Sov. Phys. Usp.* **10** 721
[4] Lee H and Rendall A D 2013 *J. Hyp. Diff. Eq* **10** 77
[5] Rendall A D 1995 *Math. Proc. Camb. Phil. Soc.* **118** 511
[6] Ringström H 2013 *On the topology and future stability of models of the universe - with an introduction to the Einstein-Vlasov system* (Oxford: Oxford University Press)
[7] Nungesser E 2012 Future non-linear stability for reflection symmetric solutions of the Einstein-Vlasov system of Bianchi types II and VI_0 *Preprint* gr-qc/1208.4231

Using Noether symmetries to specify f(R) gravity

Andronikos Paliathanasis
Department of Physics, Section of Astronomy, Astrophysics and Mechanics, University of Athens, Panepistemiopolis, Athens 157 83, Greece
E-mail: anpaliat@phys.uoa.gr

Abstract. A detailed study of the modified gravity, f(R) models is performed, using the fact that the Noether point symmetries of these models are geometric symmetries of the mini superspace of the theory. It is shown that the requirement that the field equations admit Noether point symmetries selects definite models in a self-consistent way. As an application in Cosmology we consider the Friedman -Robertson-Walker spacetime and show that the only cosmological model which is integrable via Noether point symmetries is the $(R^b - 2\Lambda)^c$ model, which generalizes the Lambda Cosmology. Furthermore using the corresponding Noether integrals we compute the analytic form of the main cosmological functions.

Keywords: General Relativity, Modified Gravity, Noether symmetries
Pacs - numbers:98.80.-k,95.35.+d,95.36.+x

1. Introduction

The recent cosmological data indicate that the universe is spatially flat and has suffered two acceleration phases. An early acceleration phase (inflation), which occurred prior to the radiation dominated era and a recently initiated accelerated expansion.

An easy way to explain this expansion is to consider an additional fluid with negative equation of state parameter, usually called dark energy, that dominates the universe at late times. In spite of that, the absence of a fundamental physical theory, regarding the mechanism inducing the cosmic acceleration, has given rise to a plethora of alternative cosmological scenarios. Most of them are based either on the existence of new fields in nature (dark energy) or in some modification of Einstein's general relativity (GR), with the present accelerating stage appearing as a sort of geometric effect ("geometrical" dark energy).

The simplest dark energy probe is the cosmological constant Λ (vacuum) leading to the ΛCDM cosmology [1–3]. However, it has been shown that ΛCDM cosmology suffers from two major drawbacks known as the fine tuning problem and the coincidence problem [4]. Besides ΛCDM cosmology, many other candidates have been proposed in the literature, such as time-varying $\Lambda(t)$ cosmologies, quintessence, $k-$essence, tachyons, modifications of gravity, Chaplygin gas and others [5–10].

There are other possibilities to explain the present accelerating stage. For instance, one may consider that the dynamical effects attributed to dark energy can be resembled by the effects of a nonstandard gravity theory. In other words, the present accelerating stage of the universe can be driven only by cold dark matter, under a modification of the nature of gravity. Such a reduction of the so-called dark sector is naturally obtained in the $f(R)$ gravity theories [11]. In the original nonstandard gravity models, one modifies the Einstein-Hilbert action with a general function

$f(R)$ of the Ricci scalar R. The $f(R)$ approach is a relatively simple but still a fundamental tool used to explain the accelerated expansion of the universe. A pioneering fundamental approach was proposed long ago with $f(R) = R + mR^2$ [12]. Later on, the $f(R)$ models were further explored from different points of view in [13–15] and indeed a large number of functional forms of $f(R)$ gravity is currently available in the literature [16–21].

The aim of the present work is to investigate which $f(R)$ models admit extra Noether point symmetries and use the first integrals of these models to determine analytic solutions of their field equations. The idea to use Noether symmetries in cosmological studies is not new and indeed a lot of attention has been paid in the literature (see [22–37]).

The main reasons for the consideration of this hypothesis is that (a) the Noether point symmetries provide integrals, which assist the integrability of the system, (b) is a geometric criterion because the Noether symmetries are associated with the geometry of the field equations. A fundamental approach to derive the Noether point symmetries of a given dynamical system moving in a Riemannian space has been proposed in [38]. A similar analysis can be found in [39, 40].

The structure of the paper is as follows. The basic theoretical elements of the problem are presented in section 2, where we also introduce the basic FRW cosmological equations in the framework of $f(R)$ models. The geometrical Noether point symmetries and their connections to the $f(R)$ models are discussed in sections 5. In section 6 we provide analytical solutions for those $f(R)$ models which are Liouville integrable via Noether point symmetries. In section 7 we study the Noether symmetries in spatially non flat $f(R)$ cosmological models. Finally, we draw our main conclusions in section 8.

2. Cosmology with a modified gravity

Consider the modified Einstein-Hilbert action:

$$S = \int d^4x\sqrt{-g}\left[\frac{1}{2k^2}f(R) + \mathcal{L}_m\right] \tag{1}$$

where $\mathcal{L}_m$ is the Lagrangian of dust-like ($p_m = 0$) matter and $k^2 = 8\pi G$. Varying the action with respect to the metric[1] we arrive at

$$(1+f')G^{\mu}_{\nu} - g^{\mu\alpha}f_{R,\alpha;\nu} + \left[\frac{2\Box f' - (f - Rf')}{2}\right]\delta^{\mu}_{\nu} = k^2 T^{\mu}_{\nu} \tag{2}$$

where the prime denotes derivative with respect to R, G^{μ}_{ν} is the Einstein tensor and T^{μ}_{ν} is the ordinary energy-momentum tensor of matter. Based on the matter era we treat the expanding universe as a perfect fluid which includes only cold dark matter with comoving observers $U^{\mu} = \delta^{\mu}_{0}$. Thus the energy momentum tensor becomes $T_{\mu\nu} = \rho_m U_{\mu}U_{\nu}$, where ρ_m is the energy density of the cosmic fluid.

Now, in the context of a flat FRW model the metric is

$$ds^2 = -dt^2 + a^2(t)(dx^2 + dy^2 + dz^2). \tag{3}$$

The components of the Einstein tensor are computed to be:

$$G^0_0 = -3H^2,\ G^a_b = -\delta^a_b\left(2\dot{H} + 3H^2\right)\ . \tag{4}$$

[1] We use the metric i.e. the Hilbert variational approach.

Inserting (4) into the modified Einstein's field equations (2), for comoving observers, we derive the modified Friedman's equation

$$3f^{'}H^2 = k^2\rho_m + \frac{f^{'}R - f}{2} - 3Hf^{''}\dot{R} \tag{5}$$

$$2f^{'}\dot{H} + 3f^{'}H^2 = -2Hf^{''}\dot{R} - \left(f^{'''}\dot{R}^2 + f^{''}\ddot{R}\right) - \frac{f - Rf^{'}}{2}. \tag{6}$$

The contraction of the Ricci tensor provides the Ricci scalar

$$R = g^{\mu\nu}R_{\mu\nu} = 6\left(\frac{\ddot{a}}{a} + \frac{\dot{a}^2}{a^2}\right) = 6(2H^2 + \dot{H})\ . \tag{7}$$

The Bianchi identity $\bigtriangledown^{\mu}T_{\mu\nu} = 0$ leads to the matter conservation law:

$$\dot{\rho}_m + 3H\rho_m = 0 \tag{8}$$

whose solution is

$$\rho_m = \rho_{m0}a^{-3}. \tag{9}$$

Note that the over-dot denotes derivative with respect to the cosmic time t and $H \equiv \dot{a}/a$ is the Hubble parameter.

If we consider $f(R) = R$ then the field equations (2) boil down to the Einstein's equations a solution of which is the Einstein de Sitter model. On the other hand, the concordance Λ cosmology is fully recovered for $f(R) = R - 2\Lambda$.

From the current analysis it becomes clear that unlike the standard Friedman equations in Einstein's GR the modified equations of motion (5) and (6) are complicated and thus it is difficult to solve them analytically.

We would like to stress here that within the context of the metric formalism the above $f(R)$ cosmological models must obey simultaneously some strong conditions [42]. These are: (i) $f^{'} > 0$ for $R \geq R_0 > 0$, where R_0 is the Ricci scalar at the present time. If the final attractor is a de Sitter point we need to have $f^{'} > 0$ for $R \geq R_1 > 0$, where R_1 is the Ricci scalar at the de Sitter point, (ii) $f^{''} > 0$ for $R \geq R_0 > 0$, (iii) $f(R) \approx R - 2\Lambda$ for $R \gg R_0$ and finally (iv) $0 < \frac{Rf^{''}}{f^{'}}(r) < 1$ at $r = -\frac{Rf^{'}}{f} = -2$

3. Modified gravity versus symmetries

In the last decade a large number of experiments have been proposed in order to constrain dark energy and study its evolution. Naturally, in order to establish the evolution of the dark energy ("geometrical" in the current work) equation of state parameter a realistic form of $H(a)$ is required while the included free parameters must be constrained through a combination of independent DE probes (for example SNIa, BAOs, CMB etc). However, a weak point here is the fact that the majority of the $f(R)$ models appeared in the literature are plagued with no clear physical basis and/or many free parameters. Due to the large number of free parameters many such models could fit the data. The proposed additional criterion of Noether point symmetry requirement is a physically meaning-full geometric ansatz.

According to the theory of general relativity, the space-time symmetries (Killing and homothetic vectors) via the Einstein's field equations, are also symmetries of the energy momentum tensor. Due to the fact that the $f(R)$ models provide a natural generalization of GR one would expect that the theories of modified gravity must inherit the symmetries of the space-time as the usual gravity (GR) does.

Furthermore, besides the geometric symmetries we have to consider the dynamical symmetries, which are the symmetries of the field equations (Lie symmetries). If the field equations are derived from a Lagrangian then there is a special class of Lie symmetries, the Noether symmetries, which lead to conserved currents or, equivalently, to first integrals of the equations of motion. The Noether integrals are used to reduce the order of the field equations or even to solve them. Therefore a sound requirement, which is possible to be made in Lagrangian theories, is that they admit extra Noether symmetries. This assumption is model independent, because it is imposed after the field equations have been derived, therefore it does not lead to conflict with the geometric symmetries while, at the same time, serves the original purpose of a selection rule. Of course, it is possible that a different method could be assumed and select another subset of viable models. However, symmetry has always played a dominant role in Physics and this gives an aesthetic and a physical priority to our proposal.

In the Lagrangian context, the main field equations (5) and (6), described in section 2, can be produced by the following Lagrangian:

$$L\left(a,\dot{a},R,\dot{R}\right)=6af^{\prime}\ \dot{a}^{2}+6a^{2}f^{\prime\prime}\ \dot{a}\dot{R}+a^{3}\left(f^{\prime}R-f\right) \tag{10}$$

in the space of the variables $\{a,R\}$. Using eq.(10) we obtain the Hamiltonian of the current dynamical system

$$E=6af^{\prime}\ \dot{a}^{2}+6a^{2}f^{\prime\prime}\ \dot{a}\dot{R}-a^{3}\left(f^{\prime}R-f\right) \tag{11}$$

or

$$E=6a^{3}\left[f^{\prime}H^{2}-\frac{1}{6f^{\prime}}\left(\left(f^{\prime}R-f\right)-6\dot{R}Hf^{\prime\prime}\right)\right]\ . \tag{12}$$

Combining the first equation of motion (5) with eq.(12) we find

$$\rho_{m}=\frac{E}{2k^{2}}\ a^{-3}\ . \tag{13}$$

The latter equation together with $\rho_m=\rho_{m0}a^{-3}$ implies that

$$\rho_{m0}=\frac{E}{2k^{2}}\Rightarrow\Omega_{m}\rho_{cr,0}=\frac{E}{2k^{2}}\Rightarrow E=6\Omega_{m}H_{0}^{2} \tag{14}$$

where $\Omega_m=\rho_{m0}/\rho_{cr,0}$, $\rho_{cr,0}=3H_0^2/k^2$ is the critical density at the present time and H_0 is the Hubble constant.

We note that the current Lagrangian eq.(10) is time independent implying that the dynamical system is autonomous hence the Hamiltonian E is conserved ($\frac{dE}{dt}=0$).

4. Noether symmetries

Before we proceed we review briefly the basic definitions concerning Lie and Noether point symmetries of systems of second order ordinary differential equations (ODEs)

$$\ddot{x}^{i}=\omega^{i}\left(t,x^{j},\dot{x}^{j}\right). \tag{15}$$

The one point parameter transformation

$$\bar{t}=t+\varepsilon\xi\left(t,x^{i}\right) \tag{16}$$

$$\bar{x}^{i}=x^{i}+\varepsilon\eta^{i}\left(t,x^{i}\right) \tag{17}$$

with generator $X = \xi\left(t, x^j\right)\partial_t + \eta^i\left(t, x\right)\partial_i$ is a Lie point symmetry of the system of ODEs (15) if the following condition is satisfied [39, 40]

$$X^{[2]}\left(\ddot{x}^i - \omega\left(t, x^j, \dot{x}^j\right)\right) = 0 \tag{18}$$

where $X^{[2]}$ is the second prolongation of X defined by the formula

$$X^{[2]} = \xi\partial_t + \eta^i\partial_i + \left(\dot{\eta}^i - \dot{x}^i\dot{\xi}\right)\partial_{\dot{x}^i} + \left(\ddot{\eta}^i - \dot{x}^i\ddot{\xi} - 2\ddot{x}^i\dot{\xi}\right)\partial_{\ddot{x}^i}. \tag{19}$$

Condition (18) is equivalent to the relation

$$\left[X^{[1]}, A\right] = \lambda\left(x^a\right)A \tag{20}$$

where $X^{[1]}$ is the first prolongation of X and A is the Hamiltonian vector field

$$A = \partial_t + \dot{x}\partial_x + \omega^i\left(t, x^j, \dot{x}^j\right)\partial_{\dot{x}^i}. \tag{21}$$

If the system of ODEs results from a first order Lagrangian $L = L\left(t, x^j, \dot{x}^j\right)$, then a Lie symmetry X of the system (15) is a Noether symmetry of the Lagrangian if the additional condition is satisfied

$$X^{[1]}L + L\frac{d\xi}{dt} = \frac{dg}{dt} \tag{22}$$

where $g = g\left(t, x^j\right)$ is the gauge function. To every Noether symmetry there corresponds a first integral (a Noether integral) of the system of equations (15) which is given by the formula:

$$I = \xi E_H - \frac{\partial L}{\partial \dot{x}^i}\eta^i + g \tag{23}$$

where E_H is the Hamiltonian of the Lagrangian

$$E_H = \dot{x}^i\frac{\partial L}{\partial x^i} - L \tag{24}$$

The vector field X in the augmented space $\{t, a, R\}$ is

$$X = \xi\left(t, a, R\right)\partial_t + \eta^{(1)}\left(t, a, R\right)\partial_a + \eta^{(2)}\left(t, a, R\right)\partial_R \tag{25}$$

and the first prolongation

$$X^{[1]} = \xi\partial_t + \eta^{(1)}\partial_a + \eta^{(2)}\partial_R + \left(\eta^{(1)} - \dot{a}\dot{\xi}\right)\partial_{\dot{a}} + \left(\dot{\eta}^{(2)} - \dot{R}\dot{\xi}\right)\partial_{\dot{R}}. \tag{26}$$

Having given the basic formula for the Noether symmetries we look for analytic solutions of the dynamical system with Lagrangian (10) with the use of Noether Integrals.

5. Noether symmetries of $f(R)$ gravity

The Noether condition (22) for the Lagrangian (10) is equivalent with the following system of eight equations

$$\xi_{,a} = 0 \tag{27}$$

$$\xi_{,R} = 0 \tag{28}$$

$$a^2 f'' \eta^{(1)}_{,R} = 0 \tag{29}$$

$$f'\eta^{(1)} + af''\eta^{(2)} + 2af'\eta^{(1)}_{,a} + a^2 f''\eta^{(2)}_{,a} - \frac{1}{2}af'\xi_{,t} = 0 \tag{30}$$

$$2af''\eta^{(1)} + a^2 f'''\eta^{(2)} + a^2 f''\eta^{(1)}_{,a} + 2af'\eta^{(1)}_{,R} + a^2 f''\eta^{(2)}_{,R} - \frac{1}{2}a^2 f''\xi_{,t} = 0 \tag{31}$$

$$-3a^2 Rf'\eta^{(1)} + 3a^2 f\eta^{(1)} - a^3 Rf''\eta^{(2)} + a^3\left(f - f'R\right)\xi_{,t} + g_{,t} = 0 \tag{32}$$

$$12af'\eta^{(1)}_{,t} + 6a^2 f''\eta^{(2)}_{,t} + a^3\left(f'R - f\right)\xi_{,a} - g_{,a} = 0 \tag{33}$$

$$6a^2 f''\eta^{(1)}_{,t} + a^3\left(f'R - f\right)\xi_{,R} - g_{,R} = 0 \tag{34}$$

The solution of the system (27)-(34) will determine the Noether symmetries.

Since the Lagrangian (10) is in the form $L = T\left(a, \dot{a}, R, \dot{R}\right) - V(a, R)$, the results of [38] can be used [2]. The kinematic term defines a two dimensional metric in the space of $\{a, R\}$ with line element

$$d\hat{s}^2 = 12af'da^2 + 12a^2 f'' da\, dR \tag{35}$$

while the "potential" is

$$V(a, R) = -a^3(f'R - f)\,. \tag{36}$$

The Ricci scalar of the two dimensional metric (35) is computed to be $\hat{R} = 0$, therefore the space is a flat space[3] with a maximum homothetic algebra. The homothetic algebra of the metric (35) consists of the vectors

$$\mathbf{K}^1 = a\partial_a - 3\frac{f'}{f''}\partial_R\ ,\ \mathbf{K}^2 = \frac{1}{a}\partial_a - \frac{1}{a^2}\frac{f'}{f''}\partial_R$$
$$\mathbf{K}^3 = \frac{1}{a}\frac{1}{f''}\partial_R\ ,\ \mathbf{H} = \frac{a}{2}\,\partial_a + \frac{1}{2}\frac{f'}{f''}\partial_R$$

where $\mathbf{K}$ are Killing vectors ($\mathbf{K}^{2,3}$ are gradients) and $\mathbf{H}$ is a gradient Homothetic vector.

Therefore applying theorem 2 of [38] we have the following cases:

Case 1: If $f(R)$ is arbitrary the dynamical system admits as Noether symmetry the $X^1 = \partial_t$ with Noether integral the Hamiltonian E.

Case 2: If $f(R) = R^{\frac{3}{2}}$ the dynamical system admits the extra Noether symmetries

$$X^2 = \mathbf{K}^2,\ X^3 = t\mathbf{K}^2 \tag{37}$$

$$X^4 = 2t\partial_t + \mathbf{H} + \frac{5}{6}\mathbf{K}^1. \tag{38}$$

with corresponding Noether Integrals

$$I_2 = \frac{d}{dt}\left(a\sqrt{R}\right) \tag{39}$$

$$I_3 = t\frac{d}{dt}\left(a\sqrt{R}\right) - a\sqrt{R} \tag{40}$$

[2] Where T is the "kinetic" term and V is the "potential

[3] All two dimensional Riemannian spaces are Einstein spaces implying that if $\hat{R} = const$ the space is maximally symmetric [41] and if $\hat{R} = 0$, the space admit gradient Homothetic vector, i.e. is flat.

$$I_4 = 2tE - 6a^2\dot{a}\sqrt{R} - 6\frac{a^3}{\sqrt{R}}\dot{R}. \tag{41}$$

the non vanishing commutators of the Noether algebra are

$$[X^1, X^3] = X^2 \quad [X^1, X^4] = 2X^1$$

$$[X^2, X^4] = \frac{8}{3}X^2 \quad [X^3, X^4] = \frac{2}{3}X^3$$

Case 3: If $f(R) = R^{\frac{7}{8}}$ the dynamical system admits the extra Noether symmetries

$$X^5 = 2t\partial_t + \mathbf{H}\ ,\ X^6 = t^2\partial_t + t\mathbf{H} \tag{42}$$

with corresponding Noether Integrals

$$I_5 = 2tE - \frac{21}{8}\frac{d}{dt}\left(a^3R^{-\frac{1}{8}}\right) \tag{43}$$

$$I_6 = t^2E - \frac{21}{8}t\frac{d}{dt}\left(a^3R^{-\frac{1}{8}}\right) + \frac{21}{8}a^3R^{-\frac{1}{8}}. \tag{44}$$

and the non vanishing commutators of the Noether algebra are

$$[X^1, X^5] = 2X^1 \quad [X^1, X^6] = X^5 \quad [X^5, X^6] = 2X^6$$

From the time dependent integrals (43),(44) and the Hamiltonian we construct the Ermakov-Lewis invariant [43, 44]

$$\Sigma = 4I_6E - I_5^2 \tag{45}$$

Case 4: If $f(R) = (R - 2\Lambda)^{\frac{3}{2}}$ the dynamical system admits the extra Noether symmetries

$$\bar{X}^2 = e^{\sqrt{m}t}\mathbf{K}^2\ ,\ \bar{X}^3 = e^{-\sqrt{m}t}\mathbf{K}^2 \tag{46}$$

with corresponding Noether Integrals

$$\bar{I}_2 = e^{\sqrt{m}t}\left(\frac{d}{dt}\left(a\sqrt{R-2L}\right) - 9\sqrt{m}a\sqrt{R-2\Lambda}\right) \tag{47}$$

$$\bar{I}_3 = e^{-\sqrt{m}t}\left(\frac{d}{dt}\left(a\sqrt{R-2L}\right) + 9\sqrt{m}a\sqrt{R-2\Lambda}\right) \tag{48}$$

where $m = \frac{2}{3}\Lambda$. The non vanishing commutators of the Noether algebra are

$$[X^1, \bar{X}^2] = \sqrt{m}\bar{X}^2 \quad [\bar{X}^3, X^1] = \sqrt{m}\bar{X}^3$$

From the time dependent integrals (47),(48) we construct the time independent integral $\bar{I}_{23} = \bar{I}_2\bar{I}_3$.

Case 5: If $f(R) = (R - 2\Lambda)^{\frac{7}{8}}$ the dynamical system admits the extra Noether symmetries

$$\bar{X}^5 = \frac{1}{\sqrt{m}}e^{2\sqrt{m}t}\partial_t + e^{2\sqrt{m}t}\ \mathbf{H} \tag{49}$$

$$\bar{X}^6 = -\frac{1}{\sqrt{m}}e^{-2\sqrt{m}t}\partial_t + e^{-2\sqrt{m}t}\ \mathbf{H} \tag{50}$$

Journal of Physics: Conference Series **453** (2013) 012009 doi:10.1088/1742-6596/453/1/012009

with corresponding Noether Integrals

$$\bar{I}_5 = e^{2\sqrt{m}t}\left[\frac{1}{\sqrt{m}}E - \frac{21}{8}\frac{d}{dt}\left(a^3\left(R-2\Lambda\right)^{-\frac{1}{8}}\right) + \frac{21}{4}\sqrt{m}a^3\left(R-2\Lambda\right)^{-\frac{1}{8}}\right] \tag{51}$$

$$\bar{I}_6 = e^{-2\sqrt{m}t}\left[\frac{1}{\sqrt{m}}E + \frac{21}{8}\frac{d}{dt}\left(a^3\left(R-2\Lambda\right)^{-\frac{1}{8}}\right) + \frac{21}{4}\sqrt{m}a^3\left(R-2\Lambda\right)^{-\frac{1}{8}}\right] \tag{52}$$

and the non vanishing commutators of the Noether algebra are

$$\left[X^1, \bar{X}^5\right] = 2\sqrt{m}\bar{X}^5 \quad \left[\bar{X}^6, X^1\right] = 2\sqrt{m}\bar{X}^6$$

$$\left[\bar{X}^5, \bar{X}^6\right] = \frac{4}{\sqrt{m}}X^1$$

From the time dependent integrals (43),(44) and the Hamiltonian we construct the Ermakov-Lewis invariant [44]

$$\phi = E^2 - \bar{I}_5\bar{I}_6 \tag{53}$$

Case 6: If $f(R) = R^n$ (with $n \neq 0, 1, \frac{3}{2}, \frac{7}{8}$) the dynamical system admits the extra Noether symmetry

$$X^7 = 2t\partial_t + \mathbf{H} + \left(\frac{4n}{3} - \frac{7}{6}\right)\mathbf{K}^1 \tag{54}$$

with corresponding Noether Integral

$$I_7 = 2tE - 8na^2R^{n-1}\dot{a}\left(2-n\right) - 4na^3R^{n-2}\dot{R}\left(2n-1\right)\left(n-1\right). \tag{55}$$

and the commutator of the Noether algebra is $\left[X^1, X^7\right] = 2X^1$.

We note that the Noether subalgebra of case 2, $\left\{X^1, X^2, X^3\right\}$ and the algebra of case 4 $\left\{X^1, \bar{X}^2, \bar{X}^3\right\}$ is the same Lie algebra but not in the same representation. The same observation applies to the subalgebra of case 3 $\left\{X^1, X^5, X^6\right\}$ and the algebra of case 5 $\left\{X^1, \bar{X}^5, \bar{X}^6\right\}$. This connection between the Lie groups is useful because it reveals common features in the dynamic systems, as is the common transformation to the normal coordinates of the systems.

For the cosmological viability of the models see [21, 45, 46]

6. Analytic Solutions

Using the Noether symmetries and the associated Noether integrals we solve analytically the differential eqs.(5), (6) and (7) for the cases where the dynamical system is Liouville integrable, that is for cases 2-5. Case 6 (i.e. $f(R) = R^n$) is not Liouville integrable via Noether point symmetries, since the Noether integral (55) is time dependent[4].

6.1. Power law model R^μ with $\mu = \frac{3}{2}$

In this case the Lagrangian eq.(10) of the $f(R) = R^{\frac{3}{2}}$ model is written as

$$L = 9a\sqrt{R}\dot{a}^2 + \frac{9a^2}{2\sqrt{R}}\dot{a}\dot{R} + \frac{a^3}{2}R^{\frac{3}{2}} \tag{56}$$

Changing the variables from (a, R) to (z, w) via the relations:

$$a = \left(\frac{9}{2}\right)^{-\frac{1}{3}}\sqrt{z} \quad R = \frac{w^2}{z} \tag{57}$$

[4] In the appendix Appendix A we present special solutions for the $f(R) = R^n$ model, using the zero order invariants.

the Lagrangian (56) and the Hamiltonian (11) become

$$L = \dot{z}\dot{w} + V_0 w^3 \tag{58}$$

$$E = \dot{z}\dot{w} - V_0 w^3 \tag{59}$$

where $V_0 = \frac{1}{9}$. The equations of motion in the new coordinate system are

$$\ddot{w} = 0 \tag{60}$$
$$\ddot{z} - 3V_0 w^2 = 0 \tag{61}$$

The Noether integrals (39),(40) in the coordinate system $\{z, y\}$ are

$$I_1' = \dot{w} \quad , \quad I_2' = t\dot{w} - w \tag{62}$$

The general solution of the system is:

$$y(t) = I_1' t - I_2' \tag{63}$$

$$z(t) = \frac{1}{36\left(I_1'\right)^2}\left(I_1' t - I_2'\right)^4 + z_1 t + z_0 \tag{64}$$

The Hamiltonian constrain gives $E = z_1 I_1'$ where $z_{0,1}$ are constants and the singularity condition results in the constrain

$$\frac{1}{36\left(I_1'\right)^2}\left(I_2'\right)^4 + z_0 = 0. \tag{65}$$

6.2. Power law model R^{μ} with $\mu = \frac{7}{8}$
In this case the Lagrangian eq.(10) is written as

$$L = \frac{21a}{4R^{\frac{1}{8}}}\dot{a}^2 - \frac{21}{16}\frac{a^2}{R^{\frac{9}{8}}}\dot{a}\dot{R} - \frac{1}{8}a^3 R^{\frac{7}{8}}. \tag{66}$$

Changing now the variables from (a, R) to (ρ, σ) via the relations:

$$a = \left(\frac{21}{4}\right)^{-\frac{1}{3}}\sqrt{\rho e^{\sigma}} \quad R = \frac{e^{12\sigma}}{\rho^4}. \tag{67}$$

The Lagrangian (94) and the Hamiltonian (11) become

$$L = \frac{1}{2}\dot{\rho}^2 - \frac{1}{2}\rho^2\dot{\sigma}^2 + V_0\frac{e^{12\sigma}}{\rho^2} \tag{68}$$

$$E = \frac{1}{2}\dot{\rho}^2 - \frac{1}{2}\rho^2\dot{\sigma}^2 - V_0\frac{e^{12\sigma}}{\rho^2}. \tag{69}$$

where $V_0 = -\frac{1}{42}$. The Euler-Lagrange equations provide the following equations of motion:

$$\ddot{\rho} + \rho\dot{\sigma}^2 + 2V_0\frac{e^{12\sigma}}{\rho^3} = 0 \tag{70}$$

$$\ddot{\sigma} + \frac{2}{\rho}\dot{\sigma}\dot{\rho} + 12V_0\frac{e^{12\sigma}}{\rho^2} = 0. \tag{71}$$

The Noether integrals (43), (44) and the Ermakov-Lewis invariant 45 in the coordinate system $\{u, v\}$ are

$$I_5' = 2tE - \rho\dot{\rho} \tag{72}$$

$$I_6' = t^2E - t\rho\dot{\rho} + \frac{1}{2}\rho^2. \tag{73}$$

$$\Sigma = \rho^4\dot{\sigma}^2 + 4V_0e^{12\sigma}. \tag{74}$$

Using the Ermakov-Lewis Invariant, the Hamiltonian (68) and equation (70) are written:

$$\frac{1}{2}\dot{\rho}^2 - \frac{1}{2}\frac{\Sigma}{\rho^2} = E \tag{75}$$

$$\ddot{\rho} + \frac{\Sigma}{\rho^3} = 0. \tag{76}$$

And the analytical solution of the system is

$$\rho(t) = \left(\rho_2t^2 + \rho_1t + \frac{\left((\rho_1)^2 - 4\Sigma\right)}{4\rho_2}\right)^{\frac{1}{2}} \tag{77}$$

$$\exp(\sigma(t)) = \left\{\frac{21}{2}\Sigma\left[\left(\tanh\left[\sigma_0\rho_2\sqrt{\Sigma} - 6\operatorname{arctan}h\left(\frac{2\rho_2t + \rho_1}{2\sqrt{\Sigma}}\right)\right]\right)^2 - 1\right]\right\}^{\frac{1}{12}} \tag{78}$$

where $B(t) = \left(\frac{1}{2}\frac{2\rho_2t+\rho_1}{\sqrt{\Sigma}}\right)$ and $\rho_{1,2}$, σ_0 are constants with Hamiltonian constrain $E = \frac{1}{2}\rho_2$. The singularity constrain gives $(\rho_1)^2 = 4\Sigma$

In the case $\Sigma = 0$ the analytical solution is

$$\rho(t) = \left(\rho_2t^2 + \rho_1t + \frac{1}{2}\frac{(\rho_1)^2}{\rho_2}\right)^{\frac{1}{2}} \tag{79}$$

$$\exp\sigma(t) = \left[\frac{1}{24\sqrt{V_0}}\frac{(2\rho_2t + \rho_1)}{(4\sigma_0\rho_2^2t + 2\sigma_0\rho_2\rho_1 - 1)}\right]^{\frac{1}{6}} \tag{80}$$

The singularity constrain gives $\rho_1 = 0$, then the solution is

$$a(t) = \frac{a_0t^{\frac{7}{6}}}{(a_2t - 1)^{\frac{1}{6}}} \tag{81}$$

In contrast with the claim of [47] this model is analytically solvable and there exists models which admit Noether integrals with time dependent gauge functions.

6.3. $\Lambda_{bc}CDM$ model with $(b, c) = (1, \frac{3}{2})$

Inserting $f(R) = (R - 2\Lambda)^{3/2}$ into eq.(10) we obtain

$$L = 9a\sqrt{R - 2\Lambda}\dot{a}^2 + \frac{9a^2}{2\sqrt{R - 2\Lambda}}\dot{a}\dot{R} + \frac{a^3}{2}(R + 4\Lambda)\sqrt{R - 2\Lambda} \tag{82}$$

Changing now the variables from (a, R) to (x, y) via the relations:

$$a = \left(\frac{9}{2}\right)^{-\frac{1}{3}} \sqrt{x} \quad R = 2\Lambda + \frac{y^2}{x} \tag{83}$$

the Lagrangian (82) and the Hamiltonian (11) become

$$L = \dot{x}\dot{y} + V_0\left(y^3 + \bar{m}xy\right) \tag{84}$$

$$E = \dot{x}\dot{y} - V_0\left(y^3 + \bar{m}xy\right) \tag{85}$$

where $V_0 = \frac{1}{9}$ and $\bar{m} = 6\Lambda$.
The equations of motion, using the Euler-Lagrange equations, in the new coordinate system are

$$\ddot{x} - 3V_0y^2 - \bar{m}V_0x = 0 \tag{86}$$

$$\ddot{y} - \bar{m}V_0y = 0. \tag{87}$$

The Noether integrals (47),(48) in the coordinate system $\{x, y\}$ are

$$I_1' = e^{\omega t}\dot{y} - \omega e^{\omega t}y \tag{88}$$

$$I_2' = e^{-\omega t}\dot{y} + \omega e^{-\omega t}y. \tag{89}$$

where $\omega = \sqrt{2\Lambda/3}$. From these we construct the time independent first integral

$$\Phi = I_1 I_2 = \dot{y}^2 - \omega^2 y^2. \tag{90}$$

The constants of integration are further constrained by the condition that at the singularity $(t = 0)$, the scale factor has to be exactly zero, that is, $x(0) = 0$.
The general solution of the system (86)-(87) is:

$$y(t) = \frac{I_2}{2\omega}e^{\omega t} - \frac{I_1}{2\omega}e^{-\omega t} \tag{91}$$

$$x(t) = x_{1G}e^{\omega t} + x_{2G}e^{-\omega t} + \frac{1}{4\bar{m}\omega^2}\left(I_2e^{\omega t} + I_1e^{-\omega t}\right)^2 + \frac{\Phi}{\bar{m}\omega^2}. \tag{92}$$

The Hamiltonian constrain gives $E = \omega\left(x_{1G}I_1 - x_{2G}I_2\right)$ where $x_{1G,2G}$ are constants and the singularity condition results in the constrain

$$x_{1G} + x_{2G} + \frac{1}{4\bar{m}\omega^2}\left(I_1 + I_2\right)^2 + \frac{\Phi}{\bar{m}\omega^2} = 0. \tag{93}$$

At late enough times the solution becomes $a^2(t) \propto e^{2\omega t}$

6.4. Λ_{bc}CDM model with $(b, c) = (1, \frac{7}{8})$
In this case the Lagrangian eq.(10) of the $f(R) = (R - 2\Lambda)^{7/8}$ model is written as

$$L = \frac{21a}{4\left(R - 2\Lambda\right)^{\frac{1}{8}}}\dot{a}^2 - \frac{21}{16}\frac{a^2}{\left(R - 2\Lambda\right)^{\frac{9}{8}}}\dot{a}\dot{R} - \frac{1}{8}a^3\frac{\left(R - 16\Lambda\right)}{\left(R - 2\Lambda\right)^{\frac{1}{8}}}. \tag{94}$$

Changing now the variables from (a, R) to (u, v) via the relations:

$$a = \left(\frac{21}{4}\right)^{-\frac{1}{3}} \sqrt{ue^v} \quad R = 2\Lambda + \frac{e^{12v}}{u^4}. \tag{95}$$

The Lagrangian (94) and the Hamiltonian (11) become

$$L = \frac{1}{2}\dot{u}^2 - \frac{1}{2}u^2\dot{v}^2 + V_0\frac{\bar{m}}{4}u^2 + V_0\frac{e^{12v}}{u^2} \tag{96}$$

$$E = \frac{1}{2}\dot{u}^2 - \frac{1}{2}u^2\dot{v}^2 - V_0\frac{\bar{m}}{4}u^2 - V_0\frac{e^{12v}}{u^2}. \tag{97}$$

where $\bar{m} = -28\Lambda$, $V_0 = -\frac{1}{42}$.
The Euler-Lagrange equations provide the following equations of motion:

$$\ddot{u} + u\dot{v}^2 - \frac{V_0\bar{m}}{2}u + 2V_0\frac{e^{12v}}{u^3} = 0 \tag{98}$$

$$\ddot{v} + \frac{2}{u}\dot{u}\dot{v} + 12V_0\frac{e^{12v}}{u^4} = 0. \tag{99}$$

The Noether integrals (51),(52) and the Ermakov-Lewis invariant (53) in the coordinate system $\{u, v\}$ are

$$I_+ = \frac{1}{\lambda}e^{2\lambda t}E - e^{2\lambda t}u\dot{u} + \lambda e^{2\lambda t}u^2 \tag{100}$$

$$I_- = \frac{1}{\lambda}e^{-2\lambda t}E - e^{-2\lambda t}u\dot{u} + \lambda e^{-2\lambda t}u^2. \tag{101}$$

$$\phi = u^4\dot{v}^2 + 4V_0e^{12v}. \tag{102}$$

where $\lambda = \frac{1}{2}\sqrt{\frac{2}{3}\Lambda}$.

Using the Ermakov-Lewis Invariant (102), the Hamiltonian (97) and equation (98) are written:

$$\frac{1}{2}\dot{u}^2 - V_0\frac{m}{8}u^2 - \frac{1}{2}\frac{\phi}{u^2} = E \tag{103}$$

$$\ddot{u} - \frac{V_0 m}{4}u + \frac{\phi}{u^3} = 0. \tag{104}$$

The solution of (104) has been given by Pinney [49] and it is the following:

$$u\left(t\right) = \left(u_1e^{2\lambda t} + u_2e^{-2\lambda t} + 2u_3\right)^{\frac{1}{2}} \tag{105}$$

where u_{1-3}. From the Hamiltonian constrain (103) and the Noether Integrals (100),(101) we find

$$E = -2\lambda u_3 \ , \ I_+ = 2\lambda u_2 \ , \ I_- = 2\lambda u_1.$$

Replacing (105) in the Ermakov-Lewis Invariant (102) and assuming $\phi \neq 0$ we find:

$$\exp\left(v\left(t\right)\right) = 2^{\frac{1}{6}}\phi^{\frac{1}{12}}e^{-A(t)}\left(4V_0 + e^{-12A(t)}\right)^{-\frac{1}{6}} \tag{106}$$

where

$$A\left(t\right) = \arctan\left[\frac{2\lambda}{\sqrt{\phi}}\left(u_1e^{2\lambda t} + u_3\right)\right] + 4\lambda^2u_1\sqrt{\phi}. \tag{107}$$

Then the solution is

$$a^2\left(t\right) = 2^{-\frac{1}{3}}\phi^{\frac{1}{12}}e^{-A(t)}\left(4V_0 + e^{-12A(t)}\right)^{-\frac{1}{6}}\left(u_1e^{2\lambda t} + u_2e^{-2\lambda t} + 2u_3\right)^{\frac{1}{2}} \tag{108}$$

where from the singularity condition $x(0) = 0$ we have the constrain $u_1 + u_2 + 2u_3 = 0$, or

$$2E - (I_+ + I_-) = 0. \tag{109}$$

At late enough time we find $A(t) \simeq A_0$, which implies $a^2(t) \propto e^{\lambda t}$.

In the case where $\phi = 0$ equations (103),(104) describe the hyperbolic oscillator and the solution is

$$u(t) = \sinh \lambda t \ , \ 2E = \lambda^2. \tag{110}$$

From the Ermakov-Lewis Invariant we have

$$\exp(v(t)) = \left(\frac{\lambda \sinh \lambda t}{\lambda v_1 \sinh \lambda t - 12\sqrt{|V_0|} e^{-2\lambda t}} \right)^{\frac{1}{6}} \tag{111}$$

where v_1 is a constant. The analytical solution is

$$a^2(t) = \left(\frac{\lambda \sinh^7 \lambda t}{\lambda v_1 \sinh \lambda t - 12\sqrt{|V_0|} e^{-2\lambda t}} \right)^{\frac{1}{6}} \tag{112}$$

7. Noether symmetries in spatially non flat $f(R)$ models

In this section we study further the Noether symmetries in non flat $f(R)$ cosmological models. In the context of a FRW spacetime the Lagrangian of the overall dynamical problem and the Ricci scalar are

$$L = 6f'a\dot{a}^2 + 6f''\dot{R}a^2\dot{a} + a^3(f'R - f) - 6Kaf' \tag{113}$$

$$R = 6\left(\frac{\ddot{a}}{a} + \frac{\dot{a}^2 + K}{a^2}\right) \tag{114}$$

where K is the spatial curvature. Note that the two dimensional metric is given by eq.(35) while the "potential" in the Lagrangian takes the form

$$V_K(a, R) = -a^3(f'R - f) + Kaf'. \tag{115}$$

Based on the above equations and using the theoretical formulation presented in section 5, we find that the $f(R)$ models which admit non trivial Noether symmetries are the $f(R) = (R - 2\Lambda)^{3/2}$, $f(R) = R^{3/2}$ and $f(R) = R^2$. The Noether symmetries can be found in section 5.

In particular, inserting $f(R) = (R - 2\Lambda)^{3/2}$ into the Lagrangian (113) and changing the variables from (a, R) to (x, y) [see section 6.3] we find

$$L = \dot{x}\dot{y} + V_0(y^3 + \bar{m}xy) - \bar{K}y \tag{116}$$

$$E = \dot{x}\dot{y} - V_0(y^3 + \bar{m}xy) + \bar{K}y \tag{117}$$

where $\bar{K} = 3(6^{1/3}K)$. Therefore, the equations of motion are

$$\begin{aligned} \ddot{x} - 3V_0y^2 - \bar{m}V_0x + \bar{K} &= 0 \\ \ddot{y} - \bar{m}V_0y &= 0 \,. \end{aligned}$$

The constant term $\bar{K}$ appearing into the first equation of motion is not expected to affect the Noether symmetries (or the integrals of motion). Indeed we find that the corresponding Noether symmetries coincide with those of the spatially flat $f(R) = (R - 2\Lambda)^{3/2}$ model. However, in the case of $K \neq 0$ (or $\bar{K} \neq 0$) the analytical solution for the x-variable is written as

$$x_K(t) \equiv x(t) + \frac{\bar{K}}{\omega^2} \tag{118}$$

where $x(t)$ is the solution of the flat model $K = 0$ (see section 6.3). Note that the solution of the y-variable remains unaltered.

Similarly, for the case of the $f(R) = R^{3/2}$ model the analytical solution is

$$z_K(t) = z(t) + \bar{K} \tag{119}$$

where $z(t)$ is the solution of the spatially flat model (see section 6.1).

8. Conclusion

In the literature the functional forms of $f(R)$ of the modified $f(R)$ gravity models are mainly defined on a phenomenological basis. In this article we use the Noether symmetry approach to constrain these models with the aim to utilize the existence of non-trivial Noether symmetries as a selection criterion that can distinguish the $f(R)$ models on a more fundamental level. Furthermore the resulting Noether integrals can be used to provide analytic solutions.

In the context of $f(R)$ models, the system of the modified field equations is equivalent to a two dimensional dynamical system moving in M^2 (mini superspace) under the constraint $\bar{E}$ =constant. Following the general methodology of [31,38], we require that the two dimensional system admits extra Noether symmetries. This requirement fixes the $f(R)$ function and the analytical solutions are computed. It is interesting that two well known dynamical systems appear: the anharmonic oscillator and the Ermakov-Pinney system. We recall that the field equations of the $\Lambda-$cosmology is equivalent with that of the hyperbolic oscillator.

Acknowledgments
This research was partially funded by the University of Athens Special Account of Research Grants no 10812.

Appendix A. Special solutions for the Power law model R^n

The case $f(R) = R^n$ is not Liouville integrable via Noether point symmetries. The zero order invariant will be used in order to find special solutions. Inserting $f(R) = R^n$, $\left(n \neq 0, 1, \frac{3}{2}, \frac{7}{8}\right)$ into eq.(10) we obtain

$$L\left(a, \dot{a}, R, \dot{R}\right) = 6naR^{n-1}\dot{a}^2 + 6n(n-1)a^2R^{n-2}\dot{a}\dot{R} + (n-1)a^3R^n \tag{A.1}$$

and the modified field equations are

$$\ddot{a} + \frac{1}{a}\dot{a}^2 - \frac{1}{6}aR = 0 \tag{A.2}$$

$$\ddot{R} + \frac{n-2}{R}\dot{R}^2 - \frac{1}{n-1}\frac{R}{a^2}\dot{a}^2 + \frac{2}{a}\dot{a}\dot{R} - \frac{(n-3)}{6n(n-1)}R^2 = 0 \tag{A.3}$$

$$E = 6naR^{n-1}\dot{a}^2 + 6n(n-1)a^2R^{n-2}\dot{a}\dot{R} - (n-1)a^3R^n. \tag{A.4}$$

The Noether symmetry (54) is also and a Lie symmetry, hence we have the zero order invariants

$$a_0 = at^{-N} \ , \ R_0 = Rt^{-2}. \tag{A.5}$$

Applying the zero order invariants in the field equations (A.2)-(A.4) and in the Noether integral (55) we have the following results.

The dynamical system admit a special solution of the form

$$a = a_0 t^N \ , \ R = 6N(2N-1)t^{-2} \tag{A.6}$$

where the constants N, E and I_7 are

$$N = \frac{1}{2} \ , \ E = 0 \ , \ I_7 = 0$$

or

$$N = -\frac{(2n-1)(n-1)}{n-2} \ , \ E = 0 \ , \ I_7 = 0$$

or

$$N = \frac{2}{3}n \ , \ E = \left(\frac{12n}{9}\right)^n (4n-3)^{n-1}\left(13n - 8n^2 - 3\right) a_0^3 \ , \ I_7 = 0.$$

Another special solution is the deSitter solution for $n = 2$

$$a = a_0 e^{H_0 t} \ , \ R = 12H_0^2 \tag{A.7}$$

where $I_7 = 0$ and the spacetime is empty i.e. $E = 0$.

References

[1] Weinberg S 1989 Rev Mod Phys **61** 1
[2] Peebles P J and Ratra B 2003 Rev. Mod. Phys. **75** 559
[3] Padmanabhan T 2003 Phys. Rept. **380** 235
[4] Perivolaropoulos L. 2008 (Preprint arxXiv.0811.4684)
[5] Ratra B and Peebles P J E 1988 Phys. Rev D. **37** 3406
[6] Overduin J M and Cooperstock F I 1998 Phys. Rev. D **58** 043506
[7] Basilakos S Plionis M and Solà S 2009 Phys. Rev. D **80** 3511
[8] Linder E V 2004 Phys. Rev. Lett. **70** 023511
[9] Lima J A S Silva F E and Santos R C 2008 Class. Quantum Grav. **25** 205006
[10] Brookfield A W et.al 2006 Phys. Rev. Lett. **96** 061301
[11] Sotiriou T P and Faraoni V 2010 Rev. Mod. Phys. **82** 451
[12] Starobinsky A A 1980 Phys. Lett. B **91** 99
[13] Carrol S Duvvuri M V Trodden M and Turner M S 2004 Phys. Rev. D. **70** 043528
[14] Amendola L Polarski D and Tsujikawa S 2007 Phys. Rev. Lett. **98** 131302
[15] Amendola L Gannouji R Polarski D and Tsujikawa S 2007 Phys. Rev. D **75** 083504
[16] Capozzielo S 2002 Int. J. Mod. Phys. D. **11** 483
[17] Nojiri S and Odintsov S D 2006 Phys. Rev. D. **73** 124038
[18] Hu W and Sawicki I 2007 Phys. Rev. D. **76** 064004
[19] Starobinsky A A 2007 JETP Lett. **86** 157
[20] Tsujikawa S 2008 Phys. Rev. D. **77** 023507
[21] Amendola L and Tsujikawa S 2008 Phys. Lett. B. **660** 125
[22] Capozziello S de Ritis R and Marino A A 1997 Class. Quantum Gravity **14** 3259
[23] Cotsakis S Leach P G L and Pantazi H 1998 Grav. Cosm. **4** 314
[24] Rubano C and Scudellaro P 2002 Gen. Relat. Grav. **34** 307
[25] Sanyal A K Modak B Rubano C and Piedipalumbo E 2005 Gen. Relat. Grav. **37** 407
[26] Szydlowski M et al. 2006 Gen. Rel. Grav. **38** 795
[27] Capozziello S Stabile A and Troisi A 2007 Class. Quant. Grav. **24** 2153
[28] Capozziello S Nesseris S and Perivolaropoulos L 2007 JCAP **0712** 009
[29] Bonanno A Esposito G Rubano C and Scudellaro P 2007 Gen. Rel. Grav. **39** 189
[30] Vakili B Phys. Lett. B. 2008 **664** 16
[31] Tsamparlis M and Paliathanasis A 2011Gen. Rel. Grav. 43 1861
[32] Basilakos S Tsamparlis M and Paliathanasis A 2011 Phys. Rev. D **83** 103512 (Preprint arXiv:1104.2980)
[33] Vakili B and Khazaie F 2012 Class. Quantum Grav. **29** 035015 (Preprint arxiv:1109.3352)

[34] Jamil M Mahomed F M and Momenti D 2011 Phys. Lett. B **702** 315 (Preprint arxiv:1105.2610)
[35] Christodoulakis T et al 2012 (Preprint arXiv:1208.0462)
[36] Capozziello S De Laurentis M and Odintsov S D 2012 Eur. Phys. J. **C72** 2086 (Preprint arxiv:1206.4842)
[37] Wei H Guo X J and Wang L F 2012 Phys. Lett. B **707** 298
[38] Tsamparlis M and Paliathanasis A 2011 J. Phys. A: Math. and Theor. **44** 175202 (Preprint arXiv:1101.5771)
[39] Olver P J 1986 Applications of Lie Groups to Differential Equations Springer - Verlag N.Y.
[40] Stephani H 1989 Differential Equations: Their Solutions using Symmetry Cambridge University Press N.Y.
[41] Barnes A 1993 Class. Quantum Grav. **10** 1139
[42] Amendola L and Tsujikawa S 2012 Dark Energy: Theory and Observations Cambridge University Press N.Y.
[43] Moyo S and Leach P G L 2002 J. Phys. A: Math. and Theor. **35** 5333
[44] Tsamparlis M and Paliathanasis A 2012 J. Phys. A: Math. and Theor. **45** 275202 (Preprint arXiv:1205.4114)
[45] Amendola L Polarski D and Tsujikawa S 2007 Int. J. Mod. Phys. D **16** 1555 (Preprint astro-ph/0605384)
[46] Paliathanasis A Tsamparlis M and Basilakos S 2011 Phys. Rev. D. **84** 123514 (Preprint arXiv:1111.4547)
[47] Nayem Sk and Sanyal A K 2012 Astrophys. Space Sci. **342** 549 (Preprint arXiv:1208.2306)
[48] Lewis H R 1967 Phys. Rev. Lett. **18** 510
[49] Pinney E 1950 Proc. Am. Math. Soc. **1** 681

Journal of Physics: Conference Series **453** (2013) 012010 doi:10.1088/1742-6596/453/1/012010

Local interior estimates for the Euler-Einstein system

Taxiarchis Papakostas[1] **and Demetrios Pliakis** [2]

[1] Department of Natural Resources and Environment, Technological Educational Institute of Crete, Chania GR 73103

[2] Department of Electronics, Technological Educational Institute of Crete, Chania GR 73103

Abstract.

We consider the spacetime of a perfect fluid described by the Einstein-Euler equations. We assume given spatial growth for the fluid and spacetime initial data and estimate the temporal evolution of the spatial growth rates through the wave equations derived for the fluid and spacetime data.

1. Introduction

We study an asymptotically flat spacetime filled with perfect fluid inside an initial geodesic ball. We assume a spatial local growth condition for the fluid and the initial slice of spacetime. We estimate the dynamic variation of these growth rates under the assumption of the existence of a local maximal foliation. Here we sketch briefly their derivation. These estimates facilitate the numerical solution of the Einstein-Euler equations that describe such gravitational systems.

Specifically, we consider initial data for the spacetime and the perfect fluid, consisting of fisrt and second fundamental forms, lapse function, fluid velocity and density:

$$(\Sigma_0, \gamma_0, k_0, \phi_0,), \qquad (v(0), \psi_0 = \log(\rho_0 + p_0)).$$

The fluid is assumed to evolve adiabatically and posses an equation of state: $p = f(\rho, s)$ We assume that Σ_0 is aymptotically flat: the fluid is contained inside a ball $\mathbf{S}^2_r = \partial \mathbf{B}^3_r$:

$$v(\Sigma_0 \setminus \mathbf{B}^3_r) = 0, \qquad \rho_0(\Sigma_0 \setminus \mathbf{B}^3_r) = 0.$$

We assume initial local growth estimates inside $\mathbf{B}^3_r$ after a decomposition of this ball in domains:

$$\mathbf{B}^3_r = \bigcup_{j=1}^{N} \mathbf{B}^0_j.$$

For cut-offs $\mathrm{supp}(\zeta) \subset \mathbf{B}^0_j$ we introduce the quantities:

$$\varrho^i(\varphi;\zeta;\Sigma_0) = \frac{\mathrm{D}^i(\varphi;\zeta;\Sigma_0)}{\mathrm{D}^0(\varphi;\zeta;\Sigma_0)}, \qquad \mathrm{D}^i(\varphi;\zeta;\Sigma_0) = \int_{\Sigma_0} \zeta^2 |\nabla^i \varphi|^2,$$

and

$$\gamma_i[\varphi] = \sup_{\zeta \in S} \varrho^i(\phi;\zeta;\Sigma_0).$$

In this paper we estimate the time evolution of the preceding constants $\gamma_i[\phi]$ for

$$\varphi = \phi_0, \quad |k_0|, \quad |v_0|, \quad \psi_0, \quad \overline{\sigma}_0 = (\overline{\nabla}_j v^j)(0), \quad |\overline{\omega}| = \left| *\frac{1}{2}\left(\overline{\nabla}_l v_m - \overline{\nabla}_m v_l\right)(0)\right|$$

assuming as given the dynamical estimates for the last two quantities. We will combine these estimates with the dynamical construction of the decomposition through the derivation of the estimates of the bump function and its derivatives through the various estimates.

These estimates will be combined with an iterative solution of the nonlinear hyperbolic equations describing the system.

Plan of the paper We present here mainly the energy estimates and appeal to the results in [9] on the Nash-DeGiorgi-Moser tecnique for the $L^\infty - L^p$ estimates on domains of the above form.

2. The deformation of the metric on the time slice

The weightfunction θ is constructed to measure the deformation of the foliation of the Euclidean ball by concentric spheres to a foliation by deformed spheres with the appearance of bumps. This deformation will be defined using a cover of the unit sphere as follows. On each time slice Σ_t - diffeomorpsic to $\mathbf{R}^3$ endowed with the metric γ we consider the deformation of the foliation by Euclidean balls though the collection of bumps $\{\mathbf{B}_{r,\ell}, \theta_\ell\}_{\ell=1,r<R}^N$ where

$$\mathbf{S}^2 = \bigcup_{\ell=1}^N \mathbf{B}_\ell$$

$$\theta_\ell : [0,R] \times \mathbf{B}_\ell \to \mathbf{R}, \qquad [0,R] \times \mathbf{B}_\ell = \mathbf{B}_{R,\ell}, \qquad \mathbf{B}_{r,\ell} = \mathbf{B}_{R,\ell} \bigcap \mathbf{S}_r^2$$

The foliation is deformed so that each leaf is deformed through the graphs of the function θ_ℓ over $\mathbf{B}_{r,\ell}$. Specifically, we have that the metric on $\mathbf{B}_R$ assumes the following form:

$$\gamma_0 = a^2 dr^2 + \mathring{\gamma}$$

The bump function are regularized on each patch $\mathbf{B}_{r,\ell}$ by the method of harmonic approximation that is described in [9].

3. Assumptions and Results

We make the following assumptions:

- The vorticity ω and divergence σ of the velocity field $\overline{v}$ satisfy estimates of the form

$$\int_{\Sigma_t} \zeta^2(|\overline{\nabla}\omega|^2 + |\overline{\nabla}\overline{\sigma}|^2) \leq \varphi(t) \int_{\Sigma_t} \zeta^2 |\overline{v}|^2.$$

- The bump and lapse functions θ, α respectively are assumed to satisfy bounds of the form, for $k = 1,2,3$:

$$\int_{\mathbf{B}_j} \zeta^2(|\overline{\nabla}^k \theta|^2 \leq (\vartheta(t))^k \int_{\mathbf{B}_j} \zeta^2 |\theta|^2$$

$$\int_{\mathbf{B}_j} \zeta^2(|\overline{\nabla}^k a|^2 \leq (\alpha(t))^k \int_{\mathbf{B}_j} \zeta^2 a^2$$

- We assume that the equation of state relating pressure and density satisfies estimates of the form:

$$p = f(\rho) \sim \rho^\gamma, \gamma \geq 1$$

- We introduce the following localized energy on our data:

$$E(\zeta;\mathrm{D})(t) = \int_{\Sigma_t} \zeta^2 \left(|\mathrm{Ric}|\phi^2|k_t|^2 + |\overline{\nabla}k|^2 + \phi_t^2 + |\overline{\nabla}\phi|^2 + \phi^2\psi_t^2 + |\overline{\nabla}\psi|^2 + |\omega|^2 + (\overline{\sigma})^2 + \alpha^2 + |\overline{\nabla}^3\theta|^2 \right)$$

where the data of our system are

$$\mathrm{D} = [(\phi,\gamma,k); (v,\psi,\omega,\overline{\sigma}); (\alpha,\theta)]$$

- Furthermore, we assume that the functions localizing in the domains $\mathbf{B}_j$ satisfy estimates of the form:

$$v(\zeta) = \pm|v|\beta_j\zeta(1-\zeta), \quad \zeta_0 = \pm\beta_j\zeta(1-\zeta)$$

where $\zeta(r,\vartheta,t) = \ell(r)\ell\left(\frac{|p(\vartheta)|}{\epsilon}\right)\ell\left(\frac{\epsilon}{|p(\vartheta)|}\right)\ell(t)$ and $\ell(t) \equiv 1$ if $\frac{1}{2} < t < 1$ and 0 if $0 < t < \frac{1}{4}$ or $T > \frac{3}{2}$.

We introduce then the domains $\mathbf{B}_j \subset \Sigma_t$

$$\mathbf{B}_j(\theta,\epsilon) = \{x \in \Sigma_t/(1-\epsilon+\epsilon^{j+1})\theta \leq |\theta(x)| \leq (1+\epsilon^{j+1})\theta\}$$

We prove then for $t \in [0,T)$-T computed explicitly that the following estimates hold:

$$\sup_{\mathbf{B}_j} a \leq P(\eta,\gamma_\theta)t^{e_a}E(\mathrm{D})(0) \quad \sup_{\mathbf{B}_j}|\overline{\nabla}^k\theta|P(\eta,\gamma_\theta)t^{e_{\theta,k}}E(\mathrm{D})(0)$$

$$\sup_{\mathbf{B}} \psi \leq P(\eta,\varphi_j)t^{e_\psi}E(\mathrm{D})(0) \quad \sup_{\mathbf{B}}|\overline{v}| \leq P(\eta,\varphi_j)t^{e_v}E(\mathrm{D})(0)$$

$$\sup_{\mathbf{B}_j}|\omega| \leq P(\eta,\varphi_j)t^{e_\omega}E(\mathrm{D})(0) \quad \sup_{\mathbf{B}_j}|\sigma| \leq P(\eta,\varphi_j)t^{e_\sigma}E(\mathrm{D})(0)$$

$$\sup_{\mathbf{B}_j}\phi \leq P(\eta,\varphi_j)t^{e_\phi}E(\mathrm{D})(0) \quad \sup_{\mathbf{B}_j}|\phi_t| \leq P(\eta,\varphi_j)t^{e_{\phi_1}}E(\mathrm{D})(0)$$

$$\sup_{\mathbf{B}_j}|\overline{\nabla}\phi| \leq P(\eta,\varphi_j)t^{e_{\phi_2}}E(\mathrm{D})(0) \quad \sup_{\mathbf{B}_j}|\overline{\nabla}^2\phi| \leq P(\eta,\varphi_j)t^{e_{\phi_3}}E(\mathrm{D})(0)$$

$$\sup_{\mathbf{B}_j}|k| \leq P(\eta,\varphi_j)t^{e_k}E(\mathrm{D})(0) \quad \sup_{\mathbf{B}_j}|\overline{\nabla}k| \leq P(\eta,\varphi_j)t^{e_{k,1}}E(\mathrm{D})(0)$$

$$\sup_{\mathbf{B}_j}|k_t| \leq P(\eta,\varphi_j)t^{e_{k,2}}E(\mathrm{D})(0)$$

while for $0 \leq t < T$ we have that:

$$\sup_{\mathbf{B}_j}\varphi \leq \varphi_j \sim \theta^{m+3}t^6, \quad \sup_{\mathbf{B}_j}\alpha \leq \alpha_j \sim \theta^{m+3}t^6$$

All these exponents are explicitly calculated. In particular the lifespan is determined through the estimates:

$$|\overline{\nabla}\theta| \geq (1+\epsilon)|\overline{v}|, \quad |\overline{v}| \leq \epsilon(1-\eta^2), \quad \eta^2 = f'(\rho)$$

The velocity estimate asserts the hyperbolic nature of the pressure wave equation.

4. The energy estimates

The wave equation for the Lorentzian metric g and source function $\mathbf{S}$:

$$\Box_g\psi = \mathbf{S}$$

where

$$g = -\lambda^2 dt + \gamma$$

We introduce the vector fields:

$$T = \lambda\frac{\partial}{\partial t}, \qquad L = T + W, \qquad \underline{L} = T - W$$

for a vector field W tangential to the time slices and constructed in the next paragraph. The energy momentum tensor, then, for the wave equation reads:

$$\mathbb{T}_{\mu\nu} = \psi_\mu\psi_\nu - \frac{1}{2}g_{\mu\nu}\psi^\lambda\psi_\lambda$$

with divergence

$$\mathbb{T}^{\mu\nu}{}_{;\nu} = \psi^{\mu}\mathbf{S}$$

The associated current

$${}^{(X)}\mathbb{J}^{\,\mu} = \mathbb{T}^{\mu\nu}X_{\nu} = X(\psi)\phi^{\mu} - \frac{1}{2}X^{\mu}\phi^{\lambda}\psi_{\lambda}$$

obeys the equation:

$${}^{(X)}\mathbb{J}^{\mu}_{;\mu} = {}^{(X)}\pi_{\mu\nu}\mathbb{T}^{\mu\nu} + \mathbf{S}\cdot X(\psi)$$

where the deformation tensor refers to:

$${}^{(X)}\pi_{\mu\nu} = \frac{1}{2}(\nabla_{\mu}X_{\nu} + \nabla_{\nu}X_{\mu})$$

5. Construction of the vector fields

The vector field W introduced above is the gradient vector field

$$W = \frac{\overline{\nabla}\theta}{|\overline{\nabla}\theta|^2}$$

for the function θ defined on the time slice.

Domains We denote by $C_u, \underline{C}_{\underline{u}}$, the level sets of the functions $u, \underline{u}$ and by Σ_t the time slices. We denote the region formed by the level sets of $t, u, \underline{u}$:

$$\mathcal{D}(u_0, 2u_0; t) = \{(t', x)/0 \le t' \le t,\ u_0 \ge -u_0,\ \text{and } \underline{u} \le 2u_0\}$$

Inside this set it is contained the cylinder $\mathcal{C}_{[0,t],R}$. Specifically let R be a regular value of $\theta(t', \cdot), t' \in [0, t]$, then:

$$\mathcal{C}_{[0,t],R} = \{(t', x)/\theta(t', x) \le R\}$$

Local optical functions We introduce the solutions of the eikonal equations $u, \underline{u}$:

$$L(\underline{u}) = 0 \quad \underline{L}(u) = 0$$

The solution of these equations proceeds as follows. First we introduce the coordinates $s, \underline{\vartheta}$ as

$$W(s) = 1 \qquad W(\vartheta_1) = W(\vartheta_2) = 0$$

then we perforem the local change of time variable

$$\tau = \alpha(t, s, \vartheta) - s$$

that we select as solution

$$\lambda\alpha_t = \alpha_s$$

through the flow out of the characteristic equation

$$\frac{ds}{d\tau} = -\frac{1}{\lambda}$$

with initial value $s = 0, \tau = 0$ and $\alpha(\tau, c(\tau)) = \tau$. Finally the optical function is

$$u = \tau - s, \qquad \underline{u} = \tau + s.$$

Journal of Physics: Conference Series **453** (2013) 012010 doi:10.1088/1742-6596/453/1/012010

The integral identities For a cut-off function ζ localizing in $\mathcal{D}(-u_0, 2u_0; t)$ we have:

$$\int_{\Sigma_t} \zeta^2 \mathbb{T}(X,T) = \int_{\underline{C}_{2u_0}} \zeta^2 \mathbb{T}(X,L) + \int_{C_{-u_0}} \zeta^2 \mathbb{T}(X,\underline{L}) + \int_{\Sigma_0} \zeta^2 \mathbb{T}(X,T) +$$

$$+ \iint_{\mathcal{D}(-u_0,u_0;t)} \mathbb{T}^{\mu\nu}\,{}^{(X)}\pi_{\mu\nu} + \mathbf{S}X(\psi) + (2X(\psi)\phi^\nu \zeta_\nu - X(\zeta)\phi^\nu \phi_\nu)\,\zeta.$$

We calculate that:

$$\begin{aligned}
\mathbb{T}(T,T) &= \frac{1}{2}\left(\lambda^2 \psi_t^2 + |\overline{\nabla}\psi|^2\right) = \frac{1}{2}\varepsilon \\
\mathbb{T}(L,L) &= |L(\psi)|^2 = \lambda^2 |\overline{\nabla}\phi|^2 + W(\psi)^2 + 2\lambda\psi W(\psi) \\
\mathbb{T}(\underline{L},\underline{L}) &= |\underline{L}(\psi)|^2 = \lambda^2 |\overline{\nabla}\psi|^2 + W(\phi)^2 - 2\lambda\psi W(\psi) \\
\mathbb{T}(L,\underline{L}) &= |\underline{L}(\psi)|^2 = \lambda^2 |\overline{\nabla}\psi|^2 - W(\psi)^2 \\
\mathbb{T}(T,L) &= \frac{1}{2}\left(\lambda^2 \psi_t^2 + |\overline{\nabla}\psi|^2\right) + \lambda\psi_t W(\psi) \\
\mathbb{T}(T,\underline{L}) &= \frac{1}{2}\left(\lambda^2 \psi_t^2 + |\overline{\nabla}\psi|^2\right) - \lambda\psi_t W(\psi)
\end{aligned}$$

These terms are majorised by ε. Similarly we have that

$$\mathbb{T}^{\mu\nu}\,{}^{(T)}\pi_{\mu\nu} = \left(\lambda\psi_t\lambda_t - 4\overline{\nabla}\lambda \cdot \overline{\nabla}\phi\right)\lambda^2\phi\zeta^2 - 4(\overline{\nabla}\psi \cdot \overline{\nabla}\zeta)\lambda\zeta \leq 4\zeta^2\varepsilon + 4\zeta^2\lambda^4\left(\lambda_t^2 + |\overline{\nabla}\lambda|^2\right).$$

Localized energy If we select as $X = T$ then the above identity suggests for the energy and the cutoff function selected so that we restrict in the interior or that the integrals on the $C, \underline{C}$ vanish then:

$$\mathrm{E}_\zeta^{\lambda,\psi}(t) = \int_{\Sigma_t} \zeta^2 \varepsilon_\lambda^\psi, \qquad \varepsilon_\lambda^\psi = \left(\lambda^2\psi_t^2 + |\nabla\psi|^2\right),$$

that

$$\mathrm{E}_\zeta^{\lambda,\psi}(t) \leq \mathrm{E}_\zeta^{\lambda,\psi}(0) + \iint_{\mathcal{D}(-u_0,2u_0;t)} \zeta^2 \varepsilon_\lambda^\psi + \mathrm{I} + \mathrm{II},$$

where

$$\mathrm{I} \leq 2 \iint_{\mathcal{D}(-u_0,2u_0;t)} \zeta^2\lambda^4\left(\lambda_t^2 + |\overline{\nabla}\lambda|^2\right),$$

and

$$\mathrm{II} = \iint_{\mathcal{D}(-u_0,2u_0;t)} \zeta^2\lambda\mathbf{S}\psi_t.$$

The coarea formula suggests that:

$$\iint_{\mathcal{D}(-u_0,2u_0;t)} \zeta^2\varepsilon = \int_0^t dt' \left(\int_{\Sigma_{t'}} \lambda\zeta^2\varepsilon d\sigma\right) \leq \int_0^t \left(\sup_{\Sigma_{t'}} \lambda\right) \mathrm{E}(t')dt'$$

The term I assumes a majorisation of the form:

$$\mathrm{I} \leq \int_0^t v(t')$$

Journal of Physics: Conference Series **453** (2013) 012010 doi:10.1088/1742-6596/453/1/012010

for a function v that will be determined while II will assume a majorisation of the form due to the structure of $\mathbf{S}$ as

$$\mathrm{II} \leq \iint_{\mathcal{D}(-u_0,2u_0;t)} \zeta^2 \lambda \mathbf{S}\psi_t \leq \int_0^t \gamma(t')\mathrm{E}(t'),$$

where γ is a function that will be determined in every case.

$$\mathrm{E}_\zeta^{\lambda,\psi} \leq \mathrm{E}_\zeta^{\lambda,\psi}(0) + \int_0^t \gamma(t')\mathrm{E}_\zeta^{\lambda,\psi}(t') + v(t').$$

Therefore Gronwall's inequality we obtain:

$$\mathrm{E}_\zeta^{\lambda,\psi}(t) \leq E(0)e^{\int_0^t \gamma(t')dt'} + \int_0^t v(t')e^{\int_t^{t''} \gamma(t'')dt''}dt'.$$

Energy radiated away We select $X = L, \underline{L}$, and in the domain $\mathcal{R}(-u_0, 2u_0; t)$:

$$\mathcal{R}(-u_0, 2u_0; t) = \{(t', x)/0 \leq t' \leq t, -2u_0 \leq u \leq -u_0 \text{ or } 2u_0 \leq \underline{u} \geq 4u_0\}$$

we have that:

$$\iint_{\mathcal{R}(-u_0,2u_0;t)} \zeta^2 \varepsilon_\lambda^\psi = \int_0^t dt' \int_{\Sigma_{t'}} \zeta^2 \lambda \varepsilon_\lambda^\psi d\sigma \leq \int_0^t \gamma_\lambda(t')\varepsilon_\zeta^\psi, \quad \gamma_\lambda(t) = \sup_{\Sigma_t} \lambda.$$

We select the cut-off to vanish on the lateral surfaces then:

$$\int_{\Sigma_t} \mathbb{T}(L,T) = \int_{\Sigma_0} \mathbb{T}(L,T) + \iint_{\mathcal{R}(-u_0,2u_0)} \zeta^2 \varepsilon_\lambda^\psi$$

Therefore we obtain similar estimates as before:

$$\mathrm{E}_\zeta^{\lambda,\psi}(t) \leq E(0)e^{\int_0^t \gamma(t')dt'} + \int_0^t v(t')e^{\int_{t''}^{t'} \gamma(t'')dt''}dt'$$

6. The Euler-Einstein system

The system is described by the Einstein equations for the energy momentum tensor:

$$\mathbf{T}_{\mu\nu} = (\rho + p)u_\mu u_\nu + pg_{\mu\nu}$$

Einstein equations then

$$\mathbf{R}_{\mu\nu} - \frac{\mathbf{R}}{2}g_{\mu\nu} = \mathbf{T}_{\mu\nu}$$

lead the Euler equations through Bianchi identity:

$$\nabla^\nu \mathbf{T}_{\mu\nu} = 0$$

Space-time We will use the space-time decomposition that requires

$$g = -\phi^2 dt^2 + \gamma$$

for the lapse function function ϕ and the induced Riemannian metric on the slices $\mathcal{H}_t$ denoted by $\gamma(t)$. The second fundamental form of the slices is $k(t)$ and we have the first variation identity

$$\partial_t \gamma = -2\phi k \quad \text{(FV)}$$

Journal of Physics: Conference Series **453** (2013) 012010 doi:10.1088/1742-6596/453/1/012010

The second variation formula and the Gauss-Codazzi (GC) and Gauss equations under the assumption of maximal slicing, i.e. $\mathrm{tr}(k) = 0$ are written as:

$$\partial_t k_{ij} = -\nabla_i \nabla_j \phi + \left(\mathrm{R}_{ij} - 2k_{im}k_j^m - \frac{1}{2}\mathrm{R}\gamma_{ij} - 2\mathbf{T}_{ij} \right) \phi \quad (\mathrm{SV})$$

$$\nabla_i k_{jm} - \nabla_j k_{im} = \mathbf{R}_{m0ij} \quad (\mathrm{GC})$$

$$\mathrm{R}_{ij} - k_{il}k_j^l = \mathbf{R}_{i0j0} + \mathbf{R}_{ij} \quad (\mathrm{G})$$

Taking the trace of (G) we obtain that:

$$\mathrm{R} - |k|^2 = \mathbf{T}_{00} \Rightarrow |k|^2 + (p+\rho)\beta^2\phi^2 = \phi^2 p + \mathrm{R}$$

We observe that

$$|k|^2 + p|\underline{v}|^2 + \rho\beta = \mathrm{R} \Rightarrow \mathrm{R} \geq p|\underline{v}|^2, |k|^2, \rho\beta$$

The trace of (GC) gives us that:

$$\mathrm{div}(k) = 2T_{j0} = 2\widetilde{p}\beta v_j$$

The lapse function The trace of (SV) suggests that

$$\phi\Delta_\gamma\phi + \left(|k|^2 + (2(p+\rho)\beta + (p-\rho))\phi^2\right)\phi^2 = 0$$

This suggests that

$$\Delta_\gamma \phi \leq 0$$

implying that $\phi \geq 1$ since this contradicts the existence of areas where $\phi \leq 1$ due to the fact that $\Delta_\gamma\phi \leq 0$ even then. This gives the integral inequality for ϕ, ζ:

$$\int_{\Sigma_t} \zeta^2 |\overline{\nabla}\phi|^2 = \int_{\Sigma_t} \zeta^2 \left[|k|^2\phi^2 + ((p+\rho)\beta + (p-\rho))\,\phi^4 - \phi\overline{\nabla}\zeta^2 \cdot \overline{\nabla}\phi \right]$$

Therefore we obtain

$$\int_{\Sigma_t} |\overline{\nabla}(\zeta\phi)|^2 \leq \left(\int_{\Sigma_t} (\beta^5 + |k|^{\frac{10}{3}} + (p+\rho)^5) \right)^{\frac{1}{5}} \left(\int_{\Sigma_t} \zeta^2\phi^5 \right)^{\frac{4}{5}}$$

Similarly we obtain for the time derivative of the lapse function that:

$$\int_{\Sigma_t} |\overline{\nabla}(\zeta\phi_t)|^2 \leq C_2 \left(\int_{\Sigma_t} (\beta^5 + k^{\frac{10}{3}} + (\rho+p)^5) \right)^{\frac{1}{5}} \left(\int_{\Sigma_t} \zeta^2\phi^5 \right)^{1/5}$$

since $\phi > 1$.

Estimates for $|\overline{\nabla}^2\phi|$ We use the formula:

$$\int_{\Sigma_t} \zeta|\overline{\nabla}^2\phi|^2 \leq C \left[\int_{\Sigma_t} (\zeta^2 + |\overline{\nabla}\zeta|^2) \left(|k|^2|\overline{\nabla}k|^2 + (1+\eta^2)^2|\overline{\nabla}\rho|^2 + (p+\rho)^2|\overline{\nabla}\beta|^2 \right) + (\rho+p)^{3/2}\beta^3 + |k|^3 \right]$$

7. The acoustical metric

The continuity equation could written in the form for the $\psi = \log(\rho + p)$:

$$\widetilde{g}^{\mu\nu}\nabla_\nu\psi = -\eta^{-2}\left(v^\nu\nabla_\nu v^\mu + v^\mu\sigma\right),$$

where

$$\widetilde{g}^{\mu\nu} = g^{\mu\nu} + \eta^{-2}u^\mu v^\nu, \qquad \sigma = \nabla_\nu v^\nu$$

We readily derive form this equation that:

$$|v_t^0| \leq \frac{\phi}{\beta}\left(\eta^2\phi^2 + \beta^2\right)\sqrt{\varepsilon_\chi^\psi} + |\phi_t| + \beta\phi|\overline{\nabla}\phi| + \phi|\mathrm{div}v|,$$

and also

$$|v_t| \leq \phi\left(\eta^2\phi^2 + \beta^2\right)\sqrt{\varepsilon_\chi^\psi} + \frac{|\phi_t|}{2\beta} + \frac{\phi}{2}|\overline{\nabla}\phi| + \frac{\phi}{2\beta}|\overline{\sigma}| + \phi(|\omega| + |\overline{\nabla}\beta|$$

We introduce the change of variables along the time-slice generated by the flow of the fluid velocity:

$$\frac{dX^j}{ds} = -\frac{\phi\beta}{\delta}v^j(x,t),\ X^j(0,x,t) = x,\ \beta = \sqrt{1 + |v|^2},\ \delta = 1 - \eta^2$$

The metric is simplified as

$$\widetilde{g} = -\chi^2 dt^2 + \tilde{\gamma}, \quad \tilde{\gamma} = X^*\gamma$$

where

$$\chi^2 = \phi^2\left(1 - \frac{\Upsilon^2}{\delta}\right) \leq \phi^2, \quad \Upsilon = \beta|v| \geq \sqrt{2}|v|^2$$

provided that

$$0 \leq |v|^2 \leq \frac{2\delta}{\sqrt{1 + 4\delta} + 1}$$

otherwise the acoustical metric then switches to Riemannian for large pressure and low velocities. Accordingly we have a Poisson equation. We introduce the acoustical connection which is related to the connection of g through the relation of the Christoffel symbols:

$$\widetilde{\Gamma}^\mu_{\nu\kappa} = \Gamma^\mu_{\nu\kappa} + F^\mu_{\nu\kappa},$$

where

$$F^\mu_{\nu\kappa} = \frac{1}{\eta^3(1-\eta^2)}\left[u^\mu u_\nu u_\kappa + u^\mu S_{\kappa\nu} - u^\mu\left(u_\kappa\nabla_\nu\eta + u_\nu\nabla_\kappa\eta\right) - \frac{1}{2}v^\mu(v^\rho\nabla_\rho)(v_\kappa v_\nu)\right] + \frac{1}{\eta^2}\left[\Omega^\mu_\kappa v_\nu + \Omega^\mu_\kappa v_\nu\right].$$

We bound this as

$$|F| \leq \frac{1}{2\delta\eta^3}\left[|v|^3 + |S||v| + 2|v|^2|\overline{\nabla}\eta| + \frac{\beta^4}{2\phi^2}\left(\eta^2\sqrt{\varepsilon} + |\sigma|\right) + \frac{2}{\eta^2}|\Omega|\beta\right],$$

if $\varepsilon_\chi^\psi = \chi^2\psi_t^2 + |\overline{\nabla}\psi|^2$. The wave equation for the acoustical metric reads as:

$$\Box_{\widetilde{g}}\psi = \mathbf{S}(v,\eta),$$

where

$$\mathbf{S}(v,\eta) = \eta^{-2}\left[\left(\widetilde{\nabla}_\mu\zeta + \widetilde{\nabla}_\mu\log(\chi\widetilde{\sigma}\gamma)\right)v^\nu\widetilde{\nabla}_\nu v^\mu - \widetilde{\nabla}_\mu v^\nu\widetilde{\nabla}_\nu v^\mu - v^\nu\nabla_\nu\sigma + v^\nu v^\mu\widetilde{\nabla}_\mu(\log(\chi\widetilde{\sigma}\gamma)) + \widetilde{\mathbf{Ric}}_{\mu\nu}v^\mu v^\nu\right],$$

where

$$\widetilde{\sigma}(x) = \det(DX)(X^{-1}(x))$$

We will apply the general method for the Lorentzian metric

$$-\chi^2 dt^2 + \mathring{\gamma}$$

where

$$\mathring{\gamma} = X^*\gamma$$

The integral identities For a cut-off function ζ localizing in $\mathcal{D}(-u_0, 2u_0; t)$ we have:

$$\int_{\Sigma_t} \zeta^2 \mathbb{T}(X,T) = \int_{\underline{C}_{2u_0}} \zeta^2 \mathbb{T}(X,L) + \int_{C_{-u_0}} \zeta^2 \mathbb{T}(X,\underline{L}) + \int_{\Sigma_0} \zeta^2 \mathbb{T}(X,T) +$$

$$+ \iint_{\mathcal{D}(-u_0,u_0;t)} \mathbb{T}^{\mu\nu}\ {}^{,(X)}\pi_{\mu\nu} + \mathbf{S}X(\psi) + (2X(\psi)\psi^\nu\zeta_\nu - X(\zeta)\psi^\nu\psi_\nu)\,\zeta,$$

We calculate that:

$$\mathbb{T}(T,T) = \frac{1}{2}\left(\chi^2\psi_t^2 + |\overline{\nabla}\psi|^2\right) = \frac{1}{2}\varepsilon$$

$$\mathbb{T}(L,L) = |L(\psi)|^2 = \chi^2|\overline{\nabla}\psi|^2 + W(\psi)^2 + 2\chi\psi W(\psi)$$

$$\mathbb{T}(\underline{L},\underline{L}) = |\underline{L}(\psi)|^2 = \chi^2|\overline{\nabla}\psi|^2 + W(\psi)^2 - 2\chi\psi W(\psi)$$

$$\mathbb{T}(L,\underline{L}) = |\underline{L}(\psi)|^2 = \chi^2|\overline{\nabla}\psi|^2 - W(\psi)^2$$

$$\mathbb{T}(T,L) = \frac{1}{2}\left(\chi^2\psi_t^2 + |\overline{\nabla}\psi|^2\right) + \chi\psi_t W(\phi)$$

$$\mathbb{T}(T,\underline{L}) = \frac{1}{2}\left(\chi^2\psi_t^2 + |\overline{\nabla}\psi|^2\right) - \lambda\psi_t W(\psi)$$

These terms are majorised by ε_χ^ψ. Similarly we have that

$$\mathbb{T}^{\mu\nu}\ {}^{,(T)}\pi_{\mu\nu} = \left(\chi\psi_t\lambda_t - 4\overline{\nabla}\chi\cdot\overline{\nabla}\psi\right)\chi^2\psi\zeta^2 - 4(\overline{\nabla}\psi\cdot\overline{\nabla}\zeta)\chi\zeta \leq 4\zeta^2\varepsilon_\chi^\psi + 4\zeta^2\chi^4\left(\chi_t^2 + |\overline{\nabla}\chi|^2\right),$$

Interior Energy estimates If we select as $X = T$ then the above identity suggests for the energy and the cutoff function selected so that we restrict in the interior or that the integrals on the $C, \underline{C}$ vanish then :

$$\mathrm{E}_\zeta^{\chi,\psi}(t) = \int_{\Sigma_t} \zeta^2\varepsilon_\chi^\psi,$$

that

$$\mathrm{E}_\zeta^{\chi,\psi}(t) \leq \mathrm{E}_\zeta^{\chi,\psi}(0) + \iint_{\mathcal{D}(-u_0,2u_0;t)} \zeta^2\varepsilon_\chi^\psi + \mathrm{I} + \mathrm{II},$$

where

$$\mathrm{I} \leq 2 \iint_{\mathcal{D}(-u_0,2u_0;t)} \zeta^2\chi^4\left(\chi_t^2 + |\overline{\nabla}\chi|^2\right),$$

and

$$\mathrm{II} = \iint_{\mathcal{D}(-u_0,2u_0;t)} \chi\zeta^2\mathbf{S}\psi_t,$$

The coarea formula suggests that:

$$\iint_{\mathcal{D}(-u_0,2u_0;t)} \zeta^2\varepsilon_\chi^\psi = \int_0^t dt'\left(\int_{\Sigma_{t'}} \chi\zeta^2\varepsilon_\chi^\psi d\sigma\right) \leq \int_0^t \left(\sup_{\Sigma_{t'}}\phi\right)\mathrm{E}_\zeta^{\chi,\psi}(t'),$$

We estimate provided that $\Upsilon \leq \epsilon\delta, \epsilon \leq \frac{1}{20}$

$$\chi^4\left(\chi_t^2 + |\overline{\nabla}\chi|^2\right) \leq \chi^4\left[(1-\epsilon)(\phi_t^2 + |\overline{\nabla}\phi|^2) + \epsilon^2\left(\Upsilon_t^2 + |\overline{\nabla}\Upsilon|^2\right) + \epsilon^4(\eta_t^2 + |\overline{\nabla}\eta|^2)\right] \leq$$

$$\leq 2\beta\left((1+\phi^2)|\overline{\nabla}v|^2 + \sigma^2\right) + 2\beta^3\phi^2\eta^2\varepsilon_\chi^\psi,$$

and also if for the equation of state $p = f(\rho)$

$$\eta_t^2 + |\overline{\nabla}\eta|^2 \leq e^{2\psi} f''^2 \varepsilon$$

Similarly we compute

$$\Upsilon_t^2 + |\overline{\nabla}\Upsilon|^2 \leq 2\beta(|v_t|^2 + |\overline{\nabla}v|^2) \leq 2\beta\left((1+\phi^2)|\overline{\nabla}v|^2 + \beta^2\phi^2\eta^2|\overline{\nabla}\psi|^2 + \sigma^2\right).$$

In conclusion we have that setting

$$v_f(t) = \int_{\Sigma_t} \zeta^2 |\overline{\nabla}v|^2$$

we bound

$$\mathrm{I} \leq 2\int_0^t \sup_{\Sigma_{t'}} \left[\phi^6\left(e^{2\psi} f''^2 + \phi\beta^2\eta^2\right)\right] E(t')dt' + \int_0^t \sup_{\Sigma_{t'}} \left(\beta\phi^7\right) v_f(t')$$

The term II is majorised as follows

$$\iint\limits_{\mathcal{D}(-u_0,2u_0;t)} \chi\zeta^2 \mathbf{S}\psi_t = \iint\limits_{\mathcal{D}(-u_0,2u_0;t)} \frac{\chi}{\eta^2}\zeta^2\psi_t\left[\widetilde{\nabla}_\mu Z + \widetilde{\nabla}_\mu \log(\chi\widetilde{\sigma}\gamma)\right) v^\nu\widetilde{\nabla}_\nu v^\mu -$$

$$- \iint\limits_{\mathcal{D}(-u_0,2u_0;t)} \frac{\chi}{\eta^2}\zeta^2\psi_t\left[\widetilde{\nabla}_\mu v^\nu \widetilde{\nabla}_\nu v^\mu + v^\nu\nabla_\nu\sigma - v^\nu v^\mu\widetilde{\nabla}_\mu(\log(\chi\widetilde{\sigma}\gamma)) - \widetilde{\mathbf{Ric}}_{\mu\nu}v^\mu v^\nu\right]$$

We estimate

$$|\nabla_\mu \log(\chi\widetilde{\sigma}\gamma)| \leq c\left|\frac{\phi_t}{\phi}\right| + \phi\left(\eta^2\phi^2 + \beta^2\right)\sqrt{\varepsilon_\chi^\psi} + \frac{|\phi_t|}{2\beta} + \frac{\phi}{2}|\overline{\nabla}\phi| + \frac{\phi}{2\beta}|\overline{\sigma}| + \phi(|\omega| + |\overline{\nabla}\beta|).$$

We get after integration by parts, $L^\infty - L^2$ and Cauchy-Schwarz that:

$$\mathrm{II} \leq \int_0^t E(t') + v_{\mathrm{II}}(t')dt',$$

where

$$v_{\mathrm{II}}(t) = \int_{\Sigma_t} \frac{\phi^2\zeta^2}{\eta^4}\left[\left(|\overline{\nabla}Z|^2 + |\overline{\nabla}\log(\chi\tilde{\sigma}\gamma)|^2\right]\beta^2|\overline{\nabla}v|^2 + |\widetilde{\mathbf{Ric}}|\beta^2 + \eta^2\sigma^4 + \frac{e^{2\psi}f''^2 + \beta^2\sigma_t^2}{\eta^2}\right].$$

We arrive at the inequality:

$$\mathrm{E}(t) \leq \mathrm{E}(0) + \int_0^t \left(\gamma(t')\mathrm{E}(t') + v(t')\right)dt',$$

where

$$\gamma = \sup_{\Sigma_{t'}}\left[\phi^6\left(e^{2\psi}f''^2 + \phi\beta^2\eta^2\right)\right] + 1,$$

$$v = \sup_{\Sigma_{t'}}\left(\beta\phi^7 v_{\mathrm{I}}\right) + v_{\mathrm{II}}.$$

Therefore Gronwall's inequality we obtain:

$$\mathrm{E}(t) \leq e^{\int_0^t \gamma(t')dt'}\left[\mathrm{E}(0) + \int_0^t e^{-\int_0^{t'}\gamma(t'')dt''}v(t')dt'\right].$$

Energy radiated away Integrating in the exterior region then

$$\mathrm{E}(t) \leq e^{\int_0^t \gamma(t')dt'}\left[\mathrm{E}(0) + \int_0^t e^{-\int_0^{t'}\gamma(t'')dt''}v(t')dt'\right].$$

The wave equation for the second fundamental form Moreover the second variation equation provides through the equation for the Ricci curvtaure the wave equation for the second fundamental form, [1];

$$\phi\Box_g k = \mathbf{S}$$

where

$$\mathbf{S}_{ij} = \mathbf{S}_{00,ij} + \mathbf{S}_{01,ij} + \mathbf{S}_{10,ij} + \mathbf{S}_{11,ij} + \mathbf{S}_{2,ij} + \mathbf{S}_{3,ij},$$

and we have that

$$\mathbf{S}_{00,ij} = -4\phi\left[k_i{}^l k_l{}^m k_{jm} - \frac{1}{2}\mathrm{R}^l{}_{ijm}k^m_l - \frac{3}{4}(\mathrm{R}_{mi}k^m_j + \mathrm{R}_{jm}k^m_i) + \frac{1}{2}\mathrm{R}k_{ij}\right] - \gamma^{rs}\left[\Gamma^n{}_{rs}\Gamma^m_{jl}k_{il} + \Gamma^n{}_{rs}\Gamma^m_{il}k_{jl}\right].$$

$$\mathbf{S}_{01,ij} = -4\phi\left[(p+\rho)(k_{il}v^l v_j + k_{jl}v^l v_i) + p\left(\gamma_{lj}k_i{}^l + \gamma_{li}k_j{}^j\right)\right].$$

$$\mathbf{S}_{10,ij} = 2(\nabla_l\phi)(\nabla_i k_j{}^l + \nabla_j k_i{}^l) - 3\nabla_l\phi\nabla^l k_{ij} + \phi^{-1}k_{ij}|\nabla\phi|^2 + \overline{\nabla}_m\log\phi\cdot\overline{\nabla}_m k_{ij} + \gamma^{rm}\Gamma^s{}_{rm}\overline{\nabla}_r k_{ij},$$

$$\mathbf{S}_{11,ij} = \phi^{-1}\left(k_{li}\nabla_j\phi + k_{lj}\nabla_i\phi\right)\nabla^l\phi - 3\phi^{-2}\phi_t(\mathrm{Ric}_{ij} - 2k_{im}k^m_j - \frac{1}{2}(\mathrm{R}+4p)\gamma_{ij} - 2(\rho+p)v_i v_j,$$

$$\mathbf{S}_{2,ij} = -\left[k_{jr}\nabla_i\nabla^r\phi + k_{ir}\nabla_j\nabla^r\phi - k_{ij}\left(|k|^2 + (2(p+\rho)\beta + (p-\rho))\phi^2\right)\phi^2\right] + 3\phi^{-3}\phi_t\overline{\nabla}_i\overline{\nabla}_j\phi,$$

$$\mathbf{S}_{3,ij} = \partial_t\mathbf{T}_{ij} + \nabla_i(\phi^2\mathbf{T}_j{}^0) + \nabla_j(\phi^2\mathbf{T}_i{}^0),$$

We use now the energy density:

$$\varepsilon^k_\phi = \phi^2|k_t|^2 + |\overline{\nabla}k|^2,$$

and the cutoff function ζ with support contained in $\mathbf{B}_j(\theta,\epsilon)$. Moreover we restrict in the interior- the integrals over $C, \underline{C}$ vanish then :

$$\mathrm{E}^{\phi,k}_\zeta(t) = \int_{\Sigma_t}\zeta^2\varepsilon^k_\phi.$$

Therefore we

$$\mathrm{E}^{\phi,k}_\zeta(t) \leq \mathrm{E}^{\phi,k}_\zeta(0) + \iint_{\mathcal{D}(-u_0,2u_0;t)}\zeta^2\varepsilon^k_\phi + \mathrm{I} + \mathrm{II},$$

where I, II are given as in the general case and we majorise below. The coarea formula suggests that:

$$\iint_{\mathcal{D}(-u_0,2u_0;t)}\zeta^2\varepsilon = \int_0^t dt'\left(\int_{\Sigma_{t'}}\phi\zeta^2\varepsilon d\sigma\right) \leq \int_0^t\left(\sup_{\Sigma_{t'}}\phi\right)\mathrm{E}.$$

Analogously

$$\mathrm{I} \leq 2\iint_{\mathcal{D}(-u_0,2u_0;t)}\zeta^2\phi^4\left[\phi^2\phi_t^2 + |\overline{\nabla}\phi|^2\right] = 2\int_0^t v_\mathrm{I}\quad v_\mathrm{I}(t) = \int_{\Sigma_t}\zeta^2\phi^5\varepsilon^\phi_\phi,$$

while

$$\mathrm{II} = \iint_{\mathcal{D}(-u_0,2u_0;t)}\frac{\zeta^2}{\phi}\mathbf{S}\cdot k_t = \int_0^t\int_{\Sigma_{t'}}\zeta^2\mathbf{S}\cdot k_t.$$

We integrate by parts and majorize by:

$$\int_{\Sigma_{t'}}(\zeta^2 + \epsilon|\overline{\nabla}\zeta|^2)\left[\mu_2|\overline{\nabla}k| + |k|\phi^2|\overline{\nabla}k|^{3/2} + \left(\phi|\overline{\nabla}\phi| + |\mathrm{Rm}| + |\mathrm{R}| + |\mathrm{Ric}|\right)|\overline{\nabla}k|^2\right] +$$

$$\int_{\Sigma_{t'}}(\zeta^2 + \epsilon|\overline{\nabla}\zeta|^2)\left[\mu_1|k| + \phi^2|\overline{\nabla}\phi|^2|k|^2 + \left(\phi^2 + |\overline{\nabla}\mathrm{Ric}||\overline{\nabla}\phi|^2 + |\mathrm{Rm}| + |\overline{\nabla}\phi|^3\right)|k|^3\right],$$

where

$$\mu_1 = \left(|\mathrm{Rm}| + e^{\psi}\beta\right)\phi^2 + (|\overline{\nabla}\mathrm{Rm}| + |\overline{\nabla}\phi|^2 + |\mathrm{Rm}|) + \left(1+\beta e^{\psi}\right)|\overline{\nabla}\phi|^3,$$

$$\mu_2 = \phi|\overline{\nabla}\phi|\left(|k|^2 + |\overline{\nabla}\mathrm{Rm}||k| + \phi|\overline{\nabla}\phi| + \beta e^{\psi} + |\mathrm{Rm}|\right).$$

The terms involving the Riemann tensor are majorised through the constraint equations as for the Ricci tensor

$$|\mathrm{Ric}| \le c_3[e^{\psi}\beta^2 + |k|^2], \qquad |\overline{\nabla}\mathrm{Ric}| \le c_4\left[e^{\psi}|\overline{\nabla}\psi| + e^{\psi}|\nabla\beta| + |\overline{\nabla}k|\right] \le c_4\left[e^{\psi}\left(\varepsilon^{\psi}_{\chi}\right)^{1/2} + |\kappa_1| + \left(\varepsilon^{k}_{\phi}\right)^{1/2}\right].$$

while the Riemmann tensor in three diemnsions is etsimated through Ricci and scalar curvature as

$$|\mathrm{Rm}| \le c_5[e^{\psi}\beta^2 + |k|^2], \qquad |\overline{\nabla}\mathrm{Rm}| \le c_6\left[e^{\psi}\left(|\overline{\nabla}\psi| + |\nabla\beta|\right) + |\overline{\nabla}k|\right] \le c_6\left[e^{\psi}\left(\varepsilon^{\psi}_{\chi}\right)^{1/2} + |\kappa_1| + \left(\varepsilon^{k}_{\phi}\right)^{1/2}\right]$$

Now we use Sobolev inequality and the $L^\infty - L^2$ estimates to get that:

$$\left|\int_{\Sigma_t} \mathbf{S}\cdot k_t\zeta^2\phi\right| \le c\gamma_{\mathrm{II}}(t)E(t) + v_{\mathrm{II}}(t),$$

if we have set that:

$$\gamma_{\mathrm{II}}(t) = \left[\int_{\Sigma_t}(\zeta^2 + |\overline{\nabla}\zeta|^2)\left(\phi^2|\overline{\nabla}\phi|^2|k|^4 + |\overline{\nabla}\mathrm{Rm}|^2|k|^2 + (1+\beta e^{2\psi} + |\mathrm{Rm}|^2 + \phi^4|\overline{\nabla}\phi|^4)\right)\right]^{1/2} +$$

$$+\left(\int_{\Sigma_t}(\zeta^2 + |\overline{\nabla}\zeta|^2)\mu_1^{6/5}\right)^{5/6}.$$

and

$$v_{\mathrm{II}}(t) = \int_{\Sigma_t}\left(|\mathrm{Rm}| + e^{\psi}(\beta+\phi^2)\right)\phi^4|\overline{\nabla}\phi|^2.$$

Then using the estimates for the lapse function we get for

$$\mathrm{E}(t) \le \mathrm{E}(0) + \int_0^t \left(\gamma(t')\mathrm{E}(t') + v(t')\right).$$

Therefore Gronwall's inequality we obtain:

$$\mathrm{E}(t) \le E(0)e^{\int_0^t \gamma(t')dt'} + \int_0^t v(t')e^{\int_{t''}^{t'}\gamma(t'')dt''}dt'.$$

Slab estimate for the lapse function We start with the propagation equation:

$$\frac{d\Phi}{dt} = 2\int_{\Sigma_t}\zeta^2\phi\phi_t \le \int_{\Sigma_t}\zeta^2\left(\phi^2 + \phi_t^2\right) \le \gamma(t)\Phi(t) + v(t),$$

if

$$\Phi = \int_{\Sigma_t}\zeta^2\phi^2,$$

$$\gamma(t) = c\left(\int_{\Sigma_t}(\beta^{5/4} + |k|^4 + (p+\rho)^5\right)^{\frac{4}{5}},$$

and

$$v(t) = \left(\int_{\Sigma_t}(\beta^5 + k^{\frac{10}{3}} + (\rho+p)^5\right)^{\frac{1}{5}}.$$

Therefore we have that

$$\int_{\Sigma_t}\zeta^2\phi^2 \le e^{\int_0^t\gamma(t')dt'}\int_{\Sigma_0}\zeta^2\phi^2 + \int_0^t v(t')e^{\int_{t'}^t\gamma(t'')dt''}.$$

Journal of Physics: Conference Series **453** (2013) 012010 doi:10.1088/1742-6596/453/1/012010

8. Estimates for fluids

Fluid currents The equation of motion for the fluid reads that:

$$\nabla_\nu \mathbb{T}^{\mu\nu} = 0.$$

We introduce the current:

$$^{(X)}\mathbb{J}^\mu = \mathbb{T}^{\mu\nu} X_\nu = (\rho + p)(X_\nu v^\nu) v^\mu + \rho X^\mu.$$

In particular if $X = v$ then

$$^{(v)}\mathbb{J}^\mu = -\rho v^\mu$$

We compute that

$$\nabla_\nu \left(\zeta^{2}\,\cdot^{(X)}\mathbb{J}^\nu\right) = \zeta^{2}\,\cdot^{(X)}\pi_{\mu\nu}\mathbf{T}^{\mu\nu} + 2\zeta^{(X)}\mathbf{J}^\nu \nabla_\nu \zeta,$$

then in the domain spanned by $\Sigma_0, \Sigma_t, C_u, \underline{C}_{\underline{u}}$ we obtain that:

$$\int_{\Sigma_t} \zeta^2 \mathbf{T}(X,T) = \int_{\Sigma_0} \zeta^2 \mathbf{T}(X,T) - \int_{C_{-u_0}} \zeta^2 \left(\mathbf{T}(X,\underline{L})\right) - \int_{\underline{C}_{2u_0}} \zeta^2 \left(\mathbf{T}(X,L)\right) +$$

$$+ \iint_{\mathcal{D}(-u_0,2u_0;t)} \zeta^2 \left({}^{(X)}\pi_{\mu\nu}\mathbb{T}^{\mu\nu}\right) + 2\zeta^{(X)}\mathbb{J}^\nu \cdot \nabla_\nu \zeta.$$

We calculate

$$\mathbf{T}(T,T) = (\rho + p)\beta^2 - p = \beta^2 \rho + p|v|^2 \geq |v|^2(\rho + p),$$

and

$$\mathbf{T}(T,L) = \phi(\rho + p)\beta\left[1 + \frac{v(\theta)}{|\overline{\nabla}\theta|^2}\right] - p \geq \phi(\rho + p)\beta\left[1 - \frac{|v|}{|\overline{\nabla}\theta|} - \frac{p}{(p+\rho)\beta\phi}\right],$$

and

$$|\mathbf{T}(T,L)| \leq \phi(\rho + p)\beta\left[1 + \frac{|v|}{|\overline{\nabla}\theta|} - \frac{p}{(p+\rho)\beta\phi}\right].$$

We will assume an estimate

$$|\overline{\nabla}\theta| \geq (1 + \epsilon)|v|.$$

and derive the conditions that assert its validity.

Velocity estimates Select $X = \alpha^2 v^j \partial_j$ then we have that:

$$\int_{\Sigma_t} \frac{\zeta^2 e^\psi \beta |v|^2 \alpha^2}{\phi} = \int_{\Sigma_0} \frac{\zeta^2 e^\psi \beta |v|^2 \alpha^2}{\phi} + \int_{C_{-u_0}} \alpha^2 \zeta^2 \left(\mathbf{T}(v,\underline{L})\right) + \int_{\underline{C}_{2u_0}} \alpha^2 \zeta^2 \left(\mathbf{T}(v,L)\right)$$

$$+ \iint_{\mathcal{D}(-u_0,2u_0;t)} \zeta^2 e^\psi \left[|v|^2 v(\alpha^2) - \alpha^2\left(\sigma + \frac{(1+2\eta^2)}{1+\eta^2} v(\psi)\right)\right] - 2\zeta\alpha^2 e^\psi \left[|v|^2\left(v(\zeta) + \frac{\beta}{\phi}\zeta_0\right) + p e^{-\psi}\beta^2 v(\zeta)\right].$$

The coarea formula implies that

$$\iint_{\mathcal{D}(-u_0,2u_0;t)} \zeta^2 \rho\alpha v(\alpha) = \int_0^t \left(\sup_{\Sigma_{t'}} \phi\right) \Gamma_0(t'),$$

where

$$\Gamma_0(t) = \int_{\Sigma_t} \zeta^2 \rho\alpha v(\alpha) \leq 2\int_{\Sigma_t} \zeta^2 \beta\rho\alpha \left(\phi^2 \alpha_t^2 + |\overline{\nabla}\alpha|^2\right)^{\frac{1}{2}}.$$

The velocity Selecting $\alpha = 1$ or $X = v$ and dropping the lateral integrals after selecting ζ so that the integrals on the conical surfaces are annihilated then we obtain:

$$\int_{\Sigma_t} \frac{\zeta^2 e^{\psi}\beta|v|^2}{\phi} = \int_{\Sigma_0} \frac{\zeta^2 e^{\psi}\beta|v|^2}{\phi} - \iint_{\mathcal{D}(-u_0,2u_0;t)} \zeta^2 e^{\psi}\left[(\sigma + \frac{(1+2\eta^2)}{1+\eta^2}v(\psi)\right] - 2\zeta e^{\psi}\left[\left(|v|^2(\overline{v}(\zeta) + \frac{\beta}{\phi}\zeta_0\right) + pe^{-\psi}\beta^2 v(\zeta)\right].$$

Therefore we set

$$\kappa_0(t) = \int_{\Sigma_t} \frac{\zeta^2 e^{\psi}\beta|v|^2}{\phi}.$$

Th coarea formula and Cauchy-Schwarz

$$\kappa_0(t) \leq \kappa_0(0) + 4\int_0^t \gamma_0(t')\kappa_0(t') + v_0(t'),$$

where

$$\gamma_0(t) = \left(\sup_{\Sigma_t}\phi\right)^2(1+\beta_j) + \varphi(t),$$

$$v_0(t) = \int_{\Sigma_t} \zeta^2 e^{\psi}\left[\frac{\phi^2}{\beta^2\chi^2}\left(\eta^2\phi^2 + \frac{\beta^2}{\phi^2} + 1\right)\varepsilon_{\phi}^{\psi} + \frac{\phi_t^2}{\beta^2} + \phi^2|\overline{\nabla}\phi|^2\right].$$

Then Gronwall's lemma suggests that:

$$\kappa_0(t) \leq \kappa_0(0)e^{\int_0^t \gamma_0(t')dt'} + \int_0^t v_0(t')e^{\int_{t''}^{t'}\gamma_0(t'')dt''}dt'.$$

The velocity gradient Similarly selecting $\alpha = |\overline{\nabla}v|^2$, $X = |\overline{\nabla}v|^2 v$ and ζ as before then:

$$\int_{\Sigma_t} \frac{\zeta^2 e^{\psi}\beta|v|^2|\overline{\nabla}v|^2}{\phi} = \int_{\Sigma_0} \frac{\zeta^2 e^{\psi}\beta|v|^2|\overline{\nabla}v|^2}{\phi} +$$

$$\iint_{\mathcal{D}(-u_0,2u_0;t)} \zeta^2 e^{\psi}\left[|v|^2 v(|\overline{\nabla}v|^2) - |\overline{\nabla}v|^2\left(\sigma + \frac{(1+2\eta^2)}{1+\eta^2}v(\psi)\right)\right] + 2\zeta|\overline{\nabla}v|^2 e^{\psi}\left[|v|^2\left(v(\zeta) + \frac{\beta}{\phi}\zeta_0\right) + pe^{-\psi}\beta^2 v(\zeta)\right].$$

The first term far more intriguing we compute with Hölder and Sobolev inequalities that

$$\int_{\Sigma_t} \zeta^2 e^{\psi}\left[|\overline{v}|^3|\overline{\nabla}^2 v|\right] \leq c\left(1+\frac{1}{\epsilon}\right)\varphi(t)^5\kappa_0(t)^{7/2} + \varphi(t)^{1/2}\int_{\Sigma_t}\varepsilon_{\chi}^{\psi}.$$

The other terms contribute that:

$$\kappa_1(t) = \int_{\Sigma_t} \frac{\zeta^2 e^{\psi}\beta|\overline{\nabla}v|^2|v|^2}{\phi}.$$

Therefore

$$\kappa_1(t) \leq \kappa_1(0) + \int_0^t \left(\gamma_1\kappa_1(t') + v_1\right)dt'$$

and for

$$\gamma_1(t) = \left(\sup_{\Sigma_t}\phi\right)^2(1+\beta_j) + \varphi(t)$$

$$v_1(t) = c\left(1+\frac{1}{\epsilon}\right)\varphi(t)^5\kappa_0(t)^{5/2} + \varphi(t)^{1/2}\int_{\Sigma_t}\varepsilon_{\chi}^{\psi}.$$

Then Gronwall's lemma suggests that:

$$\kappa_1(t) \leq \kappa_1(0)e^{\int_0^t \gamma_1(t')dt'} + \int_0^t v_1(t')e^{\int_{t''}^{t'}\gamma_1(t'')dt''}dt'.$$

Journal of Physics: Conference Series **453** (2013) 012010 doi:10.1088/1742-6596/453/1/012010

The compressibility Selecting $\alpha = \sigma = \mathrm{div}_{\Sigma_t}(v)$, $X = \sigma^2 v$ then:

$$\int_{\Sigma_t} \frac{\zeta^2 e^{\psi}\beta|v|^2\sigma^2}{\phi} = \int_{\Sigma_0} \frac{\zeta^2 e^{\psi}\beta|v|^2\sigma^2}{\phi} +$$

$$\iint_{\mathcal{D}(-u_0,2u_0;t)} \zeta^2 e^{\psi}\left[|v|^2(v(\sigma^2) - \left(\sigma + \frac{(1+2\eta^2)}{1+\eta^2}v(\psi)\right)\sigma^2\right] + 2\zeta\sigma^2 e^{\psi}\left[|v|^2\left(v(\zeta) + \frac{\beta}{\phi}\zeta_0\right) + pe^{-\psi}\beta^2 v(\zeta)\right].$$

Therefore we set

$$\kappa_{10}(t) = \int_{\Sigma_t} \frac{\zeta^2 e^{\psi}\beta|v|^2\sigma^2}{\phi}.$$

Similalry as before the first term is the most delicate we get

$$\int_{\Sigma_t} \zeta^2 e^{\psi}|v|^3|\sigma||\overline{\nabla}\sigma| \le c\left(\varphi(t)^{2+e}\kappa_0(t) + \varphi(t)^{1/2}\int_{\Sigma_t}\zeta^2 e^{\psi}\varepsilon_{\chi}^{\psi}\right).$$

Therefore

$$\kappa_{10}(t) \le \kappa_{10}(0) + \int_0^t \left(\gamma_{10}\kappa_{10}(t') + v_{10}(t')\right)dt'.$$

Then Gronwall's lemma suggests that:

$$\kappa_{10}(t) \le \kappa_{10}(0)e^{\int_0^t \gamma(t')dt'} + \int_0^t v(t')e^{\int_{t''}^{t'}\gamma(t'')dt''}dt'$$

The velocity estimates on the slice We recall the following identity from [3]:

$$\int_{\Sigma_t} |\overline{\nabla}U|^2 = \int_{\Sigma_t}\left(\frac{1}{2}|A(U)|^2 + |D(U)|^2\right) - \int_{\Sigma_t}(2p+1)\mathrm{Ric}_{mn}U^{\mathbf{I}m}U^n_{\mathbf{I}} - \frac{\mathrm{R}}{2}|U|^2 +$$

$$+2p\int_{\Sigma_t} T(U)\cdot \mathrm{R}\cdot U - \frac{p}{2}\mathrm{R}|T(U)|^2.$$

where

$$A(U)_{a_1\ldots a_p bc} = \overline{\nabla}_c U_{a_1\ldots a_p b} - \overline{\nabla}_b U_{a_1\ldots a_p c}$$
$$D(U)_{a_1\ldots a_p} = \overline{\nabla}^m U_{a_1\ldots a_p m}$$
$$T(U)_{a_1\ldots a_p} = g^{bc}U_{a_1\ldots a_{p-1}bc}$$
$$T(U)\cdot \mathrm{Ric}\cdot U = T(U)_{a_1\ldots a_{p-1}}\cdot \mathrm{Ric}_{mn}\cdot U^{a_1\ldots a_{p-1}mn}$$

We will use this for the projection of the velocity on the slice

$$\overline{v}^{\mu} = v^{\mu} - \frac{\beta}{\phi}\delta^{\mu}_0$$

then we have for $\omega = *A(\overline{v}), \sigma = D(\overline{v})$

$$\int_{\Sigma_t} |\overline{\nabla}(\zeta\overline{v})|^2 = \int_{\Sigma_t} \frac{1}{2}|\omega|^2 + |D(\overline{v})|^2 - \mathrm{Ric}(\overline{v},\overline{v}) + 2\zeta\overline{\nabla}\overline{v}\cdot\overline{\nabla}\zeta + |\overline{v}|^2|\overline{\nabla}\zeta|^2.$$

Journal of Physics: Conference Series **453** (2013) 012010 doi:10.1088/1742-6596/453/1/012010

9. Bump function estimates

The required estimates on the bumps θ_ℓ we will drop the subscript ℓ follow from estimates on the second fundamental form -which provide estimates for the hessian of $|\nabla\!\!\!/^2\theta|$ and in turn of $\theta, |\nabla\!\!\!/\theta|$. We verify readily that the traceless second fundamental form $\widehat{h}$ and the mean curvature h of the deformed spheres is related to the hessian of the bump function for positive constants c_1, c_2:

$$c_1\frac{|\nabla\!\!\!/^2\theta|}{1+|\nabla\!\!\!/\theta|^2}\leq|\widehat{h}|\leq c_2\frac{|\nabla\!\!\!/^2\theta|}{1+|\nabla\!\!\!/\theta|^2},\qquad h=\Delta\!\!\!/\theta$$

The first, second variation equations and Gauss-Codazzi equations involve $h_{AB}, h, \widehat{h}_{AB}=h_{AB}-\frac{1}{2}h\delta_{AB}$ and read as

$$\frac{d\hat{\gamma}}{dr}=-ah,\qquad \frac{dh}{dr}=-a\left(\frac{1}{2}h^2+|\widehat{h}|^2+R_{NN}\right)-\Delta\!\!\!/a,$$

$$\frac{d\widehat{h}_{AB}}{dr}=-a\left(h\widehat{h}_{AB}+S_{AB}\right)-\nabla\!\!\!/_A\nabla\!\!\!/_Ba+\frac{1}{2}\delta_{AB}\Delta\!\!\!/a,$$

$$\mathrm{d\!\!\!/iv}\,\widehat{h}_{AB}=\frac{1}{2}\nabla\!\!\!/_Ah+P_A,$$

$$K=R-2R_{NN}+\frac{1}{2}h^2-|h|^2$$

where $P_A=\mathrm{Ric}_{AN}, S_{AB}$ is the restriction of Ric on the leaves. These provide with the integral estimates for $H=\frac{1}{2}\left(\widehat{h}^2+h^2\right)$:

$$\frac{d}{dr}\int_{\mathbf{B}_j}\zeta^2H^2+\int_{\mathbf{B}_j}\zeta^2(4a-1-|\mathrm{Ric}|)hH\leq c\int_{\mathbf{B}_j}\zeta^2\left(\frac{1}{4}|\nabla\!\!\!/a|^2+\frac{1}{2}|\nabla\!\!\!/h|^2\right)$$

We assume now bounds of the form:

$$\int_{\mathbf{B}_j}\zeta^2|\nabla\!\!\!/a|^2\leq\gamma_a\int_{\mathbf{B}_j}\zeta^2a^2$$

and

$$\int_{\mathbf{B}_j}\zeta^2|\nabla\!\!\!/h|^2\leq\gamma_h\int_{\mathbf{B}_j}\zeta^2h^2$$

and obtain after integration that

$$\left(\int_{\mathbf{B}_j}\zeta^2H^2\right)(r)\leq\left(\int_{\mathbf{B}_j}\zeta^2H^2\right)(0)e^{\int_0^r g(r')dr'}+\left(\int_{\mathbf{B}_j}\zeta^2d^2e^{\int_{r'}^r g(r'')dr''}\right)$$

where

$$g(r)=\frac{1}{2}\sup_{\mathbf{B}_j}|a-1-|\mathrm{Ric}|-\gamma_h|,\qquad a(r)=\int_{\mathbf{B}_j}\zeta^2|\nabla\!\!\!/a|^2$$

Appendix A. Harmonic Approximation

We approximate harmonically functions f solving the boundary value problem:

$$\Delta\!\!\!/\widetilde{f}=0,\qquad \widetilde{f}|_{\partial\mathcal{C}_{[0,t],R}}=f$$

We will use then the initial form of this $\widetilde{f}$. We recall Lojasiewicz inequality for analytic functions near their critical value $\widetilde{f}_c$:

$$|\nabla\widetilde{f}|>c_2|\widetilde{f}-\widetilde{f}_c|^{1-\mu}$$

for some suitable rational $\mu<1$.

Appendix B. Generalized Hardy inequality

We present here the basic ingredient for the derivation of the local estimates: the generalized Hardy's inequality. In the estimates of [9] we use as homogeneous polynomial the initial form of the harmonic approxximation of the weight β, $P = \widetilde{\beta}_0$.

Let then P be a homogeneous polynomial of degree m. Then we introduce the quantities

$$H^0(P) = P^{-\frac{2}{m}}, \qquad H^1(P) = \frac{|\nabla P|^2}{P^2}, \qquad H^2(P) = \left|\frac{\Delta P}{P}\right|$$

If $f \in C_0^\infty(\{\mathbf{R}^3 \setminus \{P = 0\})$ then

$$\int_{\mathbf{I}_3} H^j(P) f^2 \leq c_j(P) \int_{\mathbf{I}_3} |\nabla f|^2$$

We start with Lagrange identity

$$\frac{|\nabla P|^2}{P^2} = \frac{|\underline{x}|^2 |\nabla P|^2}{|\underline{x}|^2 P^2} = \frac{(\underline{x} \cdot \nabla P)^2}{P^2} - \frac{1}{r^2 R^2} \sum_{i<j} \left(x_i \frac{\partial R}{\partial x_j} - x_j \frac{\partial R}{\partial x_i} \right)^2$$

and we recall that

$$\Omega_{ij} = \frac{1}{r} \left(x_i \frac{\partial}{\partial x_j} - x_j \frac{\partial}{\partial x_i} \right)$$

generate the rotation group $SO(3, \mathbf{R})$ and $\mathbf{R}^3$ and are tangent to S^2. We introduce coordinates in the conical sector

$$K_1 = \{\underline{x} \in \mathbf{R}^3 / |x_3| \geq \frac{\epsilon}{1+\epsilon} |\underline{x}|\}$$

then

$$x_1 = \frac{r\xi_1}{\sqrt{1+|\underline{\xi}|^2}}, \; x_2 = \frac{r\xi_2}{\sqrt{1+|\underline{\xi}|^2}}, \; x_3 = \frac{r}{\sqrt{1+|\underline{\xi}|^2}}, \; \underline{\xi} = (\xi_1, \xi_2)$$

Notice that $|\nabla P|^2$ is a homogeneous polynomial of degree $2m-2$ and

$$|\nabla P|^2 = r^{2(m-1)} \left(m^2 R^2 + r^2 |\not\nabla R|^2 \right) = P^2 \left(\frac{m^2}{r^2} + \left| \frac{\not\nabla R}{R} \right|^2 \right)$$

where

$$\not\nabla G = \nabla G - \left(\underline{\eta} \cdot \nabla G \right) \eta$$

We set that

$$P(\underline{x}) = r^m R(\underline{\xi}), \quad R(\underline{x}) = \frac{1}{(1+|\underline{\xi}|^2)^{\frac{m}{2}}} \widetilde{R}(\underline{\xi})$$

that:

$$\left| \frac{\nabla P}{P} \right|^2 = \frac{m^2}{r^2} + |\underline{\xi}|^2 \, ||^2 \, |\not\nabla R|^2 + (1+|\underline{\xi}|^2)(\underline{E}(R))^2, \quad \underline{E}(R) = \underline{\xi} \cdot \not\nabla R$$

Lojasiewicz inequality suggests that for rational numbers $\ell_0 = 1, \ell_1, \cdots, \ell_N$,, positive constants $c_0, \ldots, c_N, \epsilon_0 < 1, \ldots, \epsilon_N < 1$, and critical values $R_0, \ldots, R_N$:

$$|\not\nabla R| \geq \sum_{j=1}^{N} c_j |R - R_j|^{1-\ell_j}$$

if we localize in neighbourhoods:

$$|R| \geq \epsilon_j |R_j|$$

then

$$|\nabla\!\!\!/ R| \geq \sum_{j=0}^{N} c_j(1-\epsilon_j)^{1-\ell_j}|R|^{1-\ell_j}$$

For finite number of points $\underline{\eta}_1, \cdots, \underline{\eta}_N$ and for natural numbers $\mu_1, \cdots, \mu_N, \mu_j \leq m-1$:

$$\chi_j^2|\nabla\!\!\!/ R|^2 \geq c_j|\underline{\eta} - \underline{\eta}_j|^{2\mu_j}$$

We compute that

$$\int_{\mathrm{I}_3} P^{-\frac{2}{m}} f^2 = \int_{\mathrm{I}_3} \frac{P^{2(1-\frac{1}{m})}}{|\nabla P|^2} H^1(P) f^2$$

and observe that

$$M(\underline{x}) = \frac{P^{2(1-\frac{1}{m})}}{|\nabla P|^2}$$

is a real analytic function homogeneous of degree 0. The preceding estimates combined with Young's inequality for $q = m\ell_j, p = \frac{q}{q-1}$ assert that

$$M(\underline{x}) \leq \sum_{j=1}^{N} \frac{\chi_j^2}{(pm^2)^{\frac{1}{p}}(c_j q)^{\frac{1}{q}}} r^{-\frac{2}{m\ell_j}}$$

Therefore we reduce the inequality to the one for $H^1(P)$ thanks to the r^2 factor for the volume in sphrical coordinates. Integration by parts reduces the inequality for $H^2(P)$ to the one for $H^1(P)$. Applying Łojasiewicz inequality on the round spehre we deduce that:

$$\chi_j|R| \leq \frac{\chi_j}{c_j}|\nabla\!\!\!/ R|^{\frac{1}{1-\ell_j}}$$

and hence we have that:

$$\chi_j\left|\frac{\nabla\!\!\!/ R}{R}\right|^2 \leq |\underline{\eta} - \underline{\eta}_j|^{-2\frac{\ell_j\nu_j}{1-\ell_j}}$$

[1] Choquet - Bruhat Y and York J W 1976 *C. R. Acad. Sc. Paris* **3** 361
[2] Christodoulou D 2000 *The principle of least action and PDE*, (New Jersey: Princeton University)
[3] Christodoulou D 2004 *Mathematical Problems in General relativity* (European Math. Soc)
[4] Christodoulou D and Klainerman S 1993 *The global nonlinear stability of the Minkowski spactime* (New Jersey: Princeton University)
[5] Eardley D and Smarr L 1979 *Phys. Rev. D*, **19** 2293
[6] Hoffman D and Spruck D 1981 *Comm Pure Applied Math* **34** 85
[7] Landau L., Lifshitz E. M., 1960 *A course of theoretical physics: vol. 2 The classical theory of fields*, (Oxford: Pergamon Press)
[8] Pliakis D., A generalized Hardy's inequality to appear in Asian Journal of Mathematics, *arxiv:/*math-AP/0203092
[9] Pliakis D., On the volume of nodal sets, preprint, *arxiv:*math-AP/1304.7143

Journal of Physics: Conference Series **453** (2013) 012011 doi:10.1088/1742-6596/453/1/012011

The evolution of the f-mode instability and gravitational wave detection prospectives

A. Passamonti, E. Gaertig and K. D. Kokkotas

Theoretical Astrophysics, IAAT, Eberhard Karls University of Tübingen, Tübingen 72076, Germany

E-mail: andrea.passamonti@uni-tuebingen.de

Abstract. We study the dynamical evolution of the gravitational-wave driven instability of the f-mode in rapidly rotating relativistic stars with a polytropic equation of state. We use linear perturbation theory to describe the evolution of the mode amplitude and follow the trajectory of a newborn neutron star through its instability window. An unstable f-mode with a saturation energy of about $10^{-6}M_{\odot}c^2$ may generate a gravitational-wave signal which can be detected by the Einstein Telescope detector from the Virgo cluster. The effects of the magnetic field on the evolution and the detectability of the gravitational radiation are relevant when its strength is higher than 10^{12} G, while an unstable r-mode becomes dominant it reaches the maximum saturation value allowed by non-linear mode couplings. From the thermal evolution we find also that the heat generated by shear viscosity during the saturation phase completely balances the neutrinos' cooling and prevents the star from entering the regime of mutual friction. The evolution time of the instability is therefore longer and the star loses significantly larger amounts of angular momentum via gravitational waves.

1. Introduction

Rapidly rotating neutron stars born in a core collapse may develop non-axisymmetric instabilities and radiate a significant amount of gravitational waves [1]. A dissipative process, as gravitational radiation, may actually drive unstable non-axisymmetric modes via the well-known Chandrasekhar-Friedman-Schutz (CFS) mechanism [2, 3, 4]. This instability occurs when a mode which is counter-rotating with respect to the star is seen co-rotating by an inertial observer. The angular momentum radiated by gravitational waves induces an increasingly negative angular momentum of the mode which thus becomes unstable and grows up to a finite amplitude where it is saturated by some dissipative processes. In realistic neutron stars the f-mode instability may be limited by bulk and shear viscosity, non-linear interaction with other modes, superfluid effects, etc. The parameter space of the instability is therefore reduced with respect to an inviscid system and is typically described by an instability window in a temperature/star's rotation plane.

The gravitational wave signal generated during the instability can be potentially observed by the current and next generation of Earth based laser interferometers, and the identification of the oscillation modes from the spectrum will help us to unveil the properties of the dense matter at super-nuclear densities and clarify the neutron star physics by using Astereoseismology [5, 6].

In this contribution we outline our work on the evolution of the gravitational-wave driven f-mode instability [7], in which we use perturbation theory to determine a system of ordinary differential equations to describe the mode's amplitude, the star's rotation and the thermal evolution. In this formalism we incorporate the effects of viscosity and magnetic field, and the impact of an unstable r-mode on the f-mode instability.

2. Evolution of the f-mode instability

In a realistic star, an oscillation mode approximately evolves as $e^{i\omega t - t/\tau}$, where ω is the mode's frequency and τ the damping/growth timescale. A way to determine τ is to calculate the variation in time of the mode's energy given in the rotating frame E by [8]

$$\frac{dE}{dt} = -\frac{2E}{\tau}, \tag{1}$$

where the global timescale τ of equation (1) results from the individual dissipative mechanisms that act on the mode. For stars in which only gravitational waves, shear and bulk viscosity dissipate energy, we have [8]:

$$\frac{1}{\tau} = \frac{1}{\tau_{\rm gw}} + \frac{1}{\tau_{\rm s}} + \frac{1}{\tau_{\rm b}}, \tag{2}$$

where $\tau_{\rm gw}$ is the gravitational radiation time scale, while $\tau_{\rm s}$ and $\tau_{\rm b}$ are, respectively, the damping time due to shear and bulk viscosity. These quantities can be determined by using appropriate volume integrals where we need to insert the f-mode frequency and eigenfunctions [9, 10]. We determine the mode's properties by studying the time evolution of the linearised equations of rapidly rotating and relativistic stellar models. In particular, we consider uniformly rotating stars with a polytropic equations of state (EoS).

To study the evolution of the f-mode instability we can derive a system of equations for the mode amplitude and stellar rotation by using equation (1) and the evolution of the angular momentum:

$$\frac{dJ}{dt} = \frac{dJ_{gw}}{dt} + \frac{dJ_{mag}}{dt}, \tag{3}$$

where $J = J_s + J_c$ is the total angular momentum, which consists of the star's angular momentum J_s and the canonical angular momentum of the mode J_c. The latter one can be related to the mode energy by the well known equation $J_c = -mE/\omega$ [11], where m is the azimuthal number and ω the mode frequency measured in the rotating frame. The dissipative terms in equation (3) are the gravitational radiation torque,

$$\frac{dJ_{gw}}{dt} = -\frac{2J_c}{\tau_{\rm gw}} \tag{4}$$

and the magnetic torque dJ_{mag}/dt. This quantity for a standard orthogonal rotator model in vacuum with a dipolar magnetic field is given by [12]

$$\frac{dJ_{mag}}{dt} = -\frac{R^6}{6c^3} B_p^2 \Omega^3, \tag{5}$$

where Ω and R are, respectively, the star's angular velocity and radius, and $B_{\rm p}$ is the magnetic field at the magnetic pole. This formula provides a magnetic spin-down comparable to that of a standard pulsar model with magnetosphere [12, 13].

The thermal evolution of a neutron star can be studied with a global energy balance between the relevant radiative and viscous processes [12]:

$$C_v \frac{dT}{dt} = -L_\nu + H_s, \tag{6}$$

where C_v is the total heat capacity at constant volume, L_ν is the neutrino's luminosity produced by the modified Urca processes that operate in the background star, see [7], and H_s is the heating rate generated by shear viscosity:

$$H_s = \frac{2E}{\tau_s} \,. \tag{7}$$

Now, we may regard the mode energy as a function of the mode amplitude α and the stellar rotation rate Ω, i.e. we can write it as $E = \alpha\tilde{E}(\Omega)$. From equations (1) and (3) we can therefore derive a system of ordinary differential equations:

$$\frac{d\alpha}{dt} = -\frac{2\alpha}{\tau_{\rm gw}} - \frac{2\alpha}{\tau_{\rm v}}\frac{1+\alpha Q}{D} + \frac{2P}{D}\frac{\alpha}{\tau_{\rm mag}} \,, \tag{8}$$

$$\frac{d\Omega}{dt} = \frac{2F}{D}\left(\frac{\alpha}{\tau_{\rm v}} - \frac{1}{\tau_{\rm mag}}\right) \,, \tag{9}$$

where the total viscous damping time is defined as $1/\tau_{\rm v} = 1/\tau_{\rm s} + 1/\tau_{\rm b}$, $\tau_{\rm mag}$ is the timescale due to the magnetic torque, while D, F, P and Q are functions of the background stars (see [7] for their definitions).

In the non-linear study of the secular bar-mode instability [14, 15, 16, 17], it is reasonable to discern two dynamical phases of the f-mode instability, namely the mode's exponential growth and its non-linear saturation [9]. As the star cools down and enters the instability window, the mode grows exponentially while the star slowly spins down on viscous timescales. This behaviour is well described by equations (8)-(9). This initial growth phase ends either when non-linear dynamics or dissipative processes saturate the mode amplitude at a finite value. After this point, the mode amplitude is nearly constant while the star loses angular momentum via gravitational radiation [16, 17]. We can therefore approximate this evolution phase by assuming that $d\alpha/dt = 0$, which allows us to recast equations (8)-(9) into a single relation,

$$\frac{d\Omega}{dt} = -\frac{2F}{1+\alpha Q}\left(\frac{\alpha}{\tau_{\rm gw}} + \frac{1}{\tau_{\rm mag}}\right) \,. \tag{10}$$

The star is then spun down by gravitational radiation and magnetic torque. Eventually, the star will leave the instability window and the f-mode will be damped by gravitational radiation.

We normalise the mode amplitude as $E = \alpha E_{\rm rot}$, where $E_{\rm rot}$ is the stellar rotational energy. For our polytropic models an $\alpha = 1$ corresponds to $E \sim 10^{-2} M_\odot c^2$. Note that with this definition the fluid variables and the gravitational-wave strain scale as $\alpha^{1/2}$.

3. Stellar Models

We model the compact objects as relativistic neutron stars which obey a polytropic EoS,

$$p = K\rho^{1+1/N} \tag{11}$$

where p is the fluid pressure, K is the polytropic constant and N is the polytropic index. The rest-mass density ρ is related to the fluid energy density ε via the relation $\varepsilon = \rho + Np$.

We consider two sequences of uniformly rotating stars up to the mass shedding limit $\Omega_{\rm K}$. The first model represents a "standard" neutron star with gravitational mass $M = 1.4\,M_\odot$ and polytropic index $N = 1$. The non-rotating member of this sequence has a circumferential radius of $R = 14.15$ km, while the maximum rotation rate is $\nu_{\rm K} = \Omega_{\rm K}/2\pi = 673.1$ Hz. The second model is a polytrope with $N = 2/3$, and baryonic mass $M_b = 1.6\,M_\odot$ and a Kepler limit of $\nu_{\rm K} = \Omega_{\rm K}/2\pi = 1783$ Hz. This model is used to widen the parameter space of the instability, as it describes a supramassive star, i.e. an object that does not have a stable configuration in the static limit [18, 19, 20].

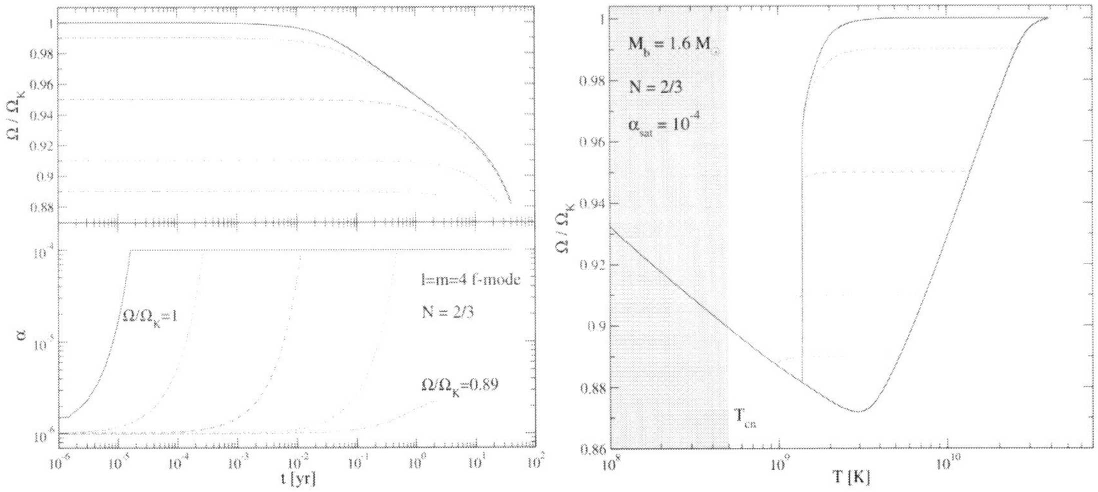

Figure 1. Evolution of the $l = m = 4$ f-mode instability for a relativistic polytrope with $N = 2/3$, baryonic mass of $M_b = 1.6\,M_\odot$ and $B_\mathrm{p} = 10^{11}$ G. The star enters the instability window at different spins *(see right panel)* and correspondingly the left panel depicts the time-evolution of the stellar rotation rate *(top left panel)* and mode amplitude *(bottom left panel)*, where the initial amplitude $\alpha = 10^{-6}$ saturates at $\alpha_{sat} = 10^{-4}$. The shaded region in the right panel represents the temperature range where the neutrons and protons of the core are expected to be in a superfluid/superconducting state in accordance with recent models for the observed cooling of Cassiopeia A [21, 22].

4. Results

We study now the dynamical evolution of the f-mode and consider how the magnetic field and a simultaneous presence of an unstable r-mode affect the instability.

The ideal conditions for the f-mode instability to work are high temperatures and rotation rates. At birth, the temperature of a neutron star is around $T \simeq 10^{11}$ K and drops down rapidly in the following few days. We therefore expect that as a rapidly rotating star cools down, it enters the f-mode instability window from the high temperature side (see figure 1). We study the evolution of the star from the onset of the gravitational-wave driven instability for different initial rotation rates, and consider an initial mode amplitude of $\alpha = 10^{-6}$. This corresponds to an energy of $E \sim 10^{-8}\,M_\odot c^2$ which is a typical value for the total energy loss in gravitational waves due to quadrupole deformations shortly after a gravitational core collapse [23, 24].

In [7], we study the $l = m = 3$ and 4 f-mode for the $N = 1$ and $N = 2/3$ models. Here we report only the most promising cases. We first consider the evolution of the $l = m = 4$ f-mode instability in a low magnetised star ($B_p \leq 10^{11}$ G), and show in figure 1 the star's spin, the mode's amplitude and the star's trajectory through the instability window for the $N = 2/3$ model. At the mass shedding limit the gravitational growth time is only $\tau_\mathrm{gw} \simeq 250\,$s in contrast to the $\tau_\mathrm{gw} \simeq 10^4\,$s for the $N = 1$ model, and the total evolution lasts for about 40 years. The exponential growth of the f-mode takes about 5 minutes (14 days) for a star with an initial $\Omega = \Omega_\mathrm{K}$ ($\Omega = 0.93\,\Omega_\mathrm{K}$). After the mode saturates, figure 1 (right panel) shows that the star spins down at nearly constant temperature ($T \simeq 10^9$ K) as a result of the heat generated by shear viscosity which completely balances the neutrino cooling. *This effect prevents the star from entering the superfluid transition zone and increases the duration of the instability.* In fact, for $T \leq T_\mathrm{cn}$ the neutrons of the core become superfluid and mutual friction damps the f-mode very efficiently on short timescales [25]. For the superfluid critical temperature we chose a value of

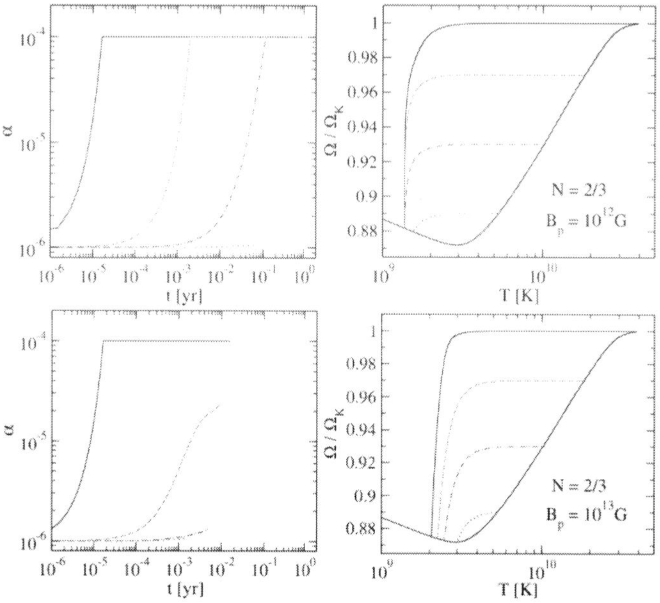

Figure 2. The impact of the magnetic torque on the evolution of the $l = m = 4$ f-mode instability for the $N = 2/3$ model. The two top panels display the mode amplitude (left panel) and the star's evolution through the instability region (right panel) for $B_p = 10^{12}$G. The same quantities are depicted in the two lower panels for $B_p = 10^{13}$G.

$T_{cn} \simeq 5\times10^8$ K which has recently been determined from the cooling curves of Cassiopeia [21, 22]. Furthermore, figure 1 shows that when a star becomes unstable at lower rotation rates near the minimum of the instability window, e.g. $\Omega \simeq 0.89\,\Omega_K$, the mode amplitude remains small and the shear viscosity heating does not balance the neutron star's cooling.

A strong magnetic field can accelerate the star's spin-down and limit the growth of the f-mode. In figure 2, we show the effects on the $l = m = 4$ f-mode for the $N = 2/3$ model and two dipolar magnetic field configurations with, respectively, $B_p = 10^{12}$G and $B_p = 10^{13}$G. For a magnetic field of $B_p = 10^{12}$ G the total duration of the instability is shorter than a factor of about 50 with respect to the $B_p = 10^{11}$G model. In fact, the evolution lasts about 1 yr (10 yr) for the $N = 2/3$ ($N = 1$) model. In slower rotating models, the magnetic torque clearly dominates the evolution even during the initial phase of the instability and limits considerably the growth of the mode amplitude. This behaviour is for instance evident for a $N = 2/3$ model with an initial rotation rate of $\Omega = 0.89\,\Omega_K$. The magnetic torque dominates completely the evolution when $B_p = 10^{13}$ G. The unstable star is quickly spun down and the mode amplitude is strongly limited even in the more massive model.

In rapidly rotating neutron stars several modes can be driven unstable by gravitational radiation at the same time. The most important ones are definitely the f- and r-modes, as they have a comparatively short growth time and therefore can generate a significant gravitational wave signal. A relevant difference between these two classes of modes is that the f-mode only gets unstable in very rapidly rotating stars while the r-mode is CFS unstable at any rotation rate. In addition, the growth time is typically shorter for the r-mode which consequently has a larger instability window. It is then reasonable to think that the r-mode should dominate the evolution of the gravitational wave driven instability. However before drawing any secure conclusion, it is necessary to know the maximum amplitude that each mode can reach during the instability. Non-linear perturbation calculations show that the r-mode may transfer energy to other inertial modes through non-linear mode coupling and has a small saturation amplitude.

As the star cools down, it is expected that the r-mode gets unstable before the f-mode as its instability window extends towards higher temperatures. The r-mode then quickly reaches its non-liner saturation value α_r^{sat} and spins down the star. However if the r-mode growth is limited by non-linear mode coupling, the star evolves towards the critical curve of the f-mode

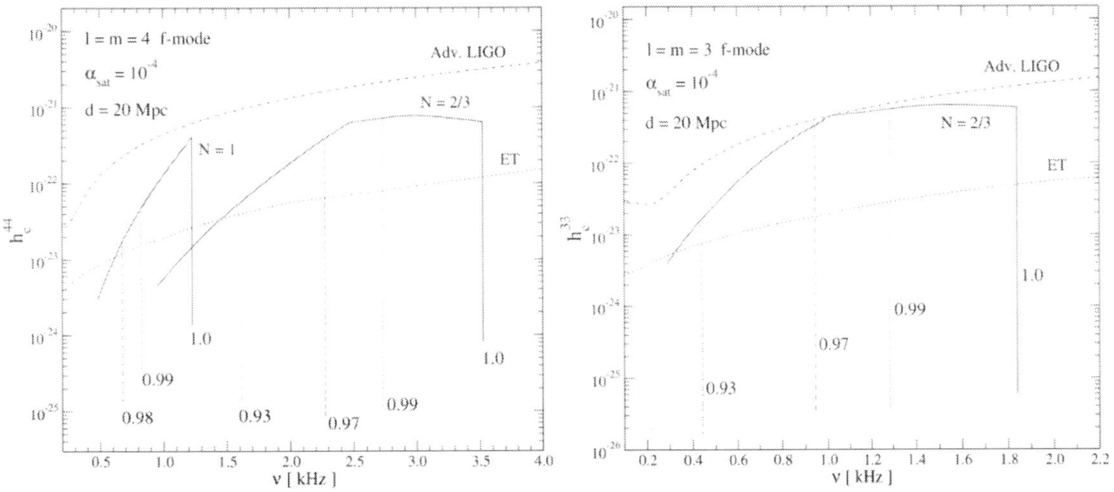

Figure 3. Characteristic strain generated by the f-mode instability for a polytrope with $B_{\rm p} = 10^{11}$G. The source is located at 20 Mpc and the saturation amplitude of the f-mode is set to $\alpha_{sat} = 10^{-4}$. *Left panel:* The gravitational-wave signal emitted by the $l = m = 4$ f-mode for the two stellar models with $N = 1$ and $N = 2/3$. The number near the vertical lines denotes the initial rotation rate $\Omega/\Omega_{\rm K}$. *Right panel:* The signal emitted the by $l = m = 3$ f-mode for the more massive model with $N = 2/3$. The sensitivity curves of Advanced LIGO and the Einstein Telescope (ET) are shown in both panels. The gravitational-wave signal generated during the f-mode instability may be detected by ET for the most part of the instability window of the $N = 2/3$ model.

and eventually also this mode is driven unstable by gravitational radiation.

We have studied several simulations with different relative saturation amplitude between the $l = m = 4$ f- and $l = m = 2$ r-modes, and found that if the r-mode reaches the maximum saturation amplitude predicted by the non-linear coupling calculations, it dominate the evolution of the instability [7]. However, it is therefore crucial to know more accurately the relative saturation amplitude between these two modes, in particular for the f-mode. This is an interesting aspect that must be clarified in a future work.

5. Gravitational Waves

The gravitational-wave signal generated during the instability of the f-mode may be potentially observed by the current and next generation of Earth-based laser interferometers. To evaluate the detection we determine the characteristic strain of the gravitational waves, which is a quantity that takes into account also the statistical amplification of the signal due to the number of oscillations accumulated in a given frequency bandwidth. It is defined by

$$h_c^{lm} \equiv \langle \left| h^{lm} \,{}_{-2}Y^{lm} \right| \rangle \sqrt{N_{cyc}} \tag{12}$$

where h^{lm} is the strain emitted by the source, ${}_{-2}Y^{lm}$ is a spin-weighted spherical harmonic and $\langle \ldots \rangle$ denotes an averaging over the angles (θ, ϕ). The number of oscillation cycles can be expressed as $N_{cyc} = \nu\, \tau_{ev}$, which is written in terms of the mode frequency $\nu = (\omega - m\Omega)/2\pi$ and the time spent near a given frequency τ_{ev}. Considering the signal integration time allowed by the detector technology, we assume that a detector may integrate the signal at most for 1 year, and consequently calculate h_c^{lm} for $\tau_{ev} \leq 1$yr. This means that in N_{cyc} we use the evolution time of our simulations whenever $\tau_{ev} < 1$yr, otherwise we set $\tau_{ev} = 1$yr.

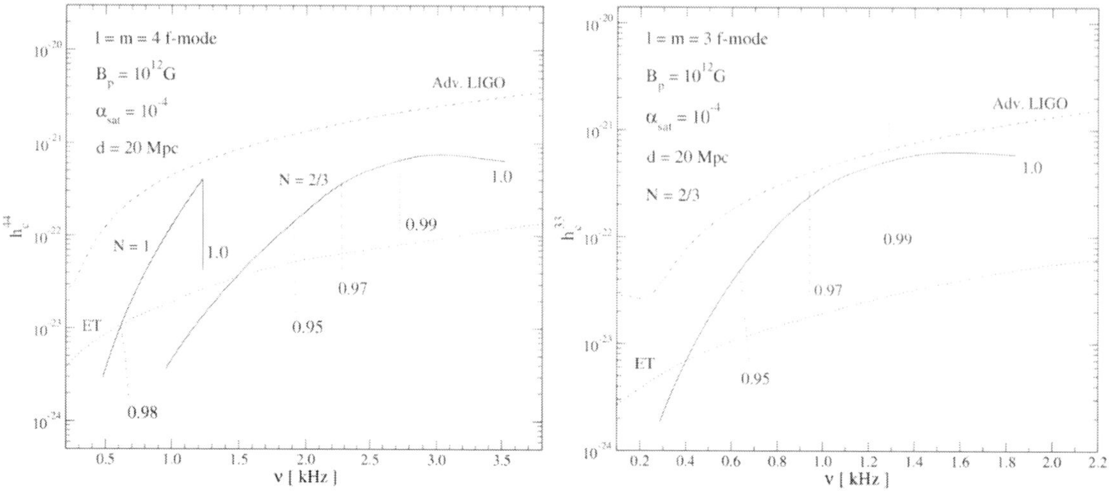

Figure 4. The characteristic strain generated by the f-mode instability for a polytropic star with $B_\mathrm{p} = 10^{12}$ G. The quantities and notation used in this figure are the same as in figure 3.

For the two polytropic models of this work, we calculate h_c^{lm} for several instability evolutionary paths with different initial stellar spin. We consider first a star with a weak magnetic fields, $B_\mathrm{p} \leq 10^{11}$ G, and focus on the $l = m = 3$ and 4 f-mode. For a saturation amplitude of $\alpha_{sat} = 10^{-4}$, figure 3 shows that the gravitational-wave signal can be detected by ET for the most part of the instability evolution for a $N = 2/3$ star located in the Virgo Cluster at a distance of 20 Mpc. For the standard $N = 1$ model an unstable $l = m = 4$ f-mode may generate a detectable signal only if the star rotates near the Kepler limit $\Omega \geq 0.98\,\Omega_\mathrm{K}$. From figure 3 we see that the gravitational-wave signal is initially monochromatic during the growth phase of the mode and evolves successively to lower frequencies as the star spins down.

A higher magnetic field may affect significantly the characteristic strain of the f-mode, as the number of accumulated cycles N_{cyc} decreases. The results for the $B_\mathrm{p} = 10^{12}$ G case are shown in figure 4 for both the $N = 1$ and $N = 2/3$ models. Although the gravitational-wave signal is slightly weaker than in the previous $B_\mathrm{p} = 10^{11}$ G case, it is still detectable by the Einstein Telescope (ET).

6. Conclusions

We have studied the evolution of the f-mode instability of rapidly rotating relativistic stars and have considered the effects of viscosity, magnetic fields and unstable r-modes. We have modelled the star with polytropic equations of state to explore the instability parameter space. The evolution of the star through the instability window may be accelerated by the presence of strong magnetic fields and a simultaneous presence of an unstable r-mode. Our results show that a magnetic field with $B_p \geq 10^{12}$ G affects the f-mode instability, and its influence is more relevant in neutron star models with smaller compactness (M/R). For the r-mode we find that it must reach the maximum value expected from non-linear mode coupling studies in order to affect considerably the f-mode evolution. However, a definitive answer to this issue cannot be given as the maximum f-mode amplitude is still unknown.

Another important result of our simulations is that the heat generated by shear viscosity during the f-mode saturation phase prevents the star from entering the regime of mutual friction. In fact, the shear viscosity re-heating balances the neutrino cooling and leads to a nearly isothermal evolution. The star therefore leaves the instability window at lower rotation rates, and even a moderate change in the superfluid transition temperature does not modify this result.

Our results show also that the gravitational-wave signal emitted during the f-mode instability may be detected by ET if the source is located within the Virgo cluster and the mode saturates at $E \sim 10^{-6} M_{\odot} c^2$. We are currently studying more realistic EoS which allow for larger masses and find that f-mode instability may generate a gravitational-wave signal which might be observed even by the advance LIGO/Virgo detectors.

Acknowledgements
AP and EG acknowledges support from the German Science Foundation (DFG) via SFB/TR7.

References

[1] Andersson N, Ferrari V, Jones D I, Kokkotas K D, Krishnan B, Read J S, Rezzolla L and Zink B 2011 *General Relativity and Gravitation* **43** 409–436
[2] Chandrasekhar S 1970 *Physical Review Letters* **24** 611–615
[3] Friedman J L and Schutz B F 1975 *Astrophys. J. Lett.* **199** L157–L159
[4] Friedman J L and Schutz B F 1978 *Astrophys. J. Lett.* **221** L99
[5] Andersson N and Kokkotas K D 1996 *Phys. Rev. Lett.* **77** 4134–4137
[6] Andersson N and Kokkotas K D 1998 *Mon. Not. R. Astron. Soc.* **299** 1059–1068
[7] Passamonti A, Gaertig E, Kokkotas K D and Doneva D 2013 *Phys. Rev. D* **87** 084010
[8] Ipser J R and Lindblom L 1991 *Astrophys. J.* **373** 213–221
[9] Owen B J, Lindblom L, Cutler C, Schutz B F, Vecchio A and Andersson N 1998 *Phys. Rev. D* **58** 084020–+
[10] Andersson N, Jones D I and Kokkotas K D 2002 *Mon. Not. R. Astron. Soc.* **337** 1224–1232
[11] Friedman J L and Schutz B F 1978 *Astrophys. J.* **221** 937–957
[12] Shapiro S L and Teukolsky S A 1983 *Black holes, white dwarfs, and neutron stars* (New York, Wiley-Interscience, 1983)
[13] Spitkovsky A 2006 *Astrophys. J. Lett.* **648** L51–L54
[14] Detweiler S L and Lindblom L 1977 *Astrophys. J.* **213** 193–199
[15] Lai D and Shapiro S L 1995 *Astrophys. J.* **442** 259–272
[16] Ou S, Tohline J E and Lindblom L 2004 *Astrophys. J.* **617** 490–499
[17] Shibata M and Karino S 2004 *Phys. Rev. D* **70** 084022
[18] Cook G B, Shapiro S L and Teukolsky S A 1992 *Astrophys. J.* **398** 203–223
[19] Cook G B, Shapiro S L and Teukolsky S A 1994 *Astrophys. J.* **422** 227–242
[20] Shibata M 2003 *Astrophys. J.* **595** 992–999
[21] Page D, Prakash M, Lattimer J M and Steiner A W 2011 *Physical Review Letters* **106** 081101
[22] Shternin P S, Yakovlev D G, Heinke C O, Ho W C G and Patnaude D J 2011 *Mon. Not. R. Astron. Soc.* **412** L108–L112
[23] Ferrari V, Miniutti G and Pons J A 2003 *Mon. Not. R. Astron. Soc.* **342** 629–638
[24] Dimmelmeier H, Ott C D, Janka H T, Marek A and Müller E 2007 *Physical Review Letters* **98** 251101
[25] Lindblom L and Mendell G 1995 *Astrophys. J.* **444** 804–809

Journal of Physics: Conference Series **453** (2013) 012012 doi:10.1088/1742-6596/453/1/012012

Testing General Relativity using the growth rate of structure

Athina Pouri[1] **& Spyros Basilakos**[2]

[1]Department of Physics, Section of Astronomy, Astrophysics & Mechanics, University of Athens, Panepistemiopolis, Athens 157 83, Greece

[2] Academy of Athens, Research Center for Astronomy & Applied Mathematics, Soranou Efessiou 4, 11-527, Athens, Greece

E-mail: [1]athpouri@phys.uoa.gr
[2] svasil@Academyofathens.gr

Abstract. We place tight constraints on the growth index γ by using the recent growth history results of 2dFGRS, SDSS-LRG, VIMOS-VLT deep Survey (VVDS), and *WiggleZ* datasets. Utilizing a standard likelihood analysis, we find that the use of the combined growth data provided by the previous mentioned galaxy surveys, puts the most stringent constraints on the value of the growth index. Assuming a constant growth index, we obtain that $\gamma = 0.602 \pm 0.055$ for the concordance ΛCDM expansion model. Based on the Dvali-Gabadadze-Porrati gravity model, we find $\gamma = 0.503 \pm 0.06$ which is lower, and almost 3σ away, from the theoretically predicted value of $\gamma_{DGP} \simeq 11/16$ implying that the present growth rate data disfavor the DGP gravity.

Keywords: cosmology: cosmological parameters

1. Introduction

In the current view of cosmology it has been shown that the model that governs the accelerated expansion of the universe is spatially flat and contains a sector of cold dark matter with some sort of dark energy associated with a large negative pressure ([4, 14, 28, 31, 33–36, 58] and references therein). Although the expansion of the universe is attributed to the so called dark energy, its nature and fundamental origin are yet to be revealed. Cosmologists have unfolded two different possible scenarios in order to describe the mechanism of acceleration. The first one involves new fields in nature (scalar fields) and the other one involves some sort of modification of Einstein's General Relativity with the present accelerating stage appearing as a sort of geometric effect (for reviews see [2, 9, 10] and references therein).

The last decade it has been proposed that measuring the growth index γ could help to test the validity of general relativity on cosmological scales because it can provide an efficient way to discriminate between scalar field dark energy (hereafter DE) models which admit to general relativity, and modified gravity models. Linder & Cahn [42] have shown that there is only a weak dependence of γ on the equation of state parameter $w(z)$, implying that one can separate the background expansion history, $H(z)$, constrained by a large body of cosmological data (SNIa, BAO, CMB), from the fluctuation growth history, given by γ. In this framework, it was theoretically found that for those DE models which adhere to general relativity, the growth index γ is well approximated by $\gamma \simeq \frac{3(w-1)}{6w-5}$ (see [42, 47, 55, 65]), which reduces to $\gamma_\Lambda \simeq 6/11$ for the traditional ΛCDM cosmology $w(z) = -1$. On the other hand, in the case of the braneworld model of Dvali-Gabadadze-Porrati [21] (hereafter DGP) the growth index becomes $\gamma_{DGP} \simeq 11/16$ (see also [22,42]), while for some $f(R)$ gravity models we have $\gamma \simeq 0.41 - 0.21$ for $\Omega_{m0} = 0.27$ [23,45]. Indirect methods to determine γ have also been proposed (mostly using a constant γ), based either on the observed growth rate of clustering [20,25,26,47,50,54] providing a wide range of γ values $\gamma = (0.58 - 0.67)^{+0.11}_{-0.11}{}^{+0.20}_{-0.17}$, or on massive galaxy clusters [64] and [51] with the latter study providing $\gamma = 0.42^{+0.20}_{-0.16}$, or even on the weak gravitational lensing [12]. Gaztanaga et al. performed a cross-correlation analysis between probes of weak gravitational lensing and redshift space distortions and found no evidence for deviations from general relativity.

The scope of the present study is to place constraints on the growth index using a single cosmologically relevant experiment, i.e., that of the recently derived growth data of the 2dFGRS, SDSS-LRG, VVDS, and *WiggleZ* galaxy surveys. We use two reference expansion models, namely flat ΛCDM and DGP, respectively, for the background evolution. The interesting aspect of the latter scenarios is that the corresponding functional forms of the Hubble parameters are affected only by one free parameter, that of the dimensionless matter density at the present time Ω_{m0}. The structure of the article is as follows. In section 2, we briefly discuss the background cosmological equations. The theoretical elements of the growth index are presented in section 3. In section 4 we briefly discuss the growth data. In section 5, we perform a likelihood analysis in order to constrain the growth index model free parameters. Finally, the main conclusions are summarized in section 6.

2. Theoretical approach

In this section we briefly discuss the main points of the background evolution for homogeneous and isotropic flat cosmologies, driven by non relativistic matter and an exotic fluid (DE models) with equation of state (hereafter EoS), $P_{DE} = w(a)\rho_{DE}$. In this case the first Friedmann equation:

$$\frac{H^2(a)}{H_0^2} \equiv E^2(a) = \Omega_{m0}a^{-3} + \Omega_{DE,0}\mathrm{e}^{3\int_a^1 d\ln y[1+w(y)]}, \tag{1}$$

where $a(z) = 1/(1+z)$ is the scale factor of the universe, $w(a)$ is the EoS parameter, Ω_{m0} is the dimensionless matter density at the present time and $\Omega_{DE,0} = 1 - \Omega_{m0}$ denotes the DE density parameter.

We can write the EoS parameter in terms of $E(a)$ [30,53] using the Friedmann equations:

$$w(a) = \frac{-1 - \frac{2}{3}a\frac{d\ln E}{da}}{1 - \Omega_m(a)} \tag{2}$$

where

$$\Omega_m(a) = \frac{\Omega_{m0}a^{-3}}{E^2(a)}\ . \tag{3}$$

Differentiating the latter and taking into account eq. (2) we obtain:

$$\frac{d\Omega_m}{da} = \frac{3}{a}w(a)\Omega_m(a)\left[1 - \Omega_m(a)\right]\ . \tag{4}$$

The important clue about this form of the DE EoS parameter is that includes our ignorance regarding the physical mechanism powering the late time cosmic acceleration since the exact nature of the DE is unknown. It is also worth noticing that for $w(a) = -1$ we have the concordance ΛCDM model.

Interestingly, the above method can be generalized to the context of modified gravity (see [38,40]). Indeed, instead of using the exact Hubble flow through a modification of the Friedmann equation, one may consider an equivalent Hubble flow somewhat mimicking eq.(1). The ingredient here is that the accelerating expansion can be attributed to a kind of "geometrical" DE contribution. Now, due to the fact that the matter density (baryonic+dark) cannot accelerate the cosmic expansion, it is useful to utilize the following parametrization [38,40]:

$$E^2(a) = \frac{H^2(a)}{H_0^2} = \Omega_{m0}a^{-3} + \Delta H^2. \tag{5}$$

It becomes clear that any modification to the Friedmann equation of general relativity is included in the last term of the above expression. Now, using eqs. (2) and (5), one can derive the effective ("geometrical") dark energy EoS parameter:

$$w(a) = -1 - \frac{1}{3}\frac{d\ln\Delta H^2}{d\ln a} = -1 - \frac{1}{3}\frac{a}{\Delta H}\frac{dlnH}{da}. \tag{6}$$

In the context of a flat DGP cosmological model, the "accelerated" expansion of the universe can be explained by a modification of the gravitational interaction in which gravity itself becomes weak at very large distances (close to the Hubble scale) due to the fact that our four dimensional brane survives into an extra dimensional manifold (see [16] and references therein). The quantity ΔH^2 is given by:

$$\Delta H^2 = 2\Omega_{bw} + 2\sqrt{\Omega_{bw}}\sqrt{\Omega_{m0}a^{-3} + \Omega_{bw}} \tag{7}$$

where $\Omega_{bw} = (1-\Omega_{m0})^2/4$. Interestingly, the quantity ΔH^2 contains only one free partameter, Ω_{m0}. From eq.(6), one can check that the geometrical (effective) DE equation of state parameter reduces to:

$$w(a) = -\frac{1}{1+\Omega_m(a)}. \tag{8}$$

In this model, due to its gravity nature, the effective Newton's parameter G_{eff} is not any more the usual constant G_N but it takes the following form [44]:

$$G_{\text{eff}}(a) = G_N Q(a) \qquad Q(a) = \frac{2+4\Omega_m^2(a)}{3+3\Omega_m^2(a)}. \tag{9}$$

3. The linear growth rate

We now discuss the basic equation which governs the evolution of the matter perturbations within the framework of any DE model (scalar or geometrical). It is very important to notice that at the sub-Hubble scales the DE component is expected to be smooth and, thus, one can use perturbations only on the matter component of the cosmic fluid [13]. In particular, following the notations of [17,37,42,44,57,61,62], we can derive the well known scale independent equation of the linear matter overdensity $\delta_m \equiv \delta\rho_m/\rho_m$:

$$\ddot{\delta}_m + 2H\dot{\delta}_m = 4\pi G_{\text{eff}}\rho_m\delta_m \tag{10}$$

a solution of which is $\delta_m(t) \propto D(t)$, with $D(t)$ denoting the linear growing mode (usually scaled to unity at the present time) For the scalar field DE models [$G_{\text{eff}} = G_N$, $Q(a) = 1$], the above equation reduces to the usual time evolution equation for the mass density contrast [48], while in the case of modified gravity models (see [23,42,44,61]), we have $G_{\text{eff}} \neq G_N$ (or $Q(a) \neq 1$). Transforming equation (10) from t to a ($\frac{d}{dt} = H\frac{d}{d\ln a}$), we simply derive the evolution equation of the growth factor $D(a)$

$$\frac{a^2}{D}\frac{d^2D}{da^2} + \left(3 + a\frac{d\ln E}{da}\right)\frac{a}{D}\frac{dD}{da} = \frac{3}{2}\Omega_m(a)Q(a). \tag{11}$$

Table 1. The growth data. The correspondence of the columns is as follows: index, redshift, observed growth rate and references. In the final column, one can find various symbols of the data appearing in Fig.2.

Index	z	A_{obs}	Refs.	Symbols
1	0.17	0.510 ± 0.060	Song & Percival 2009; Percival et al. 2004	open circles
2	0.35	0.440 ± 0.050	Song & Percival 2009; Tegmark et al. 2006	open circles
3	0.55	0.750 ± 0.180	Song & Percival 2009; Guzzo et al. 2008	open circles
4	0.25	0.351 ± 0.058	Samushia et al. 2012	open triangles
5	0.37	0.460 ± 0.038	Samushia et al. 2012	open triangles
6	0.22	0.420 ± 0.070	Blake et al. 2011	solid circles
7	0.41	0.450 ± 0.040	Blake et al. 2011	solid circles
8	0.60	0.430 ± 0.040	Blake et al. 2011	solid circles
9	0.78	0.380 ± 0.040	Blake et al. 2011	solid circles

Notice that solving eq.(10) for the concordance Λ cosmology[1], we derive the well known perturbation growth factor (see [48]):

$$D(z) = \frac{5\Omega_{m0}E(z)}{2}\int_z^{+\infty}\frac{(1+u)du}{E^3(u)} \quad . \tag{12}$$

3.1. *The evolution of the growth index*

For any type of DE, an efficient parametrization of the matter perturbations is based on the growth rate of clustering [48]:

$$f(a) = \frac{d\ln D}{d\ln a} \simeq \Omega_m^{\gamma}(a) \tag{13}$$

which implies:

$$D(a) = \exp\left[\int_1^a \frac{\Omega_m^{\gamma(x)}(x)}{x}dx\right] \tag{14}$$

where γ is the so called growth index (see [40, 42, 44, 47, 65]).

Combining eq.(13), eq.(11) and eq.(2), we find that:

$$a\frac{df}{da} + f^2 + X(a)f = \frac{3}{2}\Omega_m(a)Q(a)\,, \tag{15}$$

where

$$X(a) = \frac{1}{2} - \frac{3}{2}w(a)\left[1-\Omega_m(a)\right]\ . \tag{16}$$

If we change variables in eq.(15) from a to redshift $[\frac{d}{da} = -(1+z)^{-2}\frac{d}{dz}]$ and utilizing eqs.(13) and (4), then we can derive the evolution equation of the growth index $\gamma = \gamma(z)$ (see also [50]):

$$\begin{aligned} &-(1+z)\gamma'\ln(\Omega_m) + \Omega_m^{\gamma} + 3w(1-\Omega_m)(\gamma - \frac{1}{2}) + \frac{1}{2} \\ &= \frac{3}{2}Q\Omega_m^{1-\gamma}\,, \end{aligned} \tag{17}$$

[1] For the usual ΛCDM cosmological model we have $w(a) = -1$, $\Omega_\Lambda(a) = 1 - \Omega_m(a)$ and $Q(a) = 1$.

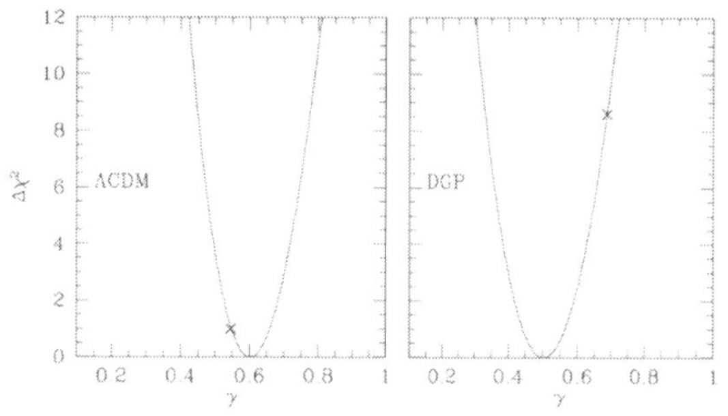

Figure 1. *Left Panel:* The variance $\Delta\chi^2 = \chi^2 - \chi^2_{min}$ around the best-fit γ value for the Λ cosmology. Note that the cross corresponds to $(\gamma_\Lambda, \Delta\chi^2_{1\sigma}) = (\frac{6}{11}, 1)$. *Right Panel:* The statistical results in the case of the DGP model. The corresponding cross is $(\gamma_{DGP}, \Delta\chi^2_{3\sigma}) = (\frac{11}{16}, 9)$.

Evaluating eq.(17) at $z = 0$, we have:

$$-\gamma'(0)\ln(\Omega_{m0}) + \Omega_{m0}^{\gamma(0)} + 3w_0(1 - \Omega_{m0})[\gamma(0) - \frac{1}{2}] + \frac{1}{2}$$
$$= \frac{3}{2}Q_0\Omega_{m0}^{1-\gamma(0)} , \qquad (18)$$

where $Q_0 = Q(z = 0)$ and $w_0 = w(z = 0)$.

It is interesting to mention here that the last few years there have been many theoretical speculations concerning the functional form of the growth index and indeed various candidates have been proposed in the literature. In this work, we decide to phenomenologically treat the functional form of the growth index $\gamma(z)$=constant.

4. The Growth data

The growth data that we utilize in this work, is based on 2dF, SDSS, and *WiggleZ* galaxy surveys, for which their combination parameter of the growth rate of structure, $f(z)$, and the redshift-dependent rms fluctuations of the linear density field, $\sigma_8(z)$, is available as a function of redshift, $f(z)\sigma_8(z)$. The $f\sigma_8 \equiv A$ estimator is almost a model-independent way of expressing the observed growth history of the universe [56]. In particular, we use:

- The 2dF [49], SDSS-LRG [59] and VVDS [26], based growth results as collected by Song & Percival [56]. This sample contains 3 entries.
- The SDSS (DR7) results (2 entries) of Samushia et al. [54], based on spectroscopic data of ~106000 LRGs in the redshift bin $0.16 < z < 0.44$.
- The *WiggleZ* results of Blake et al. [8], based on spectroscopic data of ~152000 galaxies in the redshift bin $0.1 < z < 0.9$. This dataset contains 4 entries.

In Table 1 we list the precise numerical values of the data points with the corresponding errors bars.

5. Fitting Models to the Data

In order to quantify the free parameters of the growth index, we perform a standard χ^2 minimization procedure between $N = 9$ growth data measurements, $A_{obs} = f_{obs}(z)\sigma_{8,obs}(z)$, with the growth values predicted by the models at the corresponding redshifts, $A(\mathbf{p}, z) = f(\mathbf{p}, z)\sigma_8(\mathbf{p}, z)$ with $\sigma_8(\mathbf{p}, z) = \sigma_{8,0}D(\mathbf{p}, z)$. The vector $\mathbf{p}$ represents the free parameters of

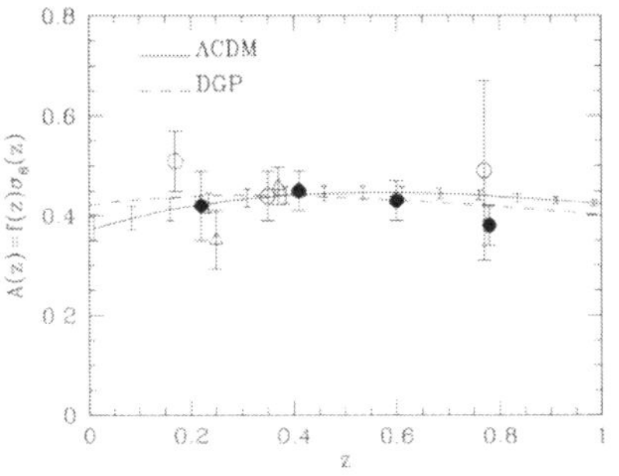

Figure 2. Comparison of the observed and theoretical evolution of the growth rate $A(z) = f(z)\sigma_8(z)$. The solid and dashed lines correspond to ΛCDM ($\gamma = 0.602$) and DGP ($\gamma = 0.503$) expansion models respectively. The different growth datasets are represented by different symbols (see Table 1 for definitions).

the model, depending on the model. In particular, the essential free parameters that enter in the statistical vector are: $\mathbf{p} \equiv (\gamma, \Omega_{m0})$. We will restrict our present analysis to the choice $(\Omega_{m0}, \sigma_{8,0}) = (0.273, 0.811)$ provided by WMAP7 [34][2]. Note that we sample $\gamma \in [0.1, 1.3]$ in steps of 0.001 and $\Omega_{m0} \in [0.1, 1]$ in steps of 0.001. The χ^2 function[3] is defined as:

$$\chi^2(z_i|\mathbf{p}) = \sum_{i=1}^{N} \left[\frac{A_{obs}(z_i) - A(\mathbf{p}, z_i)}{\sigma_i} \right]^2 \quad (19)$$

where σ_i is the observed growth rate uncertainty. A numerical summary of the statistical analysis is shown in Table 2. In general, we find that our results are in agreement, within 1σ uncertainties, with previous studies [3, 19, 20, 22, 25, 47].

In the left panel of Fig. 1, we show the variation of $\Delta\chi^2 = \chi^2(\gamma) - \chi^2_{min}(\gamma)$ around the best-fit γ value for the concordance Λ cosmology. We find that the likelihood function of the growth data peaks at $\gamma = 0.602 \pm 0.055$ with $\chi^2_{min} \simeq 7.06$ for 8 degrees of freedom. Alternatively, considering the ΛCDM theoretical value of γ ($\equiv 6/11$) and minimizing with respect to Ω_{m0}, we find $\Omega_{m0} = 0.243 \pm 0.034$ (see also [47]) with $\chi^2_{min}/dof \simeq 7.37/8$. Our growth index results are in agreement within 1σ errors, to those of Samushia et al. [54] who found $\gamma = 0.584 \pm 0.112$. However, our best-fit value is somewhat greater and almost 1σ ($\Delta\chi^2_{1\sigma} \simeq 1$) away, from the theoretically predicted value of $\gamma_\Lambda \simeq 6/11$ (see cross in the left panel of Fig. 1). It is interesting to mention here that such a small discrepancy between the theoretical ΛCDM and observationally fitted value of γ has also been found by other authors. For example, Di Porto & Amendola [18] obtained $\gamma = 0.60^{+0.40}_{-0.30}$, Gong [25] measured $\gamma = 0.64^{+0.17}_{-0.15}$ while Nesseris & Perivolaropoulos [47] found $\gamma = 0.67^{+0.20}_{-0.17}$. Recently, Basilakos [3] using a similar analysis shows that $\gamma = 0.616^{+0.088}_{-0.083}$.

Concerning the DGP model (see the right panel of Fig. 1), the best-fit value is $\gamma = 0.503 \pm 0.06$ with $\chi^2_{min}/dof \simeq 5.30/8$. If we fix the value of $\gamma (\equiv 11/16)$ to that predicted by the DGP model we find a quite large value of the dimensionless matter density at the present time, $\Omega_{m0} = 0.380 \pm 0.042$ with $\chi^2_{min}/dof \simeq 5.38/8$. It becomes clear that the best-fit γ value is much lower and almost 3σ ($\Delta\chi^2_{3\sigma} \simeq 9$) away, from $\gamma_{DGP} \simeq 11/16$ (see cross in the right panel of Fig. 1) implying that the growth data disfavor the DGP gravity. We would like to stress here that the

[2] For the DGP model, Gong [25] found $\Omega_{m0} = 0.278$.

[3] Likelihoods are normalized to their maximum values. In the present analysis, we always report 1σ uncertainties on the fitted parameters. Note that the uncertainty of the fitted parameters will be estimated, in the case of more than one such parameters, by marginalizing one with respect to the others.

Table 2. Statistical results for the combined growth data (see Table I): The 1st column indicates the expansion model, the 2nd column corresponds to the best values of γ for each cosmological model and the third column presents the reduced χ^2_{min}.

Expansion Model	γ	χ^2_{min}/dof
ΛCDM	0.602 ± 0.055	7.10/7
DGP	0.503 ± 0.060	5.32/7

above observational DGP constraints are in excellent agreement with previous studies. Indeed, Gong [25] and Dosset et al. [20] found $\gamma = 0.55^{+0.14}_{-0.13}$ and $\gamma = 0.483^{+0.113}_{-0.088}$, respectively. In Fig. 2, we plot the measured $A_{obs}(z)$ with the estimated growth rate function, $A(z) = f(z)\sigma_8(z)$ [see ΛCDM - solid line and DGP - dashed line].

The goal from the above discussion is to give the reader the opportunity to appreciate the relative strength and precision of the different methods used in order to constrain the growth index. It becomes evident that with the combined high-precision $f\sigma_8$ growth rate data of Song & Percival [56], Samushia et al. [54] and Blake et al. [8], we have achieved to place quite stringent constraints on γ. Furthermore, notice that using the cosmological parameters (Ω_m, σ_8) from Planck [1], our results remain almost the same within 1σ errors.

Finally, as we have already mentioned in Table 2, one may see a more compact presentation of our statistical results.

6. Conclusions

It is well known that the so called growth index γ plays a key role in cosmological studies because it can be used as a useful tool in order to test Einstein's general relativity on cosmological scales. We have utilized the recent growth rate data provided by the 2dFGRS, SDSS-LRG, VVDS, and *WiggleZ* galaxy surveys, in order to constraint the growth index. Performing a likelihood analysis for various $\gamma(z)$ parametrizations, we argue that the use of the above combined growth data places the most stringent constraints on the value of the growth index. Overall, considering a ΛCDM expansion model, we find that the observed growth index is in agreement, within 1σ errors, with the theoretically predicted value of $\gamma_\Lambda \simeq 6/11$. In contrast, for the DGP expansion model we find that the measured growth index is almost 3σ away from the corresponding theoretical value $\gamma_{DGP} \simeq 11/16$ which implies that the present growth data can not accomodate the DGP gravity model.

References

[1] Ade et al 2013 arXiv:1303.5076v1
[2] Amendola L & Tsujikawa S 2010 book: "Dark Energy Theory and Observations", Cambridge University Press, Cambridge UK
[3] Basilakos S 2012 arXiv:1202.1637
[4] Basilakos S & Plionis M 2010 ApJL 714 185
[5] Basilakos S & Pouri A 2012 arXiv:1203.6724v2
[6] Bean R &, Tangmatitham M 2010 Phys Rev D 81 083534
[7] Belloso A B & Garcia-Bellido J & Sapone D 2011 JCAP 1110 010
[8] Blake C et al 2011 MNRAS 415 2876
[9] Caldwell R R & Kamionkowski M 2009 Ann Rev Nucl Part Sci 59 397
[10] Copeland E J Sami M & Tsujikawa S 2006 Intern Journal of Modern Physics D 15 1753
[11] da Angela J et al 2008 MNRAS 383 565
[12] Daniel S F et al 2010 Phys Rev D 81 123508
[13] Dave R Caldwell R R & Steinhardt P J 2002 Phys Rev D 66 023516
[14] Davis T M et al 2007 ApJ 666 716
[15] de Viel M Haehnelt M G & Springel V 2004 MNRAS 354 684
[16] Deffayet C Dvali G & Cabadadze G 2002 Phys Rev D 65 044023

[17] Dent J B Dutta S & Perivolaropoulos L 2009 Phys Rev D 80 023514
[18] Di Porto C & L Amendola 2008 Phys Rev D 77 083508
[19] Di Porto & C Amendola L & Branchini E 2012 MNRAS 419 985
[20] Dosset J et al 2010 JCAP 1004 022
[21] Dvali G Gabadadze G & Porrati M 2000 Phys Lett B 485 208
[22] Fu X-y Wu P-x & H-w 2009 Phys Lett B 677 12
[23] Gannouji R Moraes B & Polarski D 2009 JCAP 62 034
[24] Gaztanaga E et al 2012 arXiv:11094852v3
[25] Gong Y 2008 Phys Rev D 78 123010
[26] Guzzo L et al 2008 Nature 451 541
[27] Hawkins E et al 2003 MNRAS 346 78
[28] Hicken M et al 2009 ApJ 700 1097
[29] Hinshaw G et al 2009 ApJ 180 225
[30] Huterer D & Turner M S 2001 Phys Rev D 64 123527
[31] Jesus J F & Cunha J V 2009 ApJL 690 L85
[32] Jesus J F Oliveira F A Basilakos S & Lima J A S 2011 arXiv:11051027
[33] Komatsu E et al 2009 ApJ 180 330
[34] Komatsu E et al 2011 Astrophys J Suplem 192 18
[35] Kowalski M et al 2008 ApJ 686 749
[36] Lima J A S & Alcaniz J S 2000 MNRAS 317 893
[37] Linder E V 2005 Phys Rev D 72 043529
[38] Linder E V 2004 Phys Rev Lett 70 023511
[39] Linder E V 2011 Phil Trans Roy Soc A 369 4985 arXiv:11030282v1
[40] Linder E V & Jenkins A 2003 MNRAS 346 573
[41] Linder E V 2008 Rep Prog Phys 71 056901
[42] Linder E V & Cahn R N 2007 Astrop Phys 2 481
[43] Lombriser L Hu W Fang W & Seljak U 2009 Phys Rev D 80 063536
[44] Lue A Scossimarro R & Starkman G D 2004 Phys Rev D 69 124015
[45] Motohashi H Starobinsky A A & Yokoyama J 2010 Progress of Theoretical Physics 123 887 arXiv:10021141
[46] McDonald P et al 2005 ApJ 635 761
[47] Nesseris S & Perivolaropoulos L 2008 Phys Rev D 77 023504
[48] Peebles P J E 1993 "Principles of Physical Cosmology" Princeton University Press Princeton New Jersey
[49] Percival W J et al 2004 MNRAS 353 1201
[50] Polarski D & Gannouji R 2008 Phys Lett B 660 439
[51] Rapetti D Allen S W Mantz A & Ebeling H 2010 MNRAS 406 179
[52] Ross N P et al 2007 MNRAS 381 573
[53] Saini T D Raychaudhury S Sahni V & Starobinsky A A 2000 Phys Rev Lett 85 1162
[54] Samushia L Percival W J & Raccanelli A 2012 arXiv:11021014
[55] Silveira V & Waga I 1994 Phys Rev D 64 4890
[56] Song Y- S Percival W J 2009 JCAP 10 4
[57] Stabenau F H & Jain B 2006 Phys Rev D 74 084007
[58] Tegmark M et al 2004 ApJ 606 702
[59] Tegmark M et al 2006 Phys Rev D 74 123507
[60] Thomas S A Abdalla F B & Weller J 2009 MNRAS 395 197
[61] Tsujikawa S Uddin K & Tavakol R 2008 Phys Rev D 77 043007
[62] Uzan P J 2007 Gen Rel Grav 39 307
[63] Verde L et al 2002 MNRAS 335 432
[64] Vikhlinin A et al 2009 ApJ 692 1060
[65] Wang L & Steinhardt J P 1998 ApJ 508 483
[66] Wang Y 2008 JCAP 0805 021

Asymptotic quasi-normal frequencies, horizon area spectra and multi-horizon spacetimes

Jozef Skákala
Dep. Matemática, Computacão e Cognicão, UFABC, Santo André, São Paulo, Brazil
E-mail: jozef.skakala@ufabc.edu.br

Abstract. There exists a long term debate about a possible link of the asymptotic quasinormal (QNM) modes to the black hole thermodynamics. The first conjecture providing a link between the classical asymptotic black hole oscillation frequencies and black hole thermodynamics was formulated more than a decade ago by Hod. The conjecture was later modified by Maggiore. We analyse the behavior of the asymptotic frequencies of the spherically symmetric multi-horizon spacetimes (in particular Reissner-Nordström, Schwarzschild-deSitter, Reissner-Nordström-deSitter spacetime) and provide some suggestions for how to interpret the results following the spirit of the modified Hod's conjecture. The interpretation suggested is in some sense analogical to the Schwarzschild black hole case, but has some new specific features. This paper[1] refers to work done over a longer period of time contained in papers [1, 2, 3, 4] and also to some extent in [5].

1. Introduction

Black holes, when perturbed, show certain characteristic damped oscillations which are called quasi-normal modes (QNMs). Consider (scalar, electromagnetic, gravitational) perturbation of a spherically symmetric black hole spacetime given by a line element

$$-f(r, c_i) \cdot dt^2 + f^{-1}(r, c_i) \cdot dr^2 + r^2 \cdot d\Omega^2. \tag{1}$$

By c_i we mean black hole spacetime parameters, such as mass, charge, or eventually in the asymptotically non-flat spacetime the cosmological constant. After decomposing the perturbation into tensor spherical harmonics and furthermore applying the normal modes decomposition, one can reduce the problem for the dynamics of the perturbations into a 1-dimensional stationary Schrödinger-like equation:

$$\frac{d^2\phi_\ell(x)}{dx^2} + \Big[\omega^2 - V(x)\Big] \cdot \phi_\ell(x) = 0. \tag{2}$$

By x we mean the tortoise coordinate defined as

$$\frac{dr}{dx} = f(r),$$

[1] Similar, but shorter paper was submitted to be published in a book by Edition Open Access as a part of conference proceedings of the conference "Relativity and Gravitation–100 years after Einstein in Prague", Prague, 2012.

and $V(x)$ is a Regge-Wheeler/Zerilli potential which depends on the black hole background parameters, the type of the perturbation, the parity of the perturbation and the wave mode number ℓ. The quasi-normal modes (QNMs) are solutions of the equation (2) such that correspond to the purely outgoing wave boundary conditions. (There are some subtleties, due to which one needs to slightly refine this simply sounding definition, but these are not going to be analysed here.) From the definition of the QNMs one can calculate for each perturbation the QNM frequencies ω. They are complex non-real frequencies, (for the $\exp(i\omega t)$ normal mode convention the frequencies have positive imaginary parts). The QNM frequencies depend on the wave-mode number and the black hole parameters. Furthermore, typically they form an infinite discrete set with unbounded imaginary part ω_I, ($\omega = \omega_R + i \cdot \omega_I$). The physical meaning of the QNM frequencies is that they are the poles of the Green function and due to this fact the QNMs dominate within a specific time scale the time evolution of an *arbitrary* compactly supported black hole perturbation. Thus QNMs can be described as the characteristic black hole oscillations, (or the "sound" of black holes).

2. The highly damped modes and conjectured connection to quantum gravity

The highly damped (asymptotic) frequencies, defined by ℓ fixed and $|\omega_I| \to \infty$, are in the known cases independent on the wave-mode number and depend only on the black hole parameters. For the Schwarzschild black hole they are given as, (κ is the surface gravity of the black hole horizon):

$$\omega_n = (\text{offset}) + i(\text{gap}) \cdot n + O(n^{-1/2}) = 4\ln(3) \cdot \kappa + i\kappa \cdot \left(n + \frac{1}{2}\right) + O(n^{-1/2}). \tag{3}$$

Hod [6] used Bohr's correspondence principle to conjecture that the asymptotic modes might be linked to the quantum black hole state transition. He linked the mass of the emitted quantum to the real part of asymptotic frequencies as, (everywhere in the paper we use the Planck units):

$$\Delta M = \lim_{n\to\infty} \omega_{nR}.$$

This suggestion plugged into the first law of black hole mechanics gives an equispaced horizon area spectrum with the area quantum $\Delta A = 4\ln(3)$.

Such quantum is of the form $\Delta A = 4\ln(k)$, $k \in \mathbf{N}$, and horizon area spectrum with the quantum of such form was already previously suggested due to the statistical interpretation of entropy [7]. Moreover, in [8] a connection was made between the asymptotic real part of the frequencies and the Immirzi parameter in Loop Quantum Gravity. The same results as in [6] can be obtained also by a different line of thinking: by using Bohr-Sommerfeld quantization condition for the adiabatic invariants and by considering the frequency asymptotic real part $\omega_{\infty R}$ to be the black hole's characteristic frequency [9]. However, Hod's conjecture led to many difficulties and as a result Maggiore suggested a modification which removed most of the complications [10]. According to Maggiore's modification of Hod's conjecture (or shorter, as we use in this paper, Maggiore's conjecture) QNMs have to be considered as a collection of damped harmonic oscillators and the mass quantum is connected to:

$$\Delta M = \lim_{n\to\infty} \Delta_{(n,n-1)} \sqrt{\omega_{nR}^2 + \omega_{nI}^2}. \tag{4}$$

Typically it holds that:

$$\lim_{n\to\infty} \Delta_{(n,n-1)} \sqrt{\omega_{nR}^2 + \omega_{nI}^2} = \lim_{n\to\infty} \Delta_{(n,n-1)} \omega_{nI}$$

and thus

$$\Delta M = \lim_{n\to\infty} \Delta_{(n,n-1)}\omega_{nI}.$$

In the case of Schwarzschild black hole this gives an equispaced area spectrum with a quantum of a form originally suggested by Bekenstein [11]:

$$\Delta A = 8\pi.$$

(The non-statistical form the of area quantum is thought not to be a problem, because of the analysis being essentially of a semi-classical nature.)

3. Spherically symmetric multi-horizon black hole spacetimes

For the multi-horizon spherically symmetric black hole spacetimes, the Reissner-Nordström spacetime (R-N), Schwarzschild-deSitter spacetime (S-dS), Reissner-Nordström-deSitter spacetime (R-N-dS), and the scalar, vector, tensor perturbations, the master equations for the asymptotic QNM frequencies were derived in [12, 13, 14, 15] and can be all written in the form:

$$\sum_{A=1}^{M} C_A \cdot \exp\left(\sum_{i=1}^{N} Z_{Ai}\frac{2\pi\omega}{|\kappa_i|}\right) = 0. \tag{5}$$

Here Z_{Ai} is an $M \times N$ matrix, such that takes one of the values $Z_{Ai} = \{0, 1, 2\}$, furthermore N is the number of horizons and κ_i are the surface gravities of the different horizons. The solutions of this formula have in general much more complicated behaviour as the asymptotic QNM frequencies of the Schwarzschild single horizon case (3). The fact that there do not generally exist analytic solutions for the formulas (5) prevented people to derive general consequences of the Maggiore's conjecture for the multi-horizon spherically symmetric cases (apart of some simplifying limits, like for example for the R-N black hole in a small charge limit [16]).

In [3, 4] we analysed the behaviour of the solutions of (5) with the following results: If the ratio of all of the surface gravities is a rational number then all the frequencies split in a finite number of equispaced families, (labeled by a), of the form:

$$\omega_{an} = (\text{offset})_a + in \cdot \ \text{lcm}(|\kappa_1|, |\kappa_2|, ..., |\kappa_N|). \tag{6}$$

Here "lcm" means the least common multiple of the numbers in the bracket, hence

$$\text{lcm}(|\kappa_1|, |\kappa_2|, ..., |\kappa_N|) = p_1 \cdot |\kappa_1| = ... = p_N \cdot |\kappa_N|,$$

where $\{p_1, ..., p_N\}$ is a set of relatively prime integers. If the ratio of two of the surface gravities is irrational, then there does *not* exist an equispaced subsequence in the sequence of asymptotic QNM frequencies. This means, although the master equations (5) cannot be in general analytically solved, one can still extract a lot of key analytic information about the behaviour of the solutions.

Moreover, the analytical insights, (without explicitly analytically knowing the solutions), are enough to prove that for the R-N black hole hold the following statements [2] (they were strictly proven for R-N black hole, but one expects them to hold for all the three cases):

1. In case the ratio of the surface gravities is irrational, the $n \to \infty$ limit for $\Delta_{(n,n-1)}\omega_{nI}$ does *not* exist.
2. For the rational ratio of the surface gravities and the R-N black hole the only case in which the limit $n \to \infty$ of $\Delta_{(n,n-1)}\omega_{nI}$ exists is if all the frequencies are given by families of the form (6) with the same (offset). But this cannot be the case when the ratio of the surface gravities is given by two relatively prime integers whose product is an odd number [2].

The previous considerations seem to suggest, that the modified Hod's conjecture has very little chance to survive the multi-horizon case. However the significantly different behaviour for the cases of rational / irrational ratios of surface gravities and the general splitting of frequencies into families (6) seem to indicate something important. Moreover, it was already observed that surface gravities rational ratios have significant consequences for the multi-horizon spacetime thermodynamics [17]. Based on this observations let us pick the R-N case where the thermodynamical interpretation is straightforward (but keep in mind that all the calculations can be repeated in an exact analogy for the other S-dS, R-N-dS cases) and consider the following [1]: Let us presuppose that both of the horizons in the R-N spacetime, the outer horizon with the area A_+ and the inner Cauchy horizon with the area A_- have equispaced area spectra given as $A_\pm = 8\pi\gamma n_\pm$.

The perturbations are supposed to carry *no* charge, so one expects that only the ADM mass of the black hole will be changed. Thus one can write the change of the areas of the black hole horizons as:

$$\Delta A_\pm = \frac{8\pi\Delta M}{\kappa_\pm}. \tag{7}$$

But $\Delta A_\pm$ can be given only as $\Delta A_\pm = 8\pi\gamma m_\pm$, which implies $\Delta M = \gamma\kappa_\pm m_\pm$. Furthermore, this implies

$$\kappa_+ m_+ = \kappa_- m_- \quad \rightarrow \quad \frac{\kappa_+}{\kappa_-} = \frac{m_-}{m_+}. \tag{8}$$

This means, that if the single ADM mass transitions have to be allowed, the surface gravities ratio must be rational. Furthermore, if one wants the emitted mass quantum to be as small as possible, such that it is still compatible with the quantization of the two horizon areas, one obtains:

$$\Delta M = \gamma \cdot \mathrm{lcm}(\kappa_+, |\kappa_-|). \tag{9}$$

Then modified Hod's conjecture suggests that

$$\lim_{n\to\infty} \Delta_{(n,n-1)}\omega_{nI} = \gamma \cdot \mathrm{lcm}(\kappa_+, |\kappa_-|). \tag{10}$$

This is indeed true if one takes the following interpretation of the frequencies (slight modification of Maggiore's conjecture): the straightforward extension of Maggiore's conjecture to the multi-horizon case is misleading, in fact only the equispaced families carry information about the quantum black hole mass transitions. (Every frequency belongs to one of the families.) Thus one has to first identify the equispaced families and then take the limit in the spacing in the imaginary part of the frequencies within each of the families. Such interpretation then fixes together with the formula (6) the γ parameter to be $\gamma = 1$. This means the area spectra of both of the horizons are given as $8\pi n$.

Note the importance of the surface gravities rational ratio condition in the preceeding argumentation: If the condition would not be fulfilled, then the absence of periodic sub-structures in case of irrational surface gravity ratio, together with the non-existence of the limit $n \to \infty$ of $\Delta_{(n,n-1)}\omega_{nI}$ means, that in such case does not seem to be even a slightly modified way how to understand the highly damped modes via Maggiore's conjecture. The fact that the rational ratios of the surface gravities are by the logic developed in this section guaranteed to exist seems to be another coincidence that "miraculously" fits into the picture. (On the other hand, since rational numbers are dense in the set of real numbers, we leave open the question of how much such rational ratios correspond to some quantum characteristics of surface gravities / horizon temperatures.)

Let us remind here, that the same analysis can be repeated for both S-dS and R-N-dS spacetimes: Assuming that all the horizons have the same equispaced area spectra, the single M parameter transitions lead to the surface gravities rational ratio condition and the QNM frequencies given by the formula (6) fix the spectra of all the horizons to be $8\pi n$, (after one considers our generalization/modification of Maggiore's conjecture). Let us also mention that the same area quantization of all the horizons in spherically symmetric multi-horizon spacetimes was obtained by a different (although not completely unrelated) way of thinking by [18].

4. Discussion

To summarize: We suppose that also in the multi-horizon case the *modified* Hod's conjecture provides information about the spacetime horizons spectra, only the way the information is encoded is more tricky than in the single horizon case. (This is hardly anything surprising as the quantization of more than one horizon might play a role in the game.) The QNM frequencies are consistent with each of the horizons being quantized with spectra given as $8\pi n$ (in Planck units). If these conclusions are accepted then still many open questions remain, for example: 1. Can be the results from this paper generalized to an arbitrary static spherically symmetric multi-horizon spacetime? (The results of [19] indicate that the behaviour of the asymptotic frequencies described in this paper holds at least in case of a general two-horizon spacetime, with one black hole horizon and one cosmological de-Sitter horizon.) 2. Could similar interpretation survive in case of charged black holes that occur after a collapse of matter? (In such case the black hole interior is very different to the extremely idealized R-N, R-N-dS cases. For example a weak mass-inflation singularity occurs at the inner horizon, but some results indicate that possibility of crossing the inner horizon might still be considered [20].)

Acknowledgments: The more recent part of this research was supported by Fundacão de Amparo a Pesquisa do Estado de São Paulo (FAPESP). In its earlier stages the research was supported by The Vice-Chancellor's Strategic Research Scholarship by Victoria University of Wellington, New Zealand.

References

[1] J. Skakala, "Quasinormal modes, area spectra and multi-horizon spacetimes", JHEP **1206** (2012) 094, gr-qc/1204.3566
[2] J. Skakala, "Non-extremal Reissner-Nordstrom black hole: Do asymptotic quasi-normal modes carry information about the quantum properties of the black hole?", JHEP **1201** (2012) 144, gr-qc/1111.4164
[3] J. Skakala and M. Visser, "Generic master equations for quasi-normal frequencies", JHEP 1011 (2010) 070, gr-qc/1009.0080
[4] J. Skakala, "Aspects of General Relativity: Pseudo-Finsler extensions, Quasi-normal frequencies and Multiplication of tensorial distributions", PhD Thesis, gr-qc/1107.2978
[5] J. Skakala and M. Visser, "Semi-analytic results for quasi-normal frequencies", JHEP **1008** (2010) 061, gr-qc/1004.2539
[6] S. Hod, "Bohr's correspondence principle and the area spectrum of quantum black holes", Phys.Rev.Lett. **81** (1998) 4293, gr-qc/9812002
[7] J.D. Bekenstein and V. Mukhanov, "Spectroscopy of the quantum black hole", Phys.Lett. B **360** (1995) 7-12, gr-qc/9505012
[8] O. Dreyer, "Quasinormal modes, the area spectrum, and black hole entropy", Phys.Rev.Lett. **90** (2003) 081301, gr-qc/0211076
[9] G. Kunstatter, "d-dimensional black hole entropy spectrum from quasinormal modes", Phys.Rev.Lett. **90** (2003) 161301, gr-qc/0212014
[10] M. Maggiore, "The Physical interpretation of the spectrum of black hole quasinormal modes", Phys.Rev.Lett. **100** (2008) 141301, gr-qc/0711.3145
[11] J. Bekenstein, "Extraction of energy and charge from a black hole", PhysRevD **7** (1973) 949953.
[12] L. Motl and A. Neitzke, "Asymptotic black hole quasinormal frequencies", Adv.Theor.Math.Phys. **7** (2003) 307-330, hep-th/0301173
[13] N. Andersson and C.J. Howls, "The Asymptotic quasinormal mode spectrum of nonrotating black holes", Class.Quant.Grav. **21** (2004) 1623-1642, gr-qc/0307020
[14] J. Natario and R. Schiappa, "On the classification of asymptotic quasinormal frequencies for d-dimensional black holes and quantum gravity", Adv.Theor.Math.Phys. **8** (2004) 1001-1131, hep-th/0411267

[15] V. Cardoso, J. Natario and R. Schiappa, "Asymptotic quasinormal frequencies for black holes in nonasymptotically flat space-times", J.Math.Phys. **45** (2004) 4698-4713, hep-th/0403132
[16] A. Lopez-Ortega, "Area spectrum of the d-dimensional Reissner-Nordstrom black hole in the small charge limit", Class.Quant.Grav. **28** (2011) 035009, gr-qc/1003.4248
[17] T.R. Choudhury and T. Padmanabhan, "Concept of temperature in multi-horizon spacetimes: Analysis of Schwarzschild-de Sitter metric", Gen.Rel.Grav. **39** (2007) 1789-1811, gr-qc/0404091
[18] Y. Kwon and S. Nam, "Entropy spectra of black holes from resonance modes in scattering by the black holes", Class.Quant.Grav. **28** (2011) 035007, hep-th/1102.3512
[19] A. Ghosh, S. Shankaranarayanan and S. Das, "High frequency quasi-normal modes for black holes with generic singularities. II. Asymptotically non-flat spacetimes", Class.Quant.Grav. **23** (2006) 1851-1874, hep-th/0510186
[20] S. Hod and T. Piran, "The Inner Structure of Black Holes", Gen.Rel.Grav. **30** (1998) 1555, gr-qc/9902008.

Induced gravity from gauge theories

R. F. Sobreiro and A. A. Tomaz
UFF – Universidade Federal Fluminense, Instituto de Física, Campus da Praia Vermelha, Avenida General Milton Tavares de Souza s/n, 24210-346, Niterói, RJ, Brasil.
E-mail: sobreiro@if.uff.br and tomaz@if.uff.br

V. J. Vasquez Otoya
IFSEMG – Instituto Federal de Educação, Ciência e Tecnologia, Rua Bernardo Mascarenhas 1283, 36080-001, Juiz de Fora, MG, Brasil.
E-mail: victor.vasquez@ifsudestemg.edu.br

Abstract. We discuss the possibility of a class of gauge theories, in four Euclidean dimensions, to describe gravity at quantum level. The requirement is that, at low energies, these theories can be identified with gravity as a geometrodynamical theory. Specifically, we deal with de Sitter-type groups and show that a Riemann-Cartan first order gravity emerges. An analogy with quantum chromodynamics is also formulated. Under this analogy it is possible to associate a soft BRST breaking to a continuous deformation between both sectors of the theory, namely, ultraviolet and infrared. Moreover, instead of hadrons and glueballs, the physical observables are identified with the geometric properties of spacetime. Furthermore, Newton and cosmological constants can be determined from the dynamical content of the theory.

1. Introduction

Since the advent of general relativity (GR) [1], gravity has been showing itself to be very different from all other fundamental interactions. The three fundamental interactions contemplated by the standard model are quantum mechanically stable gauge theories. In fact, quantum field theory (QFT) [2] describes fluctuating fields which are parameterized by a set of rigid spacetime coordinates X^μ while spacetime itself is a Minkowski space (To perform actual computations, spacetime must be Wick-rotated to the Euclidean space). On the other hand, GR is a geometrodynamical theory [3] that describes spacetime as a dynamical entity, *i.e.*, spacetime coordinates gain a dynamical character. If one declare that the metric tensor $g_{\mu\nu}(X)$ is the fundamental field of gravity, a standard quantization is automatically forbidden because its quantum nature would be transferred to spacetime. For instance, if $\hat{g}_{\mu\nu}(X)$ is the quantum version of the metric tensor then, $\hat{X}_\mu = \hat{g}_{\mu\nu}(X)X^\nu$, which implies on $\hat{X}^\mu = \hat{g}^{\mu\nu}(X)\hat{X}_\nu$. Thus, spacetime coordinates would acquire a quantum character. The inconsistency rises because a quantum field cannot be parameterized by "quantum coordinates". Axiomatic QFT states that spacetime coordinates must be a rigid set of well-defined parameters [4]. Essentially, geometrodynamics appears to be an exclusively classical description of gravity.

Another property that ruins a quantum description of gravity is the success of Einstein equations in describing macroscopic phenomena. At macroscopic scale, the *rhs* of Einstein equations is the energy-momentum tensor of the matter distribution, which is an effective object.

Journal of Physics: Conference Series **453** (2013) 012014 doi:10.1088/1742-6596/453/1/012014

Thus, the *lhs* of Einstein equations should also be an effective object and not a fundamental tensor that could be easily quantized by standard techniques.

To avoid the incompatibility between the principles of QFT and GR, the main method adopted is to expand the metric tensor around a background Minkowski metric $g_{\mu\nu} = \eta_{\mu\nu} + kh_{\mu\nu}$, where k is related to Newton's constant and carry ultraviolet dimension -1. The perturbation $h_{\mu\nu}$ is interpreted as the graviton field. However, it is widely known that the linearized Einstein-Hilbert (EH) action, known as the Pauli-Fierz action [5], does not define a stable action at quantum level [6, 7, 8]. It is necessary then to introduce higher derivative terms which, however, destroy the unitarity of the theory [9].

To get gravity closer to the rest of the fundamental interactions, it is convenient to write it as a gauge theory [10, 11, 12]. This description of gravity is known as the first order formalism (The formalism based on the metric tensor is called second order formalism), see also [13, 14]. In this approach, gravity is described by two fundamental 1-form fields, the vierbein[1] $e^{\mathfrak{a}}$ and the spin-connection $\omega^{\mathfrak{a}}{}_{\mathfrak{b}}$. The geometric properties of spacetime are obtained from specific gauge invariant composite fields [15, 16]. In particular, the metric tensor is obtained from

$$g_{\mu\nu} = \eta_{\mathfrak{ab}} e^{\mathfrak{a}}_{\mu} e^{\mathfrak{b}}_{\nu} \,, \tag{1}$$

while the affine connection is determined by

$$\Gamma^{\alpha}_{\mu\nu} = g^{\alpha\beta}\eta_{\mathfrak{ab}}\left(e^{\mathfrak{b}}_{\beta}\partial_{\mu}e^{\mathfrak{a}}_{\nu} + e^{\mathfrak{b}}_{\beta}\omega^{\mathfrak{a}}_{\mu\mathfrak{c}}e^{\mathfrak{c}}_{\nu}\right) \,. \tag{2}$$

The vierbein is a map between coordinates X^{μ} at a point X, in spacetime manifold $\mathbb{M}^4$, and coordinates $X^{\mathfrak{a}}$, in the tangent space $T_X(\mathbb{M})$, at the very same point X, *i.e.*, $dX^{\mathfrak{a}} = e^{\mathfrak{a}}_{\mu}dX^{\mu}$. The spin-connection is related to parallel transport between near tangent spaces $T_X(\mathbb{M})$ and $T_{X+dX}(\mathbb{M})$ and is recognized as the gauge field of the theory. Typically, the gauge group is the Lorentz group $SO(1,3)$ and describes the local isometries of spacetime. Nevertheless, since the fundamental fields are still related to spacetime properties, any attempt of standard quantization of these fields would lead to the same inconsistency that occur in the second order formalism. Another attempt to solve the QFT-GR incompatibility is to declare that the fundamental fields are not immediately related to spacetime. Instead, the geometric properties of spacetime emerge as effective phenomena [14, 15, 16]. For instance,

$$\begin{aligned} g_{\mu\nu} &= \eta_{\mathfrak{ab}}\left\langle e^{\mathfrak{a}}_{\mu}e^{\mathfrak{b}}_{\nu}\right\rangle \,, \\ \Gamma^{\alpha}_{\mu\nu} &= g^{\alpha\beta}\eta_{\mathfrak{ab}}\left\langle e^{\mathfrak{b}}_{\beta}\partial_{\mu}e^{\mathfrak{a}}_{\nu} + e^{\mathfrak{b}}_{\beta}\omega^{a}_{\mu\mathfrak{c}}e^{\mathfrak{c}}_{\nu}\right\rangle \,. \end{aligned} \tag{3}$$

The idea is to work with geometric variables with no specific relation to the metric tensor nor the affine connection. Although promising, it is unclear to work with a ill defined coordinate system.

Sticking to the gauge theoretical approach, many other theories have been proposed by generalizing the gauge groups and their respective actions. In particular [17, 18, 19, 20, 21], it is worth mentioning the de Sitter groups $SO(m,n)$, with $(m+n) = 5$, in four-dimensional spacetime, supplemented with a spontaneous symmetry breaking mechanism [22, 23]. In the work [24], gravity and supergravity are considered under the framework of a dynamical breaking instead of a spontaneous breaking. In any of these cases, the Yang-Mills action [25] is not considered. When the Yang-Mills action is taken into account, the starting theory is already metric dependent, see for instance [26, 27]. Only in [28], the Yang-Mills was considered

[1] The indices conventions are: $\{\mathfrak{a},\mathfrak{b},\mathfrak{c},\ldots\} \in \{0,1,2,3\}$ are frame indices; and $\{\text{greek}\} \in \{0,1,2,3\}$ are world indices.

independently of the metric. The Lorentz group was taken as the gauge group of a theory in flat spacetime. A relation with a Palatini-type gravity emerges from a color symmetry breaking generated by a condensation mechanism.

More recently [29], the Yang-Mills action for de Sitter groups was studied independently of the geometric properties of spacetime. Essentially, asymptotic freedom [30, 31], mass parameters and an Inönü-Wigner contraction [32] to Lorentz-type groups account for the emergence of a gravity theory. The absence of mass parameters in the starting action is important because it prevents the identification of the gauge field with the vierbein. Thus, it is important that the masses are dynamically generated. One possibility is the Gribov parameter [33, 34, 35] which is required for quantum consistency at low energy scales. Moreover, the presence of the Gribov parameter induces a soft BRST breaking [36, 37, 38], which continuously deforms the massless theory in flat spacetime into a gravity theory in the first order formalism. Extra dynamical masses can also contribute to the model [39, 40]. Remarkably, Newton and cosmological constants can be determined from these masses and the Yang-Mills coupling parameter. Thus, since pure Yang-Mills theory is renormalizable [41, 42], this model can be regarded as a possible theory for quantum gravity.

In the present work, we exploit some physical properties of the theory developed in [29]. In particular, we discuss the analogy between this theory and quantum chromodynamics (QCD), the soft BRST breaking, the dynamical masses contribution and the possibility of the theory in providing reliable physical results.

Section 2 is devoted to the foundations of Yang-Mills theories, BRST symmetry and confinement. Section 3 is dedicated to a QCD-Gravity analogy. In Section 4, the details of de Sitter gauge theories in four-dimensional Euclidean spacetime and their relation with gravity are discussed. In Section 5, some consistency checks of the model are provided. Our final considerations can be found in Section 6.

2. Yang-Mills theories, BRST symmetry and confinement

2.1. Preliminary concepts

The so called Yang-Mills theories [25] are gauge theories based on a gauge symmetry described by a semi-simple Lie group G. This class of theories consists of a generalization of electrodynamics, which is an Abelian gauge theory for the group $U(1)$. The fundamental field is the gauge connection 1-form $Y = Y^{\mathsf{a}} J_{\mathsf{a}}$, where J_{a} are the generators of the group and the group indices vary as $\{\mathsf{a}, \mathsf{b}, \mathsf{c}, \ldots\} \in \{1, 2, \ldots, \dim G\}$. The curvature 2-form is defined as $F = \nabla^2 = \mathrm{d}Y + \kappa YY$, where $\nabla = \mathrm{d} + \kappa Y$ is the covariant derivative, d is the exterior derivative in spacetime and κ is the coupling parameter.

The Yang-Mills action is constructed as

$$S_{\mathrm{YM}} = \frac{1}{2}\int F^{\mathsf{a}} * F_{\mathsf{a}} \;, \tag{4}$$

where $*$ is the Hodge operator. The action (4) is invariant under gauge transformations,

$$Y \longmapsto g^{-1}\left(\frac{1}{\kappa}\mathrm{d} + Y\right) g \;, \tag{5}$$

where $g \in G$. To define a path integral for the Yang-Mills action, a gauge fixing must be imposed [2, 43]. For a consistent introduction of a gauge fixing, it is convenient to adopt the BRST quantization method [42, 44, 45]. For simplicity, we adopt the Landau gauge, $\mathrm{d} * Y = 0$. In this framework, the action (4) is replaced by

$$S_0 = S_{\mathrm{YM}} + \int \left(i b^{\mathsf{a}} \mathrm{d} * Y_{\mathsf{a}} + \bar{c}_{\mathsf{a}} \mathrm{d} * \nabla^{\mathsf{ab}} c_{\mathsf{b}} \right) \;, \tag{6}$$

where the fields c^a and $\bar{c}^a$ are the ghost and anti-ghost fields, respectively, and b^a is a Lagrange multiplier which enforces the Landau gauge. The components of the covariant derivative are $\nabla^{ab} = \delta^{ab}d - \kappa f^{abc}Y_c$, where f^{abc} are the structure constants of the gauge group. The action (6), although not gauge invariant, is invariant under BRST transformations, namely,

$$\begin{aligned} sY^a &= \nabla^{ab}c_b\,, \\ sc^a &= \frac{\kappa}{2}f^{abc}c_bc_c\,, \\ s\bar{c}^a &= b^a\,, \\ sb^a &= 0\,, \end{aligned} \tag{7}$$

where s is the nilpotent BRST operator.

Let us make some important remarks about Yang-Mills theories.

- Although the fundamental fields and the action (6) are not gauge invariant, physical observables must be gauge invariant. This statement is perhaps the most sacred principle in gauge theories, it is called *gauge principle*. This principle can be extended to BRST invariant operators in order to include more general operators that might be related to physical quantities. At classical level, the BRST invariance principle reduces to the usual gauge principle.
- Together with quantum electrodynamics [46, 47, 48], and the Higgs mechanism [22, 23], Yang-Mills theories are the theoretical basis of the Standard model. Thus, it is a successful theory with a deep connection with reality.
- The action (6) is a renormalizable theory, at least to all orders in perturbation theory [41, 42]. Renormalizability states that ultraviolet divergences can be consistently eliminated from perturbative computations, *i.e.*, Yang-Mills theories are stable at quantum level. It is worth mention that BRST symmetry is very important with respect to the renormalizability of a gauge theory.
- Another feature of the action (6) is unitarity [49]. It states that the S-matrix conserves probability and, thus, asymptotic states can be defined. For pure gauge theories, unitarity requires a well defined BRST symmetry and that the gauge group is compact.
- The renormalization of the coupling parameter leads to the concept of asymptotic freedom [30, 31]. This property predicts that, at high energies, the coupling is very small and perturbation theory can be safely employed. However, as the energy decreases the coupling increases. At this strong coupling regime, the theory is non-perturbative and yet to be fully understood.
- Confinement [37, 38, 49] states that quarks and gluons cannot be observable states. Actually, the physical observables must be, not only gauge invariant, but also colorless. At high energies, confinement manifests itself as the quark-gluon plasma, a state where quarks and gluons are almost free. However, they cannot be distinguished from the plasma. At low energies, the strong coupling enforces hadronization phenomena, *i.e.*, the physical spectrum is composed by hadrons and glueballs. Although experimentally established, it is a theoretical challenge to show that the Yang-Mills action leads to an effective theory composed by hadrons and glueballs.
- Another possible effect is that, at low energies, dynamical mass generation [39, 40] takes place. It originates from the condensation of dimension-2 operators. The mass parameters that emerge drastically change the behavior of the theory at infrared scale and are relevant for confinement.
- It was shown in [33, 50] that, at low energies, gauge fixing is not possible by the simple introduction of a constraint. This problem is known as Gribov ambiguities problem. The

main point in the elimination of the Gribov ambiguities is that BRST quantization (and also the standard Faddeev-Popov quantization [43]) shows itself to be incomplete at low energy level. In fact, the implementation of a gauge fixing does not eliminate completely the gauge symmetry [33], a residual symmetry survives. Moreover, it was shown that this problem occur for any gauge choice [50]. The spurious gauge configurations are called Gribov copies and they are the kernel of the Gribov ambiguities. Remarkably, these copies gain relevance only at low energies, keeping the high energy sector untouched. To eliminate the Gribov ambiguities is a hard step that is not yet fully understood. However, at the Landau gauge, the inifinitesimal copies can be eliminated through the introduction of a soft BRST breaking[2] related to a mass parameter known as Gribov parameter. Outstandingly, the treatment of this technical problem leads to exceptional evidences of quark-gluon confinement [34, 36, 37, 38, 39].

2.2. The infrared puzzle, Gribov ambiguities and BRST soft breaking

Basically, at the infrared regime, there are several techniques leading to confinement evidences. For instance, non-trivial solutions of the Schwinger-Dyson equations [51], lattice simulations [52, 53], AdS/QCD duality [54, 55, 56], dual superconductivity and defects [57, 58, 59, 60] and dimension-2 condensates and Gribov ambiguities [33, 34, 35, 36, 37, 38, 39, 40]. The puzzle consists in merging all evidences and formalisms in order to obtain a consistent theory in terms of hadrons and glueballs. As mentioned, we confine ourselves to explore the Gribov problem and its relation to confinement.

The treatment of Gribov ambiguities has the effect of changing the action (6) to the so called Gribov-Zwanziger action,

$$S = S_0 + \int \left[\overline{\phi}_{\mathsf{ac}} \mathrm{d} * \nabla^{\mathsf{ab}} \phi_{\mathsf{bc}} - \overline{\omega}_{\mathsf{ac}} \mathrm{d} * \nabla^{\mathsf{ab}} \omega_{\mathsf{bc}} + \gamma^2 f^{\mathsf{abc}} Y_{\mathsf{a}} \left(\overline{\phi} + \phi\right)_{\mathsf{bc}} + 4M\gamma^4\right] , \qquad (8)$$

where $M = \dim G$, γ is the Gribov parameter which has dimension of a mass, the fields $\overline{\phi}_{\mathsf{ab}}$ and ϕ_{ab} are 1-form bosonic fields while $\overline{\omega}_{\mathsf{ab}}$ and ω_{ab} are fermionic 1-form fields. These fields obey the following transformation rules,

$$\begin{aligned} s\overline{\omega}_{\mathsf{ab}} &= \overline{\phi}_{\mathsf{ab}} , \\ s\overline{\phi}_{\mathsf{ab}} &= 0 , \\ s\phi_{\mathsf{ab}} &= \omega_{\mathsf{ab}} , \\ s\omega_{\mathsf{ab}} &= 0 , \end{aligned} \qquad (9)$$

and can be eliminated through field equations in favor of a nonlocal term, the horizon-function [34, 39, 40].

We now establish a few properties of the action (8).

- The action (8) is renormalizable, at least to all orders in perturbation theory, in such a way that the ultraviolet sector of the theory is unchanged. In fact, no extra renormalizations are required [34, 39]; all extra fields and Gribov parameter renormalization factors depend on the gluon and coupling parameter renormalization factors.
- The action (8) is not BRST invariant, and the breaking is proportional to the Gribov parameter, $sS \propto \gamma^2$. Since γ has dimension of a mass, the breaking is soft, and thus, harmless to the UV sector [36, 37].

[2] A soft breaking is breaking whose field operators have dimension lower than the spacetime dimension.

- The Gribov parameter is determined from the minimization of the quantum action $\delta\Sigma/\delta\gamma^2 = 0$. A simple computation at the tree level predicts that $\gamma^2 = \mu^2 \exp\{-\frac{64\pi^2}{3N\kappa^2}\}$, where μ^2 is a cutoff and N is the Casimir of the gauge group, $f^{\mathsf{acd}}f_{\mathsf{bcd}} = -N\delta^{\mathsf{a}}_{\mathsf{b}}$. Thus, at the perturbative regime, $\kappa \to 0$, we have that $\gamma \to 0$. To take this limit is equivalent to take the high energy limit. Thus, the presence of the Gribov parameter allows the action to be continuously deformed into action (6). Then, the UV and IR sectors can be continuously deformed to each other. Moreover, the BRST is asymptotically restored at the UV limit.
- One of the evidences of confinement emerges from the gauge propagator which, at finite values of γ, acquires imaginary poles. The consequence is that it violates positivity of the spectral representation of this propagator and thus, no physical particles can be associated with this propagator [39].
- More recently [40], the union of the Gribov-Zwanziger formalism and the condensation of dimension-2 operators has been developed. This is called the *refined Gribov-Zwanziger formalism* and improves the results obtained from the Gribov-Zwanziger action. For instance, the propagators of the RGZ action coincide with the results obtained from lattice simulations [53].
- We also remark that, some advances on the determination of the physical spectrum of the infrared sector of Yang-Mills theories have been made [61, 62]. Although very difficult, the main requirement is gauge invariance, *i.e.*, observables are related with gauge invariant operators. In QCD, these operators must describe its low energy spectrum, *i.e.*, hadrons and glueballs.

3. QCD-Gravity analogy

This work is based on the idea that gravity can be described at quantum level by a gauge theory in a Euclidean four-dimensional spacetime. Classical gravity emerges if one could possibly show that a geometrodynamical theory rises at the infrared sector. This is performed by applying the soft BRST breaking technique. Let us elaborate this idea in more detail.

We assume that quantum gravity can be described by a Yang-Mills theory in four-dimensional Euclidean spacetime, $\mathbb{R}^4$. This choice, instead of a Minkowski spacetime, has three main reasons: (i) first, because QFT is only solvable in Euclidean spaces. Reliable computations require a Wick rotation from the Minkowkian to the Euclidean space [2]. However, a Wick rotation is known to be valid only at perturbative level. At non-perturbative scales, there are no indications that a Wick rotation can be performed. Thus, since we are interested in non-perturbative effects, it is necessary to start with a Euclidean spacetime. (ii) The second reason, and perhaps the most important, is that the Euclidean space is the simplest geometry. Since we are constructing a theory that will determine spacetime geometry, to start with the simplest geometry is the right choice. (iii) In four dimensions, the coupling parameter of a Yang-Mills theory is dimensionless. As a consequence, pure Yang-Mills action is massless[3]. This fact is important because it prevents the gauge field, which has dimension 1, to be associated with the vierbein, which has vanishing dimension.

The gauge field is an algebra-valued 1-form and thus, carries $\dim G$ components with respect to the algebra of the gauge group. The gauge group G has to fulfill a few requirements [29, 63]: (i) the first one is that $\dim G \geq \dim ISO(4)$, in such a way that the degrees of freedom of gravity are covered. (ii) The universal principal bundle that describes this theory must be non-trivial, so the Gribov problem is inherent to the theory and has to be properly treated [50, 64, 65]. (iii) The group must decompose at least[4] as $G = H + Q$ where $Q = G/H$ must be a symmetric space

[3] In Yang-Mills theories, the four-dimensional case is the only case where the coupling parameter is dimensionless.
[4] Larger groups are also allowed. The extra decompositions will generate a matter sector in the resulting gravity theory, [67].

and H, obviously, a stability group. The stable group must share a morphism with Lorentz-type groups and Q must define a vector representation of H. Thus, H can be identified with local Lorentz transformations and Q with a sector that expand the vierbein. As a consequence, the gauge components can be identified with the spin-connection and the vierbein. (iv) It can be very convenient to have a symmetry breaking $G \rightarrow H$, in such a way that the field at the Q sector acquires a matter-type transformation with respect to H. That is why Q must be a symmetric space. It is important to mention that, depending on the gauge group, a symmetry breaking might not be needed. This occurs if the gauge transformations could be partially identified with local Lorentz transformations and partially with diffeomorphisms. A renowned example can be found in [66].

It is important that the theory develops at least one mass parameter, possibly the Gribov parameter, in order to rescale the gauge field at the sector Q. As mentioned, this field, being a connection, carries dimension 1 and thus, cannot be directly associated with the vierbein. Only after the emergence of these mass parameters is that this identification can be performed. Moreover, the Gribov parameter and the consequent BRST breaking are important concepts that must be interpreted as follows:

The theory has two sectors, the UV, which is a massless gauge theory in Euclidean four-dimensional space, and the IR sector, which presents soft BRST symmetry breaking and dynamically generated mass parameters. The UV sector is a standard, non-Abelian, asymptotic free, gauge theory of spin-1 excitations. Although the degrees of freedom coincide in number with a first order gravity, this identification is forbidden unless a mass parameter arises, so that a vierbein can be defined. Once the energy starts to decrease, the soft BRST symmetry breaking takes place, the Gribov parameter and other possible masses appear. At this stage, propagators of the fundamental fields develop complex poles, a fact that is interpreted as an evidence that these excitations are ruled out from the physical spectrum of the theory (in QCD, this is recognized as confinement). Physical observables must be defined at this point. In QCD, the observables are hadrons and glueballs. In gravity, the low energy observables must be geometry. Thus, one possibility is to identify the H gauge field with the spin-connection and the Q field with the vierbein. The consequence is a first order gravity where geometry is determined by the usual relations (1) *and* (2).

If all above requirements are fulfilled, we gain a deep analogy between gravity and quantum chromodynamics. At high energies, both theories are well defined quantum gauge theories in a four-dimensional flat space. Both theories present soft BRST breaking and have to be redefined in order to establish the physical content at the infrared regime. In the case of QCD, the physical content are confined states identified with hadrons and glueballs. In gravity, the physical content are identified with geometric properties of spacetime.

We now turn to a specific model when these ideas are realized.

4. Realization of the idea through de Sitter groups

4.1. de Sitter-Yang-Mills theory

One possible realization of the above ideas occurs in a gauge theory based on the group $SO(m,n)$ with $m+n=5$ and $m \in \{0,1,2\}$. For $m=0$, the gauge group is the orthogonal one. For $m=1$ and $m=2$, we have a de Sitter and anti de Sitter groups, respectively[5]. Spacetime is a Euclidean four-dimensional differential manifold $\mathbb{R}^4$. The algebra of the group is given by

$$[J^{AB}, J^{CD}] = -\frac{1}{2}\left[\left(\eta^{AC}J^{BD} + \eta^{BD}J^{AC}\right) - \left(\eta^{AD}J^{BC} + \eta^{BC}J^{AD}\right)\right] , \qquad (10)$$

[5] Except when necessary, we will indistinguishably call the generic $SO(m,n)$ group, with arbitrary m, by de Sitter group.

where J^{AB} are the 10 anti-hermitian generators of the gauge group, antisymmetric in their indices. Capital Latin indices are chosen to run as $\{5, 0, 1, 2, 3\}$. Comparing with the generic gauge theory described in Sect. 2.1, the indices identification is $\mathsf{a} \equiv AB = -BA$. The $SO(m,n)$ group defines a five-dimensional flat space, $\mathbb{R}^{m,n}_S$, with invariant Killing metric given by $\eta^{AB} \equiv \mathrm{diag}(\epsilon, \varepsilon, 1, 1, 1)$ with $\epsilon = (-1)^{(2-m)!}$ and $\varepsilon = (-1)^{m!+1}$. The spaces $\mathbb{R}^4$ and $\mathbb{R}^{m,n}_S$ have no dynamical relation whatsoever.

The $SO(m,n)$ Yang-Mills action is renormalizable and is defined over a non-trivial universal bundle [29, 63, 67]. Thus, BRST soft breaking takes place. Moreover, a few extra mass parameters may emerge. The presence of these masses will be forwardly used. However, we will fix our attention only at the Yang-Mills action and the gauge field Y.

The de Sitter group may be decomposed as a direct product, $SO(m,n) \equiv SO(m!-1,n) \otimes S(4)$ where $S(4) \equiv SO(m,n)/SO(m!-1,n)$ is a symmetric coset space with four degrees of freedom. This decomposition is carried out by projecting the group space in the fifth coordinate $A = 5$. Defining then $J^{5a} = J^a$, where small Latin indices vary as $\{0, 1, 2, 3\}$, the algebra (10) decomposes as

$$\begin{aligned}
\left[J^{ab}, J^{cd}\right] &= -\frac{1}{2}\left[\left(\eta^{ac}J^{bd} + \eta^{bd}J^{ac}\right) - \left(\eta^{ad}J^{bc} + \eta^{bc}J^{ad}\right)\right] , \\
\left[J^{a}, J^{b}\right] &= -\frac{\epsilon}{2}J^{ab} , \\
\left[J^{ab}, J^{c}\right] &= \frac{1}{2}\left(\eta^{ac}J^{b} - \eta^{bc}J^{a}\right) , \qquad (11)
\end{aligned}$$

where $\eta^{ab} \equiv \mathrm{diag}(\varepsilon, 1, 1, 1)$.

The gauge connection, follows the same decomposition, $Y = Y^A{}_B J_A{}^B = A^a{}_b J_a{}^b + \theta^a J_a$ and the gauge transformations in Eq. (5) are decomposed, at infinitesimal level, as

$$\begin{aligned}
A^a{}_b &\longmapsto A^a{}_b + \mathrm{D}\alpha^a{}_b - \frac{\epsilon\kappa}{4}\left(\theta^a\xi_b - \theta_b\xi^a\right) , \\
\theta^a &\longmapsto \theta^a + \mathrm{D}\xi^a + \kappa\alpha^a{}_b\theta^b . \qquad (12)
\end{aligned}$$

where the full gauge parameter splits as $\zeta = \alpha^a{}_b J_a{}^b + \xi^a J_a$ and $\mathrm{D} = \mathrm{d} + \kappa A$ is the covariant derivative with respect to the sector $SO(m!-1,n)$. The field strength also decomposes, $F = \left(\Omega^a{}_b - \frac{\epsilon\kappa}{4}\theta^a\theta_b\right) J_a{}^b + K^a J_a$, where $\Omega^a{}_b = \mathrm{d}A^a{}_b + \kappa A^a{}_c A^c{}_b$ and $K^a = \mathrm{D}\theta^a = \mathrm{d}\theta^a - \kappa A^a{}_b\theta^b$. Under this decomposition, the Yang-Mills action (4) is written as

$$S = \frac{1}{2}\int\left[\Omega^a{}_b{*}\Omega_a{}^b + \frac{1}{2}K^a{*}K_a - \frac{\epsilon}{2}\Omega^a{}_b{*}(\theta_a\theta^b) + \frac{1}{16}\theta^a\theta_b{*}(\theta_a\theta^b)\right] . \qquad (13)$$

Once the energy starts to decrease the set of mass parameters, together with BRST soft breaking, dynamically arises. At this point, it is convenient to perform the following redefinitions[6]

$$\begin{aligned}
A &\longmapsto \kappa^{-1}A , \\
\theta &\longmapsto \kappa^{-1}m\theta , \qquad (14)
\end{aligned}$$

[6] The transformations (14) are not accidental. Both sectors are rescaled with κ^{-1} in order to factor out the coupling parameter outside the action, a standard procedure in Yang-Mills theories [2]. On the other hand, the mass parameter affects only the θ-sector, transforming it in a field with dimensionless components. It turns out that this is the unique possibility if one wishes to identify θ with a vierbein field. If also A is rescaled with a mass factor, then it would never be possible to identify it with the spin connection.

where m is a mass scale depending on the mass parameters of the theory. The action (13) is then rescaled to

$$S = \frac{1}{2\kappa^2}\int\left[\overline{\Omega}^a{}_b * \overline{\Omega}_a{}^b + \frac{m^2}{2}\overline{K}^a * \overline{K}_a - \frac{\epsilon m^2}{2}\overline{\Omega}^a{}_b*(\theta_a\theta^b) + \frac{m^4}{16}\theta^a\theta_b*(\theta_a\theta^b)\right] , \tag{15}$$

where $\overline{\Omega}^a{}_b = \mathrm{d}A^a{}_b + A^a{}_c A^c{}_b$, $\overline{K}^a = \overline{\mathrm{D}}\theta^a$, and the covariant derivative is now $\overline{\mathrm{D}} = \mathrm{d} + A$. Moreover, a reparameterization of the $SO(m,n)$ generators is required due to the existence of a mass scale, *i.e.*, a stereographic projection is now allowed if one identifies the mass parameter with the radius of the gauge manifold $\mathbb{R}_S^{m,n}$, see Appendix A. The consequence for de Sitter algebra (10) is

$$\begin{aligned}
\left[J^{ab}, J^{cd}\right] &= -\frac{1}{2}\left[\left(\eta^{ac}J^{bd} + \eta^{bd}J^{ac}\right) - \left(\eta^{ac}J^{bc} + \eta^{bc}J^{ad}\right)\right] ,\\
\left[J^a, J^b\right] &= -\frac{\epsilon m^2}{2\kappa^2}J^{ab} ,\\
\left[J^{ab}, J^c\right] &= \frac{1}{2}\left(\eta^{ac}J^b - \eta^{bc}J^a\right) .
\end{aligned} \tag{16}$$

4.2. Contraction and symmetry breaking

The main trick, in order to obtain gravity, is to consider that the rate m^2/κ^2 tends to vanish at low energy scales[7]. Due to asymptotic freedom, this is a very consistent hypothesis. Thus, we can perform an Inönü-Wigner contraction [32] through $m^2/\kappa^2 \rightarrow 0$, at the algebra (16), enforcing the de Sitter group to be contracted down to the Poincaré group. The second of relations (16) deforms to $\left[P^a, P^b\right] = 0$, where $J^a \longmapsto -\kappa\gamma^{-1}P^a$ and $\theta \longmapsto -\theta^a P_a$ (See the limit of (A.7) in Appendix A). The gauge symmetry is then dynamically deformed to the Poincaré group, $SO(m,n) \longrightarrow ISO(m! - 1, n)$, for some values κ in the strong coupling regime. The details of this deformation can be understood as a stereographic projection, see Appendix A.

Typically, an Inönü-Wigner contraction is a deformation of the group to another group which is not a subgroup of the former. Thus, since the Poincaré group is not a symmetry of the action (15), a dynamical symmetry breaking takes place. In fact, the Lorentz-type group $SO(m! - 1, n)$ is a common subgroup, $ISO(m! - 1, n) \supset SO(m! - 1, n) \subset SO(m,n)$, and also a stability subgroup for both groups. Thus, the theory suffers a symmetry breaking $SO(m,n) \longrightarrow SO(m! - 1, n)$. Under the $SO(m! - 1, n)$ gauge symmetry, the transformations (12) reduce to

$$\begin{aligned}
A^a{}_b &\longmapsto A^a{}_b + \overline{\mathrm{D}}\alpha^a{}_b ,\\
\theta^a &\longmapsto \theta^a - \alpha^a{}_b\theta^b ,
\end{aligned} \tag{17}$$

where (14) was assumed. Thus, at action (15), the field A is a gauge field with respect to the Lorentz group while θ has migrated to the matter sector (it suffers only group rotations under infinitesimal Lorentz transformations, *i.e.*, it is a vector representation). Now, the theory is ready to be identified with gravity.

4.3. Effective gravity

The broken theory described in Sect. 4.2 can generate an effective geometry if [29, 63, 68]: (i) every configuration (A,θ) defines an effective geometry (ω, e); (ii) there exists a mapping from each point $x \in \mathbb{R}^4$ to a point $X \in \mathbb{M}^4$ of the deformed space. In order to preserve the algebraic

[7] We remark that, due to the hierarchy of fundamental interactions, for gravity, a low energy scale would be very high when compared to all other basic interactions. The rate m^2/κ^2 may, become small again at lower scales.

Journal of Physics: Conference Series **453** (2013) 012014 doi:10.1088/1742-6596/453/1/012014

structure already defined in $\mathbb{R}^4$, it is demanded that this mapping is an isomorphism; (iii) The local gauge group $SO(m! - 1, n)$ defines, at each point of the mapping, the isometries of the tangent space $T_X(\mathbb{M})$. Thus, θ and A can be identified with the vierbein e and spin connection ω, respectively, through

$$\begin{aligned}\omega_\mu^{\mathfrak{ab}}(X)dX^\mu &= \delta_a^{\mathfrak{a}}\delta_b^{\mathfrak{b}}A_\mu^{ab}(x)dx^\mu\;,\\ e_\mu^{\mathfrak{a}}(X)dX^\mu &= \delta_a^{\mathfrak{a}}\theta_\mu^a(x)dx^\mu\;.\end{aligned}\tag{18}$$

In expressions (18), latin indices $\{\mathfrak{a},\mathfrak{b}\ldots\}$ belong to the tangent space $T_X(\mathbb{M})$. Moreover, it is always possible to impose that the space of all p-forms in $\mathbb{R}^4$ is mapped into the space of p-forms in $\mathbb{M}^4$, namely, $\Pi^p \longmapsto \widetilde{\Pi}^p$, and the same for the Hodge spaces, $*\Pi^p \longmapsto \star\widetilde{\Pi}^p$, where $\star$ is the Hodge dual in $\mathbb{M}^4$. This mapping, together with the identifications in Eq. (18), provides

$$S = \frac{1}{8\pi G}\int\left[\frac{1}{2\Lambda^2}R^{\mathfrak{a}}{}_{\mathfrak{b}}\star R_{\mathfrak{a}}{}^{\mathfrak{b}} + T^{\mathfrak{a}}\star T_{\mathfrak{a}} - \frac{\epsilon}{2}\epsilon_{\mathfrak{abcd}}R^{\mathfrak{ab}}e^{\mathfrak{c}}e^{\mathfrak{d}} + \frac{\Lambda^2}{4}\epsilon_{\mathfrak{abcd}}e^{\mathfrak{a}}e^{\mathfrak{b}}e^{\mathfrak{c}}e^{\mathfrak{d}}\right]\;,\tag{19}$$

where $R^{\mathfrak{a}}{}_{\mathfrak{b}} = \mathrm{d}\omega^{\mathfrak{a}}{}_{\mathfrak{b}} + \omega^{\mathfrak{a}}{}_{\mathfrak{c}}\omega^{\mathfrak{c}}{}_{\mathfrak{b}}$ and $T^{\mathfrak{a}} = \mathrm{d}e^{\mathfrak{a}} - \omega^{\mathfrak{a}}{}_{\mathfrak{b}}e^{\mathfrak{b}}$ are the curvature and torsion, respectively. Newton and cosmological constants are determined by $m^2 = \kappa^2/2\pi G$ and $\Lambda^2 = m^2/4$.

Action (19) is a gravity action in the first order formalism presenting (in order of appearance) a quadratic Yang-Mills-type curvature term, a quadratic torsion term, the Einstein-Hilbert term, and the cosmological constant term. Moreover, it is easy to check that the vacuum solution is a de Sitter-type spacetime with $T^{\mathfrak{a}} = 0$ and $R^{\mathfrak{ab}} = 2\epsilon\Lambda^2 e^{\mathfrak{a}}e^{\mathfrak{b}}$.

It is remarkable that, in the present theory, Newton and cosmological constants are related through $\Lambda^2 = \kappa^2/8\pi G$. Obviously, from asymptotic freedom, κ^2 is a big quantity. And, by assumption, G must be small. Thus, Λ should be very big. In fact, if this is true, we can make two important remarks: (i) The first term in (19) can be safely neglected. Moreover, due to the absence of matter fields, torsion can be taken as very small. The resulting theory is then, the usual Einstein-Hilbert theory with cosmological constant; (ii) although astrophysical predictions [69, 70] determine that Λ^2_{obs} is very small, quantum field theory predicts [71, 72, 73, 74] a very large Λ^2_{qft}. Thus, the contribution of a pure gravitational cosmological constant, which is big in our case, can drive the cosmological puzzle to a final consistent answer. In fact, following [29, 75, 76], the renormalized cosmological constant of our model could be determined through $\Lambda^2_{ren} = \Lambda^2_{obs} - \Lambda^2_{qft}$.

5. Some consistency checks

5.1. *The explicit mapping*

The formal aspects of the mapping between a gauge theory in Euclidean spacetime and a gravity theory can be discussed in terms of fiber bundle theory. The details can be found in [29, 63, 68]. In Sect. 3, it was demanded that the mapping between the original space $\mathbb{R}^4$ and the effective space $\mathbb{M}^4$ is an isomorphism. In fact, it can be also shown that this mapping is unique and has an inverse, which ensures the absence of ambiguities and thus, its isomorphic character. The details of the proof can also be found in Appendix B.

Let us summarize this result. We suppose that, in the d-dimensional original manifold, the metric tensor is $g_{\mu\nu}(x)$, where x are the set of coordinates at a point in this space. The effective metric tensor is defined as $\widetilde{g}_{\mu\nu}(X)$, where X is the set of coordinates at a point in the target space. Thus, the matrix that defines the map $x \longmapsto X$ for all points in the effective manifold is given by

$$L^\nu{}_\mu = \left(\frac{\tilde{g}}{g}\right)^{1/2d}\tilde{g}^{\nu\alpha}g_{\alpha\mu}\;.\tag{20}$$

Its inverse is given by

$$\left(L^{-1}\right)^{\nu}_{\ \mu} = \left(\frac{g}{\tilde{g}}\right)^{1/2d} g^{\nu\alpha}\tilde{g}_{\alpha\mu}\ . \tag{21}$$

In (20) and (21), $g = \det g_{\mu\nu}$ and $\tilde{g} = \det \tilde{g}_{\mu\nu}$ are taken as non-vanishing quantities. Thus, ambiguities are absent in this mapping.

5.2. Weinberg-Witten theorems and emergent gravity

In [77], Weinberg and Witten established two very powerful theorems. Essentially, they forbid: (i) massless charged states with helicity $j > 1/2$ which have a conserved Lorentz-covariant current and (ii) massless states with helicity $j > 1$ which have conserved Lorentz-covariant energy-momentum tensor.

At first sight, the first theorem would forbid gauge theories to exist since, at high energies, vector bosons and gluons are charged massless states which carry helicity $j = 1$. However, the conserved gauge current associated with these states are not Lorentz covariant [77]. Similarly, the second theorem forbids spin-2 states with conserved Lorentz-covariant energy-momentum tensor to be defined. In traditional GR the metric tensor $g_{\mu\nu}$ is not a Lorentz covariant tensor since it is constructed over a Riemannian manifold. The linearized version of GR, on the other hand, is also a gauge theory and the same principles of the first case apply [77]. Moreover, the energy-momentum tensor of the graviton field identically vanishes.

The Weinberg-Witten theorem (ii) applies, however, to emergent gravity theories for which massless spin-2 states are generated as effective or composite states that could be associated with graviton excitations. Nevertheless, one may argue that the Weinberg-Witten theorem (ii) applies to the present mechanism. However, this is not the case here. First, the theory has a few mass parameters and the theorem holds for massless states only. Second, and more important, there are no spin-2 states in this model. The fields are identified with geometry and not with spin-2 composite fields. Gravity emerges as geometrodynamics and not as a field theory for spin-2 particles in flat space. In fact, the mapping discussed in Sects. 4.3 and 5.1 has been employed at classical level. Nevertheless, renormalizability establishes that the quantum action has the same form of its classical version. The difference between the classical and quantum actions lies on the fields and parameters, which are their respective renormalized versions. Hence, although each configuration (A, θ) can define a geometry (ω, e), the mapping should not be applied to each configuration in the path integral, but at the quantum action itself. In this way, the resulting geometrodynamical theory is obtained from the full dynamical content of the original gauge theory. Moreover, the identification of the renormalized fields are made with respect to geometric quantities and not with spin-2 states. The conclusion is that there is no violation of the Weinberg-Witten theorem.

Nevertheless, after the emergence of gravity as geometrodynamics, linearization is allowed for weak gravitation and the spin-2 states might also be considered. In that case, these states are classical states associated to propagating fields. Quantization of these states can be done since it is not the fundamental theory, but an effective theory. Clearly, this does not violate the theorems because it fits in the same category of GR, [77].

5.3. Unitarity and equivalence principle

One of the most important features of $SU(N)$ gauge theories is unitarity [2], a property that, among other features, follows from the compact character of the gauge group. In the case of the present theory, unitarity is only ensured for $m = 0$. In that case, the resulting gravity theory is an $SO(4)$ local isometric gravity. The local Euclidean character of spacetime provides a kind of incomplete prediction of the equivalence principle because it lacks the split between space and time. On the other hand, if we set[8] $m = 2$ from the beginning, the resulting theory has $SO(1,3)$ local isometries, *i.e.*, localy, spacetime is a Minkowski space. Thus, there is a split between space and time originated from the breaking that lead to the mapping. This is nothing else than the rising of the equivalence principle.

Thus, if on the one hand, we start with a unitary gauge theory, the resulting gravity is not exactly the desired one because space and time are still indistinguishable. On the other hand, by giving up unitarity, the split between space and time correctly emerges. To solve the paradox is just a mathematical problem that can be fixed by imposing a Wick rotation during the mapping between $\mathbb{R}^4$ and $\mathbb{M}^4$. See [29].

Another argument can be provided by simply saying that unitarity is irrelevant because quantum gravity is far beyond Planck scale. Above Planck scale gravity is already a classical phenomenon. Moreover, due to confinement of Higgless non-Abelian gauge theories (such as QCD and the present gravity model), unitarity might be an overestimated property, mainly because the high energy state is a plasma with no observable singlet whatsoever (in the case of QCD, this state is the quark-gluon plasma).

6. Final considerations

We have discussed that is possible to make a deep analogy between quantum chromodynamics and gravity. This analogy is realized if gravity could be described, at quantum level, by a gauge theory in a Euclidean four-dimensional spacetime while, at classical level, it is deformed to a geometrodynamical theory [29, 63]. We have established the conditions for this mechanism to be realized and an example based on de Sitter-type groups was provided. In this mechanism, gravity arises as an emergent geometric theory. The requirements are asymptotic freedom, dynamical mass generation, and BRST soft breaking due to Gribov ambiguities.

The starting action (4), for the $SO(m,n)$ gauge groups, lives in a four-dimensional Euclidean space, and thus, at high energies, it is a well defined quantum theory of spin-1 asymptotic free states. As the energy decreases, mass generation takes place, as well as the Gribov parameter and soft BRST breaking. The consequence is that the propagators acquire complex poles and are ruled out from the physical spectrum of the theory. At this point, an Inönü-Wigner contraction is assumed and a symmetry breaking $SO(m,n) \to SO(m! - 1, n)$ drives the theory to a gravity theory described by the action (19). This action defines a first order gravity which has Einstein-Hilbert and cosmological constant terms. To obtain this action, a mapping has to be imposed. The consistency of this mapping was shown and an improvement has been made in order to harmonize the concepts of unitarity and the emergence of the equivalence principle.

A remarkable feature of this mechanism is that Newton and cosmological constant could be explicitly computed from the usual QFT techniques. Moreover, there is a constraint between these constants, namely, $\Lambda^2 \propto \kappa^2/G$. Since G is supposedly small and κ is a big quantity at low energies, it is then expected that Λ is very big. This property can be combined with the predictions of QFT for the cosmological constant in order to provide a value that agrees with the observed values. Also, from the fact that Λ is big and the absence of matter in this work, it is possible to conclude that the action (19) can be safely approximated to the Einstein-Hilbert action with cosmological constant.

Let us make a comparison between the present gravity theory and the standard model. Strong and electroweak interactions are described by gauge theories. At high energies, these theories are very similar (except for the gauge groups). At low energies, however, these theories tend to behave in very different ways. Electroweak interactions suffer spontaneous symmetry breaking through the Higgs mechanism, giving rise to perturbative electrodynamics and massive

[8] The case $m = 1$ provides a non-unitary theory and also a local Euclidean spacetime with no difference between space and time.

gauge bosons. On the other hand, quark-gluon confinement shows up in chromodynamics, and hadronization phenomena take place. Specifically, confinement and the gauge principle state that physical observables must be gauge invariant and colorless. These states are recognized as hadrons and glueballs. Now, if the present theory can describe gravity, then: (i) at high energies, gravity is a gauge theory which is very similar to the other fundamental interactions; (ii) at low energies, instead of hadrons and glueballs, the physical observables are identified with geometry, and spacetime itself is affected by this theory. Thus, geometry appears as the low energy manifestation of gravity, in the same way that hadronization and spontaneous symmetry breaking are the low energy manifestations of chromodynamics and electroweak interactions.

Finally, for the moment, we can only say that a standard four-dimensional renormalizable Yang-Mills theory can generate a gravity theory at low energy regime. Obviously, many computations and tests must be performed before we recognize this theory (or some variation) as *the* quantum gravity theory or only an academic exercise.

Acknowledgements
R. F. S. is grateful to the organizing committee of *NEB 15 - Recent Developments in Gravity* for the opportunity to deliver this talk. Conselho Nacional de Desenvolvimento Científico e Tecnológico[9] (CNPq-Brazil) and the Coordenação de Aperfeiçoamento de Pessoal de Nível Superior (CAPES) are acknowledged for financial support.

Appendix A. Stereographic projection

The group $SO(m,n)$ defines a flat space, $\mathbb{R}_S^{m,n}$, with metric given by $\eta^{AB} \equiv \mathrm{diag}(\epsilon, \varepsilon, 1, 1, 1)$. In the presence of a mass scale m, we can define the radius of this space by

$$\mathcal{R} = \frac{2\epsilon}{(\epsilon\varepsilon)^{\frac{1}{2}}}\left(\frac{\kappa}{m}\right) , \tag{A.1}$$

in such a way that

$$\eta_{AB}\xi^A\xi^B = \varepsilon\mathcal{R}^2 = \eta_{ab}\xi^a\xi^b + \epsilon(\xi^5)^2 , \tag{A.2}$$

where ξ^A are Cartesian coordinates in $\mathbb{R}_S^{m,n}$. Thus, the stereographic projection is defined by

$$\begin{aligned} \xi^a &= n\overline{x}^a , \\ \xi^5 &= (\epsilon\varepsilon)^{\frac{1}{2}}\mathcal{R}(1-2n) , \end{aligned} \tag{A.3}$$

where $\eta_{ab} = diag\ (\varepsilon, 1, 1, 1)$. In order to have a projection conformally flat, *i.e.*, $g_{ab} = n^2\eta_{ab}$, it is demanded that

$$n = \frac{1}{\left(1 + \frac{\varepsilon\sigma^2}{4\mathcal{R}^2}\right)} , \tag{A.4}$$

where $\sigma^2 = \eta_{ab}\overline{x}^a\overline{x}^b$.

The effect of the projection on the group generators can also be studied. The group generators can be written as

$$J^A{}_B = \frac{1}{2}\left(\eta^{AC}\xi_C\partial_B - \eta^{BC}\xi_C\partial_A\right) . \tag{A.5}$$

Thus, it is a straightforward computation to show that

$$\begin{aligned} J^a_{\ b} &= \frac{1}{2}\left(\overline{x}^a P^b - \overline{x}^b P^a\right) , \\ J^a &= -\frac{\kappa}{m}P^a + \frac{\epsilon}{16}\frac{m}{\kappa}\left(2\overline{x}_a\overline{x}^b P_b + \sigma^2 P^a\right) , \end{aligned} \tag{A.6}$$

[9] RFS is a level PQ-2 researcher under the program *Produtividade em Pesquisa*, 304924/2009-1.

where (A.3) and (A.1) were used. Thus, from the second of relations (A.6), we have

$$\theta \longmapsto \kappa^{-1} m\theta = -\theta^a P_a + \frac{\epsilon}{16}\frac{m^2}{\kappa^2}\theta^a \left(2\overline{x}_a\overline{x}_b P^b + \sigma^2 P_a\right) . \tag{A.7}$$

Appendix B. Explicit derivation of the mapping

Let us only show how the explicit mapping between the spaces $\mathbb{R}^4$ and $\mathbb{M}^4$ can be obtained. We have considered that

$$\begin{aligned} \Pi^p &\longmapsto \tilde{\Pi}^p , \\ *\Pi^p &\longmapsto \star\tilde{\Pi}^p . \end{aligned} \tag{B.1}$$

For generality purposes, we assume a generic original metric $g_{\mu\nu}$ which, eventually, we can set as a Euclidean metric. The effective metric is denoted by $\tilde{g}_{\mu\nu}$. Moreover, we can also consider manifolds with an arbitrary dimension d. Obviously, a necessary extra condition is that both $g = |\det g_{\mu\nu}|$ and $\tilde{g} = |\det \tilde{g}_{\mu\nu}|$ are non-vanishing quantities. To find the explicit mapping, we apply the first of (B.1) to a generic p-form,

$$f_{\mu_1\ldots\mu_p}(x)dx^{\mu_1}\ldots dx^{\mu_p} = \tilde{f}_{\mu_1\ldots\mu_p}(X)dX^{\mu_1}\ldots dX^{\mu_p} , \tag{B.2}$$

where $x \in \mathbb{R}^d$ and $X \in \mathbb{M}^d$. From (B.2), one easily obtain

$$f_{\mu_1\ldots\mu_p}(x) = L^{\nu_1}{}_{\mu_1}\ldots L^{\nu_p}{}_{\mu_p}\tilde{f}_{\nu_1\ldots\nu_p}(X) , \tag{B.3}$$

where $L^\nu{}_\mu = \frac{\partial X^\nu}{\partial x^\mu}$. For the corresponding Hodge dual we have,

$$\begin{aligned} &\sqrt{g}\epsilon_{\mu_1\ldots\mu_p\nu_{p+1}\ldots\nu_d}f^{\mu_1\ldots\mu_p}(x)dx^{\nu_{p+1}}\ldots dx^{\nu_d} = \\ = \ &\sqrt{\tilde{g}}\epsilon_{\mu_1\ldots\mu_p\nu_{p+1}\ldots\nu_d}\tilde{f}^{\mu_1\ldots\mu_p}(X)dX^{\nu_{p+1}}\ldots dX^{\nu_d} , \end{aligned} \tag{B.4}$$

from which one can find that

$$f^{\mu_1\ldots\mu_p} = \left(\frac{\tilde{g}}{g}\right)^{1/2}\left(\frac{L}{d}\right)^{d-p}\tilde{f}^{\mu_1\ldots\mu_p} , \tag{B.5}$$

with $L = L^\mu{}_\mu$. A comparison of (B.3) and (B.5) leads to

$$f_{\mu_1\ldots\mu_p} = \left(\frac{\tilde{g}}{g}\right)^{1/2}\left(\frac{L}{d}\right)^{d-p}\tilde{g}^{\nu_1\alpha_1}g_{\alpha_1\mu_1}\ldots\tilde{g}^{\nu_p\alpha_p}g_{\alpha_p\mu_p}\tilde{f}_{\nu_1\ldots\nu_p} . \tag{B.6}$$

Combining (B.6) and (B.3), we achieve

$$L^\nu{}_\mu = \left(\frac{\tilde{g}}{g}\right)^{1/2p}\left(\frac{L}{d}\right)^{(d-p)/p}\tilde{g}^{\nu\alpha}g_{\alpha\mu} , \tag{B.7}$$

which is not valid for $p = 0$. In that case it is easy to find that (B.1) is valid only if

$$\left(\frac{\tilde{g}}{g}\right)^{1/2}\left(\frac{L}{d}\right)^d = 1 . \tag{B.8}$$

The constraint (B.8) implies that

$$L^\nu{}_\mu = \frac{d}{L}\,\tilde{g}^{\nu\alpha}g_{\alpha\mu} . \tag{B.9}$$

The trace L can now be calculated from (B.8) or (B.9) providing

$$\begin{aligned} L &= d\left(\frac{g}{\tilde{g}}\right)^{1/2d}, \\ L &= d^{1/2}(\tilde{g}^{\mu\nu}g_{\mu\nu})^{1/2}, \end{aligned} \tag{B.10}$$

respectively. The relations (B.10) enforce the extra constraint

$$(\tilde{g}^{\mu\nu}g_{\mu\nu})^{1/2} = d^{1/2}\left(\frac{g}{\tilde{g}}\right)^{1/2d}. \tag{B.11}$$

As a consequence, we obtain the final expression for the transformation matrix, which is given by Eq. (20). We recall that the effective metric is computed from the gravity field equations, while the original metric is a given quantity[10]. It turns out that the mapping (20) has an inverse, given by Eq. (21). The existence of the inverse ensures the non-degeneracy of the mapping.

References

[1] Einstein A 1916 The foundation of the general theory of relativity *Annalen Phys.* **49** 769
[2] Itzykson C and Zuber J B 1980 *Quantum Field Theory* (New York: Mcgraw-Hill)
[3] Hojman S A, Kuchar K and Teitelboim C 1976 Geometrodynamics Regained *Annals Phys.* **96** 88
[4] Streater R F and Wightman A F 2000 *PCT, spin and statistics, and all that* (Princeton: Princeton University Press)
[5] Fierz M and Pauli W 1939 On relativistic wave equations for particles of arbitrary spin in an electromagnetic field *Proc. Roy. Soc. Lond.* A **173** 211
[6] 't Hooft G and Veltman M J G 1974 One loop divergencies in the theory of gravitation *Annales Poincaré Phys. Theor.* **A20** 69-94
[7] Deser S and van Nieuwenhuizen P 1974 One loop divergences of quantized Einstein-Maxwell fields *Phys. Rev.* D **10** 401
[8] Deser S and van Nieuwenhuizen P 1974 Nonrenormalizability of the quantized Dirac-Einstein system *Phys. Rev.* D **10** 411
[9] Stelle K S 1977 Renormalization of higher derivative quantum gravity *Phys. Rev.* D **16** 953
[10] Utiyama R 1956 Invariant theoretical interpretation of interaction *Phys. Rev.* **101** 1597
[11] Kibble T W B 1961 Lorentz invariance and the gravitational field *J. Math. Phys.* **2** 212
[12] Sciama D W 1964 The physical structure of general relativity *Rev. Mod. Phys.* **36** 463-469
[13] Mardones A and Zanelli J 1991 Lovelock-Cartan theory of gravity *Class. Quant. Grav.* **8** 1545
[14] Zanelli J 2005 Lecture notes on Chern-Simons (super-)gravities *Preprint* hep-th/0502193
[15] Sobreiro R F and Vasquez Otoya V J 2011 On the topological reduction from the affine to the orthogonal gauge theory of gravity *J. Geom. Phys.* **61** 137-150
[16] Sobreiro R F and Vasquez Otoya V J 2011 Affine gauge theory of gravity and its reduction to the Riemann-Cartan geometry *J. Phys.: Conf. Ser.* **283** 012032
[17] Stelle K S and West P C 1979 de Sitter gauge invariance and the geometry of the Einstein-Cartan theory *J. Phys.* A **12** L205-L210
[18] Stelle K S and West P C 1980 Spontaneously broken de Sitter symmetry and the gravitational holonomy group *Phys. Rev.* D **21** 1466
[19] Pagels H R 1984 Gravitational gauge fields and the cosmological constant *Phys. Rev.* D **29** 1690
[20] Tresguerres R 2008 Dynamically broken Anti-de Sitter action for gravity *Int. J. Geom. Meth. Mod. Phys.* **5** 171-183
[21] Mielke E W 2010 Einsteinian gravity from a spontaneously broken topological BF theory *Phys. Lett.* B **688** 273-277
[22] Higgs P W 1964 Broken symmetries and the masses of gauge bosons *Phys. Rev. Lett.* **13** 508
[23] Higgs P W 1964 Broken symmetries, massless particles and gauge fields *Phys. Lett.* **12** 132
[24] MacDowell S W and Mansouri F 1977 Unified geometric theory of gravity and supergravity *Phys. Rev. Lett.* **38** 739
[25] Yang C N and Mills R L 1954 Conservation of isotopic spin and isotopic gauge invariance *Phys. Rev.* **96** 191

[10] For the present case we actually have ($d = 4$ and $g_{\mu\nu} = \delta_{\mu\nu}$) $L^{\nu}{}_{\mu} = (\tilde{g})^{1/8}\,\tilde{g}^{\nu\alpha}\delta_{\alpha\mu}$.

[26] Tseytlin A A 1982 On the Poincaré and de Sitter gauge theories of gravity with propagating torsion *Phys. Rev.* D **26** 3327
[27] Mahato P 2004 de Sitter group and Einstein-Hilbert Lagrangian *Phys. Rev.* D **70** 124024 (*Preprint* gr-qc/0603100)
[28] Sobreiro R F and Vasquez Otoya V J 2007 Effective gravity from a quantum gauge theory in Euclidean space-time *Class. Quant. Grav.* **24** 4937
[29] Sobreiro R F, Tomaz A A and Vasquez Otoya V J 2012 de Sitter gauge theories and induced gravities *Eur. Phys. J.* C **72** 1991 (*Preprint* hep-th/1109.0016)
[30] Gross D J and Wilczek F 1973 Ultraviolet behavior of non-Abelian gauge theories *Phys. Rev. Lett.* **30** 1343
[31] Politzer H D 1973 Reliable perturbative results for strong interactions? *Phys. Rev. Lett.* **30** 1346
[32] Inönü E and Wigner E P 1953 On the contraction of groups and their representations *Proc. Nat. Acad. Sci.* **39** 510-524
[33] Gribov V N 1978 Quantization of nonabelian gauge theories *Nucl. Phys.* B **139** 1
[34] Zwanziger D 1993 Renormalizability of the critical limit of lattice gauge theory by BRS invariance *Nucl. Phys.* B **399** 477-513
[35] Sobreiro R F and Sorella S P 2005 Introduction to the Gribov ambiguities in Euclidean Yang-Mills theories *Preprint* hep-th/0504095
[36] Baulieu L and Sorella S P 2009 Soft breaking of BRST invariance for introducing non-perturbative infrared effects in a local and renormalizable way *Phys. Lett.* B **671** 481 (*Preprint* hep-th/0808.1356)
[37] Baulieu L, Capri M A L, Gomez A J, Lemes V E R, Sobreiro R F and Sorella S P 2010 Renormalizability of a quark-gluon model with soft BRST breaking in the infrared region *Eur. Phys. J.* C **66** 451 (*Preprint* hep-th/0901.3158)
[38] Dudal D and Sorella S P 2012 The Gribov horizon and spontaneous BRST symmetry breaking *Phys. Rev.* D **86** 045005 (*Preprint* hep-th/1205.3934)
[39] Dudal D, Sobreiro R F, Sorella S P and Verschelde H 2005 The Gribov parameter and the dimension two gluon condensate in Euclidean Yang-Mills theories in the Landau gauge *Phys. Rev.* D **72** 014016
[40] Dudal D, Sorella S P and Vandersickel N 2011 The dynamical origin of the refinement of the Gribov-Zwanziger theory *Phys. Rev.* D **84** 065039
[41] 't Hooft G and Veltman M J G 1972 Regularization and renormalization of gauge fields *Nucl. Phys.* B **44** 189
[42] Piguet O and Sorella S P 1995 Algebraic renormalization: Perturbative renormalization, symmetries and anomalies *Springer Lect. Notes Phys.* M **28**
[43] Faddeev L D and Popov V N 1967 Feynman diagrams for the Yang-Mills field *Phys. Lett.* B **25** 29
[44] Tyutin I V 1975 Gauge invariance in field theory and statistical physics in operator formalism *Lebedev Physics Institute: Preprint* 39 (*Preprint* hep-th/0812.0580)
[45] Becchi C, Rouet A and Stora R 1976 Renormalization of gauge theories *Annals Phys.* **98** 287
[46] Tomonaga S 1946 On a relativistically invariant formulation of the quantum theory of wave fields *Prog. Theor. Phys.* **1** 27
[47] Schwinger J S 1948 Quantum electrodynamics I A covariant formulation *Phys. Rev.* **74** 1439
[48] Feynman R P 1950 Mathematical formulation of the quantum theory of electromagnetic interaction *Phys. Rev.* **80** 440
[49] Kugo T and Ojima I 1979 Local covariant operator formalism of nonabelian gauge theories and quark confinement problem *Prog. Theor. Phys. Suppl.* **66** 1
[50] Singer I M 1978 Some Remarks on the Gribov Ambiguity *Commun. Math. Phys.* **60** 7
[51] ed Binosi D, Aguilar A C, *et al* 2011 QCD Green's functions, confinement and phenomenology *Proc. Int. Workshop QCD-TNT-II* (Italy :PoS QCD-TNT-II)
[52] Greensite J 2003 The confinement problem in lattice gauge theory *Prog. Part. Nucl. Phys.* **51** 1 (*Preprint* hep-lat/0301023)
[53] Cucchieri A, Dudal D, Mendes T and Vandersickel N 2012 Modeling the gluon propagator in Landau gauge: Lattice estimates of pole masses and dimension-two condensates *Phys. Rev.* D **85** 094513 (*Preprint* hep-lat/1111.2327)
[54] Maldacena J M 1999 The Large N limit of superconformal field theories and supergravity *AIP Conf. Proc.* **484** 51 *Int. J. Theor. Phys.* **38** 1113
[55] Witten E 1988 Anti-de Sitter space and holography *Adv. Theor. Math. Phys.* **2** 253 (*Preprint* hep-th/9802150)
[56] Miranda A S, Ballon Bayona C A, Boschi-Filho H and Braga N R F 2010 Glueballs at finite temperature from AdS/QCD *Nucl. Phys. Proc. Suppl.* **199** 107 (*Preprint* hep-th/0910.4319)
[57] Nambu Y 1974 Strings, monopoles and gauge fields *Phys. Rev.* D **10** 4262
[58] 't Hooft G 1975 *High Energy Physics EPS Int. Conf.* ed. A. Zichichi
[59] Mandelstam S 1976 Vortices and quark confinement in nonabelian gauge theories *Phys. Rept.* **23** 245

[60] Faddeev L D and Niemi A J 2002 Aspects of electric magnetic duality in SU(2) Yang-Mills theory *Phys. Lett.* B **525** 195 (*Preprint* hep-th/0101078)
[61] Dudal D, Guimaraes M S and Sorella S P 2011 Glueball masses from an infrared moment problem and nonperturbative Landau gauge *Phys. Rev. Lett.* **106** 062003 (*Preprint* hep-th/1010.3638)
[62] Capri M A L, Dudal D, Guimaraes M S, Palhares L F and Sorella S P 2012 Physical spectrum from confined excitations in a Yang-Mills-inspired toy model *Preprint* hep-th/1208.5676
[63] Sobreiro R F 2012 Fiber bundles, gauge theories and gravity *Quantum Gravity* ed Sobreiro R F (Croatia: InTech)
[64] Daniel M and Viallet C M 1980 The geometrical setting of gauge theories of the Yang-Mills type *Rev. Mod. Phys.* **52** 175
[65] Nakahara M 1990 *Geometry, topology and physics* (Bristol, UK: Hilger)
[66] Witten E 1988 (2+1)-dimensional gravity as an exactly soluble system *Nucl. Phys.* B **311** 46
[67] Assimos T S, Pereira Jr A D, Santos T R S, Sobreiro R F and Tomaz A A To appear 2013
[68] Assimos T S, Pereira Jr A D, Santos T R S, Sobreiro R F, Tomaz A A and Vasquez Otoya VJ To appear 2013
[69] Perivolaropoulos L 2008 Vacuum energy, the cosmological constant, and compact extra dimensions: constraints from Casimir effect experiments *Phys. Rev.* D **77** 107301 (*Preprint* astro-ph/0802.1531)
[70] Padmanabhan T 2012 The physical principle that determines the value of the cosmological constant *Preprint* hep-th/1210.4174
[71] Nelson B L and Panangaden P 1982 Scaling behavior of interacting quantum fields in curved space-time *Phys. Rev.* D **25** 1019
[72] Toms D J 1983 The effective action and the renormalization group equation in curved space-time *Phys. Lett.* B **126** 37
[73] Buchbinder I L 1984 Renormalization group equations in curved space-time *Theor. Math. Phys.* **61** 1215 (*[Teor. Mat. Fiz.* **61** 393)
[74] Parker L and Toms D J 1985 Renormalization group and nonlocal terms in the curved space-time effective action: Weak field results *Phys. Rev.* D **32** 1409
[75] Shapiro I L and Sola J 2007 Cosmological constant problems and renormalization group *J. Phys.* A **40** 6583
[76] Shapiro I L and Sola J 2009 On the possible running of the cosmological "constant" *Phys. Lett.* B **682** 105-113
[77] Weinberg S and Witten E 1980 Limits on massless particles *Phys. Lett.* B **96** 59

Journal of Physics: Conference Series **453** (2013) 012015 doi:10.1088/1742-6596/453/1/012015

Cosmological constant and vacuum energy: old and new ideas

Joan Solà

HEP Group, Dept. Estructura i Constituents de la Matèria
and Institut de Ciències del Cosmos, Universitat de Barcelona,
Av. Diagonal 647, 08028 Barcelona, Catalonia, Spain

E-mail: `sola@ecm.ub..es`

Abstract. The cosmological constant (CC) term in Einstein's equations, Λ, was first associated to the idea of vacuum energy density. Notwithstanding, it is well-known that there is a huge, in fact appalling, discrepancy between the theoretical prediction and the observed value picked from the modern cosmological data. This is the famous, and extremely difficult, "CC problem". Paradoxically, the recent observation at the CERN Large Hadron Collider of a Higgs-like particle, should actually be considered ambivalent: on the one hand it appears as a likely great triumph of particle physics, but on the other hand it wide opens Pandora's box of the cosmological uproar, for it may provide (alas!) the experimental certification of the existence of the electroweak (EW) vacuum energy, and thus of the intriguing reality of the CC problem. Even if only counting on this contribution to the inventory of vacuum energies in the universe, the discrepancy with the cosmologically observed value is already of 55 orders of magnitude. This is the (hitherto) "real" magnitude of the CC problem, rather than the (too often) brandished 123 ones from the upper (but fully unexplored!) ultrahigh energy scales. Such is the baffling situation after 96 years of introducing the Λ-term by Einstein. In the following I will briefly (and hopefully pedagogically) fly over some of the old and new ideas on the CC problem. Since, however, the Higgs boson just knocked our door and recalled us that the vacuum energy may be a fully tangible concept in real phenomenology, I will exclusively address the CC problem from the original notion of vacuum energy, and its possible "running" with the expansion of the universe, rather than venturing into the numberless attempts to replace the CC by the multifarious concept of dark energy.

1. Introduction

Undoubtedly the most prominent performance of modern cosmology has been to provide observational evidence of the accelerated expansion of the universe [1, 2, 3] and of the existence of (other) large scale dynamical phenomena possibly caused by forms of matter beyond the usual baryonic component. To "explain" the accelerated evolution the name "dark energy" (DE) was coined; it refers to some mysterious form of diffuse (i.e. not clustering) energy presumably permeating all corners of the universe, possessing negative pressure and thus being capable of boosting the expansion of the universe as a whole. Similarly, to "explain" the anomalous dynamics of galaxies and of galaxy clusters, we have imagined that there is a large deficit of matter at different astronomical scales in the form of unknown stable particles, which are neither electrons nor protons, not even neutrinos, but some form of electrically neutral heavy stuff beyond the spectrum of the Standard Model (SM) of the strong and electroweak interactions, and referred to as "dark matter" (DM) particles. We do not yet know if any of these hypotheses is true at all, although new hints (not realities, yet) might be around recently; the only thing we know for sure is the reality of the observed physical phenomena that we are trying to explain. In the following I will not elaborate on the whereabouts of the hypothetical DM particles. I will rather focus on a few aspects of the DE problem, or more specifically the cosmological constant (CC) problem [4], which is perhaps the most intriguing of all cosmological puzzles – and is not obviously unrelated to the DM one.

It is often stated in the literature that the CC term, and its association with the notion of vacuum energy, cannot be a valid theoretical explanation for the accelerated expansion of the universe, and that we necessarily have to "go beyond Λ". The adduced reasons are manyfold, but perhaps the most brandished one is that the various contributions to the vacuum energy cannot possibly be successfully fine tuned to the measured value by any known mechanism, and therefore the idea of the vacuum energy and its connection with the CC is viewed as completely unnatural. Barring the fact that this need not be true, since there are dynamical mechanisms within modified gravity that could efficiently help here [5], it is nevertheless a curious "reason" to wield, as usually nothing more fundamental is offered as an alternative, except a defense (tooth and nail) of some particular form of DE, say from quintessence to string landscape. Unfortunately, as we know, none of these alternatives seem to improve in any practical way the fine-tuning illness – a very serious matter by the way, which for some (no less) mysterious reason is unjustly blamed to the CC option almost exclusively. In fact, the fine tuning problem in such new frameworks not only does not become milder but it gets even worse than in the CC case, simply because the traditional vacuum energy of the SM is still there, and so one has to cope with its fine tuning, plus the (no less severe) one associated to the field or string object (usually linked to some form of high energy physics) replacing the CC term. As a result the two fine tunings make the overall job even more bizarre. Let alone that, quite often, in these frameworks an extremely light new particle is predicted in the ballpark of $\sim 10^{-33}$ eV. However, we should seriously worry about the fact that such (incommensurably tiny!) mass scale is some 30 orders of magnitude smaller than the mass scale which these models aim to explain – namely the millielectronvolt ($\sim 10^{-3}$ eV) mass scale associated to the CC term! Why such strategy is not perceived as trying to solve a big problem by creating an even major one? The answer, quite likely, is that the CC problem is such a disproportionately big problem, that we are – too soon – ready to redefine dramatically the scope and limits of our physical perceptions.

The "instinctive" option of replacing the vacuum energy by alternative theoretical constructs can be counterproductive, though, because in doing so many people (consciously or unconsciously) may give up the duty of explaining why the usual vacuum energy of QFT (say the SM of Particle Physics) does not participate at all in accounting for the value of the CC. After the likely discovery of a Higgs boson at the LHC collider [6], the reality of the vacuum energy associated to the spontaneous symmetry breaking mechanism of the electroweak theory starts acquiring a palpable reality. If we wish to face the CC problem in earnest, we should somehow move on and stop leaving the vacuum energy of the SM in the most complete oblivion, literally as if the mere fact of not thinking or talking about it would make it completely disappear from our world! If we think about it, wouldn't this attitude be more typical of some "Ostrichland"?

In the following I shall dwell on the properties of the CC term in Einstein's equations and generalizations thereof. Our main aim here is to consider models where the prime driving force accelerating the universe is dynamical vacuum energy and hence a time variable CC. I will also discuss some intriguing phenomenological implications of the dynamical vacuum framework as a potential source for a mild variability of the fundamental "constants" of Nature, which could help in effectively testing these ideas. For a more detailed exposition, see e.g. [7]; and for a summarized introduction to time evolving vacuum models along these lines, see [8].

Journal of Physics: Conference Series **453** (2013) 012015 doi:10.1088/1742-6596/453/1/012015

2. Dark energy and Einstein's original "constant cosmological constant"

Historically, the Λ-term in the field equations was introduced by A. Einstein 96 years ago [9], but the "CC problem" as such was formulated 50 years later by Y. B. Zeldovich [10]. The latter is the realization that the quantum theory applied to the world of the elementary particles seems to predict an effective value for Λ which is much larger than the critical density of the universe (to which the vacuum energy density is found to be comparable) – see the reviews [4]. The first models trying to circumvent this tough difficulty from a fundamental quantum field theory (QFT) point of view – confer e.g. [11, 12, 13, 14, 15] – tried to use dynamical scalar fields; they tried to explain the small value of the CC density $\rho_\Lambda = \Lambda/(8\pi G)$, assuming that the latter was actually the energy density value of a cosmic scalar field (called "cosmon" in one of the formulations [12]) starting with a huge value in the early universe and then eventually settling it down dynamically (hence without fine tuning) to the present current value [3] $\rho_\Lambda^0 \simeq 2.5\times 10^{-47}$ GeV^4. No model of this kind ever succeeded in achieving that main aim. Later on the first quintessence models bearing this name appeared in the market [16], but with a much more modest aim: they did not try to explain the value of ρ_Λ^0, but just the cosmic coincidence problem, i.e. the reason why the current value of the CC density is so close to the present matter density – see [17] for a comprehensive exposition and more references. However, not even this lower target was ever reached satisfactorily. Nor could be reached either by the many phenomenological proposals for a time dependent cosmological term or other alternative notions [18] – see e.g. the reviews [19] and [20]. Subsequently (in the years after the discovery of the accelerated expansion of the universe [1]) a new and powerful wave of proposals invaded all corners of the cosmological literature: the multifarious notion of dark energy (DE) was born, its main purpose being to replace (in infinitely many different disguises) the function made by the Λ-term in Einstein's equations. And here we are, still fighting with the phenomenal cosmological problems triggered by Einstein's first idea to introduce the Λ term in his almost centennial General Relativity framework! Somehow Einstein knew he was going to raise such a pandemonium, as he felt that by introducing Λ in the field equations he was in danger to be interned in a madhouse: "Ich habe wieder etwas verbrochen in der Gravitationstheorie, was mich ein wenig in Gefahr bringt, in ein Tollhaus interniert zu werden" (A. Einstein, Letter to P. Ehrenfest, February 4th 1917).

Almost a century ago the gravitational field equations (*Die Feldgleichungen der Gravitation*) were first introduced by A. Einstein in 1915, with no cosmological term at all [21]:

$$G_{\mu\nu} \equiv R_{\mu\nu} - \frac{1}{2}\, g_{\mu\nu}\, R = 8\pi G\, T_{\mu\nu}\,. \tag{1}$$

It was only two years later when Einstein, in an attempt to describe the largely accepted idea, at that time, of a finite, static and closed universe hypothetical fulfilling Mach's principle, introduced the Λ term [9] and modified his field equations in the form we still write them today:

$$G_{\mu\nu} - g_{\mu\nu}\,\Lambda = 8\pi G\, T_{\mu\nu}\,. \tag{2}$$

The reason why he was able to introduce that term is because it is perfectly consistent with the idea of general covariance, which was the building principle of GR. This is mathematically expressed by the fact that the covariant derivative on both sides of the original field equations (1) gives zero: indeed $\nabla^\mu G_{\mu\nu} = 0$ is always (automatically) satisfied on account of the Bianchi identity of the Riemann tensor; whereas the covariant derivative on the right-hand-side can be satisfied in different ways, the simplest one being perhaps to assume that G (Newton's gravitational coupling) is a fundamental constant, and that matter is covariantly conserved (i.e. $\nabla^\mu T_{\mu\nu} = 0$). Under these conditions, if we compute once more the covariant derivative on both sides, but this time on the modified field equations (2), and under the same set of assumptions (on fixed G and matter conservation), we are immediately led to

$$\nabla_\mu \Lambda = \partial_\mu \Lambda = 0 \quad \Rightarrow \quad \Lambda = \text{const}\,. \tag{3}$$

This is of course what justifies the name "cosmological constant" to the parameter Λ introduced by Einstein in Eq. (2). But the justification is only partial, as it should be clear from our carefully stating the conditions under which it has been derived. If G is not constant and/or matter is not covariantly conserved (both of them being assumptions which should not be rejected too fast) then the canonical conclusion (3) is not guaranteed at all. A time dependent Λ, or more precisely, a spatially homogeneous function of the cosmic time, $\Lambda = \Lambda(t)$, would still be perfectly compatible with the Cosmological Principle. However, in order to still fulfill the Bianchi identity we would need either a time dependent gravitational coupling, $G = G(t)$, or to admit the possibility that matter exchanges energy with vacuum (hence that matter is not self-conserved, in a covariant sense), or a combination of the two possibilities. While some of these possibilities may look bizarre at first sight, it is no less bizarre that the fundamental "constants" of Nature might not be constant after all, as there are some hints that this could be the case – and also that the two surprising options could be deeply related [22] – see Sect. 9.

Coming back to the "simplest" possibility, namely a strictly constant Λ and the assumed uniformity and spherical symmetry of the space, Einstein could easily derive the explicit form of the field equations in terms of just the constant matter density ρ_m, the newly introduced Λ-term and the scale factor a of the metric. After imposing equilibrium (static universe) a simple relation ensues between these three quantities: $4\pi G\rho_m = 1/a^2 = \Lambda$, or equivalently

$$\rho_m = \frac{1}{4\pi G\, a^2} = 2\,\rho_\Lambda\,. \tag{4}$$

Clearly, it is thanks to the assumed nonvanishing, and positive, Λ-term that such relation is possible – recall that $\Lambda = 8\pi G\rho_\Lambda$. This made Einstein happy, but his happiness was ephemeral. It is not only that E. Hubble soon provided evidence that our universe is actually expanding, but the fact (immediately noticed by A. Eddington) that even in the absence of this information the relation (4) is completely unrealistic, for it corresponds to an unstable position of equilibrium! It means that the slightest perturbation in the value of ρ_m around the one satisfying (4) triggers a runaway solution that kicks forever the "marble off the hill's top"! This result is perfectly intuitive. Indeed, the gravitational law corrected with the Λ-term reads

$$\mathbf{g}(r) = -G\frac{M}{r^2}\mathbf{u}_r + \frac{1}{3}\,\Lambda\, r\,\mathbf{u}_r\,, \tag{5}$$

where $\mathbf{u}_r$ is a unit vector directed radially outwards with respect to the position of the body of mass M creating the field. Thus, if for some reason this universe would expand slightly ($r \to r + \delta r$), this would diminish the gravitational attraction but at the same time would enhance the repulsive Λ-force because the latter is larger the larger is the separation between particles, if $\Lambda > 0$ – as implicit in Eq. (4). So there is no possible compensation between the two. As a result the original seed expansion, no matter how small it is, would destabilize the universe into a runaway expansion. Similarly, an initial seed contraction ($r \to r - \delta r$) would cause the universe to shrink indefinitely into the "Big Crunch".

Paradoxically, ten years later after Λ was introduced by Einstein to insure a static non-evolving universe, G. Lemaître [23] used the form (2) with nonvanishing Λ to discuss his dynamical models of the expansion of the universe, strongly motivated by E. Hubble's observations prior their publication in 1929. The same models had already been independently discussed (without Λ) a few years before by A. Friedmann [24] on pure mathematical grounds and with no connection whatsoever with observations. In 1931, fourteen years after Einstein had introduced the Λ-term, he finally rejected it [25], as if the observational evidence collected by Hubble against his original static model of the universe should have any impact at all on the theoretical status of the Λ term. As we know, 96 years after its introduction in the gravitational field equations, Λ is still there, "alive and kicking", we like it or not!

3. The electroweak Higgs vacuum in classical field theory

In the following we summarize the old CC problem and perform a preliminary discussion of the associated fine-tuning problem, leaving the more sophisticated effects for the subsequent sections. We wish to illustrate the problem within the context of the standard model (SM) of particle physics, and more specifically within the Glashow-Weinberg-Salam model of electroweak interactions [26]. This is the most successful QFT we have at present (together with the QCD theory of strong interactions [27]), both theoretically and phenomenologically, and therefore it is the ideal scenario where to formulate the origin of the problem. As is well-known, the unification of weak and electromagnetic interactions into a renormalizable theory requires to use the principle of local gauge symmetry in combination with the phenomenon of spontaneous symmetry breaking (SSB) [28]. It is indeed the only known way to generate all the particle masses by preserving the underlying gauge symmetry. In the SM, one must introduce a fundamental complex doublet of scalar fields. However, in order to simplify the discussion, let us just consider a field theory with a real single scalar field ϕ, as this does not alter at all the nature of the problem under discussion. To trigger SSB, one must introduce a potential for the field ϕ, which in renormalizable QFT takes the form (the tree-level Higgs potential):

$$V(\phi) = \frac{1}{2}\, m^2\, \phi^2 + \frac{1}{4!}\,\lambda\,\phi^4 \qquad (\lambda > 0)\,. \tag{6}$$

Since we are dealing with a problem related with the CC, we must inexcusably consider the influence of gravity. To this effect, we shall conduct our investigation of the CC problem within the semiclassical context, i.e. from the point of view of quantum field theory (QFT) in curved space-time [29, 30]. It means that we address the CC problem in a framework where gravity is an external gravitational field and we quantize matter fields only [31]. The potential in equation (6) is given at the moment only at the classical level, but it will eventually acquire quantum effects generated by the matter fields themselves (cf. Sect.6). In this context, we need to study what impact the presence of such potential may have on Einstein's equations both at the classical and at the quantum level.

Einstein's field equations for the classical metric in vacuo are derived from the Einstein-Hilbert (EH) action with a cosmological term $\Lambda^{(b)}$ (hereafter the CC *vacuum term*). The EH action in vacuo reads [32]:

$$S_{\rm EH} \;=\; \frac{-1}{16\pi\, G}\int d^4x\sqrt{-g}\;(\,R + 2\,\Lambda\,) = -\int d^4x\sqrt{-g}\,\left(\frac{1}{16\pi\, G}\,R + \rho_{\Lambda{\rm vac}}\right). \tag{7}$$

Here we have defined $\rho_{\Lambda{\rm vac}}$, the energy density associated to the CC vacuum term:

$$\rho_{\Lambda{\rm vac}} = \frac{\Lambda}{8\pi\, G_N}\,. \tag{8}$$

The classical action including the scalar field ϕ with its potential (6) is

$$S = S_{EH} + \int d^4x\,\sqrt{|g|}\,\left[\frac{1}{2}\,g^{\mu\nu}\,\partial_\mu\phi\,\partial_\nu\phi - V(\phi)\right]. \tag{9}$$

Due to the usual interpretation of Einstein's equations as an equality between geometry and a matter-energy source, it is convenient to place the $\rho_{\Lambda{\rm vac}}$ term as a part of the matter action, $S[\phi]$. Then the total action (9) can be reorganized as

$$S = \frac{1}{16\pi\, G_N}\int d^4x\sqrt{|g|}\,R + S[\phi]\,, \tag{10}$$

with

$$S[\phi] = \int d^4x\, \sqrt{|g|}\, \left[\frac{1}{2}\, g^{\mu\nu}\, \partial_\mu\phi\, \partial_\nu\phi - \rho_{\Lambda\rm vac} - V(\phi)\right] \equiv \int d^4x\, \sqrt{|g|}\, \mathcal{L}_\phi\,, \tag{11}$$

where $\mathcal{L}_\phi$ is the matter Lagrangian for ϕ. For the moment, we will treat the matter fields contained in $\mathcal{L}_\phi$ as classical fields, and in particular the potential V is supposed to take the classical form (6) with no quantum corrections. If we compute the energy-momentum tensor of the scalar field ϕ in the presence of the vacuum term $\rho_{\Lambda\rm vac}$, let us call it $\tilde{T}^\phi_{\mu\nu}$, we obtain

$$\tilde{T}^\phi_{\mu\nu} = \frac{2}{\sqrt{|g|}}\, \frac{\delta\, S[\phi]}{\delta\, g^{\mu\nu}} = 2\,\frac{\partial\mathcal{L}_\phi}{\partial\, g^{\mu\nu}} - g_{\mu\nu}\, \mathcal{L}_\phi = g_{\mu\nu}\, \rho_{\Lambda\rm vac} + T^\phi_{\mu\nu}\,. \tag{12}$$

where we have used $\partial\sqrt{|g|}/\partial g^{\mu\nu} = -(1/2)\sqrt{|g|}\, g_{\mu\nu}$. Here

$$T^\phi_{\mu\nu} = \left[\partial_\mu\phi\, \partial_\nu\phi - \frac{1}{2}\, g_{\mu\nu}\, \partial_\sigma\phi\, \partial^\sigma\phi\right] + g_{\mu\nu}\, V(\phi) \tag{13}$$

is the ordinary energy-momentum tensor of the scalar field ϕ.

In the vacuum (i.e. in the ground state of ϕ) there is no kinetic energy, so that the first term on the *r.h.s* of (13) does not contribute in that state. Only the potential can take a non-vanishing vacuum expectation value, which we may call $\langle V(\phi)\rangle$. Thus, the ground state value of (12) is

$$\langle\tilde{T}^\phi_{\mu\nu}\rangle = g_{\mu\nu}\, \rho_{\Lambda\rm vac} + \langle T^\phi_{\mu\nu}\rangle = g_{\mu\nu}\, (\rho_{\Lambda\rm vac} + \langle V(\phi)\rangle) \equiv \rho^{\rm cl}_{\rm vac}\, g_{\mu\nu} \qquad (m^2 < 0)\,, \tag{14}$$

where $\rho^{\rm cl}_{\rm vac}$ is the *classical vacuum energy* in the presence of the field ϕ.

If $m^2 > 0$ in equation (6), then $\langle\phi\rangle = 0 \;\Rightarrow \langle V(\phi)\rangle = 0$, and there is no SSB. The classical vacuum energy is just the original $\rho_{\Lambda\rm vac}$ term,

$$\langle\tilde{T}^\phi_{\mu\nu}\rangle = g_{\mu\nu}\, \rho_{\Lambda\rm vac} \qquad (m^2 > 0)\,. \tag{15}$$

This result also applies in the free field theory case. However, if the phenomenon of SSB is active, which precisely occurs when $m^2 < 0$, we have a non-trivial ground-state value for ϕ, or vacuum expectation value (VEV):

$$v \equiv \langle\phi\rangle = \sqrt{\frac{-6\, m^2}{\lambda}}\,. \tag{16}$$

In this case, there is an *induced part* of the vacuum energy at the classical level owing to the electroweak phase transition generated by the Higgs potential. This transition induces a non-vanishing contribution to the cosmological term which is usually called the "induced CC". At the classical level, it is given by

$$\rho_{\Lambda\rm ind} \equiv \langle V(\phi)\rangle = -\frac{3\, m^4}{2\lambda} = \frac{1}{4}\, m^2\, v^2 = -\frac{1}{8}\, M^2_{\mathcal{H}}\, v^2 = -\frac{1}{8\sqrt{2}}\, M^2_{\mathcal{H}}\, M^2_F\,, \tag{17}$$

In the last equation we have used the physical Higgs mass squared:

$$M^2_{\mathcal{H}} = \left.\frac{\partial^2\, V(\phi)}{\partial\phi^2}\right|_{\phi=v} = m^2 + \frac{1}{2}\,\lambda\, v^2 = -2m^2 > 0\,. \tag{18}$$

Indeed, if we redefine the Higgs field as $\mathcal{H} = \phi - v$ then its value at the minimum will obviously be zero. This is the standard position for the ground state of the field before doing perturbation

Journal of Physics: Conference Series **453** (2013) 012015 doi:10.1088/1742-6596/453/1/012015

theory. The physical mass is just determined by the oscillations of $\mathcal{H}$ around this minimum, i.e. it follows from the second derivative of V at $\phi = v$, as in Eq. (18). In the last equality of that equation we have also introduced the so-called Fermi's scale $M_F \equiv G_F^{-1/2} \simeq 293$ GeV, which is defined from Fermi's constant obtained from muon decay, $G_F \simeq 1.166 \times 10^{-5}$ GeV^{-2}. The relation of G_F with the $W^\pm$ gauge boson mass and the $SU(2)_L$ weak gauge coupling, g, reads (at the lowest order):

$$\frac{G_F}{\sqrt{2}} = \frac{g^2}{8M_W^2} = \frac{1}{2\,v^2}\,, \tag{19}$$

where in the second equality we have used the formula for the $W^\pm$ mass in the SM, namely $M_W = (1/2)\,g\,v$. In this way, Eq. (19) provides a direct determination of the Higgs VEV in terms of Fermi's constant:

$$v = 2^{-1/4}\,G_F^{-1/2} \simeq 246\,\text{GeV}\,, \tag{20}$$

and this relation has been used in the last equality of Eq. (17).

In view of the SSB phenomenon, it is clear that we must replace $g_{\mu\nu}\,\rho_{\Lambda\text{vac}} \to g_{\mu\nu}\,\rho_{\text{vac}}^{\text{cl}}$, given by Eq. (14), in the expression of Einstein's equations in vacuo. This modifies the effective cosmological constant contribution in Einstein's equations. Furthermore, in the presence of incoherent matter contributions (e.g. from dust and radiation) described by a perfect fluid we have the additional term $T_{\mu\nu} = (\rho + p)U_\mu U_\nu - p\,g_{\mu\nu}$. Therefore, the final Einstein's equations in terms of coherent and incoherent contributions of matter, plus the vacuum energy of the fields, finally read

$$R_{\mu\nu} - \frac{1}{2}g_{ab}R = 8\pi\,G_N\,\left(\langle \tilde{T}^{\phi}_{\mu\nu}\rangle + T_{\mu\nu}\right) = 8\pi\,G_N\,\left[g_{\mu\nu}\left(\rho_{\Lambda\text{vac}} + \rho_{\Lambda\text{ind}}\right) + T_{\mu\nu}\right]\,. \tag{21}$$

We conclude that the "physical value" of the CC, at this stage, is not just the original term $\rho_{\Lambda\text{vac}}$, but

$$\rho_{\Lambda\text{ph}} = \rho_{\Lambda\text{vac}} + \rho_{\Lambda\text{ind}}\,, \tag{22}$$

where the induced part is given by (17). On the face of this result, it is pretty obvious that when we compare theory and experiment a severe fine tuning problem is conjured in equation (22). Indeed, the lowest order contribution from the Higgs potential, as given by equation (17), is already much larger than the observational value of the CC. Using the recent LHC measurement of the mass of the Higgs-like particle, suggesting the value $M_{\mathcal{H}} \simeq 125$ GeV [6], equation (17) yields $\rho_{\Lambda\text{ind}} \simeq -1.2 \times 10^8$ GeV^4. Thus, being the CC observed value of order $\rho_\Lambda^0 \sim 10^{-47}$ GeV^4, the electroweak vacuum energy density is predicted to be 55 orders of magnitude larger than the currently measured ρ_Λ^0:

$$\left|\frac{\rho_{\Lambda\text{ind}}^{\text{EW}}}{\rho_\Lambda^0}\right| = \mathcal{O}(10^{55})\,. \tag{23}$$

Suppose that the induced result would exactly be $\rho_{\Lambda\text{ind}} = -10^8$ GeV^4 and that the vacuum density would exactly be $\rho_\Lambda^0 = +10^{-47}$ GeV^4. In such case one would have to choose the vacuum term $\rho_{\Lambda\text{vac}}$ in equation (22) with the rather bizarre precision of 55 decimal places in order to fulfill the equation

$$10^{-47}\,\text{GeV}^4 = \rho_{\Lambda\text{vac}} + \rho_{\Lambda\text{ind}} = \rho_{\Lambda\text{vac}} - 10^8\,\text{GeV}^4\,. \tag{24}$$

This is of course the famous fine-tuning problem.

Let us note that this problem is in no way privative of the cosmological constant approach to the DE, but is virtually present in *any* known model of the DE, in particular also in the quintessence approach [17]. Indeed, the quintessence scalar field potential $V(\varphi)$ is supposed to precisely match the value of the measured DE density at present, ρ_Λ^0, starting from a high energy scale, usually some grand unified theory (GUT) scale $\varphi = M_X$ between $\sim 10^{16}$ GeV and $M_P \sim 10^{19}$ GeV. Even in the simplest case $V(\varphi) \sim m_\varphi^2 \varphi^2$, one finds $m_\varphi^2 \sim \rho_\Lambda^0/M_X^2$. Defining the mass scale associated to the current CC value, $m_\Lambda \equiv (\rho_\Lambda^0)^{1/4} = \mathcal{O}(10^{-3})$ eV, we have the ratio

$$\frac{m_\varphi}{m_\Lambda} \sim \frac{m_\Lambda}{M_X} \sim 10^{-30}\,, \tag{25}$$

which tells us that the mass of the quintessence field should be some thirty orders of magnitude smaller than the CC mass scale that one tries to explain. Apart from the numerical mass value that this implies ($m_\varphi \sim 10^{-33}$ eV), such situation is of course preposterous. Therefore, the quintessence approaches, in addition from being plagued with fine-tuning problems in no lesser degree than the original CC problem, they introduce extremely unnatural small mass scales.

However, this is not quite the end of the story yet. In QFT the induced value of the vacuum energy is much more complicated than just the simple result (17), and the fine-tuning problem is much more cumbersome than the one expressed in equation (24), see Sect.6 for some clues about this additional complication.

4. Vacuum energy: ZPE and some cosmic numerology

As we know, a nonvanishing Λ leads to a nonvanishing value of the vacuum energy density, or CC density, $\rho_\Lambda = \Lambda/(8\pi G)$. In the previous section we have seen that even at the classical level we can get a large contribution to ρ_Λ from the spontaneous symmetry breaking (SSB) of the electroweak symmetry, i.e. from the ground state of the Higgs potential. Furthermore, quantum corrections to the this potential can be quite significant, as we will discuss later. The possible confirmation of the Higgs finding at the LHC collider at CERN certainly strengthens the case for the vacuum energy in QFT, but the quantum effects which we are going to refer now do not depend on the existence of the Higgs potential as they have a very generic character. They were noted much before the Higgs potential and the SSB phenomenon entered the stage. We pay now some preliminary attention to this important (and generic) phenomenon of the quantum vacuum, only to come back to it in subsequent sections from a more rigorous point of view.

The quantum effects we are now referring to are those emerging from the the vacuum-to-vacuum fluctuations of the quantum matter fields. They occur already in the free field theory, in contrast to the SSB phenomenon. These quantum fluctuations correspond to closed loop diagrams without external tails ("blobs"). They describe the infinite number of oscillators with all possible frequencies $\omega_{\mathbf{k}}$ to which we attach to any free quantum matter field. The sum of all the (nonvanishing) ground state energies of these oscillators constitutes the zero-point energy (ZPE) of the corresponding quantum field: $E_0 = (1/2)\sum_{\mathbf{k}} \hbar\omega_{\mathbf{k}}$. Let us recall that the historical origin of the ZPE emerges from Planck's theory for the black-body radiation in 1900. For the average energy of an oscillator of frequency ω in equilibrium with radiation at temperature T, Planck obtained

$$E_\omega = \frac{\hbar\omega}{e^{\hbar\omega/kT} - 1} + \frac{1}{2}\,\hbar\omega\,. \tag{26}$$

He realized immediately that $E_\omega \to (1/2)\,\hbar\omega$ for $T \to 0$. This is the *Nullpunktsenergie* or zero-point energy of the oscillator, therefore a nonvanishing one. Despite he soon tried to normalize it away treating it as an unphysical effect, thirteen years later Einstein and Stern paid clever attention to the fact that if one expands the formula (26) in the classical limit $kT \gg \hbar\omega$ one finds $E_\omega \simeq kT - (1/2)\,\hbar\omega + (1/2)\,\hbar\omega = kT$, as easily checked. In other words, thanks to the isolated ZPE term on the *r.h.s.* of (26) one can recover the expected classical limit for the average energy of an oscillator in thermal equilibrium at temperature T. This firmly convinced Einstein that the ZPE could have some physical meaning after all.

But things are not so easy. The electromagnetic field oscillates with all frequencies, and therefore one has to compute the infinite sum $E_0 = (1/2)\sum_{\mathbf{k}} \hbar\omega_{\mathbf{k}}$. The unrenormalized result

Journal of Physics: Conference Series **453** (2013) 012015 doi:10.1088/1742-6596/453/1/012015

is of course infinite, but the renormalized quantity is perfectly finite (cf. Sect. 5) and its value may be of concern. The first considerations on the ZPE of the quantum vacuum dates back to the theoretical discussions made by W. Nernst as early as 1916 and by W. Pauli in the 1920s (and again in a more formal way in 1933, see Sect. 5). However, all these discussions were mainly focused on the electromagnetic field. As, however, the renormalized form of the ZPE is proportional to the quartic power of the renormalized mass of the corresponding field quantum ($\rho_\Lambda \propto m^4$, see below), in the electromagnetic case we have no such physical contribution to the ZPE (at least in the way it was originally conceived by the first pioneers discussing these matters) owing to the massless nature of the photon.

It was probably Y.B. Zeldovich who first raised a serious concern on the contribution to the ZPE from a "typical" massive particle [10] and expressed the necessity to subtract its leading – or "first order" – effect in the computation of the physical value of the CC [1]. In particular he noticed that the leading proton contribution to the ZPE ($\rho_\Lambda \propto m_p^4 \sim 1\mathrm{GeV}^4$) is overwhelmingly large as compared to any cosmic density, say the current critical density ρ_c^0 expressed in typical particle physics units:

$$\rho_c^0 = \frac{3\,H_0^2}{8\pi\,G} = \frac{3}{8\pi}\,H_0^2\,M_P^2 \sim 10^{-47}\,\mathrm{GeV}^4\,, \tag{27}$$

where $H_0 \sim 10^{-42}$ GeV is the current value of the Hubble rate and $M_P = 1/\sqrt{G} \sim 10^{19}$ GeV is Planck's mass (both expressed in natural units). The current matter and vacuum energy densities (ρ_m^0 and ρ_Λ^0) are proportional to (27) up to factors of order one, i.e. $\rho_m^0 = \Omega_M^0\,\rho_c^0$ and $\rho_\Lambda^0 = \Omega_\Lambda^0\,\rho_c^0$, with $\Omega_M^0 \sim \Omega_\Lambda^0 = \mathcal{O}(1)$. After subtracting the exceedingly large leading term in the theoretical ZPE estimate, Zeldovich realized that starting once more from m_p as the "typical" mass scale of particle physics one could produce a much more reasonable order of magnitude estimate of the CC density by means of a "second order" (or "next-to-leading") formula involving the natural presence of the gravitational coupling, G. To this purpose he concocted the dimensionally consistent "construct"

$$\rho_\Lambda \simeq G\,m_p^6 = \frac{m_p^6}{M_P^2} \sim 10^{-38}\,\mathrm{GeV}^4\,. \tag{28}$$

The previous estimate still strays off the modern value, but by "only" nine orders of magnitude – in contrast to the demolishing 47 orders of magnitude that one has to face if keeping the leading term $\rho_\Lambda \sim m_p^4$. Needless to say at the time of Zeldovich there was no real measurement of Λ, although of course an upper bound estimate of the order of the critical density was in force. Therefore the order of magnitude discrepancy was not that different from the present one. Even so, formula (28) could somehow still be considered as a respectable estimate. An even more intriguing result obtains if one replaces the proton by the pion (whose mass $m_\pi \simeq 0.1$ GeV is roughly ten times smaller than that of the proton). Then one gets a better approximation which differs now by "only" three orders of magnitude from the correct order of magnitude result (27). It turns out that a kind of "cosmic prediction" of the pion mass was proposed by Weinberg in 1972 [36] through the curious numerical relation

$$m_\pi^3 \sim \frac{H_0}{G} = H_0\,M_P^2 \sim 10^{-4}\,\mathrm{GeV}^3\,, \tag{29}$$

which indeed leads to $m_\pi \simeq 0.1$ GeV. Amusingly, if we now substitute Eq. (29) into (28) — after first replacing $m_p \to m_\pi$, according to our prescription – we find

$$\rho_\Lambda \sim H_0^2\,M_P^2 \sim 10^{-46}\,\mathrm{GeV}^4\,, \tag{30}$$

[1] It is not my intention to make full justice, in this summarized account, to the extensive historical literature on the CC problem – see e.g. [4] and [33, 34], and references therein. It will suffice to say that the connection between vacuum energy and CC had already been glimpsed by Lemaître in 1934 a few years after he found the cosmological expanding solution in the presence of a Λ-term [35].

which, according to (27), is very close to the correct order of magnitude of the current CC density, since $\rho_\Lambda^0 = \Omega_\Lambda^0\,\rho_c^0$, with $\Omega_\Lambda^0 \simeq 0.7$. From (29) and (30) it also ensues the (no less "cabalistic") relation $\rho_\Lambda \sim m_\pi^3\,H_0$, somehow suggesting a possible link between the meson world of particle physics and cosmology. Unfortunately such relation is untenable within GR, and therefore such link is impossible if following that pathway. Equally untenable is a (subtlety disguised) form of the previous relation, which has iteratively appeared in the literature in more recent times, to wit:

$$\rho_\Lambda \sim H_0\,\Lambda_{\rm QCD}^3\,. \tag{31}$$

where $\Lambda_{\rm QCD} \simeq 200$ MeV is the QCD scale of the strong interactions (not far away from the pion mass). Some people has tried hard to seek for a fundamental reason behind a formula like (31) [37]. Numerically it is much worse than (30), as it yields $\rho_\Lambda \sim 10^{-44}$ GeV^4, thus failing by "only" three orders of magnitude. But it is not this numerical failure which is most worrisome (as numbers are comparable to the situation with the previous pion formula – thought of as an improved form of Zeldovich's one); the problem here is of theoretical nature. As has been emphasized in [31], an equation like (31) – or, for that matter, any other relation where ρ_Λ is extracted from an odd power of the Hubble rate – is incompatible with the general covariance of the effective action in QFT in curved spacetime[2]. One possibility would be to resort to fractional powers of the invariants, but are we ready for such an eccentric possibility while other, more amenable ones, are still there to be fully exploited? For example, the numerically successful relation (30) depends quadratically on H and therefore is compatible with the aforementioned covariance. It follows that the argument that led to Eq. (30) cannot be (without invoking contrived assumptions) the one that we followed above through $\rho_\Lambda \sim m_\pi^3\,H_0$ or (31) since the latter cannot be accepted in a natural way. It means that if there is any truth around Eq. (30) there must exist some completely independent pathway leading to it. We will see that there are indeed quite different paths pointing to (30), or in general to the more general "affine" quadratic relation

$$\rho_\Lambda(H) = c_0 + \beta\,M_P^2\,H^2\,, \tag{32}$$

with a non-vanishing c_0 term, β being here a dimensionless coefficient. The presence of $c_0 \neq 0$ is crucial for a realistic implementation of the model, as it enables the transition from deceleration to acceleration in this kind of models. The strict model (30), understood as a model of the kind (32) with $c_0 = 0$, is ruled out [38, 39] as it cannot satisfy the aforementioned transition condition. This is the situation with entropic force models [40], for example, and many other models previously presented on purely phenomenological grounds. In contrast, the class of affine models (32) is perfectly safe in this respect, and in fact it has been successfully tested against the recent cosmological data [38, 39, 41].

In principle, we have no fundamental reason behind any of these numerical games. The only "fundamental" thing to be done here is to make sure that we subtract the leading ("first order") contribution from the ZPE, which for all kinds of known elementary particles (except perhaps for a light neutrino) is much larger than the measured ρ_Λ. For a hypothetical neutrino of a few meV$= 10^{-3}$ eV, we have the suggestive result [42]

$$\rho_\Lambda \sim m_\nu^4 \sim 10^{-11}\,{\rm meV}^4 \sim 10^{-47}\,{\rm GeV}^4\,, \tag{33}$$

that falls in the right ballpark of the cosmic densities (27). In the next sections we will go in a summarized way thorough models that could justify some of the more successful evolution laws for the cosmological term, say of the form (32), on more fundamental grounds.

[2] In Sect. 9 we will see that the $\Lambda_{\rm QCD}$ scale could play a relevant cosmological role, but for quite a different reason, namely on account of its potential cosmic time dependence linked to that of the vacuum energy, and within a fully covariant formulation [22].

5. Zero-point energy in quantum field theory in flat spacetime

If we wish to go beyond the previous numerical games, things can get a bit harder, even if we still try to keep them as simple as possible. Let us therefore first follow a very naive formulation. Formally the ZPE of a given quantum field, say a scalar field ϕ, is obtained by selecting that part of the effective potential which does not depend on the external tails of ϕ (i.e. that part which is *not* a function of ϕ). For example, the Higgs potential (cf. Section 3) is in general *not* a part of the ZPE because it has a classical part (one which does *not* vanish when we set $\hbar = 0$). It means that it consists of all the bubble-type (vacuum-to-vacuum) diagrams at all orders in perturbation theory. The ZPE is thus a pure quantum effect: it vanishes if there is no quantum theory, $\hbar = 0$. Indeed, vacuum-to-vacuum diagrams can only exist in a field theory with vacuum fluctuations: QFT. The final result therefore can only depend on a list of parameters P (masses and coupling constants), but not on ϕ itself, which can only enter virtually in the loop propagators. If we count the loop order of perturbation theory with the corresponding power of $\hbar$, the loopwise expansion can be presented as a power series in $\hbar$:

$$V_{\rm ZPE}(P) = \hbar\, V_P^{(1)} + \hbar^2\, V_P^{(2)} + \hbar^3\, V_P^{(3)} + ... \tag{34}$$

(A contribution to the 21th term of this pure vacuum-to-vacuum series within the SM can be seen in Fig. 1.) From Eq. (34) it is obvious that even the first term depends on $\hbar$, just linearly. This is the one-loop approximation. Since we promised to keep things simple, let us evaluate the ZPE of a QFT for a free scalar field at one loop only. For fermion field of spin $s_f = 1/2$ an additional factor of $4\,(= {\rm Tr}\,\hat{\mathbf{1}})$ and an overall minus sign should both be inserted.

A naive calculation of the coefficient $V_P^{(1)}$ in (34) is obtained by ignoring for the moment gravity and regularizing the infinite sum $(1/2)\sum_{\mathbf{k}} \hbar\omega_{\mathbf{k}} = (1/2)\sum_{\mathbf{k}} \hbar\sqrt{\mathbf{k}^2+m^2}$ by means of a cutoff $\mathcal{M}$. Moving to the continuum and trivially integrating the solid angle, it gives

$$V_P^{(1)} = \frac{1}{2}\int \frac{{\rm d}^3k}{(2\pi)^3}\sqrt{\mathbf{k}^2+m^2} = \frac{1}{4\pi^2}\int_0^{\mathcal{M}} {\rm d}k\, k^2\sqrt{k^2+m^2} = \frac{\mathcal{M}^4}{16\pi^2}\left(1+\frac{m^2}{\mathcal{M}^2}-\frac{1}{4}\frac{m^4}{\mathcal{M}^4}\ln\frac{M^2}{m^2}+\cdots\right), \tag{35}$$

where in the expansion in powers of $m/\mathcal{M}$ we have explicitly kept the important $\sim m^4 \ln m^2$ term because this one does not depend on any power of the cutoff and therefore is the term that should remain after we attempt to remove the cutoff by some renormalization procedure. In the 1930's Pauli applied this naïf approach to the electromagnetic field ($m = 0$ for the photon) and choosing the inverse of the classical electron radius for the cutoff, i.e. $\mathcal{M} = 2\pi\, m_e/\alpha$ (here α standing for the fine structure constant). Then he plugged the result into Einstein's universe formula (4), i.e. he replaced ρ_Λ there by the previous one-loop estimate of the ZPE, and obtained the "radius" of the ensuing universe:

$$a = \sqrt{\frac{2\pi}{G}\frac{1}{\mathcal{M}^2}} = (2\pi)^{-3/2}\left(\frac{M_P}{m_e}\right)\frac{\alpha^2}{m_e}\,. \tag{36}$$

It would not even reach the Moon! (as a matter of fact the result is appallingly small, some twenty six kilometers only!)[3]. As we can see, the larger is the vacuum density the smaller is the equilibrium "radius". This is intuitively obvious from the fact that the negative pressure associated to the CC value is very big even for a cutoff as small as the electron mass, so that if gravity has to compensate such outwards vacuum pressure the universe must be small enough to produce a gravitational (inwards) effect of the same size. He was dismayed, but the argument is itself actually non-rigorous. Indeed, the results (35)-(36) explicitly depend on the value we can

[3] Strictly speaking one has to include a factor of 2 in Eq. (35) to account for the two helicities of the photon, so the result for the "radius" is $1/\sqrt{2}$ smaller , i.e. some $\sim$ 18 kilometers, but of course this nicety is irrelevant here.

Journal of Physics: Conference Series **453** (2013) 012015 doi:10.1088/1742-6596/453/1/012015

arbitrarily assign to the regulator $\mathcal{M}$; in other words, it is merely an unphysical result obtained in the "bare theory". Sometimes a particular value can better approximate the final result, but in actual fact the physical result should be completely independent of $\mathcal{M}$. It means that one must first renormalize the theory before jumping to conclusions, as without renormalization the result (36) means very little. This implies, among other things, to get rid of the regulator before extracting meaningful conclusions on the ZPE contribution from the electromagnetic field. However, after doing so we are left with nothing since the photons have no mass and hence no contribution to the ZPE could remain from them. Indeed, upon removing the cutoff we expect that the renormalized result should be of order

$$V_P^{(1)\text{renorm}} = -\frac{m^4}{64\pi^2}\ln\frac{\mu^2}{m^2} + \cdots . \tag{37}$$

where μ is some subtraction scale that must remain after renormalization. From the above formula it is clear that the ZPE becomes zero if we set $m = 0$ in it. If we would instead compute the contribution, not from the electromagnetic field, but from a scalar field of mass m, the renormalized contribution should be of the order (35), thus proportional to m^4. Unfortunately, we find that even after trying to make some sense out of the ZPE within QFT we are back to the huge quartic contributions to the CC which we tried to avoid in the previous section. They are back here and this becomes of course worrisome. As for Pauli's quantitative result, despite its roughness and severe limitations it transmits the correct qualitative idea that any typical choice for the cutoff scale within the particle physics domain leads to a very unsatisfactory value for the cosmological constant as compared to the observational measurements. We thus realize that the ZPE calculation becomes conflictive with observations even after renormalization. Despite we did not use a concrete renormalization scheme to guess at the renormalized form (37) from the regularized expression (35), it turns out that the latter is indeed what is obtained e.g. using dimensional renormalization (see below).

Although every scheme is in principle valid, not every scheme has the property that the renormalized quantities are in good correspondence with the physical quantities in the particular framework of the calculation. For example, it makes no much sense to renormalize QCD (the gauge theory of strong interactions) in the on-shell scheme because we never find quarks and gluons on mass shell! One is forced to use an off-shell renormalization scheme. This can complicate the physical interpretation, for there can indeed be a nontrivial gap between the renormalized parameters and the physical ones. And if all that is not enough, the CC problem [4] is of course a problem where gravity should play a role somewhere in the calculation of the ZPE, shouldn't it? So, at the end of the day, we realize that the result (35) is in itself far from having any physical meaning!

In an attempt to smooth out some of these problems, suppose we adopt the MS-scheme (or Minimal Subtraction scheme) in n-dimensional regularization. Then, the one-loop approximation to the ZPE renders

$$V_P^{(1)} = \frac{1}{2}\,\mu^{4-n}\int\frac{d^{n-1}k}{(2\pi)^{n-1}}\sqrt{\mathbf{k}^2+m^2} = \frac{1}{2}\,\beta_\Lambda^{(1)}\left(-\frac{2}{4-n} - \ln\frac{4\pi\mu^2}{m^2} + \gamma_E - \frac{3}{2}\right), \tag{38}$$

where γ_E is Euler's constant, and

$$\beta_\Lambda^{(1)} = \frac{m^4}{2\,(4\pi)^2} \tag{39}$$

is the one-loop β-function for the vacuum term (see below). Notice that μ is the characteristic 't Hooft mass unit of dimensional regularization, and $n \to 4$ is understood in the final result. So obviously the ZPE is UV-divergent once more, as it could not be otherwise without renormalization. In the MS scheme one introduces, as usual, a counterterm killing the "bare

Journal of Physics: Conference Series **453** (2013) 012015 doi:10.1088/1742-6596/453/1/012015

bone" UV-part (the pole at $n \to 4$). In the slightly modified $\overline{\text{MS}}$ scheme one collects also some additive constants. Specifically, the counterterm reads:

$$\delta\rho_\Lambda^{\overline{\text{MS}}} = \frac{m^4\,\hbar}{4\,(4\pi)^2}\left(\frac{2}{4-n} + \ln 4\pi - \gamma_E\right). \tag{40}$$

How to get now a finite (if still not physical) value of the ZPE? We just have to follow the renormalization program. Recall that the Einstein-Hilbert action from which the field equations (2) are derived reads as in Eq. (7) [32]:

$$S_{\rm EH} = \frac{-1}{16\pi\, G^{(b)}}\int d^4x\sqrt{-g}\,\left(R + 2\,\Lambda^{(b)}\right) = -\int d^4x\sqrt{-g}\,\left(\frac{1}{16\pi\, G^{(b)}}\,R + \rho_\Lambda^{(b)}\right). \tag{41}$$

Here (and hereafter) we have rewritten the parameters $\Lambda \to \Lambda^{(b)}$, $\rho_{\Lambda\rm vac} \to \rho_\Lambda^{(b)}$ and $G \to G^{(b)}$ so as to emphasize it is the bare action, i.e. the action before any renormalization program is applied to account for the UV-divergences related to the quantum matter contributions. In particular, the CC term and the gravitational coupling are the bare ones – and so have been denoted with the superindex "b" . Even though this action is not renormalizable, we know that by including the higher order invariant terms (i.e. R^2, $R_{\mu\nu}R^{\mu\nu}$..., only relevant for the short-distance behavior of the theory) the extended action is renormalizable in the context of QFT in curved spacetime (where gravity is not quantized but the matter fields yes), provided (!) we keep the Λ-term in it. So let us maintain this term in the low-energy part of the full renormalizable action, i.e. the long-distance EH action (41), as this part is both necessary for the renormalization program and is accessible to observation (as we indeed know from the observed accelerated expansion).

The key point now, at the theoretical level, is that the CC-term in (41) is *not* yet the physical quantity, it is only the bare parameter of the bare EH action. When we include the ZPE as a part of the full action, the overall additive term is no longer $\rho_\Lambda^{(b)}$ but the sum $\rho_\Lambda^{(b)} + V_{\rm ZPE}^{(\rm bare)}$, where $V_{\rm ZPE}^{(\rm bare)}$ is given by (34). Next we split $\rho_\Lambda^{(b)}$ into a renormalized part plus a counterterm, $\rho_\Lambda^{(b)} = \rho_\Lambda(\mu) + \delta\rho_\Lambda$, where the renormalized part depends on the arbitrary renormalization scale μ and the counterterm depends on the regularization and renormalization scheme. Adopting $\overline{\text{MS}}$ in dimensional regularization we are led to use the explicit form (40) for $\delta\rho_\Lambda$. Finally, since the (effective action of the) bare theory must equal (that of) the renormalized theory (in whatever renormalization scheme), we must have $\rho_\Lambda^{(b)} + V_{\rm ZPE}^{(\rm bare)} = \rho_\Lambda(\mu) + V_{\rm ZPE}(\mu)$. This condition defines the $\overline{\text{MS}}$-renormalized one-loop value of the ZPE. Using the above equations, at one-loop it reads:

$$V_{\rm ZPE}^{(1)}(\mu) = \hbar\, V_P^{(1)} + \delta\rho_\Lambda^{\overline{\text{MS}}} = \frac{m^4\,\hbar}{4\,(4\,\pi)^2}\left(\ln\frac{m^2}{\mu^2} - \frac{3}{2}\right), \tag{42}$$

which is perfectly finite. Incidentally, with this result we have checked that the one-loop correction in dimensional regularization with minimal subtraction indeed provides the kind of finite correction that we guessed in Eq. (37) from the bare result (35), up to additive parts related with the subtraction procedure. This does not mean that the obtained expression is the physical result, but at least is a renormalized and hence finite one. The price for the finiteness of the renormalized result is its dependence on the arbitrary mass scale μ. However, while both pieces (42) and $\rho_\Lambda(\mu)$ separately depend on the scale μ, the sum

$$\rho_{\rm vac}^{(1)} = \rho_\Lambda(\mu) + V_{\rm ZPE}^{(1)}(\mu) = \rho_\Lambda(\mu) + \frac{m^4\,\hbar}{4\,(4\,\pi)^2}\left(\ln\frac{m^2}{\mu^2} - \frac{3}{2}\right), \tag{43}$$

does *not* since by construction we started from the bare theory, which is of course μ-independent. The sum (43) represents the $\overline{\rm MS}$-renormalized vacuum energy of the free field at one loop. As this quantity is the same starting bare expression, just rewritten in terms of renormalized parameters, it is overall μ-independent. This is actually the main message of the renormalization group (RG): the sum of the various μ-dependencies must cancel in the renormalized effective action, and also in the renormalized S-matrix elements (when they can be defined). We can now check explicitly that (39) is indeed the one-loop β-function for the running of the renormalized CC. Computing the logarithmic derivative $d/d\ln\mu = \mu d/d\mu$ on both sides of (43) and taking into account that this gives $d\rho^{(1)}_{\rm vac}/d\ln\mu = 0$ on the $l.h.s$ (for the reasons just mentioned above), we immediately find the desired result:

$$\mu\frac{d\rho_\Lambda(\mu)}{d\mu} = \frac{\hbar\, m^4}{2\,(4\pi)^2} = \beta_\Lambda^{(1)}\,. \tag{44}$$

The RG tells us something useful about the explicit μ-dependence affecting incomplete structures of the effective action. Indeed, while we may not know the full structure of the effective action in a particular complicated situation, our educated guess in associating μ with some relevant dynamical variable of the system can give relevant information on physics, similarly as we do in particle physics. This is e.g. the case with an effective charge (or "running coupling constant"), say the QED or QCD renormalized gauge coupling $g = g(\mu)$, which is explicitly μ-dependent even though the full effective action or S-matrix element is not.

For a more rigorous connection with the curved space-time case discussed in the next section, it is convenient to approach Eq. (38) from a more formal point of view, namely from the notion of effective action [43]. The desired form for the effective action at one-loop reads:

$$\Gamma_{\rm eff}[\phi_c] = S[\phi_c] + \frac{i\,\hbar}{2}\,Tr\,\ln\mathcal{K}(x,x')\,. \tag{45}$$

where $S[\phi_c]$ is the classical action

$$S[\phi_c] = \int d^4x\,\mathcal{L} = \int d^4x\,\left[\frac{1}{2}\,g^{\mu\nu}\,\partial_\mu\phi_c\,\partial_\nu\phi_c - V_c(\phi_c)\right]\,. \tag{46}$$

and

$$\mathcal{K}(x,x') = \left[\Box_x + V''(\phi_c)\right]\,\delta(x-x')\,, \tag{47}$$

is essentially the inverse propagator in the presence of the background matter field ϕ_c. Here V_c is the tree-level or classical part of the potential for that field, typically of the form (6). For a free field ($\lambda = 0$) it contains only the mass term: $V(\phi) = (1/2)\,m^2\,\phi^2$ and Eq. (47) simplifies to

$$\mathcal{K}(x,x') = \left[\Box_x + m^2\right]\,\delta(x-x') \qquad \text{(free QFT)}\,. \tag{48}$$

The result (45) was expected, namely the effective action at this order is the sum of the classical action (46) and the one-loop term (47) – notice the presence of $\hbar$ in front of it in Eq. (45). The quantum correction term, call it $\Gamma^{(1)}$, is generated exclusively by the vacuum diagrams, and therefore represents the ZPE. We can understand that indeed $\Gamma^{(1)}$ is associated to vacuum-to-vacuum diagrams (i.e. closed loop diagrams without external tails of quantum matter) by the fact that we have to compute a trace over all indices, including the spacetime ones:

$$\Gamma^{(1)} = \frac{i\,\hbar}{2}\,Tr\,\ln\mathcal{K}(x,x') = \frac{i\,\hbar}{2}\,\int d^4x\,\lim_{x\to x'}\,\ln[\mathcal{K}(x,x')]\,. \tag{49}$$

Diagrammatically we can interpret we are following a closed line starting at one point and ending at the same point, i.e. a vacuum-to-vacuum diagram or "blob". Furthermore, we have

Journal of Physics: Conference Series **453** (2013) 012015 doi:10.1088/1742-6596/453/1/012015

to sum these blobs over all of the spacetime points, as indicated by the above integral. Setting ϕ_c =const. the classical action (46) just boils down to (minus) the classical potential times the spacetime volume Ω, i.e. $S[\phi_c] = -\int d^4x V(\phi_c) = -\Omega V(\phi_c)$. The *l.h.s.* of (45) can then be written as $-V_{\rm eff}\,\Omega$, where $V_{\rm eff}$ is the so-called effective potential. Clearly $V_{\rm eff} = V_c + \hbar V^{(1)}$, where the one-loop correction reads

$$V^{(1)} = -\frac{i}{2}\,\Omega^{-1}\,Tr\ln\mathcal{K}\,. \tag{50}$$

Next the integral can be conveniently worked out in momentum space. For the free field case (48) it immediately leads to

$$Tr\,\ln\mathcal{K} = \Omega\int\frac{d^nk}{(2\pi)^n}\,\ln\left(-k^2+m^2\right)\,, \tag{51}$$

where we moved to n dimensions to regularize the result. The spacetime volume finally cancels in (50) and after a simple calculation one finally retrieves the formula (38) and the corresponding $\overline{\rm MS}$-renormalized final result(43).

6. The ZPE and the full effective Higgs potential for QFT in flat spacetime

The ZPE calculation for free fields considered in the previous section is the simplest kind of quantum vacuum effect one can deal with. However, the general renormalized effective potential extending to the quantum domain the classical potential, $V_c \to V_{\rm eff}$, takes on the form

$$V_{\rm eff} = V_c + \hbar\,V_1 + \hbar^2\,V_2 + \hbar^2\,V_3 + ... \tag{52}$$

The quantum effects to all orders of perturbation theory arrange themselves in the form of a loopwise expansion where the number of loops is tracked by the powers of $\hbar$. Thus, at one loop we have only one power of $\hbar$, at two loops we have two powers of $\hbar$ etc. For $\hbar = 0$, however, there are no loops and the effective potential just reduces to the classical potential, V_c, given by equation (6) in the electroweak standard model. On the other hand, each of the loop terms in (52) can be split into two independent contributions, one consisting of loops with no external legs (vacuum-to-vacuum parts $V_P^{(i)}$, i.e. the ZPE contribution at ith-order) and the other involving loops with external legs of the Higgs field ϕ (i.e. the ith-loop correction $V_{\rm scal}^{(i)}(\phi)$ to the classical Higgs potential):

$$V_1 = V_P^{(1)} + V_{\rm scal}^{(1)}(\phi)\,,\qquad V_2 = V_P^{(2)} + V_{\rm scal}^{(2)}(\phi)\,,\qquad V_3 = V_P^{(3)} + V_{\rm scal}^{(3)}(\phi)....\tag{53}$$

As a result, the effective potential (52) at the quantum level splits naturally into two parts, one which is ϕ-independent and another that is ϕ-dependent:

$$V_{\rm eff}(\phi) = V_{\rm ZPE} + V_{\rm scal}(\phi)\,, \tag{54}$$

where

$$V_{\rm ZPE} = \hbar\,V_P^{(1)} + \hbar^2\,V_P^{(2)} + \hbar^3\,V_P^{(3)} + ... \tag{55}$$

is the full *zero-point energy* (ZPE) contribution. It is a number, it only depends on the set of parameters $P = m, \lambda, ...$ of the classical potential, not at all on the fields. As mentioned, the latter consists in the sum of all the vacuum-to-vacuum parts of the effective potential. The ZPE part is sourced exclusively from closed loops of matter fields (i.e. vacuum loops without external ϕ-legs). In the previous section we have computed the one-loop contribution to the ZPE and for free fields, i.e. just the term $V_P^{(1)}$. The ZPE receives in general contributions to all orders of perturbation theory, except at zero loop level since $V_{\rm ZPE}$ is a pure quantum effect that vanishes for $\hbar = 0$. Now, besides the ZPE there is the scalar field dependent part of the effective potential:

$$V_{\rm scal}(\phi) = V_c(\phi) + \hbar\, V^{(1)}_{\rm scal}(\phi) + \hbar^2\, V^{(2)}_{\rm scal}(\phi) + \hbar^3\, V^{(3)}_{\rm scal}(\phi) + ... \tag{56}$$

This one is not purely quantum (i.e. it does not vanish for $\hbar = 0$) as the first term (the classical potential) is, of course, *not* proportional to $\hbar$. The above ϕ-dependent part of $V_{\rm eff}$ receives in general also contributions to all orders of perturbation theory, and vanishes for $\phi = 0$ since in this case all the loops have external ϕ legs, including the tree-level part. Thus, in the absence of SSB the effective potential boils down to just the ZPE.

$$V_{\rm eff}(\phi = 0) = V_{\rm ZPE} \tag{57}$$

which is a number entirely constructed from a power series of $\hbar$. On the other hand, the expression (56) is the effective potential excluding that ZPE number. The full effective potential (54) contains both contributions.

In the above formulae, all the field theoretical ingredients (m, λ, ϕ and $V_{\rm eff}$) are in fact bare quantities (m_0, λ_0, ϕ_0 and $V_{\rm eff\,0}$) that require renormalization, as the loopwise expansion is UV-divergent order by order. Renormalization means that we replace all the bare quantities with renormalized ones (in some given renormalization scheme with a specific set of renormalization conditions) plus counterterms (which are also scheme dependent and are partially fixed by the condition of canceling the UV-divergences): $m_0 = m + \delta m$, $\lambda_0 = \lambda + \delta\lambda$, $\phi_0 = Z_\phi^{1/2}\,\phi = (1 + \delta Z_\phi/2)\,\phi$... Of course, a similar splitting occurs with the vacuum term, which was originally a bare term $\rho_\Lambda^{\rm (b)}$. We must also split it into a renormalized piece plus a counterterm: $\rho_\Lambda^{\rm (b)} = \rho_\Lambda(\mu) + \delta\rho_\Lambda$. The full set of counterterms is essential to enable the loop expansion to be finite order by order in perturbation theory. For instance, if we would renormalize the theory in the $\overline{\rm MS}$ scheme in dimensional regularization, the suitable counterterm for the vacuum parameter was given in Sect. 5, see Eq. (40).

For a practical calculation of the first (one-loop) quantum correction to the Higgs potential (6), we have to compute the one-loop term (49) in the presence of the $\lambda \neq 0$ term in the potential. In this case,

$$\mathcal{K}(x,x') = \left[\Box_x + V''(\phi_c)\right]\,\delta(x-x') = \left[\Box_x + m^2 + \frac{1}{2}\,\lambda\,\phi_c^2\right]\,\delta(x-x')\,. \tag{58}$$

For ϕ_c =const. this will lead us to determine the explicit form for $V^{(1)}_{\rm scal}(\phi)$ in the above language. This introduces some complications but the calculation can still be carried out without much problems. In the constant mean field limit we may equate the bare and renormalized effective action and we obtain

$$\rho_\Lambda^{\rm (b)} + V_{\rm eff}(\phi_0, m_0, \lambda_0) = \rho_\Lambda(\mu) + V_{\rm eff}(\phi(\mu), m(\mu), \lambda(\mu); \mu)\,. \tag{59}$$

As always the renormalized result depends on an arbitrary mass scale μ The overall μ-dependence, however, must eventually cancel. In fact, the renormalized parameters are finite quantities which are also functions of μ: $\phi = \phi(\mu)$, $m = m(\mu)$, $\lambda = \lambda(\mu)$, $\rho_\Lambda = \rho_\Lambda(\mu)$, and since the vacuum energy cannot depend on the arbitrary scale μ, the sum of the renormalized vacuum term and the renormalized potential must be globally scale-independent (i.e. μ-independent). This is obviously so because the bare vacuum term and bare effective potential were scale-independent to start with. Thus, from (59) we have

$$\mu\frac{d}{d\mu}\left[\rho_\Lambda(m(\mu), \lambda(\mu); \mu)) + V_{\rm eff}(\phi(\mu); m(\mu), \lambda(\mu); \mu)\right] = 0\,. \tag{60}$$

This relation implies that the full effective potential is actually *not* renormalization group (RG) invariant (contrary to some inaccurate statements in the literature), but it becomes so only after we add up to it the renormalized CC vacuum part ρ_Λ. In reality, the structure of the effective potential (54) is such that the previous relation splits into two independent RG equations:

$$\mu\frac{d}{d\mu}\left[\rho_\Lambda(m(\mu),\lambda(\mu);\mu)) + V_{\rm ZPE}(m(\mu),\lambda(\mu);\mu)\right] = 0 \tag{61}$$

and

$$\mu\frac{d}{d\mu}V_{\rm scal}(\phi(\mu);m(\mu),\lambda(\mu);\mu) = 0\,. \tag{62}$$

Equation (61) shows that it is only the strict vacuum-to-vacuum part (i.e. the ZPE) the one that needs the renormalized vacuum term ρ_Λ to form a finite and RG-invariant expression, whereas the renormalized ϕ-dependent part of the potential (i.e. the tree-level plus the loop expansion with external ϕ-tails) is finite and RG-invariant by itself. This is of course the essential message from the renormalization group. Explicitly, equation (62) reads

$$\left\{\mu\frac{\partial}{\partial\mu} + \beta_P\,\frac{\partial}{\partial P} - \gamma_\phi\,\phi\,\frac{\partial}{\partial\phi}\right\} V_{\rm scal}\left[P(\mu),\phi(\mu);\mu\right] = 0\,, \tag{63}$$

where as usual $\beta_P = \mu\partial P/\partial\mu$ ($P = m, \lambda, ...$) and $\gamma_\phi = \mu\partial\ln Z_\phi^{1/2}/\partial\mu$. Similarly, equation (61) can be put in the form (63), except that the ϕ term is absent.

Plugging equation (42) in the general RG equation (61), we find immediately that the renormalized vacuum term $\rho_\Lambda(\mu)$ obeys the one-loop RG-equation which we have found previously, i.e. Eq. (44).

The next step is the one-loop renormalization of the effective potential. This is standard [43], of course, although the usual discussions on this subject rarely pay much attention to disentangle the ZPE part from it. Let us do it. Once more we have to compute the one-loop expression (49), but in this case for the operator (58), so we have

$$Tr\,\ln\mathcal{K} = \Omega\int\frac{d^nk}{(2\pi)^4}\,\ln\left(-k^2 + V''(\phi_c)\right)\,. \tag{64}$$

Substituting this expression in (50), we may split the result in the suggestive form

$$V_1(\phi_c) = -\frac{i}{2}\int\frac{d^nk}{(2\pi)^n}\,\ln\left[-k^2+m^2\right] - \frac{i}{2}\int\frac{d^nk}{(2\pi)^n}\,\ln\left(\frac{k^2 - V''(\phi_c)}{k^2-m^2}\right)\,. \tag{65}$$

Notice that the first term is independent of ϕ, i.e. it only depends on the parameters of the potential. In the one-loop case, it depends only on m, but at higher loops it would also depend on λ (in the interactive theory). Diagrammatically, it corresponds to a closed one-loop diagram without external ϕ-tails, i.e. to a vacuum-to-vacuum one-loop diagram. As could be expected, this term can be identified with the one-loop contribution for the $V_{\rm ZPE}$ in Eq. (55), previously addressed in Sect. 5. Similarly, the second integral on the *r.h.s.* of Eq. (65) just gives the one-loop correction to the ϕ-dependent part of the potential, i.e. the second term on the *r.h.s.* of Eq. (56). Therefore Eq. (65) contains the full one-loop correction, which includes the ZPE part and the ϕ-dependent contribution to the potential. To it we still have to add the classical potential, cf. Eq. (56).

Sticking to the $\overline{MS}$ scheme in dimensional regularization to fix the counterterms, one finds the final result for the renormalized full effective potential (54) up to one-loop:

$$V_{\rm eff}(\phi) = \frac{1}{2}\,m^2(\mu)\,\phi^2 + \frac{1}{4!}\,\lambda(\mu)\,\phi^4 + \hbar\,\frac{(m^2+\frac{1}{2}\,\lambda\,\phi^2)^2}{4(4\pi)^2}\left(\ln\frac{m^2+\frac{1}{2}\,\lambda\,\phi^2}{\mu^2} - \frac{3}{2}\right)\,, \tag{66}$$

Notice the implicit μ-dependence of the masses and couplings. Together with the explicit μ-dependent parts of the effective potential, this insures the full RG-invariance of the effective action in the constant mean field limit, i.e. the fulfilment of Eq. (60).

As shown in Eq. (65), the two kind of one-loop effects are built in the calculation. Therefore, the expression (66) must boil down to the renormalized ZPE for $\phi = 0$, and indeed we verify that in this limit we recover the one-loop term on the $r.h.s.$ of Eq. (43):

$$V_{\rm eff}(\phi = 0) = \frac{\hbar\, m^4}{4(4\pi)^2}\left(\ln\frac{m^2}{\mu^2} - \frac{3}{2}\right). \tag{67}$$

This is accordance with the expectation in Eq. (57). We are now ready for addressing the CC fine-tuning problem in the context of a well-defined, renormalized, and RG-invariant vacuum energy density in flat space.

Once the full effective potential has been renormalized, the two loopwise expansions (55) and (56) become finite to all orders of perturbation theory. Furthermore, the basic equation (22) remains formally the same in the quantum theory, i.e. the physical energy density associated to the CC is the sum of the vacuum part plus the induced part. The only difference is that the induced part now contains all the quantum effects, i.e. it reads $\rho_{\Lambda{\rm ind}} = \langle V_{\rm eff}^{\rm ren}(\phi)\rangle$, where $V_{\rm eff}^{\rm ren}(\phi) \equiv V_{\rm eff}(\phi(\mu); m(\mu), \lambda(\mu); \mu)$ is the renormalized effective potential. Notice that the latter includes the (renormalized) ZPE part, which was absent in the classical theory. Thus, the physical CC emerging from the renormalization program (in any given subtraction scheme) reads

$$\rho_{\rm vac} = \rho_\Lambda^{\rm ren} + \langle V_{\rm eff}^{\rm ren}(\phi)\rangle = \rho_\Lambda^{\rm ren} + V_{\rm ZPE}^{\rm ren} + \langle V_{\rm scal}^{\rm ren}(\phi)\rangle\,. \tag{68}$$

The formal structure of this renormalized result is valid to all orders of perturbation theory. For simplicity we have obviated the μ-dependence, which appears implicitly in all couplings and fields, and explicitly in the structure of the terms beyond the tree-level. RG-invariance of physical quantities, initially formulated in the bare theory, tells us that such μ-dependence must finally cancel among all terms in the renormalized theory. Thus $d\rho_{\rm vac}/d\ln\mu = 0$. This cancelation, however, does not mean that the role played by the scale μ is necessarily irrelevant. In fact, since the entire structure of the renormalized $\rho_{\rm vac}$ is parameterized through μ, this fact can help us to unveil the form of the quantum effects in $\rho_{\rm vac}$ and their dependence on relevant physical scales of the problem. In the cosmological case this is only possible if the theory is formulated in curved spacetime and when we consider the renormalization effects on the effective action, not just the effective potential. We will come back to this point later on.

We are now ready to formulate the extreme severity of the fine-tuning problem, which we address here only in QFT in flat spacetime. Indeed, being the expression (68) the precise QFT prediction of the physical value of the vacuum energy density to all orders of perturbation theory, it must be equal to the observational measured value [3], i.e. $\rho_{\rm vac} = \rho_\Lambda^0 \simeq 2.5\times 10^{-47}$ GeV^4. We have already seen in section 3 that the lowest order contribution from the Higgs potential is 55 orders of magnitude larger than ρ_Λ^0, and that this enforces us to choose the vacuum term ρ_Λ with a precision of 55 decimal places such that the sum $\rho_\Lambda + \rho_{\Lambda{\rm ind}}$ gives a number of order 10^{-47} GeV^4. The problem is that the fine-tuning game, ugly enough already at the classical level, becomes devastating at the quantum level. Indeed, recall that we have the all order expansions (55) and (56). Therefore, the quantity that must be equated to ρ_Λ^0 is not just (17) but the full $r.h.s.$ of (68), which as we said is a well-defined (finite and RG-invariant) expression. It means that, instead of the "simple" equation (24), we must now fulfill the much more gruesome one:

$$\begin{aligned} 10^{-47}\,{\rm GeV}^4 = \rho_\Lambda - 10^8\,{\rm GeV}^4 \;&+\; \hbar\, V_P^{(1)} + \hbar^2\, V_P^{(2)} + \hbar^3\, V_P^{(3)}.... \\ &+\; \hbar\, V_{\rm scal}^{(1)}(\phi) + \hbar^2\, V_{\rm scal}^{(2)}(\phi) + \hbar^3 V_{\rm scal}^{(3)}(\phi)... \end{aligned} \tag{69}$$

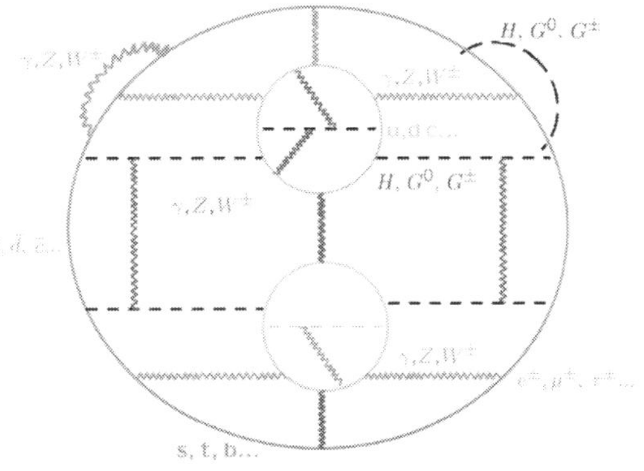

Figure 1. One of the many thousand 21th-loop vacuum-to-vacuum diagram still contributing to the value of the cosmological constant in the Standard Model of Electroweak interactions.

As compared to Eq. (24), on the *r.h.s.* of the equation we now have, in addition, two independent perturbatively renormalized series contributing to the observed value of the vacuum energy density to one-loop, two-loops, three-loops etc... up to some nth-loop order (both for the ZPE series and the series associated to the quantum corrected Higgs potential). It follows that the numerical value for the "renormalized vacuum counterterm" ρ_Λ must be changed accordingly order by order in perturbation theory. Specifically, the number ρ_Λ on the *r.h.s.* of equation (69) must be re-tuned with 55 digits of precision as many times as the number of diagrams (typically thousands) that contribute to the highest nth-loop still providing a contribution to the CC that is at least of the order of the experimental number placed on the *l.h.s.* of that equation. For example, let us roughly assume that each electroweak loop contributes on average a factor $g^2/(16\pi^2)$ times the fourth power of the electroweak scale $v \equiv \langle\phi\rangle \sim 250$ GeV (see section 3), where g is either the $SU(2)$ gauge coupling constant or the Higgs self-coupling, or a combination of both. It follows that the order, n, of the highest loop diagram that may contribute to the measured value of the vacuum energy density, and that therefore could still be subject to fine-tuning, can approximately be derived from

$$\left(\frac{g^2}{16\,\pi^2}\right)^n v^4 = 10^{-47}\, GeV^4\,. \tag{70}$$

Take e.g. g equal to just the $SU(2)$ gauge coupling constant of the electroweak SM, which satisfies $g^2/(16\pi^2) = \alpha_{\rm em}/(4\pi\sin^2\theta_W) \simeq 2.5\times 10^{-3}$. This is actually a conservative assumption because in practice there are larger contributions in the SM associated to the big top quark Yukawa coupling, but it will suffice to illustrate the situation. Since $v \equiv \langle\phi\rangle \sim 250$ GeV, we find $n \simeq 21$. Therefore all of the (many thousand) loop diagrams pertaining to the 21th electroweak order (see Fig. 1 for a typical example) are still contributing sizeably to the value of the CC, and must therefore be readjusted by an appropriate choice of the renormalized value of the vacuum term ρ_Λ. This is of course preposterous and completely unacceptable. Even though this is nonsense, it is implicitly accepted by everyone that admits that such "technical trick" is a viable solution to the CC problem. It goes without saying that this situation worsens even more for higher energy extensions of the SM, such as in GUT's. Furthermore, replacing the cosmological term by a cosmic scalar field with some peculiar potential just iterates the same kind of fine tuning

Journal of Physics: Conference Series **453** (2013) 012015 doi:10.1088/1742-6596/453/1/012015

problem, let alone that it does not explain why e.g. the electroweak vacuum energy can be hidden under the rug with no relation to the CC value.

Special symmetries (such as Supersymmetry (SUSY), for example) are of little help to solve the CC problem, despite some early hopes [44], since SUSY is necessarily broken, and hence all the above problems just replicate very similarly to the SM case. Only dynamical mechanisms could really help here, namely mechanisms capable of automatically adjusting the CC to the present tiny value even starting from an arbitrarily large one in the early universe. As we know, they were first attempted using scalar fields, but suffer from a "no-go theorem" (see Weinberg in [4]). However, they are technically possible within generalized dynamical vacuum models with modified gravity (cf. the intriguing proposal in Ref. [5]). Unfortunately, we still don't know which fundamental theory could naturally accommodate them.

7. Zero-point energy in curved space-time

So far so good, but where does the curvature of spacetime enter the previous discussion on the ZPE in Sect. 5 or the Higgs potential in Sect. 6? Nowhere! However, it should be natural to discuss the vacuum energy in a curved background if we aim at elucidating its possible connection to the value of the cosmological term, shouldn't it? In this section, we are going to summarize the effects of curvature for the calculation of the ZPE. The discussion of the Higgs potential part in curved spacetime does not change in any fundamental way the kind of problems that we will meet for the discussion of the ZPE in a curved background, so we shall limit here to briefly consider this last aspect of the vacuum problem and leave the (somewhat bulky) details for a more comprehensive exposition elsewhere [7].

The first and most important observation is that in the presence of vacuum energy (and hence with a nonvanishing value of the CC in Einstein's equations) the constant Minkowski metric $\eta_{\mu\nu} = (+,-,-,-)$ obviously is not a solution of the field equations (2). It means that, with $\Lambda \neq 0$, the metric $g_{\mu\nu}$ solving them is a nontrivial one and therefore the spacetime is intrinsically curved. However, this is not what we have assumed in the calculation leading to the result (43). Therefore, that calculation is, in principle, not appropriate to reflect any dynamical aspect of the expanding spacetime. In particular, the arbitrary mass scale μ has no immediate connection with any physical quantity of relevant interest, in contrast to the situation of a cross-section calculation in, say, QCD, where one tries to make an association with some characteristic energy scale of the colliding or decaying process in order to estimate the quantum corrections with the help of the RG.

This is certainly a relevant issue for QFT in curved spacetime [29, 30], i.e. the theory of quantized fields on a curved classical background, where the gravitational field itself does not participate of the quantization process. In the functional formulation this means that the generating functional of the theory (from which we can derive the Green's functions from the functional derivatives of it with respect to a classical source $J(x)$) can be written as a path integral over only the quantum matter fields (collectively represented here by ϕ):

$$Z[J] = \int [D\phi] \exp\left\{\frac{i}{\hbar}\left[S\left[\phi, g_{\mu\nu}\right] + \int d^4x\sqrt{-g(x)}\, J(x)\,\phi(x)\right]\right\}. \tag{71}$$

Once more $S[\phi_c]$ in that formula is the classical action for the scalar field, but adapted to the curved spacetime:

$$S[\phi, g_{\mu\nu}] = \int d^4x\,\sqrt{-g(x)}\,\left[\frac{1}{2}\,g^{\mu\nu}\,\partial_\mu\phi\,\partial_\nu\phi + \frac{1}{2}\,\xi\,\phi_c^2\,R - V_c(\phi)\right]. \tag{72}$$

A first complication is that this matter action generalizes the original form (46) in that we have included in it the generally invariant integration measure $d^4x\sqrt{-g(x)}$; and also a possible non-minimal coupling term $\xi\,\phi_c^2\,R$, where ξ is a dimensionless coefficient, which is necessary for renormalizability (see below).

In principle, the problem under discussion should also be a focus of attention for Quantum Gravity [45], the theory where the gravity field is also quantized together with the matter fields. Unfortunately, QG is essentially non-perturbative and it does not exist as a consistent theory yet. String theory seems to be able to address some of the fundamental problems posed by QG, but it is not obvious that it can solve all of them, not even a significant part. In any case we do not wish to discuss them here. Our objective right now is much more modest, and yet not trivial. We deal with the problems of QFT in curved spacetime. In contrast to QG, the former exists as a renormalizable theory in the perturbative sense. It can be considered a fairly successful theory of the external gravitational field in interaction with quantum matter. That beautiful theory has already a long time-honored tradition, collected in several books and reviews [29, 30, 46, 47]. However, not even at present such semiclassical approach to gravity is completely understood. In fact, it is far from being so [31]. A serious hint is the fact that the ZPE contribution to the vacuum energy density in QFT in curved spacetime turns out to be exactly the same as the one we have obtained for flat space, i.e. Eq. (43), up to the renormalization of the higher order operators R^2, $R_{\mu\nu}R^{\mu\nu}$... in the vacuum action, including of course the low energy part represented by the Einstein-Hilbert term with cosmological constant (41). To see this, let us first recall that in order that the QFT theory becomes renormalizable in curved spacetime, the classical vacuum action must also contain the standard higher derivative (HD) part:

$$S_{\rm HD} = \int d^4x\sqrt{-g}\,\left\{\alpha_1^{\rm (b)}C^2 + \alpha_2^{\rm (b)}\,R^2 + \alpha_3^{\rm (b)}E + \alpha_4^{\rm (b)}\nabla^2 R\right\}\,, \tag{73}$$

where, $\alpha_{1,2,3,4}^{\rm (b)}$ are bare parameters, $C^2 = R_{\mu\nu\rho\sigma}R^{\mu\nu\rho\sigma} - 2R_{\mu\nu}R^{\mu\nu} + (1/3)\,R^2$ is the square of the Weyl tensor and $E = R_{\mu\nu\rho\sigma}R^{\mu\nu\rho\sigma} - 4R_{\mu\nu}R^{\mu\nu} + R^2$ is the topological Euler's density. The coefficients α_i of these terms are then renormalized by the quantum effects, which have the same structure, and in this way we can absorb the new infinities. The theory is thus one-loop renormalizable in curved spacetime. The HD terms are short distance effects which have no impact at low energies, i.e. they do not lead to significant corrections to the Einstein equations (2), valid at long distances (the situation of the present, low-energy, universe). In the FLRW metric all these terms are of order $R^2 \sim H^4$ (including $\Box R \sim H^4$), which are negligible for the entire post-inflationary history of the universe. The terms we are interested in are those which are at least of order $R \sim H^2$.

To achieve a finite renormalized theory we proceed as follows. As usual we split each bare coefficient into a renormalized term plus a counterterm:

$$\alpha_1^{\rm (b)} = \alpha_1(\mu) + \delta\alpha_1\,, \qquad\qquad \alpha_2^{\rm (b)} = \alpha_2(\mu) + \delta\alpha_2 \tag{74}$$

and similarly,

$$\frac{1}{G^{\rm (b)}} = \frac{1}{G(\mu)} + \delta\left(\frac{1}{G}\right)\,, \qquad\qquad \rho_\Lambda^{\rm (b)} = \rho_\Lambda(\mu) + \delta\rho_\Lambda\,, \tag{75}$$

We disregard here the details on the topological term and the total derivative part of the HD action. In doing these splittings the renormalized quantities are μ-dependent because they are supposed to be defined at a given renormalization point μ in RG-space, where they should hypothetically make contact with some experimental input. The point μ is arbitrary, and in fact the wisdom of the RG is that the physics should not depend on its election. However, μ can have a more or less physical meaning depending on the renormalization scheme that is used. Finally, the counterterms can be chosen to cancel the UV divergences. These have been dimensionally regularized in our approach (they appear as poles at $n = 4$) and therefore the

Journal of Physics: Conference Series **453** (2013) 012015 doi:10.1088/1742-6596/453/1/012015

cancelation can be easily arranged. We just require that $\delta\alpha_{1,2}$, as well as $\delta(1/G)$ and $\delta\rho_\Lambda$, are chosen so as to make finite the values of the final one-loop parameters. The use of dimensional regularization is quite standard and relatively simple, but on the other hand it is a source of ambiguity when we face the physical interpretation of the results. This ambiguity reaches a climax when we next use minimal subtraction to renormalize the theory. Again, minimal subtraction (whether in the $\overline{\rm MS}$ form or pole subtraction followed with whatever finite part we like) is simple enough, but the renormalized quantities (which become functions of the arbitrary mass scale μ) suffer from a direct physical interpretation. Let us note that we can do all this without still mentioning what is the physical system we have behind, we only know that we are renormalizing the classical vacuum action. The system can be the universe or some other gravitational framework in which we may be interested to study the quantum matter effects. It is obvious that at some point we have to put our system in context, and then assess what could be the physical quantity with which we could dream establishing a possible association with μ. This would, of course, be unnecessary if μ had been a physical subtraction point from the very beginning (say some momentum subtraction point [48]), although this has a calculational price. But if we want to still stay with the mathematical simplicity of dimensional regularization with MS-like subtraction, we are forced to look for such a physical interpretation in the last stage of the calculation. Unfortunately, not even solving this problem should mean the end of our troubles; there are still some other headaches in the list, maybe the most severe ones.

Let us however move on and see. Following the aforementioned procedure, the total effective action (classical plus one-loop corrections) can be conveniently organized as follows:

$$\Gamma = S[\phi_c] + S_{\rm HD} + S_{\rm EH} + \Gamma^{(1)}_{\rm eff} = S[\phi_c] + S^{(1)}_{\rm HD} + S^{(1)}_{\rm EH}\,, \tag{76}$$

where

$$S^{(1)}_{\rm HD} == \int d^4x\sqrt{-g}\,\left(\alpha_1^{(1)}C^2 + \alpha_2^{(2)}\,R^2 + ...\right) \tag{77}$$

and

$$S^{(1)}_{\rm EH} = -\int d^4x\sqrt{-g}\left(\frac{1}{16\pi G^{(1)}}R + \rho^{(1)}_{\rm vac}\right)\,. \tag{78}$$

In the above equations we have absorbed the bare terms and the quantum matter effects in the one-loop coefficients $\alpha_1^{(1)}$ and $\alpha_2^{(1)}$ of the HD action and the one-loop Newton's constant and CC term, $G^{(1)}$ and $\rho^{(1)}_{\rm vac}$, of the low energy EH action [4]. The final result reads

$$\alpha_1^{(1)} = \alpha_1(\mu) - \frac{\hbar}{2(4\pi)^2}\frac{1}{120}\left(\ln\frac{m^2}{\mu^2} + \text{finite const.}\right) \tag{79}$$

$$\alpha_2^{(1)} = \alpha_2(\mu) - \frac{\hbar}{4(4\pi)^2}\left(\frac{1}{6} - \xi\right)^2\left(\ln\frac{m^2}{\mu^2} + \text{finite const.}\right) \tag{80}$$

$$\frac{1}{16\pi G^{(1)}} = \frac{1}{16\pi G(\mu)} + \frac{\hbar\, m^2}{2(4\pi)^2}\left(\frac{1}{6} - \xi\right)\left(\ln\frac{m^2}{\mu^2} + \text{finite const.}\right) \tag{81}$$

$$\rho^{(1)}_{\rm vac} = \rho_\Lambda(\mu) + \frac{m^4\,\hbar}{4\,(4\,\pi)^2}\left(\ln\frac{m^2}{\mu^2} + \text{finite const.}\right)\,. \tag{82}$$

[4] These are the quantum effects from the matter field ϕ encoded in the explicit computation of the one-loop part of the effective action, Eq. (49), in the curved space-time case. This is of course the hardest part of the calculation, that I am fully sparing to the reader here – see e.g. [29, 30]; confer also [7] for the details along the present lines.

The parameters $P_i^{(1)} = \alpha_1^{(1)}, \alpha_2^{(1)}, 1/G^{(1)}, \rho_{\rm vac}^{(1)}$ on the *l.h.s.* of these expressions are the final "one-loop parameters"; they are finite at one loop because we have used the counterterms to cancel the divergences coming from the one-loop contributions to them. The very fact that such redefinition of the original coefficients of (73) can be made shows the practical possibility to renormalize the theory. It does not mean, however, that we can get an immediate physical interpretation of the renormalized expressions. The arbitrariness of the finite constant terms in the expressions (82) is only a small hint of the ambiguity and lack of direct interpretation of these formulas.

If we would continue the calculation at 2-loops, the parameters $P_i^{(1)}$ would play the role of bare parameters, in a similar way as the bare parameters (74)-(75) of the classical action, and therefore they would be UV-divergent quantities that split once more into a renormalized parameter and a corresponding counterterm. These counterterms would then be used to cancel the 2-loop divergences etc. As parameters of the bare action, all of the $P_i^{(1)}$ are of course independent of the arbitrary renormalization point μ. The explicit μ-dependence of the various terms in each $P_i^{(1)}$ must cancel in the overall expression. One source of μ-dependence comes from the corresponding "renormalized parameters", $P_i(\mu)$, which therefore "run" with the scale μ. However, as already warned before, at this point such "running" has no obvious relation with the variation of any physical quantity; whether such connection is possible cannot be unambiguously decided in the absence of a concrete physical framework. In the next section, however, we place these formulae in the cosmological context and only then some chance exists for a possible physical interpretation.

Mathematically, the μ-running is determined by the corresponding renormalization group equation (RGE) for each of the parameters, and follows from setting the total derivatives of the one-loop parameters $P_i^{(1)}$ with respect to μ to zero. For convenience we compute the logarithmic derivatives of each one of them, i.e. $dP_i^{(1)}/d\ln\mu = 0$. The explicit RGE's are immediately obtained from (79)-(82):

$$\frac{d\alpha_1(\mu)}{d\ln\mu} = -\frac{\hbar}{120(4\pi)^2} \equiv \beta_1^{(1)} \tag{83}$$

$$\frac{d\alpha_2(\mu)}{d\ln\mu} = -\frac{\hbar}{2(4\pi)^2}\left(\frac{1}{6}-\xi\right)^2 \equiv \beta_2^{(1)} \tag{84}$$

$$\frac{d}{d\ln\mu}\left(\frac{1}{16\pi G(\mu)}\right) = \frac{\hbar\, m^2}{(4\pi)^2}\left(\frac{1}{6}-\xi\right) \equiv \beta_{G^{-1}}^{(1)} \tag{85}$$

$$\frac{d\rho_\Lambda(\mu)}{d\ln\mu} = \frac{\hbar\, m^4}{2\,(4\pi)^2} \equiv \beta_\Lambda^{(1)}\,. \tag{86}$$

Notice that the obtained RGE's do not depend at all on the unspecified finite terms indicated in equations (79)-(82). The *r.h.s*'s of these expressions define the corresponding one-loop β-functions that control the running of these parameters in the renormalization scheme we have used, an MS-based one. Note, in particular, the RGE for ρ_Λ, Eq. (86). It coincides exactly with Eq. (44), the flat spacetime result. Integration of (86) immediately furnishes the one-loop finite result

$$\rho_{\rm vac}^{(1)} = \rho_\Lambda(\mu) + \frac{m^4\,\hbar}{4\,(4\,\pi)^2}\left(\ln\frac{m^2}{\mu^2} + \text{finite const.}\right)\,. \tag{87}$$

The finite additive constant in this formula is not very important, as it depends on the kind of MS-based subtraction scheme we use. For exact $\overline{\rm MS}$, the constant must of course be the same one as in Eq. (43), but this is quite irrelevant as it does not change at all the worrisome aspect of the result, namely the fact that we get a quantum "correction" to the vacuum energy density growing as $\sim m^4$, for a particle of mass m. So we are back to the numerological disaster first struck by Zeldovich (cf. Sect.4), which forced him to change his strategy to estimate the vacuum energy in particle physics. Somehow we have not advanced a single step since then, despite much QFT in curved spacetime! This is the result we mentioned before, and it comes as a kind of an "unexpected" surprise. It certainly does not help shed light on our poor understanding of the vacuum energy density in QFT in curved spacetime. We will come back to it in the next section.

In view of the situation, it would be unwise to rush into conclusions at this point. For example, it would make little sense to evaluate the found formula for the vacuum energy density by using some value of μ and, say, the full collection of particle masses of the SM of strong and electroweak interactions, or whatever extension of it. First, because on the face of the obtained result we feel that we are somehow in deep water and we don't know what is the next surprise we'll come across in this story; and, second, because it is obvious that the contribution from a single mass, being proportional to the quartic power of it, is completely out of range (cf. our "numerology discussion" in Sect. 4). The only possible exception is perhaps a very light neutrino mass, see Eq. (33), or the existence of new degrees of freedom at a similar mass scale that would determine the behavior of the cosmological term [42].

It goes without saying that neither QG nor string theory have been able to solve that problem either. No theoretical framework at present is capable to provide a fully consistent and realistic account for the vacuum energy density in the gravitational context. Although some of the technical problems can be glimpsed [31], the situation is far from being completely (not even substantially) understood. We feel that if the problem cannot be minimally dealt with at the level of unquantized gravity, the work of QG may not be such decisive to settle this issue. In this sense, to appeal directly to QG as an indispensable approach for finding the solution to the CC problem, maybe disputable, especially if we limit ourselves to the scale of the SM (which is extremely far away from the Planck scale). We should judiciously expect that QFT in curved spacetime is competent enough in this energy domain so as to provide a first reliable hint of the possible solution, something that goes beyond the "cul de sac" situation we have now.

8. Dynamical vacuum energy in an expanding universe

Let us focus here on the vacuum part of the effective cosmological term, i.e. we are now mainly concerned with the ZPE in curved spacetime. With the traditional approach the metric expansion is usually made around flat space, i.e. $g_{\mu\nu} = \eta_{\mu\nu} + h_{\mu\nu}$. If one traces the details of the calculation [7], one can easily see that this is the reason why the result (87) is the same as in flat space, Eq. (42). In fact, the result is entirely dependent on the free propagator in flat spacetime, which appears as the first term of the adiabatic expansion of the Green's function in the curved case. This can be troublesome since we are dealing with the renormalization of the cosmological term, namely a term which does *not* exist in Minkowski space. Therefore, in those situations when the spacetime background is unavoidably curved, sticking to the expansion $g_{\mu\nu} = \eta_{\mu\nu} + h_{\mu\nu}$ misses the precise dynamical correction to the vacuum energy density that we are looking for. Although the root of the problem has been identified [31], the remedy for it is not available yet. One needs to compute not only the divergences, but also the first finite correction using a physical renormalization scheme (beyond $\overline{\text{MS}}$ or the like); and, above all, one has to learn how to cope with the technical limitations of the current QFT methods for computing the renormalization corrections around a non-trivial background. For de Sitter background, for example, one expects $\rho_{\text{vac}} \sim H^4$ by pure dimensional analysis, and indeed particular calculations reach that kind of conclusion [49, 50]. But no realistic attempt has ever been made for a FLRW background, where we should expect both $M_i^2 H^2$ and H^4 terms, where M_i are the particle masses. Such calculation should be a rather nontrivial one, perhaps unreachable by the presently known methods. However we expect that it should lead to a RGE of the form:

$$\frac{d\rho_\Lambda}{d\ln H^2} = \frac{1}{(4\pi)^2}\sum_i\left[a_i M_i^2\, H^2 + b_i\, H^4 + c_i\frac{H^6}{M_i^2} + ...\right], \tag{88}$$

in which $\mu = H$ should sensibly act as the natural running scale in the cosmological context. Obviously the $\mathcal{O}(H^6/M_i^2)$ terms represent the decoupling contributions that should appear in a physical renormalization scheme. Notice that the scheme used in Sect. 7 was off-shell and it could not trace the decoupling effects. The above equation should describe the "change in the value of the vacuum energy density" (triggered by the quantum matter effects) associated to a "change in the curvature of the spacetime", this curvature being of order $R \sim H^2$ in the FLRW metric. It is well known that renormalization theory can only be predictive if we first input the value of the relevant parameters at a given renormalization point. It is only then when we can predict the value at another point. For instance, the e.m. fine structure constant at the scale of the electron mass, $m_e \simeq 0.5$ MeV $= 5\times 10^{-4}$ GeV, is $\alpha(m_e) \simeq 1/137$, whereas its value at the scale of the Z-boson mass, $M_Z \simeq 90$ GeV, is $\sim 7\%$ larger: $\alpha(M_Z) \simeq 1/128$. The RG approach cannot pretend to "compute" $\alpha(m_e)$ or $\alpha(M_Z)$, only the change from one value to the other when moving from $\mu = m_e$ to $\mu = M_Z$, where here we make a direct association of μ with the physical scale of the particle masses. Similarly, renormalization theory cannot aim at computing "the value" itself of the vacuum energy density and the CC, but only the evolution or running of this value after we have measured it at some cosmic energy scale. This should suffice for an effective QFT study of the running vacuum energy density, quite different from the more ambitious attempts at predicting its current value (the old CC problem [4]). These attempts are probably doomed to fail until we can first cope with the prediction, or a more fundamental understanding, of the fundamental parameters of the SM, if that is possible at all.

8.1. Running gravitational coupling and vacuum energy

While it is not possible to derive the general RGE for ρ_Λ in QFT in curved spacetime of the form given in (88) yet, we may be able to hint at the expected dynamical effects induced by the quantum corrections in some indirect way.

If we look at the vacuum diagrams of Fig. 2, we notice the following. The simplest diagram there is the classic "blob" with just a closed loop line of the scalar field. This one is already present in the flat spacetime case. However, the presence of the external gravitational field introduces "hair" (i.e. $h_{\mu\nu}$-tails of the classical field departing from the flat space structure). This modifies the originally "bald" quantum vacuum blob, and for this reason we have many other blob diagrams in Fig. 2, the "haired" ones. The infinitely many tails are induced on all possible diagrams preexisting in flat space, and are generated by the non-polynomial expansion of the factor $\sqrt{-g} = 1 + \frac{1}{2}h + \mathcal{O}\left(h^2\right) + \mathcal{O}\left(h_{\mu\nu}\, h^{\mu\nu}\right) + ...$ in the action (72) of the scalar field, which we take as a free field here, i.e. $V(\phi) = (1/2)\, m^2\, \phi^2$. For the particular case of the vacuum diagrams under consideration, the expansion of $\sqrt{-g}$ is performed in Eq. (49) after we replace $d^4x \to \sqrt{-g}\; d^4x$ in it. The previously computed renormalization effects from the matter field ϕ on the parameters of the purely geometric vacuum action $S_{EH} + S_{HD}$ (conf. Sect. 7) can now be viewed diagrammatically in Fig. 2, which leads more intuitively to a possible physical interpretation after we put our external gravity system in context – the cosmological one, in this case.

What is important from this diagrammatic representation is to realize that the first "blob" does not interact with the curved background at all, as it has no external tails of the gravitational field. So any result that depends solely on that "bald" diagram contains the same information as in flat space. This is what happens with the unsuccessful Eq. (87). In contrast, the 'haired blobs" do interact with the background geometry. Let us concentrate on the first "haired blob".

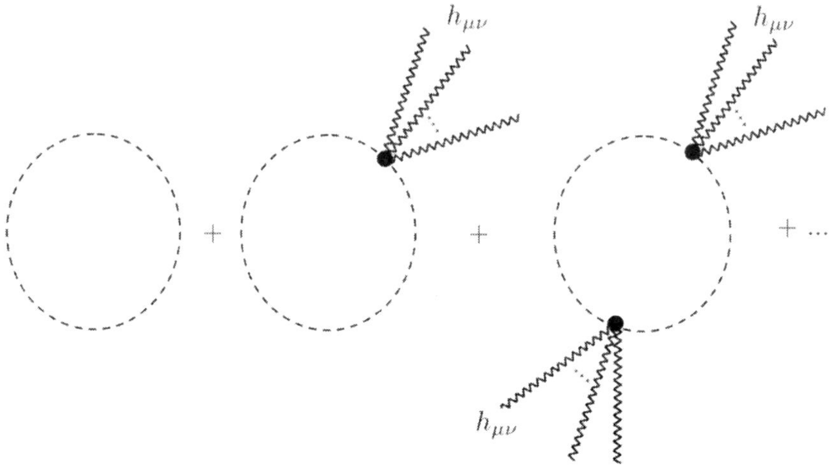

Figure 2. The one-loop vacuum-to-vacuum diagrams of the scalar matter field in the presence of an external gravitational field. The first type of contribution is the "bald blob" with no external tails, which is quartically divergent. The other contribution are the "haired blobs" with one or more external field insertion, where an arbitrary number of $h_{\mu\nu}$-tails of the background field are attached to one or more points. They appear from the expansion of the metric in the form $g_{\mu\nu} = \eta_{\mu\nu} + h_{\mu\nu}$, and the corresponding determinant in the action: $\sqrt{-g} = 1 + \frac{1}{2}h + \mathcal{O}\left(h^2\right) + \mathcal{O}\left(h_{\mu\nu}h^{\mu\nu}\right) + ...$ with $(h \equiv g_{\mu\nu}h^{\mu\nu})$. The first haired vacuum diagram with one insertion is quadratically divergent, as there is one propagator of the scalar field. The bunch of tails organize themselves in a covariant way to generate an action term of the form $\sqrt{-g}\, m^2 R$, and hence renormalize the inverse gravitational Newton's coupling $1/G$ in the EH action (78). The third type of diagram is the "doubly haired blob" and contains two insertion points. With two propagator lines, it is only logarithmically divergent; it renormalizes the coefficients $\alpha_{1,2}$ of the HD-action (77). The diagrams with three or more insertions of the external field are finite. In the flat spacetime case only the "bald blob" contributes, so the dynamics of the vacuum energy in an expanding universe must come from the "haired blobs", which may pump in energy from the expanding background into the vacuum matter loops.

It contains one insertion point where infinitely many tails of the external gravitational field $h_{\mu\nu}$ can be attached. That blob, as all the other'haired" ones, is pumping external energy-momentum from the FLRW background. The typical magnitude of the momentum should be of the order of H, which has dimension of energy in natural units. In this sense the association $\mu = H$ looks quite reasonable and physically intuitive, but only after we have placed the system in the cosmological context. This is similar to the correspondence of μ with a momentum variable q in particle physics processes, although in the cosmological case the interpretation is less straightforward, specially in the low energy domain.

The bunch of $h_{\mu\nu}$-tails in the first "haired" diagram organize themselves in a covariant way to generate an action term of the form $\sqrt{-g}\, m^2 R$, and hence renormalize the inverse gravitational Newton's coupling $1/G$ in the EH action (78). The precise RGE for this coupling was given before in Eq. (85). At this point this does not yet lead to a renormalization of the vacuum energy, but it can be related to it in an indirect way, as we shall see. Whereas in the absence of a physical context the μ-dependence would not be of much help, if we now follow the aforementioned ansatz $\mu = H$ within the cosmological context, we can immediately integrate (85) to find (in natural units):

$$\frac{1}{G(H)}=\frac{1}{G_0}+\frac{m^2}{2\pi}\left(\frac{1}{6}-\xi\right)\ln\frac{H^2}{H_0^2}\,, \tag{89}$$

where $G_0 \equiv G(H_0)=1/M_P^2$ is the current value of the gravitational coupling. Equivalently,

$$G(H)=\frac{G_0}{1+\nu\ln\left(H^2/H_0^2\right)}\,, \tag{90}$$

where $\nu=(1/2\pi)\left(\frac{1}{6}-\xi\right)m^2/M_P^2$ is a dimensionless coefficient which acts as the reduced β-function for the running of the gravitational coupling with the physical scale H. Since the coefficient ξ in the previous equation is not determined and the number of participating matter fields is arbitrary, we can generalize ν in the form

$$\nu=\frac{1}{2\pi}\sum_i\left(\frac{1}{6}-\xi_i\right)\frac{m_i^2}{M_P^2}\,, \tag{91}$$

It is important to realize that if the running gravitational coupling as a function of $\mu=H$ is given by Eq. (90), we cannot continue with Eq. (87) as a valid RGE for the running vacuum energy density, since there is a link between the running of $G(H)$ and the running of $\rho_\Lambda(H)$ which must be preserved. That link is enforced by the Bianchi identity satisfied by the Einstein tensor on the *l.h.s.* of Eq. (1), namely $\nabla^\mu G_{\mu\nu}=0$. At low energies we can disregard the $\mathcal{O}(H^4)$ contributions provided by the HD terms (the effects from the "double haired" diagrams in Fig. 2), and in this way we obtain the following relation for the source term on the *r.h.s.* of Einstein's equations:

$$\nabla^\mu\left(G\,\tilde{T}_{\mu\nu}\right)=\nabla^\mu\left[G\left(T_{\mu\nu}+g_{\mu\nu}\,\rho_\Lambda\right)\right]=0\,. \tag{92}$$

Using the FLRW metric and the standard energy-momentum tensor for matter in the form of a perfect fluid, the last equation provides the following "mixed" local conservation law:

$$\frac{d}{dt}\left[G(\rho_m+\rho_\Lambda)\right]+3\,G\,H\,(\rho_m+p_m)=0\,. \tag{93}$$

Therefore, if $G(t)=G(H(t))$ is evolving with the expansion in some particular way, the previous identity enforces $\rho_\Lambda=\rho_\Lambda(H(t))$ to evolve accordingly. We can compute the way ρ_Λ evolves as a function of H if we assume that $G(t)=G(H(t))$ evolves as in Eq. (90). Let us assume at this point that matter is covariantly conserved, i.e.

$$\dot\rho_m+3\,H\,(\rho_m+p_m)=0\quad\Rightarrow\quad\rho_m(a)=\rho_m^0\,a^{3(1+\omega_m)}\,, \tag{94}$$

where $p_m=\omega_m\rho_m$ is the equation of state. It is then easy to show that (93) boils down to

$$(\rho_m+\rho_\Lambda)\dot{G}+G\dot{\rho}_\Lambda=\left[(\rho_m+\rho_\Lambda)\frac{dG}{dH}+G\frac{d\rho_\Lambda}{dH}\right]\dot{H}=0\,. \tag{95}$$

Obviously $\dot{H}\neq 0$. Therefore, combining (95) with Friedman's equation for flat space,

$$H^2=\frac{8\pi G}{3}(\rho_m+\rho_\Lambda)\,, \tag{96}$$

in which G is given in the present instance by (90), it is straightforward to solve for ρ_Λ as a function of the Hubble rate, the result being a simple "affine" quadratical law:

$$\rho_\Lambda(H)=c_0+\frac{3\nu}{8\pi}\,M_P^2\,H^2\,, \tag{97}$$

Journal of Physics: Conference Series **453** (2013) 012015
doi:10.1088/1742-6596/453/1/012015

ν being here, of course, the same coefficient defined in (91). This is a nice equation. With it we have found an explicit realization of the kind of dreamed law for the vacuum energy density suggested in Eq. (32), with $\beta = 3\nu/8\pi$.

Interestingly enough the obtained equation (97) is precisely of the same form as the one that would follow from solving the previously mentioned general RGE (88) provided we restrict the latter to the current universe, namely when the $\mathcal{O}(H^4)$ terms can be neglected. Integrating (88), and comparing with (97) it follows that ν must also be given by

$$\nu = \frac{1}{6\pi} \sum_{i=f,b} a_i \frac{M_i^2}{M_P^2}, \tag{98}$$

in which only the coefficients a_i of the first term on the *r.h.s* of (88) are involved. We shall adopt this equation as it is more general and assumes independent contributions from bosons and fermions with different multiplicities. Normalizing the obtained result (97) with respect to the current CC value $\rho_\Lambda(H_0) = \rho_\Lambda^0$, we can rewrite it as

$$\rho_\Lambda(H) = \rho_\Lambda^0 + \frac{3\nu}{8\pi} M_P^2 (H^2 - H_0^2). \tag{99}$$

where c_0 is related with the current value of the vacuum energy as $\rho_\Lambda^0 = c_0 + (3\nu/8\pi)\, M_P^2 H_0^2$.

As explained around Eq. (32) in Sect. 4, the presence of the affine term $c_0 \neq 0$ is crucial for a realistic implementation of the model. A vacuum energy evolving only as $\sim H^2$ (with $c_0 = 0$) would be incompatible with the transition from deceleration to acceleration [39]. Remarkably, the $c_0 \neq 0$ term appears in this context quite naturally as an integration constant of the RGE, which is fixed by the boundary conditions as in Eq. (99).

Let us emphasize that the affine quadratic law (99) insures a mild evolution of the vacuum energy density, specially if the parameter $|\nu| < 1$. We shall consider observational limits on this parameter in Sect. 8.3, but it is obvious from its definition (91) or (98) that the natural theoretical range expected for it is $|\nu| \ll 1$. This expectation will be confirmed by the phenomenological analysis. What is utmost important to note is that no term of the form $\sim M_i^4$ drives now the evolution of the vacuum energy. Indeed, this kind of terms are incompatible with the mild running of the gravitational coupling, as determined by Eq. (90). This equation, together with the covariant constraint imposed by the Bianchi identity (95), imply that the cosmological term must evolve in the form (99). Furthermore, from the renormalization group point of view the $\sim M_i^4$ terms are not expected on the *r.h.s.* of the general RGE (88) either, as otherwise this would mean that a particle with mass M_i is an active degree of freedom for the running of ρ_Λ. Since, however, the running is parameterized by the scale $\mu = H$, this would entail $H > M_i$, which is impossible for any known particle mass at any time in the matter dominated epoch (recall that $H_0 \sim 10^{-42}$ GeV) [5].

To summarize, while a direct calculation of the dynamical renormalization effects on the cosmological term is not possible at the moment without expanding around a non-flat background in the presence of massive fields and within a physical renormalization scheme, at least an indirect hint of the result should be glimpsed by requiring the consistency of the renormalization effects on the different terms of the effective action. As these terms are linked

[5] Even if we go to the radiation dominated epoch, at temperature T, we find from Friedman's equation (with $\rho_m \sim T^4$) that to satisfy the condition $H > M_i$ roughly means $T^2/M_P > M_i$, or equivalently $M_i^4/T^4 < M_i^2/M_P^2 \ll 1$ (for any particle). Hence, at the time when the $\sim M_i^4$ vacuum contributions start being active they are negligible as compared to the radiation contribution $\sim T^4$. We conclude that, in the RG formulation, the terms $\sim M_i^4$ remain innocuous throughout the entire cosmic history: for, at low energies, these terms are not allowed since the condition $H > M_i$ can never be fulfilled, whilst the effect becomes completely irrelevant at high energies (when that condition is possible).

by the Bianchi identity, the possible quantum effects are tied to the general covariance of the theory. This requirement can give us an indirect clue, which can be read off more physically from the diagrams of Fig. 2, to wit: while for flat spacetime only the "bald blob" contributes, the dynamics of the vacuum energy in an expanding universe should emerge from the "haired blobs", which may pump in energy from the expanding background into the vacuum matter loops. The renormalization effects impinged upon the other terms of the vacuum action should be covariantly consistent with those affecting the cosmological term. At low energies it is enough to collect the $\mathcal{O}(H^2)$ effects from this method, as reflected in Eq. (97).

As long as a direct calculation is not feasible, such roundabout path gives us a handle on the type of effects we are looking for. It is encouraging to see that the obtained result by this procedure is free from the unwanted $\sim m^4$ contributions that plague the traditional approach. Somehow the $\mathcal{O}(H^2)$ terms are the real effects that remain in the expanding spacetime after we subtract the flat spacetime result — quite in the same manner as when in the calculation of the Casimir effect (see e.g. [51, 52] for a review) one removes the the divergence of the result upon subtracting the vacuum energy density when there are no plates [6]. Similarly as the Casimir effect is related to the disturbance caused on the vacuum state in the presence of boundary conditions (the plates), the dynamical evolution of the vacuum energy in an expanding background is what remains of the disturbed vacuum energy density after we remove the flat spacetime result – which is also contained in the curved spacetime calculation, Eq. (87) — and with it also disappear the unwanted $\sim m^4$ contributions. At the end of the day only the $\mathcal{O}(H^2)$ disturbance, along with the additive c_0 term in Eq. (97), are left at low energy as the final measurable output. Let us emphasize once more that the additive term c_0, which is essential for a correct phenomenological description, appears in a natural way in the present framework as required by the integration of the general RGE (88) for the vacuum energy in an expanding spacetime.

Interestingly, a kind of equation such as (99) was suggested in [31], and previously in [54, 55, 56, 57] on more phenomenological grounds. But the first hint that an expression of that sort could be somehow derived from a specific QFT framework was provided in Ref. [58] in a conformal field theory framework. More recently there have appeared alternative QFT frameworks suggesting a similar kind of evolution of the vacuum energy leading once more to dynamical terms $\sim H^2$ [59]. See also [60] for other consdierations along these lines.

We expect that the low energy running vacuum model (97) can be generalized in the form (88) so as to include the important effects from the higher powers of H at the early inflationary times. A unified model of this kind, where the dark energy emerges at late times as a "fossil" of the early inflationary universe was presented in [58] in the framework of the generalized anomaly induced inflation [61]. These models of the early universe, combined with the low energy theory, suggest that the effective expression of the vacuum energy density should be a combination of even powers of H, essentially H^2 and H^4. Recently, in Ref. [62, 63, 64] a detailed analysis of the complete cosmic history of the universe has been presented by considering the class of models of the form

$$\rho_\Lambda(H) = c_0 + \frac{3\nu}{8\pi}\,M_P^2 H^2 + \frac{3\alpha}{8\pi}\,M_P^2 \frac{H^{2n}}{H_I^{(2n-2)}}\,, \tag{100}$$

in which $n \geqslant 2$. Clearly, this expression is an extension of Eq. (97) along the lines of the general RGE (88). The higher power of H should obviously be operative only for large values of H near the inflationary scale H_I (presumably a GUT scale not very far from the Planck scale) [7].

[6] Recall that the Casimir effect is the force between a pair of neutral, parallel conducting planes due to the disturbance of the quantum electrodynamics (QED) vacuum. Since at zero temperature there are no real photons in between the plates, it should be the vacuum alone, i.e., the ground state of QED which causes the plates to attract each other. The Casimir effect is usually advocated as experimental evidence for the ZPE, although this matter is far from being settled, see e.g. [53].

[7] See also Ref. [65] for interesting thermodynamical considerations around these dynamical vacuum models.

Journal of Physics: Conference Series **453** (2013) 012015 doi:10.1088/1742-6596/453/1/012015

Typically we expect $n = 2$, i.e. a high energy behavior $\sim H^4$ [62]. At present, the H^4 term is of course negligible and we are effectively left with the low energy theory (97). Notice that the dimensionless coefficient α (enabling the running of the vacuum energy near the GUT scale) can be related to the coefficients of the general RGE (88) as follows:

$$\alpha = \frac{1}{12\pi}\frac{H_I^2}{M_P^2}\sum_{i=f,b} b_i\,. \tag{101}$$

We point out that while we use only even powers of the Hubble rate, odd powers have also been considered phenomenologically in the old literature for the treatment of the inflationary stages [66]. More recently one can also find models where odd powers are used for describing the CC evolution for the present universe, or in combination with even powers for the early universe [37, 67]. However, the general form that we propose in (100) involves only even powers of the expansion rate, at all stages of the cosmological evolution, as a most fundamental requirement from the general covariance of the effective action of QFT in curved spacetime [31].

8.2. Different scenarios for running cosmological parameters

We discuss now some possible scenarios for running cosmological parameters. We shall focus here on the implications for the low energy regime of the cosmological evolution and therefore we do not consider the highest powers of H introduced in Eq. (100), it will be enough to consider the H^2 dynamical effects. Furthermore, we consider the spatially flat FLRW metric only, i.e. $ds^2 = dt^2 - a^2(t)d\vec{x}^2$, where $a(t)$ is the time-evolving scale factor. After all this seems to be the most plausible possibility in view of the present observational data [3] and the natural expectation from the inflationary universe. In such conditions, the relevant Friedmann's equation providing the Hubble rate reads as in Eq. (96). As stated, we assume that $\rho_\Lambda = \rho_\Lambda(t)$ and $G = G(t)$ can be functions of the cosmic time t. The current value of the Hubble rate is $H_0 \equiv 100\,\mathrm{h}\,Km/s/Mpc = 1.0227\,h \times 10^{-10}\ \mathrm{yr}^{-1}$. The observations give $h \simeq 0.70$ (e.g. $h = 0.673 \pm 0.012$ from PLANCK [3]). Furthermore, the dynamical equation for the acceleration of the universe is:

$$\frac{\ddot{a}}{a} = -\frac{4\pi\,G}{3}\left(\rho_m + 3p_m - 2\rho_\Lambda\right) = -\frac{4\pi\,G}{3}\left(1+3\omega_m\right)\rho_m + \frac{8\pi\,G}{3}\,\rho_\Lambda\,. \tag{102}$$

In the late universe ($\rho_m \to 0$) the vacuum energy density ρ_Λ dominates. It accelerates the cosmos for $\rho_\Lambda > 0$. This may occur either, because ρ_Λ is constant, and for a sufficiently old universe one finally has $\rho_m(t) < 2\,\rho_\Lambda$, or because $\rho_\Lambda(t)$ evolves with time, and the situation $\rho_\Lambda(t) > \rho_m(t)/2$ is eventually reached sooner or later than expected.

Let us come back to the general covariant conservation law, Eq. (93), and consider various possibilities [8]:

- **Model I**: G =const. and ρ_Λ =const.:
  Under these conditions and in the absence of other components in the cosmic fluid, apart from matter and a strictly constant Λ-term, the local covariant conservation law of matter-radiation is strictly fulfilled, i.e. Eq. (94). If, in addition, we have zero spatial curvature, then Model I becomes the almost thirty years old flat ΛCDM, or concordance cosmological model [68], viz. the currently reigning standard cosmological model.
- **Model II**: G =const and $\dot{\rho}_\Lambda \neq 0$:

[8] This equation is not independent from equations (96) and (102), as it is actually a first integral of the system formed by the last two. For example, it is easy to check that the acceleration equation for the scale factor takes the usual form (102) even for time evolving G and ρ_Λ, as it follows from combining equations (93) and (96).

Journal of Physics: Conference Series **453** (2013) 012015 doi:10.1088/1742-6596/453/1/012015

Here Eq.(93) leads to the mixed conservation law:

$$\dot{\rho}_\Lambda + \dot{\rho}_m + 3\,H\,(\rho_m + p_m) = 0\,. \tag{103}$$

An exchange of energy between matter and vacuum takes place. The model can be solved only if more information is provided on e.g. the cosmic evolution of ρ_Λ, see the next section.

- **Model III**: $\dot{G} \neq 0$ and ρ_Λ =const.:

$$\dot{G}(\rho_m + \rho_\Lambda) + G[\dot{\rho}_m + 3H(\rho_m + p_m)] = 0\,. \tag{104}$$

Since G does not stay constant here, this equation implies matter non-conservation. It could be solved e.g. for G, if the (anomalous) cosmic evolution of ρ_m would be given by some particular ansatz.

- **Model IV**: $\dot{G} \neq 0$ and $\dot{\rho}_\Lambda \neq 0$:
Although several possibilities are available here, the simplest one is of course the framework that has motivated our analysis in the previous section, where matter is covariantly conserved and the dynamical interplay occurs between G and ρ_Λ through Eq. (95). The cosmological equations for this model were solved in Sect. 8.1. In the next section we solve also the "running" Models II and III and compare with the present one.

8.3. Solving the cosmological equations for Models II and III

The class of Models II and III is quite general and we cannot solve them unless we provide some more information, similarly to the situation with Model IV. Let us next concentrate on solving the class of scenarios denoted as Model II. Let ρ_M^0 be the total matter density of the present universe, which is essentially non-relativistic ($\omega_m \simeq 0$). The corresponding normalized density is $\Omega_M^0 = \rho_M^0/\rho_c^0 \simeq 0.3$, where ρ_c^0 is the current critical density. Similarly, $\Omega_\Lambda^0 = \rho_\Lambda^0/\rho_c^0 \simeq 0.7$ is the current normalized vacuum energy density, for flat space. If ρ_Λ evolves with the Hubble rate in the form indicated in Eq. (99), the non-relativistic matter density and vacuum energy density evolve with the redshift as follows [38, 55]:

$$\rho_M(z;\nu) = \rho_M^0\,(1+z)^{3(1-\nu)}\,, \tag{105}$$

and

$$\rho_\Lambda(z) = \rho_\Lambda^0 + \frac{\nu\,\rho_M^0}{1-\nu}\left[(1+z)^{3(1-\nu)} - 1\right]\,. \tag{106}$$

The corresponding Hubble function reads

$$H^2(z) = \frac{8\pi\,G}{3\,(1-\nu)}\left[\rho_\Lambda^0 - \nu\,\rho_c^0 + \rho_m^0\,(1+z)^{3(1-\nu)}\right]\,. \tag{107}$$

The crucial parameter is ν, which we have introduced in sect. 8.1. It is responsible for the time evolution of the vacuum energy. From Eq. (105) we confirm, that it accounts also for the non-conservation of matter. For $\nu = 0$ it leads to the exact local covariant conservation, which for non-relativistic matter reads

$$\rho_M(z) = \rho_M^0\,(1+z)^3\,. \tag{108}$$

Next we note that $\delta\rho_M \equiv \rho_M(z;\nu) - \rho_M(z)$ is the net amount of non-conservation of matter per unit volume at a given redshift. This expression must be proportional to ν, since we subtract the conserved part. At this order we have $\delta\rho_M = -3\,\nu\,\rho_M^0(1+z)^3\ln(1+z)$. We differentiate it with respect to time and expand in ν, and finally divide the final result by ρ_M. This provides the relative time variation:

$$\frac{\delta\dot{\rho}_M}{\rho_M} = 3\nu\,\left(1 + 3\ln(1+z)\right)\,H + \mathcal{O}(\nu^2)\,. \tag{109}$$

Here we have used $\dot{z} = (dz/da)\dot{a} = (dz/da)aH \doteq -(1+z)H$. Assuming relatively small values of the redshift, we may neglect the log term and are left with:

$$\frac{\delta\dot{\rho}_M}{\rho_M} \simeq 3\nu\, H\,. \tag{110}$$

From (106) we find:

$$\frac{\dot{\rho}_\Lambda}{\rho_\Lambda} \simeq -3\nu\,\frac{\Omega_M^0}{\Omega_\Lambda^0}\,(1+z)^3\, H + \mathcal{O}(\nu^2)\,. \tag{111}$$

It is of the same order of magnitude as (110) and has the opposite sign. From a detailed analysis of the combined data on type Ia supernovae, the Cosmic Microwave Background (CMB), the Baryonic Acoustic Oscillations (BAO) and the structure formation data a direct cosmological bound on ν has been obtained in the literature [38]:

$$|\nu|^{\rm cosm.} \lesssim \mathcal{O}(10^{-3})\,, \qquad \text{(Model II)}\,. \tag{112}$$

It is consistent with the theoretical expectations [58].

Let us now analyze Model III, which can also accommodate matter non-conservation in the form (105), but at the expense of a time varying G. We compare it with a similar model where G is also running, Model IV, but where matter is conserved.

Within the class of scenarios indicated as Model III the parameter ρ_Λ remains constant ($\rho_\Lambda = \rho_\Lambda^0$) and G is variable. This is possible due to the presence of the *non* self-conserved matter density (105). Trading the time variable by the scale factor, we can rewrite Eq. (104) as follows:

$$G'(a)\left[\rho_M(a) + \rho_\Lambda^0\right] + G(a)\left[\rho_M'(a) + \frac{3}{a}\,\rho_M(a)\right] = 0\,. \tag{113}$$

The primes indicate differentiation with respect to the scale factor. We insert equation (105) in (113), integrate the resulting differential equation for $G(a)$ and express the final result in terms of the redshift:

$$G(z) = G_0\left[\Omega_M^0\,(1+z)^{3(1-\nu)} + \Omega_\Lambda^0\right]^{\nu/(1-\nu)}\,. \tag{114}$$

Here $G_0 = 1/M_P^2$ is the current value of the gravitational coupling. The previous equation is correctly normalized: $G(z=0) = G_0$, due to the cosmic sum rule in flat space: $\Omega_M^0 + \Omega_\Lambda^0 = 1$. For $\nu = 0$ the gravitational coupling G remains constant: $G = G_0$. Since ρ_Λ is constant in the current scenario, the small variation of G is entirely due to the non-vanishing value of the ν-parameter in the matter non-conservation law (105). This leads to the dynamical feedback of G with matter [9]. For the present model Friedmann's equation (96) becomes:

$$H^2(z) = \frac{8\pi G(z)}{3}\left[\rho_M^0\,(1+z)^{3(1-\nu)} + \rho_\Lambda^0\right] = H_0^2\frac{G(z)}{G_0}\left[\Omega_M^0\,(1+z)^{3(1-\nu)} + \Omega_\Lambda^0\right]. \tag{115}$$

Combining (114) and (115), we find the Hubble function of this model in terms of z:

$$H^2(z) = H_0^2\left[\Omega_M^0\,(1+z)^{3(1-\nu)} + \Omega_\Lambda^0\right]^{1/(1-\nu)}\,, \tag{116}$$

and we obtain:

$$\frac{G(z)}{G_0} = \left[\frac{H^2(z)}{H_0^2}\right]^\nu\,. \tag{117}$$

[9] The matter non-conservation law (105) was first suggested and analyzed in [55], and later on in [69].

Journal of Physics: Conference Series **453** (2013) 012015 doi:10.1088/1742-6596/453/1/012015

Since ν is presumably small in absolute value (as in the previous section), we can expand (117) in this parameter:

$$G(H) \simeq G_0 \left(1 + \nu \ln \frac{H^2}{H_0^2} + \mathcal{O}(\nu^2)\right) . \tag{118}$$

At leading order in ν this expression for the variation of G is identical to the one found for Model IV, see Eq. (90), except for the sign of ν. The equation(118) allows us to estimate the value of the parameter ν by confronting the model with the experimental data on the time variation of G. Differentiating (118) with respect to the cosmic time, we find in leading order in ν:

$$\frac{\dot{G}}{G} = 2\nu \frac{\dot{H}}{H} = -2\,(1+q)\,\nu\, H\,, \tag{119}$$

where we have used the relation $\dot{H} = -(1+q)H^2$, in which $q = -\ddot{a}/(aH^2)$ is the deceleration parameter. From the known data on the relative time variation of G the bounds indicate that $|\dot{G}/G| \lesssim 10^{-12}\,\mathrm{yr}^{-1}$ [70]. If we take the present value of the deceleration parameter, we have $q_0 = 3\Omega_M^0/2 - 1 = -0.595 \simeq -0.6$ for a flat universe with $\Omega_M^0 = 0.27$. It follows:

$$\left|\frac{\dot{G}}{G_0}\right| \lesssim 0.8|\nu|\,H\,. \tag{120}$$

Taking the current value of the Hubble parameter: $H_0 \simeq 7 \times 10^{-11}\,\mathrm{yr}^{-1}$ (for $h \simeq 0.70$), we obtain $|\nu| \lesssim 10^{-2}$. The real value of $|\nu|$ can be smaller, but to compare the upper bound that we have obtained with observations makes sense in view of the usual interpretation of ν in sect. 8.1 and the theoretical estimates indicated there. The constraints from Big Bang nucleosynthesis (BBN) for the time variation of G are more stringent and lead to the improved bound. Since Models III and IV share a similar kind of running law for the gravitational coupling (except for the sign of ν) we can extract the same bound for $|\nu|$ in the two models following the method of Ref. [71] and references therein, particularly [72]. The final result is

$$|\nu|^{\mathrm{BBN}} \lesssim 10^{-3} \qquad (\text{Models III and IV})\,. \tag{121}$$

The cosmological data from different sources furnish about the same upper bound on $|\nu|$ for the two running models where matter is non-conserved, i.e. Models II and III. In both cases the upper bound on $|\nu|$ is $\sim 10^{-3}$, as shown by equations (112) and (121).

Although the order of magnitude of the bounds on $|\nu|$ are sometimes coincident for different models, the interpretation can be quite different. For example, Model IV cannot – in contrast to Models II and III – be used to explain the possible time variation of the fundamental constants of the strong interactions and the particle masses (see next section). It can only be used to explain the time variation of the cosmological parameters ρ_Λ and G in a way which is independent, in principle, from the microphysical phenomena in particle physics and nuclear physics.

9. Dynamical vacuum energy and the time variation of the fundamental constants

As we have seen in Sect. 8.1, in an expanding universe the vacuum energy density ρ_Λ is expected to be a dynamical quantity, and should exhibit a slow evolution determined by the expansion rate of the universe H. While a formal and completely rigorous proof of this contention is at the moment not feasible, the hints that it must be so are sufficiently encouraging to further spur these investigations at least from the phenomenological side [8], let alone the theoretical studies pointing to that possibility in particular QFT frameworks [58, 73]. If so, the dynamical vacuum in QFT in curved spacetime could provide an alternative scenario (beyond the usual quintessence kind of approaches) for implementing dynamical dark energy (DE) as a general paradigm for curing or alleviating the old CC problem, as well as the so-called cosmic coincidence problem. Interestingly enough, this possibility has been phenomenologically tested and profusely confronted with the latest accurate data on type Ia supernovae (SNIa), the Baryonic Acoustic Oscillations (BAOs), and the anisotropies of the Cosmic Microwave Background (CMB), see particularly [38, 39, 74, 41] for the most recent and comprehensive studies. Furthermore, a unified vacuum framework capable of describing the complete cosmic history as of the early inflationary times to the present dark energy epoch has also recently been proposed within this same kind of approach [62, 63].

If that is not enough, quite intriguingly the idea could also be tested from an entirely different vein, namely one capable of providing a new and completely independent insight into the whole subject. This novel and recent approach was first suggested in [22], and it is related with the precise measurements of the so-called Fundamental Constants of Nature and their possible time variation, see e.g. [70] (and the long list of references therein). I would like to devote some time in this review to this particular phenomenological approach to dynamical vacuum energy.

Recent measurements on the time variation of the fine structure constant and of the proton-electron mass ratio suggest that basic quantities of the Standard Model, such as the QCD scale parameter $\Lambda_{\rm QCD}$, may not be conserved in the course of the cosmological evolution [75]. The masses of the nucleons m_N and of the atomic nuclei would also be affected. Matter is not conserved in such a universe. In the framework of ideas that we have developed in Sect. 8.2, these measurements could be interpreted as a leakage of matter into vacuum or vice versa. In the following we wish to show that the amount of leakage necessary to explain the measured value of the time variation $\dot{m}_N/m_N$ could be of the same order of magnitude as the observationally allowed value of $\dot{\rho}_\Lambda/\rho_\Lambda$, with a possible contribution from the dark matter particles [22]. Therefore, the dark energy in our universe could be the dynamical vacuum energy in interaction with ordinary baryonic matter as well as with dark matter.

The QCD scale parameter is related to the strong coupling constant $\alpha_s = g_s^2/(4\pi)$. To lowest (1-loop) order one finds:

$$\alpha_s(\mu_R) = \frac{1}{\beta_0 \ln\left(\Lambda_{\rm QCD}^2/\mu_R^2\right)} = \frac{4\pi}{(11 - 2\,n_f/3)\ln\left(\mu_R^2/\Lambda_{\rm QCD}^2\right)}\,, \tag{122}$$

where μ_R is the renormalization point and $\beta_0 \equiv -b_0 = -(33-2\,n_f)/(12\,\pi)$ (n_f being the number of quark flavors) is the lowest order coefficient of the β-function. If we include all the know flavors, we obviously have $\beta_0 < 0$ and therefore when the renormalization scale μ_R increases the strong coupling constant diminishes (this is the famous asymptotic freedom property of QCD). However, for lower scales, namely for μ_R near $\Lambda_{\rm QCD} \sim \mathcal{O}(100)$ MeV, the strong coupling constant increases and in practice it becomes arbitrarily large. This is just the regime which we cannot explore with perturbation theory, and as we see the two regimes are roughly separated by the $\Lambda_{\rm QCD}$ scale. Not surprisingly this scale is the one that rules the calculation of the strongly interacting hadronic bound states like the proton, and in fact it dominates the calculation of its mass, see Eq. (126) below. The QCD scale parameter $\Lambda_{\rm QCD}$ has actually been experimentally measured with $\sim 10\%$ precision: $\Lambda_{\rm QCD} = 217 \pm 25$ MeV.

The next observation is crucial for our discussion [22]. When we embed QCD in the FLRW expanding background, the value of $\Lambda_{\rm QCD}$ need not remain rigid anymore. The value of $\Lambda_{\rm QCD}$ could change with H, and this would mean a change in the cosmic time. If $\Lambda_{\rm QCD} = \Lambda_{\rm QCD}(H)$ is a function of H, the coupling constant $\alpha_s = \alpha_s(\mu_R; H)$ is also a function of H (apart from a function of μ_R). The relative cosmic variations of the two QCD quantities are related (at one-loop) by:

$$\frac{1}{\alpha_s}\frac{d\alpha_s(\mu_R; H)}{dH} = \frac{1}{\ln\left(\mu_R/\Lambda_{\rm QCD}\right)}\left[\frac{1}{\Lambda_{\rm QCD}}\frac{d\Lambda_{\rm QCD}(H)}{dH}\right]. \tag{123}$$

Journal of Physics: Conference Series **453** (2013) 012015 doi:10.1088/1742-6596/453/1/012015

The potential significance of this relation is out of discussion: if the QCD coupling constant α_s or the QCD scale parameter $\Lambda_{\rm QCD}$ undergo a small cosmological time shift, the nucleon mass and the masses of the atomic nuclei of the universe would also change in proportion to $\Lambda_{\rm QCD}$.

The cosmic dependence of the strong coupling $\alpha_s(\mu_R; H)$ can be generalized to the other couplings $\alpha_i = \alpha_i(\mu_R; H)$ [76], in particular the electromagnetic fine structure constant $\alpha_{\rm em}$. In a grand unified theory (GUT) these couplings converge at the unification point. Let $d\alpha_i$ be the cosmic variation of α_i with H. Each of the α_i is a function of μ_R, but the expression $\alpha_i^{-1}\,(d\alpha_i/\alpha_i)$ is easily seen to be independent of μ_R. As a result we can see, using Eq. (123), that the running of $\alpha_{\rm em}$ is related to the corresponding cosmic running of $\Lambda_{\rm QCD}$ as follows:

$$\frac{1}{\alpha_{\rm em}}\,\frac{d\alpha_{\rm em}(\mu_R; H)}{dH} = \frac{8}{3}\,\frac{\alpha_{\rm em}(\mu_R; H)/\alpha_s(\mu_R; H)}{\ln\left(\mu_R/\Lambda_{\rm QCD}\right)}\left[\frac{1}{\Lambda_{\rm QCD}}\,\frac{d\Lambda_{\rm QCD}(H)}{dH}\right]. \tag{124}$$

At the renormalization point $\mu_R = M_Z$, where both $\alpha_{\rm em}$ and α_s are well-known, one finds:

$$\frac{1}{\alpha_{\rm em}}\,\frac{d\alpha_{\rm em}(\mu_R; H)}{dH} \simeq 0.03\left[\frac{1}{\Lambda_{\rm QCD}}\,\frac{d\Lambda_{\rm QCD}(H)}{dH}\right]. \tag{125}$$

Thus the QCD scale $\Lambda_{\rm QCD}$ runs more than 30 times faster with the cosmic expansion than the electromagnetic fine structure constant. Searching for a cosmic evolution of $\Lambda_{\rm QCD}$ is thereby much easier than searching for the time variation of $\alpha_{\rm em}$. It is important to note that the cosmic running properties of the QCD scale parameter could have nontrivial implications on the physics of the fundamental constants of Nature. This is sooner said than done: how do we search for a possible cosmic running of the QCD scale? The answer is: searching for a possible running of the proton mass, a parameter which has been (and is being) monitored by many astrophysical and laboratory experiments (or more specifically, what is traced is the ratio of the proton mass to the electron mass, where the latter is of course independent of QCD) [70]. Let us take for instance the proton mass $m_p \simeq 938$ MeV, supposed to be a fundamental constant. It can be computed from $\Lambda_{\rm QCD}$, the quarks masses and the electromagnetic contribution. The precise formula reads

$$\begin{aligned} m_p &= c_{\rm QCD}\Lambda_{\rm QCD} + c_u\, m_u + c_d\, m_d + c_s\, m_s + c_{\rm em}\Lambda_{\rm QCD} \\ &= (860 + 21 + 19 + 36 + 2)\ {\rm MeV}\,, \end{aligned} \tag{126}$$

from which we learn that it is largely dominated by the QCD scale, namely the bulk ($\sim 92\%$) of the proton mass is given by $m_p \simeq c_{\rm QCD}\Lambda_{\rm QCD}$. We will take advantage of this fact in what follows.

Let us focus on the impact of the cosmological Models II and III of Sect. 8.2 on the non-conservation of matter in the universe. Recall that we have considered bounds on the "leakage parameter" ν within the class of these models based on the non-conservation matter density law (105). We must be careful in interpreting such a non-conservation law. For example, if we take the baryonic density in the universe, which is essentially the mass density of protons, we can write $\rho_M^B = n_p\, m_p$, where n_p is the number density of protons and $m_p^0 = 938.272013(23)$ MeV is the current proton mass. If this mass density is non-conserved, either n_p does not exactly follow the normal dilution law with the expansion, i.e. $n_p \sim a^{-3} = (1+z)^3$, but the anomalous law:

$$n_p(z) = n_p^0\,(1+z)^{3(1-\nu)} \qquad (\text{at fixed proton mass } m_p = m_p^0)\,, \tag{127}$$

and/or the proton mass m_p does not stay constant with time and redshifts with the cosmic evolution:

$$m_p(z) = m_p^0\,(1+z)^{-3\nu} \qquad (\text{with normal dilution } n_p(z) = n_p^0\,(1+z)^3)\,. \tag{128}$$

In all cases it is assumed that the vacuum absorbs the difference (i.e. $\rho_\Lambda = \rho_\Lambda(z)$ "runs with the expansion"). The first possibility implies that during the expansion a certain number of particles (protons in this case) are lost into the vacuum (if $\nu < 0$; or ejected from it, if $\nu > 0$), whereas in the second case the number of particles is strictly conserved. The number density follows the normal dilution law with the expansion, but the mass of each particle slightly changes (decreases for $\nu < 0$, or increases for $\nu > 0$) with the cosmic evolution.

For the following considerations we adopt the second point of view, i.e. Eq. (128). Being the matter content of the universe dominated by the dark matter (DM), we cannot exclude that the particle masses of which is made also vary with cosmic time. Let us denote the mass of the average DM particle m_X, and let ρ_X and n_X be its mass density and number density, respectively. The overall matter density of the universe can be written as follows:

$$\begin{aligned} \rho_M &= \rho_B + \rho_L + \rho_R + \rho_X = (n_p\, m_p + n_n\, m_n) + n_e\, m_e + \rho_R + n_X\, m_X \\ &\simeq n_p\, m_p + n_n\, m_n + n_X\, m_X\,. \end{aligned} \tag{129}$$

Here n_p, n_n, n_e, n_X (m_p, m_n, m_e, m_X) are the number densities (and masses) of protons, neutrons, electrons and DM particles. The baryonic and leptonic parts are $\rho_B = n_p\, m_p + n_n\, m_n$ and $\rho_L = n_e\, m_e$ respectively. The small ratio $m_e/m_p \simeq 5\times 10^{-4}$ implies that the leptonic contribution to the total mass density is negligible: $\rho_L \ll \rho_B$. We have also neglected the relativistic component ρ_R (photons and neutrinos).

If we assume that the mass change through the cosmic evolution is due to the time change of m_p, m_n and m_X, we can compute the mass density anomaly per unit time, i.e. the deficit or surplus with respect to the conservation law, by differentiating (129) with respect to time and subtracting the ordinary (i.e. fixed mass) time dilution of the number densities. The result is:

$$\delta\dot{\rho}_M = n_p\, \dot{m}_p + n_n\, \dot{m}_n + n_X\, \dot{m}_X\,. \tag{130}$$

The relative time variation of the mass density anomaly can be estimated as follows:

$$\frac{\delta\dot{\rho}_M}{\rho_M} = \frac{n_p\, \dot{m}_p + n_n\, \dot{m}_n + n_X\, \dot{m}_X}{n_p\, m_p + n_p\, m_p + n_X\, m_X} \simeq \frac{n_p\, \dot{m}_p + n_n\, \dot{m}_n + n_X\, \dot{m}_X}{n_X\, m_X}\left(1 - \frac{n_p\, m_p + n_n\, m_n}{n_X\, m_X}\right)\,. \tag{131}$$

The current normalized DM density $\Omega^0_{\rm DM} = \rho_X/\rho_c \simeq 0.23$ is significantly larger than the corresponding normalized baryon density $\Omega^0_B = \rho_B/\rho_c \simeq 0.04$. Therefore $n_X\, m_X$ is larger than $n_p\, m_p + n_n\, m_n$ by the same amount. If we assume $\dot{m}_n = \dot{m}_p$, we find approximately:

$$\frac{\delta\dot{\rho}_M}{\rho_M} = \frac{n_p\, \dot{m}_p}{n_X\, m_X}\left(1 + \frac{n_n}{n_p} - \frac{\Omega_B}{\Omega_{\rm DM}}\right) + \frac{\dot{m}_X}{m_X}\left(1 - \frac{\Omega_B}{\Omega_{\rm DM}}\right)\,. \tag{132}$$

In the approximation $m_n = m_p$ we can rewrite the prefactor on the *r.h.s* of Eq. (132) as follows:

$$\frac{n_p\, \dot{m}_p}{n_X\, m_X} = \frac{\Omega_B}{\Omega_{\rm DM}}\frac{\dot{m}_p}{m_p}\left(1 - \frac{n_n/n_p}{1 + n_n/n_p}\right) \simeq \frac{\Omega_B}{\Omega_{\rm DM}}\frac{\dot{m}_p}{m_p}\left(1 - \frac{n_n}{n_p}\right)\,. \tag{133}$$

The ratio n_n/n_p is of order 10% after the primordial nucleosynthesis. Since $\Omega_B/\Omega_{\rm DM}$ is also of order 10%, we can neglect the product of this term with n_n/n_p. When we insert the previous equation into (132), the two n_n/n_p contributions cancel each other. The expression $1 - \Omega_B/\Omega_{\rm DM}$ factorizes in the two terms on the *r.h.s* of Eq. (132). The final result is:

$$\left(1 - \frac{\Omega_B}{\Omega_{\rm DM}}\right)^{-1}\frac{\delta\dot{\rho}_M}{\rho_M} = \frac{\Omega_B}{\Omega_{\rm DM}}\frac{\dot{m}_p}{m_p} + \frac{\dot{m}_X}{m_X} = \frac{\Omega_B}{\Omega_{\rm DM}}\frac{\dot{\Lambda}_{\rm QCD}}{\Lambda_{\rm QCD}} + \frac{\dot{m}_X}{m_X}\,. \tag{134}$$

As mentioned, we have approximated $m_p \simeq c_{\rm QCD}\, \Lambda_{\rm QCD}$.

The expression $\delta\dot{\rho}_M/\rho_M$ in Eq. (134) must be the same as the one we have computed in (109), if we consider the models based on the generic matter non-conservation law (105). Therefore the two expressions should be equal, and we obtain approximately:

$$3\nu_{\rm eff}\, H = \frac{\Omega_B}{\Omega_{\rm DM}}\frac{\dot{\Lambda}_{\rm QCD}}{\Lambda_{\rm QCD}} + \frac{\dot{m}_X}{m_X}\,, \tag{135}$$

where we have defined

$$\nu_{\rm eff} = \frac{\nu}{1-\Omega_B/\Omega_{\rm DM}}\,, \tag{136}$$

and numerically $\nu_{\rm eff} \simeq 1.2\,\nu$. The differential equation (135) describes approximately the relationship between the matter non-conservation law (105), the evolution of the vacuum energy density ρ_Λ (and/or G) and the time variation of the nuclear and particle physics quantities.

Various scenarios are possible. Suppose that the dark matter particles do not vary with time, i.e. $\dot{m}_X = 0$, and only the cosmic evolution of $\Lambda_{\rm QCD}$ accounts for the non-conservation of matter. Trading the cosmic time for the scale factor through $\dot{\Lambda}_{\rm QCD} = (d\Lambda_{\rm QCD}/da)\; a\,H$ and integrating the resulting equation, we can express the final result in terms of the redshift:

$$\Lambda_{\rm QCD}(z) = \Lambda^0_{\rm QCD}\;(1+z)^{-3\,(\Omega^0_{\rm DM}/\Omega^0_B)\,\nu_{\rm eff}}\,. \tag{137}$$

Since the contribution of the quark masses m_u,m_d and m_s to the proton mass is small – cf. Eq. (126) – we can approximate the proton mass by $m_p \simeq c_{\rm QCD}\,\Lambda_{\rm QCD}$. Therefore, for the protons we have

$$m_p(z) = m_p^0\;(1+z)^{-3\,(\Omega^0_{\rm DM}/\Omega^0_B)\,\nu_{\rm eff}}\,. \tag{138}$$

Here $\Lambda^0_{\rm QCD}$ and m_p^0 are the QCD scale and proton mass at present ($z=0$); $\Omega^0_{\rm DM}$ and Ω^0_B being the current values of these cosmological parameters.

The presence of the factor $\Omega^0_B/\Omega^0_{\rm DM}$ in the power law makes eq. (138) more realistic than eq. (128). In the case $\nu=0$ the QCD scale and the proton mass would not vary with the expansion of the universe, but for non-vanishing ν it describes the cosmic running of $\Lambda_{\rm QCD} = \Lambda_{\rm QCD}(z)$ and $m_p = m_p(z)$. For $\nu>0$ ($\nu<0$) the QCD scale and proton mass decrease (increase) with the redshift. This is consistent, since for $\nu>0$ ($\nu<0$) the vacuum energy density is increasing (decreasing) with the redshift – cf. Eq. (106) –, and it is smaller (larger) now than in the past.

We can write down the variation of the QCD scale in terms of the Hubble rate H. With the help of Eq. (107) it is easy to see that Eq. (137) can be turned into an expression for $\Lambda_{\rm QCD}$ given explicitly in terms of the primary cosmic variable H:

$$\Lambda_{\rm QCD}(H) = \Lambda^0_{\rm QCD}\left[\frac{1-\nu}{\Omega^0_M}\frac{H^2}{H_0^2} - \frac{\Omega^0_\Lambda-\nu}{\Omega^0_M}\right]^{-(\Omega^0_{\rm DM}/\Omega^0_B)\,\nu_{\rm eff}/(1-\nu)}\,, \tag{139}$$

with $\Omega^0_M = \Omega^0_B + \Omega^0_{\rm DM}$. ν and $\nu_{\rm eff}$ are involved in (139), since they come from different sources. This equation satisfies the normalization condition $\Lambda_{\rm QCD}(H_0) = \Lambda^0_{\rm QCD}$ due to the cosmic sum rule for flat space: $\Omega^0_M + \Omega^0_\Lambda = 1$.

Obviously the cosmic time variation of the $\Lambda_{\rm QCD}$ scale is very small in our framework. This can be more easily assessed if we use Eqs. (137) and (139) to compute the relative time variation of the QCD scale with respect to the present value. Since ν is small it it easy to show that

$$\frac{\Lambda_{\rm QCD}(z)-\Lambda^0_{\rm QCD}}{\Lambda^0_{\rm QCD}} = -\frac{\Omega^0_{\rm DM}}{\Omega^0_B}\frac{\nu_{\rm eff}}{1-\nu}\ln\left[\frac{1-\nu}{\Omega^0_M}\frac{H^2(z)}{H_0^2} - \frac{\Omega^0_\Lambda-\nu}{\Omega^0_M}\right]. \tag{140}$$

As a concrete example, let us consider the studies made in Ref. [77] on comparing the H_2 spectral Lyman and Werner lines observed in the Q 0347-383 and Q 0405-443 quasar absorption systems. The comparison with the corresponding spectral lines at present may be sensitive to a possible evolution of these lines in the last twelve billion years and involves redshifts in the range $z \simeq 2.6 - 3.0$. A positive result could be interpreted as a small variation of the proton to electron mass ratio between two widely separated epochs of the cosmological evolution [77] . Assuming that $|\nu| = \mathcal{O}(10^{-3})$, as suggested by Eq. (112), it follows from the previous formulae that the relative variation of $\Lambda_{\rm QCD}$ in this lengthy time interval is only at the few percent level with respect to its present day value. From Eq. (106) we can then easily check that the corresponding variation of $\rho_\Lambda(z)$ with respect to the current value ρ_Λ^0 is also of a few percent. As expected, the two scales undergo tiny variations over very long periods of time, in fact cosmological periods, and therefore the large hierarchy between them at present – namely $\Lambda_{\rm QCD} = \mathcal{O}(100)$ MeV$= \mathcal{O}(10^8)$ eV and $\rho_\Lambda^{1/4} = \mathcal{O}(10^{-3})$ eV – is essentially preserved over the cosmological evolution. However, even this small crosstalk between these two widely separated scales could be sufficient for being eventually detected by the aforementioned high precision experiments aiming at measuring very tiny variation of the proton to electron mass ratio.

Indeed, this is suggested by the fact that the expected range of values of ν is within the scope of the precision of these experiments. Consider the state of the art in the current laboratory tests, using atomic clocks. According to our estimate (125), the largest effect is expected to be a cosmological redshift (hence time variation) of the nucleon mass, which can be observed by monitoring molecular frequencies. These are precise experiments in quantum optics, e.g. obtained by comparing a cesium clock with 1S-2S hydrogen transitions. In a cesium clock the time is measured by using a hyperfine transition [10]. Since the frequency of the clock depends on the magnetic moment of the cesium nucleus, a possible variation of the latter is proportional to a possible variation of $\Lambda_{\rm QCD}$. A hyperfine splitting is a function of $Z\,\alpha_{\rm em}$ (Z being the atomic number) and is proportional to $Z\,\alpha_{\rm em}^2(\mu_N/\mu_B)(m_e/m_p)\,R_\infty$, where R_∞ is the Rydberg constant, μ_N is the nuclear magnetic moment and $\mu_B = e\hbar/2m_pc$ is the nuclear magneton. We have $\dot{\mu}_N/\mu_N \propto -\dot{\Lambda}_{\rm QCD}/\Lambda_{\rm QCD}$. The hydrogen transitions are only dependent on the electron mass, which we assume to be constant. The comparison over a period of time between the cesium clock with hydrogen transitions provides an atomic laboratory measurement of the ratio $\mu_{\rm pe} \equiv m_p/m_e$. The most sophisticated atomic clock experiments aim soon to reach a sensitivity limit of $\left|\dot{\Lambda}_{\rm QCD}/\Lambda_{\rm QCD}\right| < 10^{-14}\,{\rm yr}^{-1}$ [22]. Since the proton mass is given essentially by $\Lambda_{\rm QCD}$, as indicated by Eq. (126), we have $\dot{m}_p \simeq c_{\Lambda_{\rm QCD}}\,\dot{\Lambda}_{\rm QCD}$ and the corresponding time variation of the ratio of the proton mass would be:

$$\left|\frac{\dot{\mu}_{pe}}{\mu_{\rm pe}}\right| = \left|\frac{\dot{m}_p}{m_p}\right| \simeq \left|\frac{\dot{\Lambda}_{\rm QCD}}{\Lambda_{\rm QCD}}\right| < 10^{-14}\,{\rm yr}^{-1}\,. \tag{141}$$

The atomic clock result (141) would indicate a time variation of the ratio $\mu_{\rm pe}$, which is consistent (in absolute value) with the astrophysical measurements [77].

Using the above equations and Eq. (122), we can obtain the corresponding evolution of the strong coupling constant α_s with the redshift:

$$\frac{1}{\alpha_s(\mu_R; z)} = \frac{1}{\alpha_s(\mu_R; 0)} + 6\,b_0\,\frac{\Omega_{\rm DM}^0}{\Omega_B^0}\,\nu_{\rm eff}\,\ln\left(1+z\right), \tag{142}$$

[10] Recall that the cesium hyperfine clock provides the modern definition of time. In SI units, the second is defined to be the duration of 9.192631770×10^9 periods of the transition between the two hyperfine levels of the ground state of the ^{133}Cs atom

where $\alpha_s(\mu_R;0)$ is the value of $\alpha_s(\mu_R;z)$ today ($z=0$). Since $b_0>0$, we observe that for $\nu>0$ ($\nu<0$) the strong interaction $\alpha_s(\mu_R;z)$ decreases (increases) with z, i.e. with the cosmic evolution. The corresponding variation of the strong coupling with the Hubble rate can also be determined [11]:

$$\frac{1}{\alpha_s(\mu_R;H)}=\frac{1}{\alpha_s(\mu_R;H_0)}+2\,b_0\,\frac{\Omega^0_{\rm DM}}{\Omega^0_B}\,\frac{\nu_{\rm eff}}{1-\nu}\,\ln\left[\frac{1-\nu}{\Omega^0_M}\,\frac{H^2}{H_0^2}-\frac{\Omega^0_\Lambda-\nu}{\Omega^0_M}\right]. \qquad (143)$$

Here $\alpha_s(\mu_R;H_0)$ is the current value of $\alpha_s(\mu_R;H)$.

Remarkably, the above expression displays the logarithmic running of the strong coupling as a function of *two* energy scales: one is the ordinary QCD running scale μ_R, the other is the cosmic scale defined by the Hubble rate H, which has dimension of energy in natural units. The second term on the *r.h.s.* depends on the product of the two β-function coefficients, the one for the ordinary QCD running (b_0) and the one for the cosmic running ($\nu\propto\nu_{\rm eff}$).

The following remarks are in order:

i) for $\nu=0$ there is no cosmic running of the strong interaction,

ii) for $\nu>0$ the strong coupling $\alpha_s(\mu_R;H)$ is "doubly asymptotically free". It decreases for large μ_R and also for large H, whereas for $\nu<0$ the cosmic evolution drives the running of α_s opposite to the normal QCD running,

iii) the velocity of the two runnings is very different, because H is slowly varying with time and $|\nu|\ll 1$ and $|\nu|\ll b_0\lesssim 1$. The cosmic running only operates in the cosmic history and is weighed with a very small β-function. But it may soon be measured in the experiments with atomic clocks and through astrophysical observations.

We should not overlook the fact that the previous equations describe not only the leading cosmic evolution of the QCD scale and the proton mass with the redshift and the expansion rate H of the universe, but they can account for the redshift evolution of the nuclear masses. For the neutron we can write approximately: $m_n\simeq c_{\rm QCD}\,\Lambda_{\rm QCD}$. For an atomic nucleus of current mass M_A and atomic number A we have $M_A=Z\,m_p+(A-Z)\,m_n-B_A$, where Z is the number of protons and $A-Z$ the number of neutrons, and B_A is the binding energy. Although B_A may also change with the cosmic evolution, the shift should be less significant, since at leading order the binding energy relies on pion exchange among the nucleons. The pion mass has a softer dependence on $\Lambda_{\rm QCD}$: $m_\pi\sim\sqrt{m_q\,\Lambda_{\rm QCD}}$, due to the chiral symmetry.

In the previous approximations we have neglected the light quark masses m_q. We can assume that the binding energy has a negligible cosmic shift as compared to the masses of the nucleons. In the limit where we neglect the proton-neutron mass difference and assume a common nucleon mass m_N^0 at present, the corresponding mass of the atomic nucleus at redshift z is given at leading order by:

$$M_A(z)\simeq A\,m_N^0\,(1+z)^{-3\,(\Omega^0_{\rm DM}/\Omega^0_B)\,\nu_{\rm eff}}-B_A\,. \qquad (144)$$

Although the chemical elements redshift their masses, a disappearance or overproduction of nuclear mass (depending on the sign of ν) is compensated by a running of the vacuum energy ρ_Λ, which is of opposite in sign, see (111).

Above we have described a simplified case, in which the nuclear matter evolves with the cosmic evolution as a result of the evolution of the fundamental QCD scale. In this scenario the light quark masses are neglected, and the DM does not participate in the cosmic time evolution.

Alternatively we can assume that the nuclear matter does not vary with time, i.e. $\dot\Lambda_{\rm QCD}=0$, and only the DM particles account for the non-conservation of matter. In general we expect a

[11] We point out that a similar running of the strong coupling constant with the cosmic expansion was pointed out in a different context by J.D. Bjorken in [78].

Journal of Physics: Conference Series **453** (2013) 012015 doi:10.1088/1742-6596/453/1/012015

mixed situation, in which the temporal rates of change for nuclear matter and for DM particles are different:

$$\frac{\dot{\Lambda}_{\rm QCD}}{\Lambda_{\rm QCD}} = 3\,\nu_{\rm QCD}\,H\,, \qquad \frac{\dot{m}_X}{m_X} = 3\,\nu_X\,H\,. \tag{145}$$

We have defined the QCD time variation index, which is characteristic of the redshift rate of the QCD scale, while ν_X is the corresponding one for the DM. In this more general case we find:

$$\Lambda_{\rm QCD}(z) = \Lambda^0_{\rm QCD}\,(1+z)^{-3\,\nu_{\rm QCD}}\,, \qquad m_X(z) = m^0_X\,(1+z)^{-3\,\nu_X}\,. \tag{146}$$

We introduce the effective baryonic redshift index ν_B:

$$\nu_B = \frac{\Omega_B}{\Omega_{\rm DM}}\,\nu_{\rm QCD}\,. \tag{147}$$

The equations (146) satisfy the relation (135), provided the coefficients ν_B and ν_X are related by

$$\nu_{\rm eff} = \nu_B + \nu_X\,. \tag{148}$$

$\nu_{\rm QCD}$ is the intrinsic cosmic rate of variation of the strongly interacting particles. The effective index ν_B weighs the redshift rate of these particles taking into account their relative abundance with respect to the DM particles. Even if the intrinsic cosmic rate of variation of $\Lambda_{\rm QCD}$ would be similar to the DM index (i.e. if $\nu_{\rm QCD} \gtrsim \nu_X$), the baryonic index (147) would still be suppressed with respect to ν_X, because the total amount of baryon matter in the universe is much smaller than the total amount of DM.

In this mixed scenario the mass redshift of the dark matter particles follows a similar law as in the case of protons (138), except now we have $\nu_{\rm eff} \to \nu_B$. The proton would have the index $\nu_{\rm QCD}$ characteristic of the free (and bound) stable strongly interacting matter:

$$m_p(z) = m^0_p\,(1+z)^{-3\,(\Omega_{\rm DM}/\Omega_B)\,\nu_B} = m^0_p\,(1+z)^{-3\nu_{\rm QCD}}\,. \tag{149}$$

The DM particles have another independent index ν_X. The sum (148) must reproduce the original index $\nu_{\rm eff} \propto \nu$, which we associated with the non-conservation of matter.

Finally we consider the possible quantitative contribution to the matter density anomaly from the dark matter. The global mass defect (or surplus) is regulated by the value of the ν parameter, but the contribution of each part (baryonic matter and DM) depends on the values of the individual components ν_B and ν_X. We can obtain a numerical estimate of these parameters by setting the expression (134) equal to (110). The latter refers to the time variation of the matter density ρ_M without tracking the particular way in which the cosmic evolution can generate an anomaly in the matter conservation. The former does assume that this anomaly is entirely due to a cosmic shift in the masses of the stable particles. Taking the absolute values, we obtain:

$$3|\nu_{\rm eff}|\,H \simeq \left|\frac{4}{23}\,\frac{\dot{\Lambda}_{\rm QCD}}{\Lambda_{\rm QCD}} + \frac{\dot{m}_X}{m_X}\right| < \frac{4}{23}\times 10^{-14}\,{\rm yr}^{-1} + \left|\frac{\dot{m}_X}{m_X}\right|\,. \tag{150}$$

Here we have used the experimental bound (141) on the time variation of $\Lambda_{\rm QCD}$.

Several cases can be considered, depending on the relation between the intrinsic cosmic rates of variation of the strongly interacting particles and DM particles, $\nu_{\rm QCD}$ and ν_X. Since these indices can have either sign, we shall compare their absolute values:

- 1) $|\nu_X| \ll |\nu_B|$:
  This condition implies $|\nu_X| \ll |\nu_{\rm QCD}|$. By demanding the stronger condition $|\nu_X| \ll |\nu_B|$, we insure that the intrinsic QCD cosmic rate $|\nu_{\rm QCD}|$ is much larger than the corresponding

	$\lvert\nu\rvert^{\rm cosm}$	$\lvert\nu\rvert^{\rm lab} = \lvert\nu_B\rvert$	$\lvert\nu_X\rvert^{\rm cosm}$
Model II	10^{-3} (SNIa+BAO+CMB)	10^{-5} (Atomic clocks+Astrophys.)	10^{-3}
Model III	10^{-3} (BBN)	10^{-5} (Atomic clocks+Astrophys.)	10^{-3}
Model IV	10^{-3} (SNIa+BAO+CMB)+BBN	0	0

Table 1. Upper bounds on the running index $|\nu|$ for the various models defined in sect. 8.2 – cf. [22]. Only for Models II and III a non-vanishing value of $|\nu|$ is related to non-conservation of matter and a corresponding time evolution of ρ_Λ and G, respectively. For these models, the baryonic part of ν (denoted ν_B) can be accessible to accurate lab experiments – cf. Eq. (141) – whereas the DM part (ν_X) can only be bound indirectly from cosmological observations (same cosmological bound as for the overall ν). For Model IV matter is conserved, and a non-vanishing value of $|\nu|$ (only accessible from pure cosmological observations) is associated to a simultaneous time evolution of ρ_Λ and G – with no microphysical implications.

DM rate $|\nu_X|$. We can neglect the $\dot{m}_X/m_X$ term on the *r.h.s.* of (150), and we recover the equations (137)-(143) with $\nu_{\rm eff} \simeq \nu_B$. Using $H_0 \simeq 7\times 10^{-11}\,{\rm yr}^{-1}$, we find:

$$|\nu_X| \simeq 0\,, \qquad |\nu_{\rm eff}| \simeq |\nu_B| < 10^{-5}\,, \qquad |\nu_{\rm QCD}| < 5\times 10^{-5}\,. \tag{151}$$

- 2) $|\nu_X| \simeq |\nu_B|$:
Here we still have $|\nu_X|$ smaller than $|\nu_{\rm QCD}|$, but the requirement is weaker. It follows: $|\nu_{\rm eff}| \simeq 2|\nu_X| \simeq 2|\nu_B| = 2(\Omega_B/\Omega_{\rm DM})\,|\nu_{\rm QCD}|$, and we find

$$|\nu_{\rm eff}| < 2\times 10^{-5}\,, \qquad |\nu_X| \simeq |\nu_B| < 10^{-5}\,, \qquad |\nu_{\rm QCD}| < 5\times 10^{-5}\,. \tag{152}$$

- 3) $|\nu_X| \simeq |\nu_{\rm QCD}|$:
The two intrinsic cosmic rates for strongly interacting and DM particles are similar, i.e. $\dot\Lambda_{\rm QCD}/\Lambda_{\rm QCD}$ and $\dot{m}_X/m_X$ do not differ significantly. In this case Eq. (150) leads to

$$3|\nu_{\rm eff}|\,H < \left(\frac{4}{23}+1\right)\times 10^{-14}\,{\rm yr}^{-1}\,. \tag{153}$$

There are two sign possibilities ($\nu_{\rm QCD} = \pm\nu_X$), and we take the absolute value:

$$|\nu_{\rm eff}| \lesssim \left(\frac{\Omega_B}{\Omega_{\rm DM}}+1\right)|\nu_{\rm QCD}| \simeq |\nu_{\rm QCD}|\,. \tag{154}$$

We find:

$$|\nu_{\rm eff}| \lesssim |\nu_{\rm QCD}| \simeq |\nu_X| < 5\times 10^{-5}\,. \tag{155}$$

- 4) $|\nu_{\rm QCD}| \ll |\nu_X|$:
Here the nuclear part is frozen. The non-conservation of matter is entirely due to the time variation of the DM particles. Eq. (150) gives:

$$3\nu\,H \simeq \frac{\dot{m}_X}{m_X}\left(1-\frac{\Omega_B}{\Omega_{\rm DM}}\right)\,. \tag{156}$$

We have written this expression directly in terms of the original ν parameter. In this case we cannot get information from any laboratory experiment on $\dot{m}_X/m_X$, but we do have independent experimental information on the original ν value (irrespective of the particular contributions form the nuclear and DM components). It comes from the cosmological data on type Ia supernovae, BAO, CMB and structure formation. The analysis of this data [38, 39] leads to the bound (112), which applies to all models, in which matter follows the generic non-conservation law (105) and the running vacuum law (99) — or the same matter non-conservation law and the running gravitational coupling law (118), as shown in Eq. (121). Since it depends on the cosmological effects from all forms of matter, it applies to the DM particles in particular. We find:

$$|\nu_X|^{\rm cosm} \lesssim 10^{-3}\,. \tag{157}$$

This bound is significantly weaker than any of the bounds found for the previous scenarios in which the nuclear matter participated of the cosmic time variation. It cannot be excluded that the matter non-conservation and corresponding running of the vacuum energy in the universe is mainly caused by the general redshift of the DM particles. In this case only cosmological experiments could be used to check this possibility. If the nuclear matter also participates in a significant way, it could be analyzed with the help of experiments in the laboratory. For a summary of the bounds, see Table 1.

If in the future we could obtain a tight cosmological bound on the effective $\nu_{\rm eff}$-parameter (148), using the astrophysical data, and an accurate laboratory (and/or astrophysical) bound on the baryonic matter part ν_B, we could compare them and derive the value of the DM component ν_X. If $\nu_{\rm eff}$ and ν_B would be about equal, we should conclude that the DM particles do not appreciably shift their masses with the cosmic evolution, or that they do not exist. If, in contrast, the fractional difference $|\,(\nu_{\rm eff}-\nu_B)/\nu_{\rm eff}\,|$ would be significant, the DM particles should exist to compensate for it.

10. Discussion and conclusions

We have discussed a few old and new aspects of the Λ-term in Einstein's equations and its relation with the vacuum energy. After some brief historical remarks, we have dealt with the cosmological constant (CC) problem and assessed its significance and possible phenomenological implications. In doing so we have focused on the notion of vacuum energy in quantum field theory (QFT) both in flat and curved spacetime, and we have dwelled on the fact that the term which is usually interpreted as the vacuum energy density (originating from the quantum vacuum fluctuations of the matter fields) is the same both in flat and curved spacetime. This is a well-known fact, but this does not make it less worrisome, since particles with mass m provide huge quartic contributions to the zero-point energy (ZPE), $V_{\rm ZPE}\sim m^4$, and these do not seem to be acceptable from the phenomenological point of view.

In the SM of particle physics there are other, no less preoccupying, contributions to the vacuum energy that come from the spontaneous symmetry breaking (SSB) of the electroweak gauge symmetry, specifically from the Higgs mechanism. These contributions are associated to the vacuum expectation value of the Higgs potential and increase as $\langle V\rangle\sim v^2\,M_H^2\sim 10^8$ GeV^4, where $v=\mathcal{O}(100)$ GeV is the vacuum expectation value that defines the electroweak scale, and $M_H\simeq 125$ GeV is the presumed physical mass of the Higgs boson [6]. While every single vacuum energy source alone is already vastly worrisome, the combination of them all amounts to a devastating fine tuning problem, which is further aggravated at the quantum level when we consider the many higher loop effects involved. We have illustrated quite vividly this fact in Sect. 6 by considering what are the highest loop diagrams still contributing to the CC value (and to the awful fine tuning process).

Of the two main sources of difficulties with the vacuum energy in QFT, namely the ZPE and the SSB, the reality of the former has been disputed sometimes by the inconclusive interpretations about the origin of the Casimir effect as being a pure QFT vacuum effect or something else; whereas the latter also remained in the limbo of the theoretical ideas as long as the Higgs mechanism could not be fully substantiated on the experimental side. Because of this the CC problem could remain dormant for a long time in the ethereal world of the theoretical conundrums, which are that kind of problems whose dangerousness and threatening power on the physical world is only potential, not yet factual. Such situation, however, may have given signs of changing quite dramatically in recent times. Needless to say, not because the physical world changes an inch every time the human knowledge gives a jerk in its perception of the reality, but because we might now be reaching a situation where the two giant paradigms of the fundamental physics knowledge, viz. the SM of particle physics and the SM of cosmology, have finally been put furiously face to face for a very serious parley. Indeed, we have recently heard of the exciting news from CERN about the 5σ evidence on the discovery of a bosonic Higgs-like resonance at the LHC collider [6]. If fully confirmed, this can be considered as one of the greatest triumphs of particle physics ever. But, at the same time, we should not overlook that such discovery could be certifying the very existence of the electroweak vacuum energy, with all its potential consequences for (theoretical) cosmology.

On the other hand, we should not forget that apart from the classical and quantum Higgs vacuum energy we have other sorts of vacuum fluctuations in the electroweak domain, which are perfectly "alive and kicking" since long ago. Recall that quantum corrections to high precision electroweak observables have reached a level of certainty that is beyond any possible doubt. A simple example should suffice. Take the famous Δr parameter from electroweak theory [12]. This is the famous parameter that allows to compute the $W^{\pm}$ gauge boson mass, M_W, in the on-shell renormalization scheme with quantum precision, i.e. including the quantum output from radiative corrections. We may ask ourselves, to which extent we can attest that the "genuine" electroweak quantum effects (i.e. those beyond the pure QED running of $\alpha_{\rm em}$) have been measured when we compare the theoretical value of M_W and the experimentally measured one. To ask (and answer!) this question is important since the electroweak radiative corrections are a manifestation of the properties of the electroweak vacuum at the quantum level, namely the same vacuum that is supported by the likely existence of the Higgs particle. The answer is known, and is astounding: we know unmistakably that they are there with a confidence level of $\sim 25\sigma$! The formerly proclaimed 5σ evidence of the Higgs-like particle pales in comparison. But if we put both the direct and indirect signatures together they reinforce each other; and, overall, the overwhelming evidence of the electroweak vacuum effects becomes even more defiant and challenging for cosmology. Somehow time has come to try to find a final solution which comes to grips with the basic notion of vacuum energy in QFT, rather than constantly eschewing the issue in an almost non-denumerable number of ways.

Although the problem is huge, and far from being solved, some avenues for its eventual solution might be looming in the horizon. We have discussed a possible reinterpretation of the results obtained in the calculation of the vacuum energy density ρ_Λ in QFT in curved spacetime, and suggested that although the resolution of the CC problem cannot be addressed at present from a rigorous computation of ρ_Λ in an expanding FLRW universe, at least some consistency relations seem to hint at the possible form of the correct dynamical dependence of that important quantity as a function of the Hubble rate. For example, if both the vacuum energy and the gravitational coupling are evolving with the Hubble rate, H, the Bianchi identity leads to the possible form for the running of the vacuum density $\rho_\Lambda = \rho_\Lambda(H)$, given the logarithmic running suggested for $G = G(H)$ in the expanding spacetime. One is led to the following behavior for the low energy regime: $\rho_\Lambda(H) = c_0 + \beta\, M_P^2\, H^2$, where M_P is the Planck mass, c_0 a constant (close to the current CC density value ρ_Λ^0) and β is a dimensionless coefficient parameterizing the CC running. This form for the evolving vacuum energy density is perfectly tenable and has been

[12] See e.g. Ref.[79], and references therein, for contextual explanations on the Δr parameter, and for a comprehensive and updated study of this important electroweak parameter within the SM and beyond.

profusely tested against the latest cosmological data [38, 39]. At the same time, a nonvanishing value of β leads to a dynamical vacuum behavior which may effectively appear as quintessence, or as phantom energy, without need of invoking the existence of fundamental quintessence and phantom fields [41, 57]. This could eventually provide a strong phenomenological evidence in favor of the vacuum energy being a serious candidate for dynamical dark energy (DE).

We believe that the CC problem can only be solved as a physical problem, not just as a theoretical conundrum. This means that only through phenomenological tests it should be possible to disentangle the most difficult theoretical aspects of the CC problem. In this respect, another potentially interesting aspect of the dynamical vacuum models is the fact that they could provide an explanation for a possible variation of the so-called fundamental constants of Nature [22]. There is currently plentiful of experimental activity, both in the lab and from observations in the astrophysical domain, that will provide sooner or later interesting news on this field [70, 77]. Future data from these experiments should be very helpful for effectively testing the proposed vacuum ideas in the near future. For this reason we have discussed this issue at length in Sect. 9.

The idea of a running vacuum energy in an expanding universe, i.e. $\rho_\Lambda = \rho_\Lambda(H)$, appears as a most natural one. It would be very difficult to admit the existence of a tiny and immutable constant from the early times to the present days, maybe even ruling the entire future evolution of the universe. Notwithstanding, the ultimate value that $\rho_\Lambda(H)$ takes at present, i.e. ρ_Λ^0, cannot be predicted within these models and hence can only be extracted from observations. Notice that if we would have the ability to predict this value it would be tantamount to solve the old CC problem [4]. This is of course the toughest part of the job. In our discussion, however, the running vacuum paradigm ascribes a new look to the problem, one that could perhaps make it more amenable for an eventual solution; namely it conceives the cosmological term as a time evolving variable that underwent a dramatic reduction from the inflationary time till the present days. A generic model that implements this idea is the following: $\rho_\Lambda(H) = c_0 + c_2\,H^2 + c_4\,H^4$, in which the highest power of the Hubble rate, H^4, would only be relevant for the early universe, i.e. for values of H of order of the inflationary expansion rate H_I. Models of this kind can be motivated by the general running vacuum framework that we have described, and they should have a real chance to provide a complete description of the cosmic history [62, 63, 64]. In contrast to the standard ΛCDM model, in which the two opposite poles of the cosmic history (inflation and DE) are completely disconnected, the running vacuum models offer a clue for interconnecting them and let the present DE appear as a "quantum fossil" from the inflationary universe [58]. In this way the two de Sitter epochs, viz. the primordial inflationary one and the late DE epoch, can be thought of as two vacuum dominated stages of the cosmic evolution smoothly interpolated (within a single unified model) by the radiation and matter dominated epochs.

In the kind of decaying vacuum framework that we have proposed, we can better understand why the present vacuum energy density has to be small as compared to its primordial value at the inflationary scale, i.e. $\rho_\Lambda^0 \ll \rho_\Lambda(H_I)$. Therefore, the main task that remains should be to understand why ρ_Λ^0 has the concrete value we have measured. In other words, we are left with the question of why the mass scale $m_\Lambda \equiv (\rho_\Lambda^0)^{1/4}$ associated to the cosmological term is of order of a millielectronvolt ($m_\Lambda = \mathcal{O}(10^{-3})$ eV) rather than, say, ten times or a hundred times bigger. While such values cannot be admitted, as they would obviously be incompatible with the observations, here we are addressing a matter of principle, i.e. we are asking: is there a fundamental physical theory which can explain the value of the vacuum energy that we measure at present? Put another way: should the millielectronvolt energy scale m_Λ be ultimately predictable, for example from the value of the Planck scale M_P? This is a very interesting question, but is not at all an obvious one [80].

When we face the possibility to explain the mass scale m_Λ in our universe, we should note

Journal of Physics: Conference Series **453** (2013) 012015 doi:10.1088/1742-6596/453/1/012015

that in particle physics essentially all physical scales remain unexplained. For example, we cannot explain why the value of the electron mass is $m_e \simeq 0.511$ MeV in our universe, since we do not understand why its Yukawa coupling, λ_e, takes the value it takes in order to yield the precise value of the electron mass from its product with the vacuum expectation value of the Higgs doublet: $m_e = \lambda_e\, v$. Similarly, we do not understand why we have so many fermion flavors and with widely different mass scales (or family of Yukawa couplings). In particular, why the masses of the neutrinos are much lighter, and why one species of neutrinos is possibly not far away from the same $\sim$ meV scale m_Λ associated to the cosmological constant.

Whether we can ultimately predict or not the scale of the vacuum energy in the present universe can be a debatable question, but what is perhaps less debatable is that there is, in principle, no reason why we should be able to bypass all the foreseeable difficulties well before we can minimally understand the origin of the scales of the fermion and boson masses in the SM of particle physics, and of course the values of the two basic "order parameters" that set the fundamental scales of the model, to wit: i) the electroweak scale $v \simeq G_F^{-1/2}$ (associated to the Higgs mechanism and linked to Fermi's constant); and ii) the value of the QCD scale $\Lambda_{\rm QCD}$ (associated to the strong interactions and determined by the non-perturbative confinement dynamics of QCD). In both cases an experimental input is needed to make contact with physics. For the electroweak sector (which provides the dominant contribution to the vacuum energy of the SM) we need to perform a precise measurement of Fermi's constant from muon decay. Similarly, in the domain of strong interactions we have to measure $\Lambda_{\rm QCD}$ to account for the QCD vacuum contribution. It is fascinating that the possible cosmic time dependence of $\Lambda_{\rm QCD}$ could also play a decisive role in elucidating the nature of the CC problem, as we have amply emphasized in Sect. 9.

Owing to its especial position in the realm of the physical quantities, and because of the general covariance of Einstein's equations, we should expect a tight connection of the CC with the remaining scales of the universe. For this reason the help of the phenomenological input appears as indispensable. Is this not a most fundamental physical requirement, even for the glory of the CC problem? In this sense, if the vacuum energy density is treated as a running quantity in an expanding universe, it would be natural to input its value at a given cosmic time (say, now) and then focus our efforts on finding its past and future evolution. After all it is not granted that we can reach a purely theoretical solution, unless e.g. all scales of the universe should come from a single one, say the Planck mass M_P, and all the others be referred to it through dimensionless (and computable) ratios. But it is far from obvious that we can happily make such an aprioristic assumption. For the time being quantum gravity is not a consistent theory, and we *cannot* know for certain if M_P – which is nothing but a shorthand for $(\hbar\, c/G)^{1/2}$ (G being Newton's constant) – is truly a physical scale!

The bare truth is that we still need the help from the phenomenological input so as to set the scale for the vacuum energy in our low energy universe, and there is no foreseeable change in this situation. But this does not mean we cannot hope for some progress. Take for example the RG approach we have discussed here; after we input m_Λ (or, equivalently, c_0) what remains of the CC problem is still a hard enough challenge for our intellect! The CC problem is then formulated as the problem of explaining why the dynamical term in the low energy expression $\rho_\Lambda(H) = c_0 + \beta M_P^2\, H^2$ is just the soft (and completely harmless) $\sim H^2$ contribution in the context of QFT in an expanding spacetime. Such term endows the vacuum energy density of a very mild (albeit non-negligible) time evolution, which causes no tension with the observations. It is therefore a very attractive option, in which the time effect could eventually surface in the form of a smooth dynamical DE. To check this is a task reserved for the future observations. On the theoretical side, while we have given some indirect clues on how to effectively achieve an appealing scenario for $\rho_\Lambda(H)$ within QFT in curved spacetime, much more work (and thought) is of course needed to tackle the most difficult angles of the CC problem face to face.

Acknowledgments
I have been partially supported by DIUE/CUR Generalitat de Catalunya under project 2009SGR502, by the research Grant FPA 2007-20807, and by the Consolider CPAN project SD2007-00042. It is my pleasure to thank S. Basilakos, H. Fritzsch, J. Grande, J.A.S. Lima, D. López-Val, M. Plionis, and D. Polarski for their nice and stimulating collaboration in recent times on some of the topics discussed here. Last but not least, I am also very grateful to the organizers of NEB 15, in particular to E. Vagenas, for the invitation to review the cosmological constant problem in the incomparable scenario of the hellenic conference on Recent Developments in Gravity, held in the town of Chania, in the beautiful island of Crete, Greece.

References
[1] R. Knop et al., Astrophys. J. **598** (2003) 102; A. Riess *et al.* Astrophys. J. **607** (2004) 665.
[2] D.N. Spergel, et al., Astrophys. J. Suplem., **170**, 377, (2007); E. Komatsu et al. [arXiv:1001.4538].
[3] P.A.R. Ade *et al.*, Planck Collab. *Cosmological parameters*, arXiv:1303.5076.
[4] S. Weinberg, Rev. Mod. Phys., **61** (1989) 1; V. Sahni, A. Starobinsky, Int. J. of Mod. Phys. **A9** (2000) 373; S.M. Carroll, *Living Rev. Rel.* **4** (2001) 1; T. Padmanabhan, Phys. Rept., **380** (2003) 235; J. Martin, Comptes Rendus Physique 13 (2012) 566.
[5] F. Bauer, J. Solà, H. Stefancic, JCAP 1012 (2010) 029 [arXiv:1006.3944]; S. Basilakos, F. Bauer, J. Solà, JCAP 1201 (2012) 050 [arXiv:1109.4739].
[6] G. Aad *et al.* (ATLAS Collaboration), Phys.Lett. B 716 (2012) 1; S. Chatrchyan *et al.*. (CMS Collaboration), Phys. Lett. B 716 (2012) 30.
[7] J. Solà, *The cosmological constant and the vacuum energy in quantum field theory: a review with reflections* (to appear).
[8] J. Solà, *Cosmologies with a time dependent vacuum*, J. Phys. Conf. Ser. 283 (2011) 012033 [arXiv:1102.1815].
[9] A. Einstein, Sitzungsber. Konigl. Preuss. Akad. Wiss. (1917) 142.
[10] Y. B. Zeldovich, *Cosmological constant and elementary particles*, Sov. Phys. JETP Lett **6** (1967) 3167.
[11] A.D. Dolgov, in: *The very Early Universe*, Ed. G. Gibbons, S.W. Hawking, S.T. Tiklos (Cambridge U., 1982); Y. Fujii, Phys. Rev. **D26** (1982) 2580; L. Abbott, *Phys. Lett.* **B150** (1985) 427.
[12] R.D. Peccei, J. Solà and C. Wetterich, Phys. Lett. **B195** (1987) 183.
[13] S. M. Barr, Phys. Rev. **D36** (1987) 1691; L.H. Ford, Phys. Rev. **D 35** (1987) 2339.
[14] C. Wetterich, Nucl. Phys. **B302** (1988) 668; B. Ratra, P.J.E. Peebles, Phys. Rev. **D37** (1988) 3406.
[15] J. Solà, Phys.Lett. B228 (1989) 317.
[16] R.R. Caldwell, R.Dave, P.J. Steinhardt, Phys. Rev. Lett. **80** (1998) 1582.
[17] P.J. Peebles, B. Ratra, Rev. Mod. Phys. **75** (2003) 559; E.J. Copeland, M. Sami, S. Tsujikawa, Int. J. Mod. Phys. D15 2006 1753.
[18] O. Bertolami, Nuovo Cimento, **93B**, 36, (1986); M. Ozer M. and O. Taha, Nucl. Phys., **B287**, 776, (1987); O. K. Freese K., et al., Nucl. Phys., **287**, 797, (1987); J. C. Carvalho, J. A. S. Lima and I. Waga, Phys. Rev. **D46**, 2404, (1992).
[19] J. M. Overduin and F. I. Cooperstock, Phys. Rev. D., **58**, 043506, (1998) [arXiv: astro-ph/9805260]; R.G. Vishwakarma, Class. Quant. Grav. 18 (2001) 1159 [arXiv: astro-ph/0012492]; and references therein.
[20] J. A. S. Lima, Braz. J. Phys. 34 (2004) 194 [arXiv:astro-ph/0402109].
[21] A. Einstein, *Grundgedanken der allgemeinen Relativitätstheorie und Anwendung diese Theorie in der Astronomie, Sitzber. Preuss. Akad. Wiss. Berlin,* (1915) 315; *Zur allgemeinen Relativitätstheorie, Sitzber. Preuss. Akad. Wiss. Berlin,* (1915) 778-786; *Zur allgemeinen Relativitätstheorie (Nachtrag) Sitzber. Preuss. Akad. Wiss. Berlin,* (1915) 799-801.
[22] H. Fritzsch, J. Solà, Class. Quant. Grav., **29** (2012) 215002 [arXiv:1202.5097].
[23] G. Lemaître, Ann. Soc. Sci. Brux. **A47** (1927) 49.
[24] A. Friedmann, Z. Phys. **10** (1922) 377; **21** (1924) 326.
[25] A. Einstein, Sitzungsber. Preuss. Akad. Wiss. (1931) 235.
[26] S. Weinberg, Phys. Rev. Lett. **19** (1967) 1264; A. Salam, Elementary particle Theory; Nobel Symposium, No.8, ed. N. Svartholm (Almqvist and Wiksell, Stockholm 1969); S.L. Glashow, Nucl. Phys. **22** (1961) 579.
[27] H. Fritzsch, M. Gell-Mann, proc. of the XVI Int. Conf. on High Energy Physics, Chicago, 1972, vol 2, 135; H. Fritzsch, Murray Gell-Mann, H. Leutwyler, Phys.Lett. B47 (1973) 365.
[28] P.W. Higgs, Phys. Lett. **12** (1964) 132; Phys. Rev. Lett. **13** (1964) 508; G.S. Guralnik, C.R. Hagen and T.W.B. Kibble, Phys. Rev. Lett. **13** (1964) 585; F. Englert and R. Brout, Phys. Rev. Lett. **13** (1964) 321; T. W. B. Kibble, Phys. Rev. 155 (1967) 1554.

[29] L.E. Parker and D.J. Toms, *Quantum Field Theory in Curved Spacetime: quantized fields and gravity* (Cambridge U. Press, 2009).
[30] N.D. Birrell and P.C.W. Davies, *Quantum Fields in Curved Space* (Cambridge U. Press, 1982).
[31] I. L. Shapiro and J. Solà, Phys.Lett. B **682** (2009) 105, arXiv:0910.4925; arXiv:0808.0315.
[32] P.J.E. Peebles, *Principles of Physical Cosmology* (Princeton Univ. Press, 1993); T. Padmanabhan, *Structure Formation in the Universe* (Cambridge Univ. Press, 1993).
[33] S. E. Rugh, H. Zinkernagel, Studies in History and Philosophy of Modern Physics, **33** (2002) 663.
[34] N. Straumann, proc. of the 3rd Int. Conf. on Dark Matter in Astro and Particle Physics (Dark 2000) [arXiv:astro-ph/0009386].
[35] G. Lemaître, *Evolution of the Expanding Universe*, Proc. of the National Academy of Sciences of the USA **20** (1934) 12-17.
[36] S. Weinberg, *Gravitation and Cosmology: Principles and Applications of the General Theory of Relativity* (Wiley, New York, 1972).
[37] R. Schutzhold, Phys. Rev. Lett. **89** (2002) 081302, gr-qc/0204018.
[38] S. Basilakos, M. Plionis and J. Solà, Phys. Rev. **D 80**, (2009) 3511, arXiv:0907.4555 [astro-ph.CO]; J. Grande, J. Solà, S. Basilakos, and M. Plionis, J. of Cosmology and Astrop. Phys. JCAP **08** (2011) 007 [arXiv:1103.4632].
[39] S. Basilakos, D. Polarski, and J. Solà, Phys. Rev. **D 86** (043010) 2012 [arXiv:1204.4806].
[40] D. A. Easson, P. H. Frampton, G. F. Smoot, Phys. Lett. **B 696** (2011) 273.
[41] S. Basilakos, and J. Solà, *Effective equation of state for running vacuum: "mirage" quintessence and phantom dark energy* (to appear).
[42] I. L. Shapiro and J. Solà, Phys. Lett. B **475** (2000) 236, hep-ph/9910462.
[43] P. Ramond, *Field Theory. A Modern Primer* (Benjamin Cummings , Publish. Company, 1981); T.P. Cheng, L. Fong Li, *Gauge Theory of Elementary particle Physics* (Clarendon Press, Oxford, 1986); L.H. Ryder, *Quantum Field Theory* (Cambridge U. Press, 1985); S. Coleman, *Aspects of Symmetry* (Cambridge U. Press, 1985); L.S. Brown, *Quantum Field Theory* (Cambridge U. Press, 1992).
[44] J. Wess, B. Zumino, Nucl.Phys. **B78** (1974) 1.
[45] B. S. DeWitt, G. Esposito, Int. J. Geom. Meth. Mod. Phys. **5** (2008) 101 [arXiv:0711.2445]; L. Smolin, hep-th/0303185; S. B. Giddings, arXiv:1105.6359.
[46] B.S. DeWitt, *Quantum Field Theory in Curved Space-Time*, Phys. Rept. 19 (1975) 295; L. Parker, *Aspects of quantum field theory in curved space-time: Effective action and energy-momentum tensor*, in: Recent Developments in Gravitation, Cargèse 1978, ed. M. Lévy and S. Deser (Plenum Press, New York, 1978) 219-273; T. S. Bunch, L. Parker, Phys. Rev. **D 20** (1979) 2499.
[47] I.L. Buchbinder, S.D. Odintsov, I.L. Shapiro, *Effective Action in Quantum Gravity* (IOP, Bristol, 1992).
[48] E.V. Gorbar, I.L. Shapiro, JHEP 0302 (2003) 021.
[49] R.P. Woodard, Phys.Rev.Lett. 101 (2008) 081301.
[50] A. Kaya e-Print: arXiv:1303.5459 [hep-th].
[51] M. Bordag, U. Mohideen, V.M. Mostepanenko Phys.Rept. **353** (2001) 1.
[52] E. Elizalde, *Ten physical applications of spectral zeta functions* (Springer, 1995).
[53] R. L. Jaffe, Phys.Rev. D72 (2005) 021301.
[54] I.L. Shapiro, J. Solà, JHEP 02(006), 2002 [arXiv:hep-th/0012227].
[55] I.L. Shapiro, J. Solà, Nucl. Phys. Proc. Suppl. **127** (2004) 71 [hep-ph/0305279]; I. L. Shapiro, J. Solà, C. España-Bonet, P. Ruiz-Lapuente, Phys. Lett. B 574 (2003) 149 [astro-ph/0303306]; JCAP 0402 (2004) 006 [hep-ph/0311171]; I.L. Shapiro, J. Solà, JHEP proc. AHEP2003/013 [astro-ph/0401015].
[56] A. Babić, B. Guberina, R. Horvat, H. Štefančić, Phys. Rev. **D 65** (085002) 2002; Phys. Rev. **D 71** (124041) 2005.
[57] J. Solà, H. Štefančić, Phys. Lett. B **624**, 147 (2005) [arXiv:astro-ph/0505133]; Mod. Phys. Lett. A **21**, 479 (2006) [arXiv:astro-ph/0507110].
[58] J. Solà, J. of Phys. **A 41** (2008) 164066 [arXiv:0710.4151].
[59] M. Maggiore, Phys. Rev. D83 (2011) 063514, arXiv:1004.1782; L. Hollenstein, M. Jaccard, M. Maggiore, E. Mitsou, Phys. Rev. D85 (2012), arXiv:1111.5575; N. Bilic, Phys.Rev. D83 (2011) 105003, arXiv:1104.1349; Phys.Lett. B707 (2012) 221, arXiv:1110.2393; N. Bilic, S. Domazet, B. Guberina, Phys.Lett. B707 (2012) 221 arXiv:1110.2393.
[60] T. Padmanabhan, Class. Quant. Grav. **22** (2005) L107-L110 [arXiv:hep-th/0406060].
[61] I. L. Shapiro and J. Solà, Phys.Lett. B **530** (2002) 10 [arXiv:hep-ph/0104182].
[62] J. A. S. Lima, S. Basilakos, and J. Solà, Mon. Not. Roy. Astron. Soc. **431** (2013) 923, [arXiv:1209.2802].
[63] S. Basilakos, J. A. S. Lima, and J. Solà, Int. J. of Mod. Phys. D (to appear).(Work awarded Honorable Mention by the Gravity Research Foundation, 2013).
[64] E. L. D. Perico, J.A.S. Lima, S. Basilakos, and J. Solà, *Complete Cosmic History with* $\Lambda = \Lambda(H)$ (to appear).

[65] J. P. Mimoso, D. Pavón, Phys. Rev. **D 87**, 047302 (2013) [arXiv:1302.1972].
[66] J.A.S. Lima, J.M.F. Maia, Phys.Rev. **D49** (1994) 5597; J.A.S. Lima, M. Trodden, Phys. Rev. **D53** (1996) 4280 [astro-ph/9508049].
[67] F.R. Klinkhamer, G.E. Volovik, Phys.Rev. D79 (2009) 063527; E. C. Thomas, F. R. Urban, A. R. Zhitnitsky, JHEP 0908 (2009) 043; S. Carneiro, R. Tavakol, Phys.Rev. D80 (2009) 043528; N. Ohta, Phys. Lett. **B 695** (2011) 41.
[68] P.J.E. Peebles, Astrophys. J. 284 (1984) 439.
[69] P. Wang, X.-H. Meng, Class.Quant.Grav. **22** (2005) 283 [arXiv:astro-ph/0408495]; J.S. Alcaniz, J.A.S. Lima Phys.Rev. D72 (2005) 063516 [arXiv: astro-ph/0507372].
[70] J-P. Uzan, Living Rev. Rel. **14** (2011) 2 [arXiv:1009.5514]; *Rev. Mod. Phys.* 752003403 [arXiv: hep-ph/0205340]; T. Chiba, Prog. Theor. Phys. 126 (2011) 993 [arXiv:1111.0092]; and references therein.
[71] J. Grande, J. Solà, J.C. Fabris, I.L.Shapiro, Class. Quant. Grav. **27** (2010) 105004 [arXiv:1001.0259]; I.L. Shapiro, J. Solà, H. Štefančić, JCAP **0501** (2005) 012, [arXiv:hep-ph/0410095]; J. C. Fabris, I. L. Shapiro, J. Solà, JCAP 0702 (2007) 016 [arXiv:gr-qc/0609017].
[72] F. Iocco, G. Mangano, G. Miele, O. Pisanti and P. D. Serpico, Phys. Rept. **472** (2009) 1.
[73] J. Grande, and J. Solà (in preparation).
[74] J. Grande, A. Pelinson, and J. Solà, Phys. Rev. **D79**, (2009) 043006 [arXiv:0809.3462]; J. Grande, R. Opher, A. Pelinson, and J. Solà, JCAP **0712** (2007) 007 [arXiv:0709.2130]; arXiv:0904.3293; J. Grande, J. Solà and H. Štefančić, JCAP **08** (2006) 011 [arXiv:gr-qc/0604057].
[75] H. Fritzsch, Prog. Part. Nucl. Phys. **66** (2011) 193;, Nucl.Phys. Proc. Suppl. **203-204** (2010) 3.
[76] X. Calmet, H. Fritzsch, Europhys. Lett. **76** (2006) 1064; Phys. Lett. **540B** (2002) 173; Eur. Phys. J **C24** (2002) 639.
[77] E. Reinhold, R. Buning, U. Hollenstein, A. Ivanchik, P. Petitjean, W. Ubachs, Phys. Rev. Lett. **96** (2006) 151101; W. Ubachs, E. Reinhold, Phys. Rev. Lett. **92** (101302) 2004; A. Ivanchik, P. Petitjean, D. Varshalovich, B. Aracil, R. Srianand, H. Chand, C. Ledoux, P. Boisse, Astron. Astrophys. **440** (2005) 45 [arXiv:astro-ph/0507174].
[78] J. D. Bjorken, Phys. Rev. **D67** (2003) 043508 [arXiv:hep-th/0210202].
[79] D. López-Val, J. Solà, Eur. Phys. J. **C73** (2013) 2393 [arXiv:1211.0311].
[80] T. Padmanabhan [arXiv:1210.4174]; T. Padmanabhan, H. Padmanabhan [arXiv:1302.3226].

Journal of Physics: Conference Series **453** (2013) 012016 doi:10.1088/1742-6596/453/1/012016

Constraints on the nuclear symmetry energy via asteroseismology

Hajime Sotani[1], Ken'ichiro Nakazato[2], Kei Iida[3], and Kazuhiro Oyamatsu[4]

[1]Yukawa Institute for Theoretical Physics, Kyoto University, Kyoto 606-8502, Japan
[2]Faculty of Science & Technology, Tokyo University of Science, 2641 Yamazaki, Noda, Chiba 278-8510, Japan
[3]Department of Natural Science, Kochi University, 2-5-1 Akebono-cho, Kochi 780-8520, Japan
[4]Department of Human Informatics, Aichi Shukutoku University, 9 Katahira, Nagakute, Aichi 480-1197, Japan

E-mail: sotani@yukawa.kyoto-u.ac.jp

Abstract. We examine the fundamental torsional oscillations of the crust in the neutron star. Especially, we focus on the crystalline properties obtained from macroscopic nuclear models. As a result, we find that the frequencies of fundamental torsional oscillations are sensitive to the density dependence of the symmetry energy, but almost independent of the incompressibility of symmetric nuclear matter. Identifying the lowest frequency among the quasi-periodic oscillations observed in the giant flares as the fundamental $\ell = 2$ torsional oscillations, we put constraints on the density derivative of the symmetry energy as $L \simeq 50$ MeV.

1. Introduction

One of the most promising evidences about the oscillations of neutron stars are the quasi-periodic oscillations (QPOs) in giant flares observed in soft gamma repeaters (SGRs). The SGRs are considered as the candidate of magnetars, which are neutron stars with strong magnetic fields [1]. In addition to the sporadic X- and gamma-ray bursts, SGRs rarely emit much stronger gamma-rays called "giant flares". Up to now, at least three giant flares have been detected, i.e., SGR 0526-66, SGR 1900+14, and SGR 1806-20. Through the timing analysis of the decaying tail of giant flares, the existence of QPOs has been discovered, whose frequencies are in the range from tens Hz up to a few kHz [2]. To explain the observed QPO frequencies theoretically, many attempts have made with the torsional oscillations in the crust region and/or the magnetic oscillations (e.g., [3, 4, 5, 6, 7, 8, 9, 10, 11]). Associated with the QPOs in giant flares, there also exist the suggestions of the importance to consider the microscopic properties of crust region [12] and the possibility to obtain the information about nonuniform nuclear structure in the crust [13].

The structure of neutron stars can be described as follows. The ocean of melted iron exists in the vicinity of stellar surface for density lower than $10^6 - 10^8$ g/cm^3. Subsequently, the elastic crust region extends up to a density of the order of saturation density of nuclear matter, $\rho_s \sim 3 \times 10^{14}$ g/cm^3. At last, the inner region where the density is higher than $\sim \rho_s$ corresponds to the fluid core of neutron stars. The nuclei in the greater part of the crust form a body center cubic (bcc) lattice due to Coulomb interactions. But the recent studies suggest the existence of a nonuniform nuclear structure in the bottom of the crust, i.e., the shape of nuclear matter could be changing from sphere (bcc lattice) to cylinder, slab, cylinder hole, spherical hole, and uniform matter (inner fluid core) with increasing the density [14, 15]. This variation in the shape of nuclear matter is known as "nuclear pasta". It should be noticed that not only the density region of pasta phase but also the charge number of spherical nuclei depend strongly on the symmetry energy density derivative coefficient, L (see below for the definition of L) [16]. However, to see the properties of crust in neutron stars is not so easy. One way to solve this difficulty might be the constraint via the observations of the neutron star oscillations, which is well-established as "asteroseismology". In such a way, one could be able to see the stellar properties and/or to restrict on the stellar equation of state (EOS) (e.g., [17, 18, 19, 20]). In this article, we will examine the torsional oscillations in the neutron star crust and set make a restriction on the curst EOS via observations. The more details of this study can be found in Ref. [21].

2. Crust Equilibrium Configuration

The bulk energy per nucleon near the saturation point of symmetric nuclear matter can be written as a function of number density, n, and the neutron excess, α [22];

$$w = w_0 + \frac{K_0}{18n_0^2}(n-n_0)^2 + \left[S_0 + \frac{L}{3n_0}(n-n_0)\right]\alpha^2, \tag{1}$$

where w_0, n_0, and K_0 are the saturation energy, saturation density, and incompressibility of symmetric nuclear matter, respectively. Additionally, the parameters L and S_0 are associated with the symmetry energy coefficient, $S(n)$, which corresponds to the coefficient of α^2 in Eq. (1). That is, S_0 is the symmetry energy coefficient at $n = n_0$, while $L = 3n_0(dS/dn)_{n=n_0}$ is the symmetry energy density derivative coefficient. Three parameters of them, w_0, n_0, and S_0, can be comparatively easier to determine via experiments in the laboratory [23]. On the other hand, the remaining two parameters, L and K_0, are more difficult to determine experimentally. Now, instead of L, we introduce a new parameter, y, defined as $y = -K_0S_0/(3n_0L)$, which denotes the slope of the saturation line in the vicinity of $\alpha = 0$ [16, 23]. This is because y can express the characteristic of EOS, although y is not a parameter determined from experiments directly. Hence, the curst EOS can be parametrized by two parameters, y (or L) and K_0. Similar to [16], in this article, we adopt that $0 < L < 160$ MeV and 180 MeV $\leq K_0 \leq 360$ MeV, as long as $y < -200$ MeV fm^3. In practice, these parameter ranges are consistent with the mass and radius data for stable nuclei and effectively cover even extreme cases [16]. With such parameters, using a Thomas-Fermi model to agree with the experimental results of stable nuclei, the EOS in the crust region can be obtained. The adopted values of parameters are shown in Table 1, where n_1 corresponds to the baryon number density at which the spherical nuclear matter changes to cylindrical one, while n_2 corresponds to the density at which the nuclear matter becomes uniform. Namely, the interval between n_1 and n_2 is corresponding to the nuclear pasta phase.

The equilibrium structure of nonrotating neutron star becomes the static, spherically symmetric solution of the Tolman-Oppenheimer-Volkoff (TOV) equations. In this case, the metric can be described as

$$ds^2 = -e^{2\Phi}dt^2 + e^{2\Lambda}dr^2 + r^2d\theta^2 + r^2\sin^2\theta\, d\phi^2, \tag{2}$$

where Φ and Λ are functions with respect to r. In order to be closed the equation system, one needs to prepare the additional equation, i.e., EOS. Then, if knowing the EOS inside the neutron star, one can construct the neutron star models with one parameter, such as the central density of stars. However, the EOSs for inner core are still quite uncertain, in spite of many suggestions

Journal of Physics: Conference Series **453** (2013) 012016 doi:10.1088/1742-6596/453/1/012016

Table 1. Several sets of crust EOS parameters.

y (MeV fm^3)	K_0 (MeV)	L (MeV)	n_1 (fm^{-3})	n_2 (fm^{-3})
-220	180	52.2	0.060	0.079
-220	230	73.4	0.064	0.073
-220	360	146.1	0.066	0.066
-350	180	31.0	0.058	0.091
-350	230	42.6	0.063	0.086
-350	360	76.4	0.072	0.076
-1800	180	5.7	0.058	0.134
-1800	230	7.6	0.058	0.127
-1800	360	12.8	0.058	0.118

based on the assumptions of nuclear interactions. In order to avoid such an uncertainty of EOSs for inner core, as in Ref. [24], we construct the crust region with the crust EOS mentioned the above by integrating from the stellar surface inward to the boundary between the curst and core regions. In this way, one needs two parameters to determine the stellar model, where we adopt the stellar mass and radius, M and R, as those parameters.

In the limit of zero temperature, the shear modulus can be described as

$$\mu = 0.1194 \times n_i (Ze)^2/a, \tag{3}$$

where n_i and $+Ze$ are corresponding to the ion number density and the ion charge, while a is defined as $a^3 = 3/(4\pi n_i)$, which corresponds the average ion spacing [25, 26]. It should be noticed that this approximate formula for the shear modulus is derived with the assumption of bcc crystalline state, which is averaged over all directions. So, if one takes into account the effect of nonuniform nuclear structure known as nuclear pasta, the shear modulus might be modified [13]. Actually, it is suggested that the elastic properties in pasta phase could be softer than that in bcc lattice [27]. However, since the realistic shear modulus in pasta phase is not revealed yet, in this article we will deal with the pasta phase as liquid, i.e., $\mu = 0$ for $n \geq n_1$. With this assumption, the shear modulus may be effectively estimated smaller. As a result, the corresponding frequencies of torsional oscillations also become smaller, i.e., the obtained frequencies in this article could be corresponding to lower limit. But, as mentioned later, the effect of pasta phase might not be so important in the torsional oscillations.

3. Torsional Oscillations and Numerical Results

Since the torsional oscillation on the spherically symmetric star is incompressible, this type of oscillation does not induce the stellar deformation and density variation. Thus, one can determine the frequencies of torsional oscillations with satisfactory accuracy even if the metric perturbations would be neglected by setting $\delta g_{\mu\nu} = 0$, which is known as the relativistic Cowling approximation. With this approximation, the torsional oscillations can be described with one perturbation variable, i.e., the angular displacement of the stellar matter, $\mathcal{Y}$, which is associated with the ϕ-component of the perturbed 4-velocity of fluid, δu^ϕ, as $\delta u^\phi = \mathrm{e}^{-\Phi}\partial_t \mathcal{Y}(t,r)\partial_\theta P_\ell(\cos\theta)/\sin\theta$. Here, ∂_t and ∂_θ denote the partial derivative with respect to t and θ, while P_ℓ is the Legendre polynomial of order ℓ. Assuming that $\mathcal{Y}(t,r) = \mathrm{e}^{\mathrm{i}\omega t}\mathcal{Y}(r)$, one can obtain the perturbation equation from the linearlized equation of motion [28];

$$\mathcal{Y}'' + \left[\left(\frac{4}{r} + \Phi' - \Lambda'\right) + \frac{\mu'}{\mu}\right]\mathcal{Y}' + \left[\frac{\epsilon + p}{\mu}\omega^2 \mathrm{e}^{-2\Phi} - \frac{(\ell+2)(\ell-1)}{r^2}\right]\mathrm{e}^{2\Lambda}\mathcal{Y} = 0, \tag{4}$$

where ϵ and p correspond to the energy density and pressure, and the prime denotes the derivative with respect to r. In order to determine the eigenfrequencies, ω, we adopt the following boundary conditions: a zero-traction condition at $n = n_1$ and zero-torque condition at the stellar surface. The both conditions can reduce to $\mathcal{Y}' = 0$ [6, 28].

First, to see the dependence of the fundamental torsional oscillations on the EOS parameters, we calculate their frequencies on the typical stellar model with $M = 1.4M_\odot$ and $R = 12$km. Fig. 1 shows the frequencies of $\ell = 2$ fundamental torsional oscillations as a function of L, where the lines with circles, diamonds, and squares are corresponding to the results for $K_0 = 180$, 230, and 360 MeV. In addition to the numerical results, for comparison, the dot-dashed line denotes the lowest observed frequency in the SGR 1806-20, which is 18 Hz. From this figure, one can observe that the frequencies of fundamental torsional oscillations are almost independent from the value of K_0 at least in the parameter range of our calculations. We make sure of this dependence with different stellar models, such as $R = 10$, 12, and 14km for $M = 1.4M_\odot$ and $1.8M_\odot$. It could be because the charge number of spherical nuclei, Z, which contributes largely to the shear modulus as in Eq. (3), depends strongly on not K_0 but L [16]. As a consequence, one can derive the fitting formula of the frequencies of $\ell = 2$ fundamental torsional oscillations as

$$ {}_0t_2 = c_0 - c_1 L + c_2 L^2, \tag{5} $$

where c_0, c_1, and c_2 are constants depending on the stellar properties.

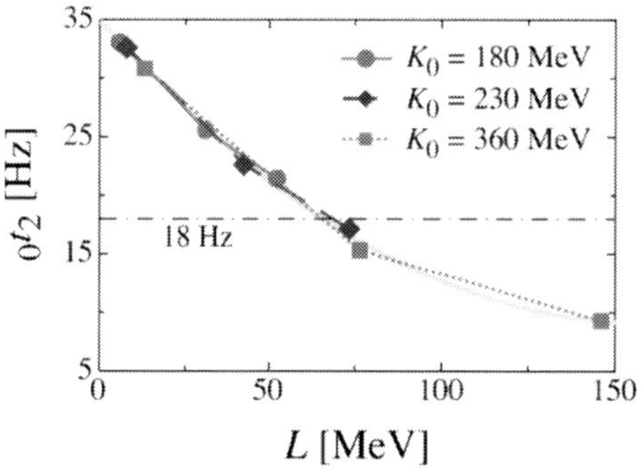

Figure 1. Frequency of $\ell = 2$ fundamental torsional oscillations plotted as a function of L for the stellar model with $M = 1.4M_\odot$ and $R = 12$ km. For comparison with the observational data, we also add the horizontal dot-dashed line, which is the lowest QPO frequency observed from SGR 1806-20 [2]. Additionally, the solid thick line denotes the fitting formula (5).

With the obtained fitting formula, one can draw Fig. 2, where the stellar mass is fixed to be $M = 1.4M_\odot$. It is found from this figure that the dependence of the frequency of $\ell = 2$ fundamental torsional oscillation on the stellar radius is relatively small. Namely, the frequency with the fixed value of L is determined within the accuracy of $\sim 20\%$, if the stellar radius is in the range of $10 \leq R \leq 14$km. Additionally, we find that the dependence of $_0t_2$ on the stellar mass becomes smaller than that on the radius, i.e., the frequencies for $R = 10$, 12, and 14 km decrease ~ 14, 10, and 9% if the stellar mass increases from $1.4M_\odot$ up to $1.8M_\odot$. In practice, the frequencies of $\ell = 2$ fundamental torsional oscillations for the stellar models with 10 km $\leq R \leq$ 14 km and $1.4M_\odot \leq M \leq 1.8M_\odot$ can be confined in the colored region in Fig. 3. Thus, almost independent of the stellar models, one could be possible to restrict on the value of L via the observations of torsional oscillations.

It has been suggested that the observed QPOs in giant flares can be due to the torsional oscillations in neutron star crust. On the other hand, among many frequencies of torsional oscillations, the frequency of $\ell = 2$ fundamental oscillations is the lowest one. So, if the observed QPO frequencies would be corresponding to the torsional oscillations in neutron star crust, the frequency of $\ell = 2$ fundamental torsional oscillation should become lower than the lowest frequency in the observed QPOs. Based on this statement, one can restrict on L as $L \gtrsim 50$ MeV independent of the stellar models (see Fig. 3). It should be noticed that as mentioned the above, the calculated frequencies might be underestimated because we neglect the effect of pasta phase in our calculations. In other words, the restriction that $L \gtrsim 50$ MeV might be severer than that including the effect of pasta phase. However, the region of past phase become narrower as L increases, and that region could vanish for $L \gtrsim 100$ MeV [16]. Thus, even if the effect of pasta phase will be taken into account, the restriction that $L \gtrsim 50$ MeV may not be modified.

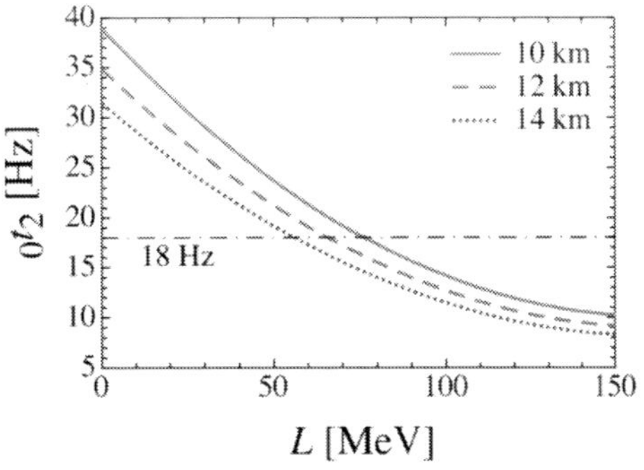

Figure 2. For the stellar models with $M = 1.4M_\odot$ and $R = 10$, 12, 14km, the frequency of ${}_0t_2$ given by the fitting formula (5) is shown as a function of L.

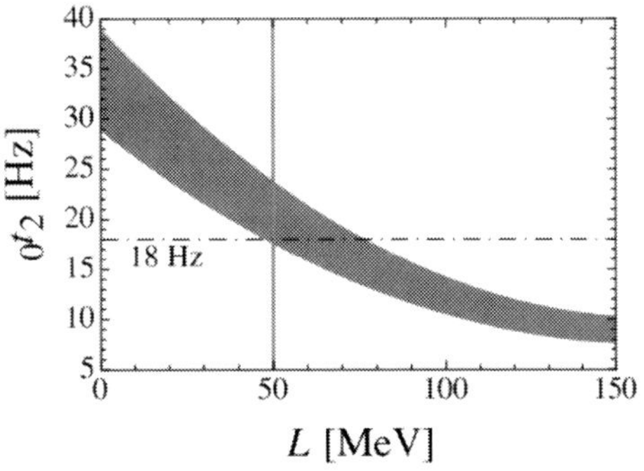

Figure 3. The colored region is corresponding to the expected frequencies of ${}_0t_2$ for the stellar models with $R = 10 - 14$ km and $M = 1.4M_\odot - 1.8M_\odot$.

4. Conclusion

To verify completely the EOS of nuclear matter with the ground experiments is not so easy and many uncertainties still exist. In contrast to this, the asteroseismology might provide the additional constraints in the EOS. This is an approach from totally another direction. In fact, the QPOs observed in the giant flares, which are considered as the results of oscillations in neutron stars, could become the appropriate evidence to reveal the interior properties. In this article, we show that the frequency of fundamental torsional oscillation depends only on L. Compared with the QPO frequencies observed in the giant flares, one can expect that $L \gtrsim 50$ MeV independent of the stellar models. We point out that our restriction on L is fairly stringent because experimental constraints on L have yet to converge (e.g., [31, 32]). Additionally, in this article, we omit the effect of neutron superfluidity, which can affect on the enthalpy density and the torsional oscillations [29, 30]. Including such an effect, one can make further constraint in the EOS parameters. Moreover, since the existence of nonuniform structure in the core of neutron star is also suggested, the torsional oscillations in such phase might be important to theoretically explain the observed QPO [33].

This work was supported in part by Grants-in-Aid for Scientific Research on Innovative Areas through No. 23105711 and No. 24105008 provided by MEXT and in part by Grant-in-Aid for Young Scientists (B) through No. 24740177 and for Research Activity Start-up through No. 23840038 provided by JSPS.

References

[1] Duncan R C and Thompson C 1992 *Astrophys. J.* **392** L9
[2] Watts A L and Strohmayer T E 2006 *Adv. Space Res.* **40** 1446
[3] Levin Y 2006 *Mon. Not. R. Astron. Soc.* **368** L35
[4] Lee U 2007 *Mon. Not. R. Astron. Soc.* **374** 1015
[5] Samuelsson L and Andersson N 2007 *Mon. Not. R. Astron. Soc.* **374** 256
[6] Sotani H, Kokkotas K D and Stergioulas N 2007 *Mon. Not. R. Astron. Soc.* **375** 261
[7] Sotani H, Kokkotas K D and Stergioulas N 2008 *Mon. Not. R. Astron. Soc.* **385** L5
[8] Sotani H, Colaiuda A and Kokkotas K D 2008 *Mon. Not. R. Astron. Soc.* **385** 2161
[9] Sotani H and Kokkotas K D 2009 *Mon. Not. R. Astron. Soc.* **395** 1163
[10] Gabler M, Cerdá-Durán P, Font J A, Müller E and Stergioulas N 2011 *Mon. Not. R. Astron. Soc.* **410** L37
[11] Colaiuda A and Kokkotas K D 2011 *Mon. Not. R. Astron. Soc.* **414** 3014
[12] Steiner A W and Watts A L 2009 *Phys. Rev. Lett.* **103** 181101
[13] Sotani H 2011 *Mon. Not. R. Astron. Soc.* **417** L70
[14] Lorenz C P, Ravenhall D G and Pethick C J 1993 *Phys. Rev. Lett.* **70** 379
[15] Oyamatsu K 1993 *Nucl. Phys.* A **561** 431
[16] Oyamatsu K and Iida K 2007 *Phys. Rev.* C **75** 015801
[17] Andersson N and Kokkotas K D 1996 *Phys. Rev. Lett.* **677** 4134
[18] Sotani H, Tominaga K and Maeda K I 2001 *Phys. Rev.* D **65** 024010
[19] Sotani H, Kohri K and Harada T 2004 *Phys. Rev.* D **69** 084008
[20] Sotani H, Yasutake N, Maruyama T and Tatsumi T 2011 *Phys. Rev.* D **83** 024014
[21] Sotani H, Nakazato K, Iida K and Oyamatsu K 2012 *Phys. Rev. Lett.* **108** 201101
[22] Lattimer J M 1981 *Annu. Rev. Nucl. Part. Sci.* **31** 337
[23] Oyamatsu K and Iida K 2003 *Prog. Theor. Phys.* **109** 631
[24] Iida K and Sato K 1997 *Astrophys. J.* **477** 294
[25] Ogata S and Ichimaru S 1990 *Phys. Rev.* A **42** 4867
[26] Strohmayer T, van Horn H M, Ogata S, Iyetomi H and Ichimaru S 1991 *Astrophys. J.* **375** 679
[27] Pethick C J and Potekhin A Y 1998 *Phys. Lett.* B **427** 7
[28] Schumaker B L and Thorne S K 1983 *Mon. Not. R. Astron. Soc.* **203** 457
[29] Pethick C J, Chamel N and Reddy S 2010 *Prog. Theor. Phys. Suppl.* **186** 9
[30] Sotani H, Nakazato K, Iida K and Oyamatsu K 2013 *Mon. Not. R. Astron. Soc.* **428** L21
[31] Li B A, Chen C W and Ko C M 2008 *Phys. Rep.* **464** 113
[32] Tsang M B *et al* 2009 *Phys. Rev. Lett.* **102** 122701
[33] Sotani H, Maruyama T and Tatsumi T 2013, *Nucl. Phys.* A **906** 37

Journal of Physics: Conference Series **453** (2013) 012017 doi:10.1088/1742-6596/453/1/012017

Thermodynamic phase structure of charged anti-de Sitter scalar-tensor black holes

Ivan Zh. Stefanov[1], **Daniela D. Doneva**[2], **Stoytcho S. Yazadjiev**[3], **Kostas D. Kokkotas**[2] **and Michail D. Todorov**[4]

[1] Department of Applied Physics, Technical University of Sofia, 8, Kliment Ohridski Blvd., 1000 Sofia, Bulgaria

[2] Theoretical Astrophysics, Eberhard-Karls University of Tübingen, Tübingen 72076, Germany

[3] Department of Theoretical Physics, Faculty of Physics, St.Kliment Ohridski University of Sofia 5, James Bourchier Blvd., 1164 Sofia, Bulgaria

[4] Dept of Diff. Equations, Faculty of Applied Math. and Informatics, Technical University of Sofia, Sofia, 1000, Bulgaria

E-mail: izhivkov@tu-sofia.bg

Abstract. When electromagnetic field with nonlinear lagrangian acts as a source of gravity the no-scalar-hair theorems can be eluded and black holes with non-trivial scalar field can be found in scalar tensor theories [1]. Black holes with secondary scalar hair exist also when a cosmological constant is added in the theory. The thermodynamics of black holes in anti-de Sitter (AdS) space-time has attracted considerable interest due to the AdS/CFT conjecture. A natural question that arises is whether the non-trivial scalar field would alter the black-hole thermodynamical phase structure. In the current work we present the phase structure of charged hairy black holes coupled to nonlinear Born-Infeld electrodynamics in canonical ensemble which is naturally related to AdS space-time. In certain regions of the parameter space we find the existence of a first-order phase transition between small and very large black holes. An unexpected result is that for a small subinterval of charge values two phase transitions are observed – one of zeroth and one of first order.

1. Introduction

In the current paper we present the thermodynamical phase structure in canonical ensemble of charged anti-de Sitter black holes coupled to nonlinear Born-Infeld electrodynamics within a certain class of scalar-tensor theories. The thermodynamics of black holes in AdS space-time has attracted considerable research interest in the last decade especially in the context of the AdS/CFT conjecture according to which the thermodynamics of the AdS black holes is related to the thermodynamics of the dual conformal field theory (CFT) residing on the boundary of the AdS space [3, 4]. In their pioneering work Hawking and Page [5] showed the existence of a phase transition between the AdS black hole and thermal AdS space. The AdS/CFT duality provides us with a new tool to study the phase transitions in the dual CFT theories on the base of studying the phase transitions of the AdS black holes.

On the other side, scalar tensor theories (STT) attract significant research interest since they are among the most natural generalizations of General Relativity (GR). Fundamental scalar fields are a possible remnant of a fundamental unified theory like string theory or higher dimensional gravity theories [6]. STT and GR can hardly be distinguished in the weak field limit. Significant

Journal of Physics: Conference Series **453** (2013) 012017 doi:10.1088/1742-6596/453/1/012017

differences between these theories are more likely to occur in strong fields, hence the interest in black holes in STT. According to the no scalar-hair conjecture, the black-hole solutions in the STT coincide with the solutions from GR. No-scalar-hair theorems treating the cases of static, spherically symmetric, asymptotically flat, electrically neutral black holes and charged black holes in the Maxwell electrodynamics have been proved for a large class of scalar-tensor theories (see [7, 8] and papers referring to them for more details). The no-scalar-hair conjecture can be circumvented when nolinear matter models, e.g. non-linear electrodynamics (NLED), are considered. In [1, 9] black holes with non-trivial scalar field which are significantly different from the corresponding Einstein-Born-Infeld solutions in GR have been constructed numerically. The solutions obtained in [1] have been generalized in [2] where black holes with AdS asymptotic have been numerically studied. Here we present in details the thermodynamical structure in canonical ensemble of the latter.

2. Black holes in STT coupled to Born-Infeld electrodynamics

The general action of STT in Jordan frame has the following form

$$S = \frac{1}{16\pi G_*}\int d^4x\sqrt{-\tilde{g}}\left(F(\Phi)\tilde{R} - Z(\Phi)\tilde{g}^{\mu\nu}\partial_\mu\Phi\partial_\nu\Phi - 2U(\Phi)\right) + S_m.$$

The action of matter which acts as a gravitational source is denoted by S_m and it is constructed as in General relativity. STT are often studied in Einstein frame in which the action takes the form

$$S = \frac{1}{16\pi G_*}\int d^4x\sqrt{-g}\left(R - 2g^{\mu\nu}\partial_\mu\varphi\partial_\nu\varphi - 4V(\varphi)\right) + S_m[\Psi_m;\mathcal{A}^2(\varphi)g_{\mu\nu}].$$

In this frame the scalar field in not directly coupled to the Ricci scalar curvature but appears explicitly in the action of the sources. In the particular class of theories that we have considered the potential of the scalar field and the coupling function are given by

$$V(\varphi) = \frac{1}{2}\Lambda \quad \text{and} \quad \mathcal{A}(\varphi) = e^{\alpha\varphi},$$

where Λ is a negative constant that ensures the AdS asymptotic and α is the coupling parameter between the electromagnetic field and the scalar field. This choice of potential for the scalar field allows solutions which are asymptotically AdS to be found. The coupling function is the same as in Brans-Dicke theory. The action of the nonlinear electrodynamics is

$$S_m = \frac{1}{4\pi G_*}\int d^4x\sqrt{-g}\mathcal{A}^4(\varphi)L(X,Y),$$

where $L(X,Y)$ denotes the Lagrangian of the nonlinear electrodynamics. The equations defining the functions X and Y are

$$X = \frac{\mathcal{A}^{-4}(\varphi)}{4}F_{\mu\nu}F^{\mu\nu}, \tag{1}$$

$$Y = \frac{\mathcal{A}^{-4}(\varphi)}{4}F_{\mu\nu}(\star F)^{\mu\nu}, \tag{2}$$

where $F_{\mu\nu}$ is the electromagnetic field strength tensor and $\star$ stands for the Hodge dual with respect to the metric $g_{\mu\nu}$.

In the particular case of Born-Infeld electrodynamics the nonlinear lagrangian has the following form

$$L = 2b\left[1 - \sqrt{1 + \frac{X}{b} - \frac{Y^2}{4b^2}}\right], \tag{3}$$

where b is the Born-Infeld parameter. The black holes that we have constructed numerically in that class of theories [2] are

- asymptotically anti-de Sitter
- have a single, non-degenerate event horizons (no extremal black holes exist)
- purely magnetically charged (the metric, the scalar field and the thermodynamical variables are preserved under electric-magnetic duality rotations)
- have three free parameters: α, b, P (Due to invariance of the solution with respect to rigid rescaling the parameter Λ can be excluded from the list of free parameters.)

The major difference between them and the pure Einstein-Born-Infeld AdS black holes is that for the latter as the mass M and, respectively, the radius of the event horizon r_H are decreased an extremal black hole is reached, while for the former – a naked singularity.

3. Thermodynamics of scalar-tensor black holes coupled to Born-Infeld electrodynamics

For the temperature of the event horizon we apply a standard definition

$$T = \left.\frac{f'(r)\, e^{-\delta(r)}}{4\pi}\right|_{r=r_H}.$$

The formula for the entropy in Jordan frame needs to be generalized in the following way

$$S_J = \frac{1}{4G_*}\int d^2x\sqrt{-^{(2)}\tilde{g}}F(\Phi) = \frac{1}{4G_*}\int d^2x\sqrt{-^{(2)}g} = S_E = S.$$

This definition of entropy is invariant with respect to conformal transformations of the metric.

The thermodynamic potential that is naturally related to the canonical ensemble is the free energy (see Appendix B in [2] for more details concerning the derivation of the thermodynamical quantities and the first law of thermodynamics for the system):

$$F(T,P) = M - TS.$$

The phase structure is usually presented on conjugate diagrams. The two conjugate variables are

$$-S(T) = \left(\frac{\partial F}{\partial T}\right)_P \quad \text{and} \quad \Psi(P) = \left(\frac{\partial F}{\partial P}\right)_T,$$

where S is the entropy and Ψ is the magnetic potential. We will consider only the case with fixed magnetic charge P. Since the entropy is a monotonous function of the radius of the event horizon r_H the thermodynamic phase structure of the black holes is usually presented on the $T(r_H)$ diagram instead of the $T(S)$. Our aim is to study the effect that the different free parameters have on the thermodynamical phase structure. First, we will briefly discuss the role of the parameters b and α. When b is lower than some critical value $b < b_{\rm crit}$ the phase structure is trivial – there is just one stable phase so no phase transitions can occur. For $b > b_{\rm crit}$ the number of phases is either two, one stable and one unstable, or four – two stable and two unstable, depending on the value of the magnetic charge. Phase transitions occur between the two stable phases. The parameter coming from the coupling function of the STT α has a negligible effect on the value of $b_{\rm crit}$.

Of all three parameters the magnetic charge has the most significant effect on the thermodynamical phase structure of the solutions. The function $T = T(r_H, P)$ has two inflection points $P^{(1)}_{\rm crit}$ and $P^{(2)}_{\rm crit}$ defined by

$$\frac{\partial T}{\partial r_H} = \frac{\partial^2 T}{\partial r_H^2} = 0.$$

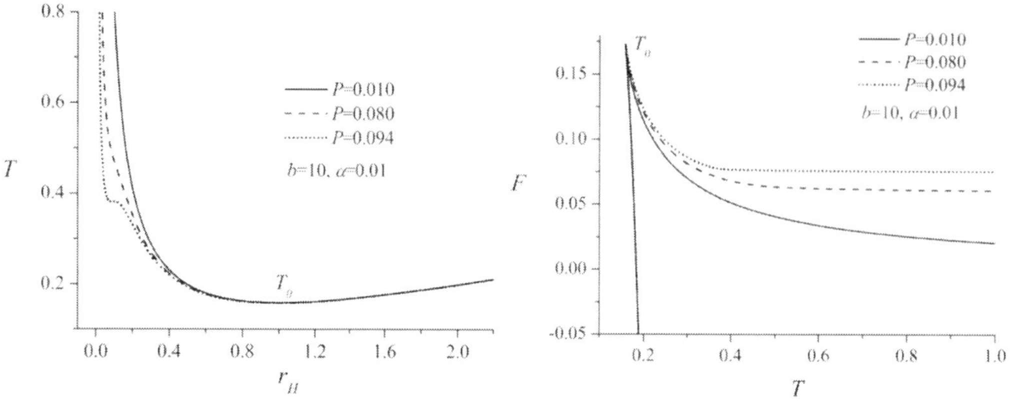

Figure 1. The $T(r_H)$ and the $F(T)$ diagrams for low magnetic charge interval.

They determine three major intervals of values of P: low charge $P < P^{(1)}_{\rm crit}$, middle charge $P^{(1)}_{\rm crit} < P < P^{(2)}_{\rm crit}$ (this one has three subintervals) and high charge $P^{(2)}_{\rm crit} < P$. Each of them will be discussed separately below.

3.1. Low magnetic charge interval
For the case of low magnetic charge the conjugate diagram $T(r_H)$ and the free energy as function of the temperature $F(T)$ are presented, respectively, on the left and on the right panel of Fig. (1). In that interval of the magnetic charge the thermodynamical system has two phases. The phase with negative slope of the temperature and higher free energy is unstable. The presence of only one stable phase means that no phase transitions occur in that interval of values of the magnetic charge.

3.2. Middle magnetic charge interval
For all values of the magnetic charge between $P^{(1)}_{\rm crit}$ and $P^{(2)}_{\rm crit}$ the thermodynamical system has four phases which we term: very small black holes (VSBH), small black holes (SBH), large black holes (LBH) and very large black holes (VLBH). The temperature T has three extrema – two local minima at $T_{\rm min}$ and T_0 and one local maximum at $T_{\rm max}$. The branch of VSBH spans from $T_{\rm min}$ to infinite temperatures, the branch of SBH spans between $T_{\rm min}$ and $T_{\rm max}$, the branch of LBH spans between $T_{\rm max}$ and T_0 and the last branch corresponding to VLBH – from T_0 to infinity.

For two of the phases, the SBH and the VLBH, the slope of the temperature is positive which means that these phases are stable. The other two phases are unstable. The thermodynamical system is dominated by the stable phase which has the lowest value of the free energy. As T is varied while P is kept fixed the system passes from one stable phase to another stable phase, i.e. a phase transition is observed.

The middle magnetic charge interval is divided in the following three subintervals. The number and the order of the observed phase transition is different in each of them.

I) The first subinterval is presented on Fig. (2). For this subinterval $T_0 < T_{\rm min}$ so when $T_0 < T < T_{\rm min}$ the only stable phase is the VLBH. For $T_{\rm min} < T < T_{\rm max}$ there is a second stable phase – the SBH – but since for all values of the temperature $F_{\rm VLBH} < F_{\rm SBH}$ no phase transitions occur.

II) The second subinterval is presented on Fig. (3). This is the subinterval with the largest number of phase transitions. Here still $T_0 < T_{\rm min}$. Again for $T_0 < T < T_{\rm min}$ the system is in the

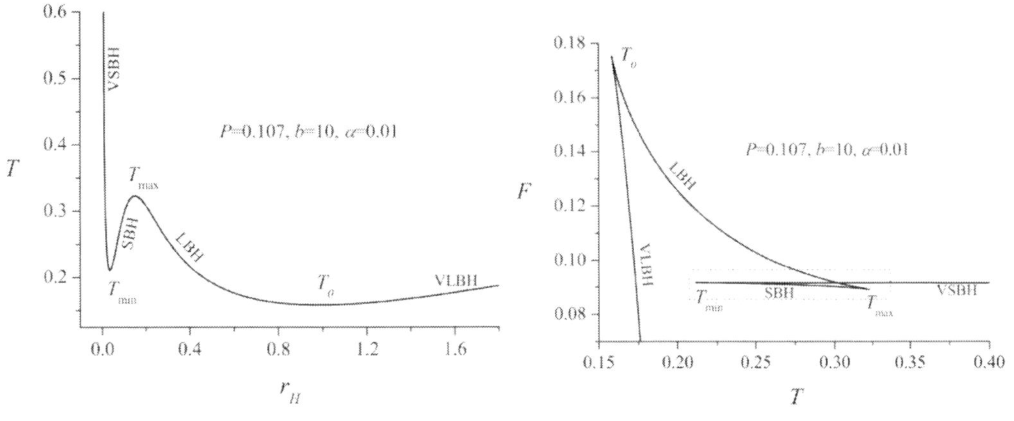

Figure 2. Middle magnetic charge interval (first subinterval).

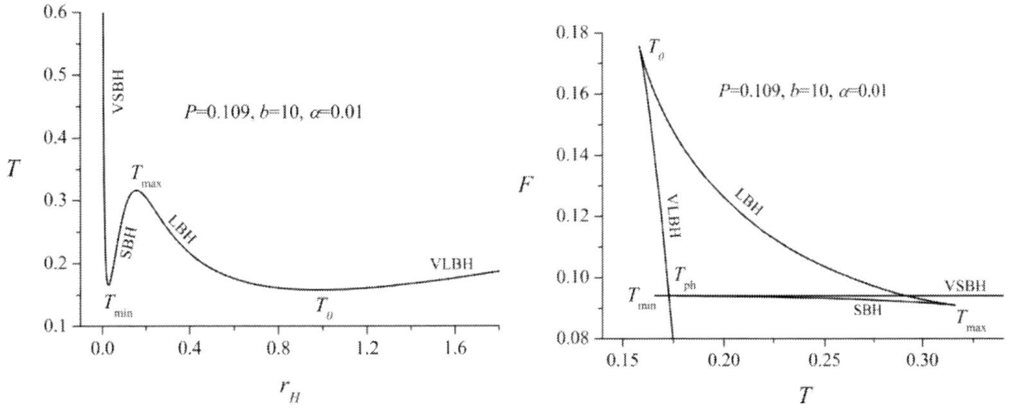

Figure 3. Middle magnetic charge interval (second subinterval).

phase of VLBH since that is the only stable phase. What happens when $T_{\rm min} < T < T_{\rm max}$ and the number of stable phases is two? As it can be seen on the right panel of (3) for $T_{\rm min} < T < T_{\rm ph}$ the system is dominated by the SBH since $F_{\rm SBH} < F_{\rm VLBH}$. At $T = T_{\rm min}$ a phase transition between VLBH and SBH is observed. The free energy is discontinuous at that point so the order of the phase transition zeroth.

When the temperature rises above $T_{\rm ph}$, however, $F_{\rm VLBH} < F_{\rm SBH}$ so a first order phase transition between SBH and VLBH occurs.

III) The last subinterval is presented on Fig. (4). Here $T_0 > T_{\rm min}$. When $T_{\rm min} < T < T_0$ the system is in the only stable state – that of the SBH. For temperatures between T_0 and $T_{\rm ph}$, $F_{\rm SBH} < F_{\rm VLBH}$ so no phase transition is observed at T_0 where the second stable phase comes into existence. The first order phase transition at $T_{\rm ph}$ which was present in the previously discussed subinterval is observed also in the current subinterval of the magnetic charge.

3.3. High magnetic charge interval

The graphics for the interval of high magnetic charges are presented on Fig. (5). Again only two phases exist – one unstable and one stable, and no phase transitions are observed.

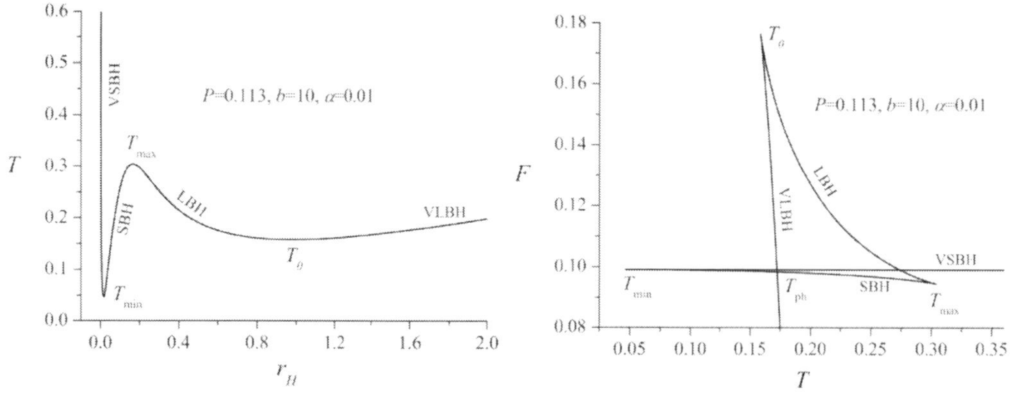

Figure 4. Middle magnetic charge interval (third subinterval).

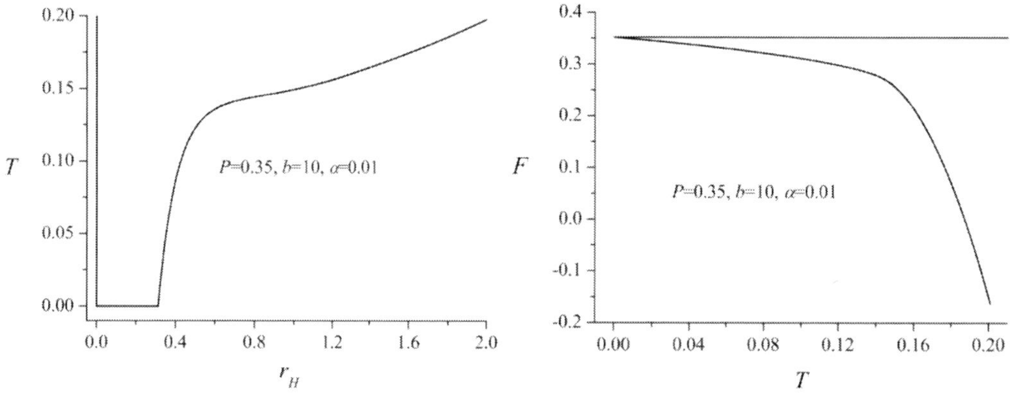

Figure 5. The $T(r_H)$ and the $F(T)$ diagrams and for high magnetic charge interval.

4. Off-shell considerations

Conclusions for the phase structure and the stability of the phases of a thermodynamical system can be made from the off-shell thermodynamic potentials. This method has been applied in [10] for the case of Einstein-Born-Infeld black holes. We will apply it here to give another perspective to the zeroth order phase transition that is observed in the middle charge interval $P^{(1)}_{\rm crit} < P < P^{(2)}_{\rm crit}$. In canonical ensemble the off-shell free energy is defined as

$$F_{\rm off} = M - T_{\rm th} S. \tag{4}$$

Here $T_{\rm th}$ is the temperature of the thermostat and varies independently from S unlike the temperature of the event horizon T. According to the off-shell formalism, equilibrium states occur at the extrema of the off-shell free energy, with minima corresponding to stable states and maxima – to unstable [11]. The on-shell free energy intersects the curves of the off-shell free energy in the extrema corresponding to the equilibrium phases. Equilibrium black holes have temperature equal to the that of the thermostat $T = T_{\rm th}$.

Let us demonstrate the relation between the on-shell and the off-shell approach. As we can see on the left panel of Fig. (3) for temperatures $T \in (T_{\rm min}, T_{\rm max})$ there are four black holes with different radii of the event horizon and different entropies S, respectively, that have the same temperature. What is the structure of the $F_{\rm off}(S)$ diagram? On Fig. (6) we have presented the off-shell free energy for three values of the temperature of the thermostat: $T_0 < T = 0.16 < T_{\rm min}$,

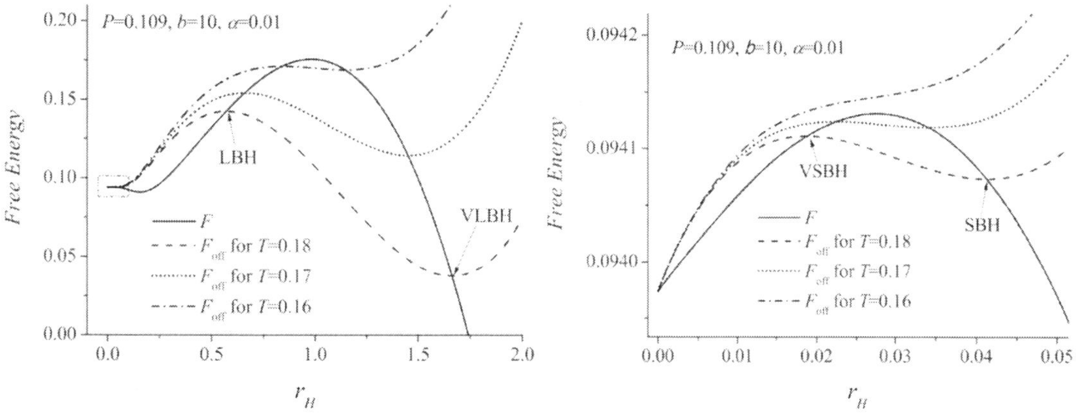

Figure 6. The off-shell free energy for three different values of the temperature and the equilibrium free energy as functions of the entropy. The right panel is a magnification of the enclosed region from the left panel.

$T_{\rm min} < T = 0.17 < T_{\rm ph} < T_{\rm max}$ and $T_{\rm ph} < T = 0.18 < T_{\rm max}$. We can clearly see the presence of the VLBH and the LBH on the left panel of Fig. (6). Due to the scale of the figure we cannot get sufficient information for the VSBH and SBH. A magnification of the enclosed region from the left panel of the figure is shown on the right panel. As it can be seen on the figure, in the latter region for the first temperature interval, $T_0 < T < T_{\rm min}$, the off-shell free energy is monotonous and the VSBH and SBH do not exist. Those phases, however, are present in the two other temperature intervals – $T_{\rm min} < T < T_{\rm ph} < T_{\rm max}$ and $T_{\rm ph} < T < T_{\rm max}$.

What does the $F_{\rm off}(S)$ diagram tell us? It is usually accepted that the system is in the state which realizes not only a local but also a global minimum of the off-shell free energy.

For low enough temperatures, $T_0 < T < T_{\rm min}$, the thermodynamical system is in the only stable phase – the VLBH. If the temperature rises above $T_{\rm min}$ but is still lower than $T_{\rm ph}$ the ensemble will be dominated by the SBH since the minimum corresponding to SBH is lower than the one of the VLBH. This temperature interval is presented by the dotted line on Fig. (6). In other words at $T_{\rm min}$ a phase transition between VLBH and SBH is observed. It could be classified as zeroth order since the on-shell free energy is discontinuous at this point (see Fig. (3)).

What happens if the temperature is further raised? For $T_{\rm ph} < T < T_{\rm max}$, presented by the dashed line, the global minimum of $F_{\rm off}$ is realized by the VLBH which means that a phase transition occurs at $T_{\rm ph}$. The phase transition could be classified as first order since the first derivative of the on-shell free energy is discontinuous at this point (see Figs. (2)–(4)).

5. Sum-up

Let us summarize our results for the rich thermodynamical phase structure of the AdS-scalar-tensor-Born-Infeld black holes in canonical ensemble.

- At low values of the magnetic charge $P < P^{(1)}_{\rm crit}$ the thermodynamical system has just two phases so no phase transitions are observed when the temperature T is varied while P is kept fixed.
- In the middle charge interval $P^{(1)}_{\rm crit} < P < P^{(2)}_{\rm crit}$ there are four thermodynamical phases – two of them stable. The middle charge interval has to be divided in three subintervals. The number and the order of the phase transition is different in each of the subintervals.
  - subinterval I: no phase transitions are observed;

- subinterval II: two phase transitions occur with the increase of the temperature.
  The phase transition VLBH $\rightarrow$ SBH is of 0-th order,
  The phase transition SBH $\rightarrow$ VLBH is of 1-st order;
- subinterval III: one phase transition is observed,
  The phase transition SBH $\rightarrow$ VLBH is classified as 1-st order phase transition;

- For high values of the magnetic charge $P^{(2)}_{\rm crit} < P$ again no phase transitions are present.

Similar phase transitions are observed also in the case with zero scalar field, namely, the Einstein-Born-Infled black holes.

Acknowledgements
This work was partially supported by the Bulgarian National Science Fund under Grant No DMU 03/6. The authors would also like to thank the Research Group Linkage Programme of the Alexander von Humboldt Foundation for the support.

References

[1] I. Stefanov, S. Yazadjiev and M. Todorov, *Phys. Rev.* D **75**, 084036 (2007).
[2] Daniela D. Doneva, Stoytcho S. Yazadjiev, Kostas D. Kokkotas, Ivan Zh. Stefanov, Michail D. Todorov, *Phys. Rev.* D **81**, 104030 (2010).
[3] J. Maldacena, *Adv. Theor. Math. Phys.* **2**, 231 (1998); arXiv: hep-th/9711200.
[4] J. L. Petersen, *Int. J. Mod. Phys.* **A 14**, 3597 (1999) ; arXiv: hep-th/9902131.
[5] S. W. Hawking and D. N. Page, *Commun. Math. Phys.* **87**, 577 (1983).
[6] T. Damour and A. Polyakov, *Nucl. Phys.* **B423**, 532 (1994).
[7] M. Heusler, *Class.Quant.Grav.* **12**, 2021 (1995).
[8] A. Mayo and J. Bekenstein, *Phys. Rev.* **D54**, 5059 (1996).
[9] I. Stefanov, S. Yazadjiev and M. Todorov, *Mod. Phys. Lett.* **A 23** (34), 2915 (2008), arXiv: 0708.4141.
[10] Yun Soo Myung, Yong-Wan Kim and Young-Jai Park, *Phys. Rev.* **D 78**, 084002 (2008); arXiv: 0805.0187.
[11] G. Arcioni and E. Lozano-Tellechea, *Phys.Rev.* D **72**, 104021 (2005).

Journal of Physics: Conference Series **453** (2013) 012018 doi:10.1088/1742-6596/453/1/012018

MTV-G experiment : probing a non-standard strong gravitational field at nuclear scale using geodetic precession

Saki Tanaka[1], Yusuke Nakaya[1], Kazufumi Ninomiya[1,2], Hironori Nishio[1], Junichi Onishi[1], Robert Openshaw[3], Matthew Pearson[3], Yumi Totsuka[1] and Jiro Murata[1]

[1]Department of Physics, Rikkyo University, Nishi-Ikebukuro, Tokyo 171-8501, JAPAN
[2]RIKEN Nishina Center, 2-1 Hirosawa, Wako, Saitama 351-0198, JAPAN
[3]TRIUMF, Vancouver, BC V6T 2A3, CANADA

E-mail: `saki_t@rikkyo.ac.jp`

Abstract. The MTV-G project has started in 2011 to explore a strong gravitational field around nuclei utilizing an experimental technique developed to search time reversal symmetry violation in nuclear beta decay experiment at a radioactive beam facility. A large electron spin-precession due to a possible strong gravitational field, which has been predicted by a large extra dimension model, is investigated in an electron-nuclear scattering experiment at TRIUMF. The experimental design, which use a spin polarized electron source and a Mott-spin analyzer, the commissioning experiment, the preliminary results, together with an introduction to the next generation device, are described in this article.

1. Introduction

Gravitational phenomena have been completely ignored in particle and nuclear physics since the Newtonian gravity is about 10^{-38} weaker than the other three gauge interactions. However, if we seriously think about the possibility of a strong gravitational field predicted by the so called large extra dimension model which is known as the ADD (N. Arkani-Hamed, S. Dimopoulos, G. Dvali) model [1], a small correction in nuclear phenomena coming from the enforced gravitational field should be carefully investigated. In fact, gravity is the most mysterious interaction among the four fundamental interactions. Its extreme weakness prevents particle theorists from building unified theories. Recently, a possible existence of a very strong gravitational field at a microscopic scale was discussed based on the large extra dimension model. According to the ADD model, the gravitational inverse square law can be modified due to the existence of additional spatial dimensions. In order to naturally resolve the hierarchy problem unifying the Planck energy at around 1 TeV in the higher dimensional world, at least two extra dimensions with sizes of the order a millimeter are required. In this case, however, no precision test of the inverse square law has been performed. Two possible ways to test the large extra dimension scenario have been proposed. One way is to perform a direct laboratory test of the gravitational law below the millimeter scale, using a torsion balance pendulum or a similar Cavendish-type device [2]. Another way is based on a high energy collider experiment trying to search for quantum gravity related phenomena such as mono jet events and micro black hole creation [1, 3].

In the present study, we aim to investigate a possible strong gravitational field around nuclei as a new approach to search for the large extra dimension. If there are two large extra dimensions of the order of 0.1mm, we should expect to see a 10^{22} times stronger gravitational field comparing to the original Newtonian prediction. It can be shown that in a $4+n$ dimensional model the gravitational potential is:

$$V_{4+n}(r) = -\lambda^n G \frac{Mm}{r^{1+n}} = \left(\frac{\lambda}{r}\right)^n V_{Newton}(r). \tag{1}$$

If we assume that the size of the extra dimension is $\lambda = 0.1$ mm and the number of extra dimensions are $n = 2$, 3, and 4, then the gravitational potential strength is modified as:

$$\begin{aligned} V_{4+2}(1fm) &= \left(\frac{\lambda = 0.1mm}{r = 1fm}\right)^{n=2} V_{Newton}(1fm) = 10^{22} \times V_{Newton}(1fm) \qquad (n=2) \\ V_{4+3} &= 10^{33} \times V_{Newton}(1fm) \qquad (n=3) \\ V_{4+4} &= 10^{44} \times V_{Newton}(1fm) \qquad (n=4) \end{aligned} \tag{2}$$

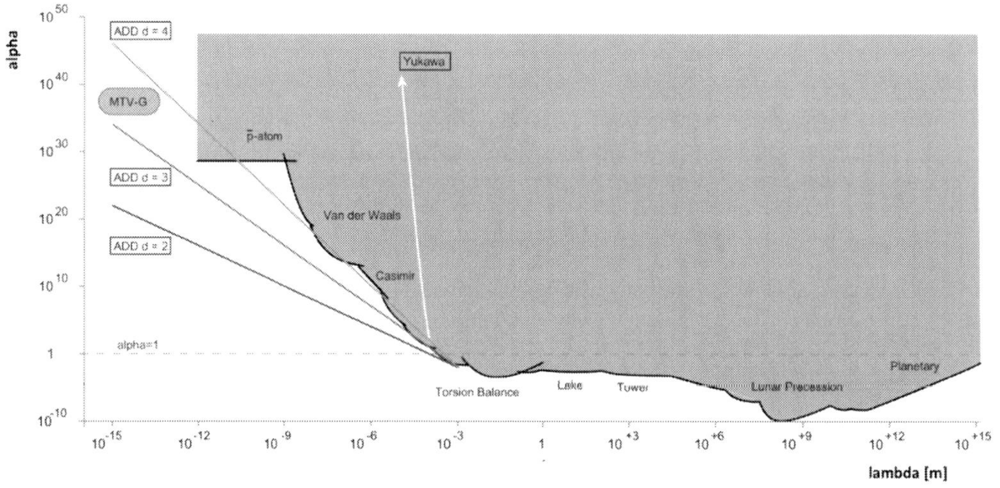

Figure 1. A summary of the experimental search for the Yukawa term. The experimentally excluded region is illustrated as a shaded area. The region which the present study aims to explore is indicated at around 100 fm.

Figure 1 shows a summary of experimental tests of the Newtonian inverse square law at various length scales λ [4]. The vertical axis is the coupling constant α of an additional Yukawa term defined in a modified gravitational potential with a Yukawa term as:

$$V(r) = -G\frac{Mm}{r}\left(1 + \alpha e^{-r/\lambda}\right). \tag{3}$$

The shaded area of this figure indicates the experimentally excluded region in the $\alpha - \lambda$ parameter space. It can be noticed that very little is known below the atomic scale. The length scale where $\alpha(\lambda) > 1$ indicates that even the existence of the Newtonian gravity has not been get confirmed in the current experimental precision. If we assume that the inverse square law is tested with

100% relative precision above 0.1 mm, i.e. $\alpha < 1$ at $\lambda > 0.1$ mm, the gravity for the $n = 2, 3, 4$ cases can respectively be $10^{22}, 10^{33}, 10^{44}$ times greater than the original Newtonian predictions. In the present study, we are aiming to test the inverse square law at a precision of $\alpha \sim 10^{37}$ at around 100 fm scale.

2. Principles

In order to probe the possible strong gravitational field around nuclei, we performed an electron scattering experiment. Electrons emitted from a beta-unstable radiation source are naturally polarized in their longitudinal direction because of the parity violating nature of the weak interaction. The spin precession effects, due to the gravitational geodetic precession, from the strong gravitational field around nuclei are examined in this experiment. The geodetic precession is a precessing effect of a spinning particle travelling in a warped spacetime produced by a gravitational field, which is predicted by the general relativity theory [5]. The existence of the geodetic precession phenomena has been confirmed in 2011 by the NASA satellite Gravity Probe B as a precession of a gyroscope on an orbit around the Earth [6]. As shown in Figure 2, in our experiment, we regard the nuclei as the Earth, and the polarized electron as the gyroscope on the satellite.

We utilize an experimental device designed for the MTV experiment (Mott polarimetry for T-Violation experiment), which aims to search for a large time reversal symmetry violation in the nuclear beta decay [7], to measure a tiny transverse polarization of electrons using the Mott scattering analyzing power. The MTV experiment is measuring a transverse polarization of the electrons emitted from spin-polarized ^{8}Li nuclei, which should be negligible in the standard model, in a 0.1% polarization precision. In this Mott-analyzer, the backscattering left-right asymmetry from a Mott scattering at a thin lead foil is used as a measurement of the transverse polarization.

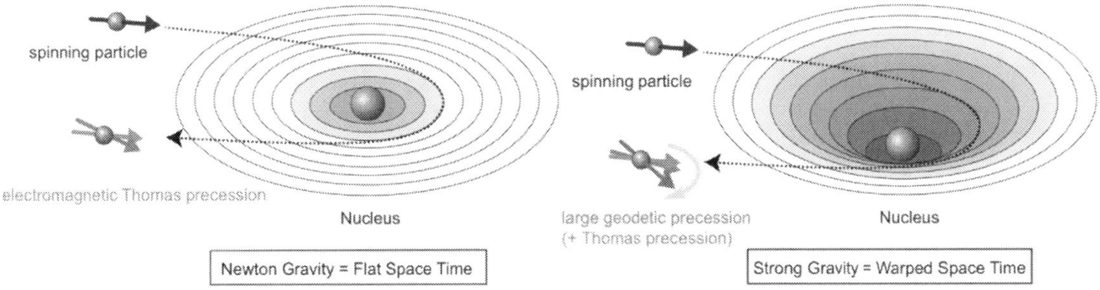

Figure 2. Probing principle using the geodetic precession of a spinning electron scattered by a nuclei.

Because The present project aims to examine gravitational phenomena utilizing the MTV experimental device, it is called the MTV-G (MTV-Gravity) experiment. As shown in Figure 3, the MTV-G experiment consists of a ^{90}Sr radiation source, a primary scattering lead foil, and the MTV polarimeter with a secondary scattering foil and an electron tracking chamber. The existence of strong precession effects at the primary scattering foil is examined with the MTV polarimeter by checking the secondary scattering asymmetry.

3. The experiment and the results

The experiment was performed at TRIUMF in 2011 (Figure 3 and 4) at the MTV experimental beam line with a 37 MBq ^{90}Sr source for about two weeks during which data were

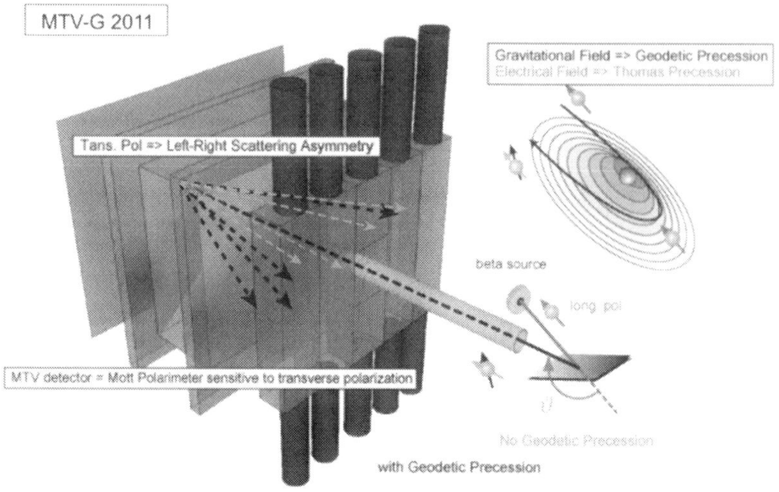

Figure 3. The setup of the MTV-G experiment.

Figure 4. Experimental setup of the MTV-G experiment at TRIUMF in 2011.

collected. The relative setting angle θ between the radiation source and the primary scattering foil can be changed in order to see the scattering angular dependence. By changing this scattering angle, we can measure the distance dependence from the nuclei. The secondary scattering left-right asymmetry, defined as $Asymmetry = (N_{left} - N_{right})/(N_{left} + N_{right})$, is measured as a function of the primary scattering angle θ. In order to cancel out the detector intrinsic efficiency deference, the source configuration flipping between *UP*/*DOWN* position settings was performed. In Figure 5, typical counting yield distributions are plotted as functions of the secondary Mott scattering angle, for *UP* and *DOWN* configurations. The shape difference between the *UP* and the *DOWN* distributions (Figure 5) indicates the pure scattering asymmetry. This difference is not suffering from the detector efficiency difference. Here, we can see a clear evidence of a transverse polarization because of the non-zero asymmetry. The left-right asymmetry can be interpreted as the transverse polarization P in the $Asymmetry = P \times A$ by using the known analyzing power A of the Mott scattering. The analyzing power A includes de-polarization effects inside the scattering foils.

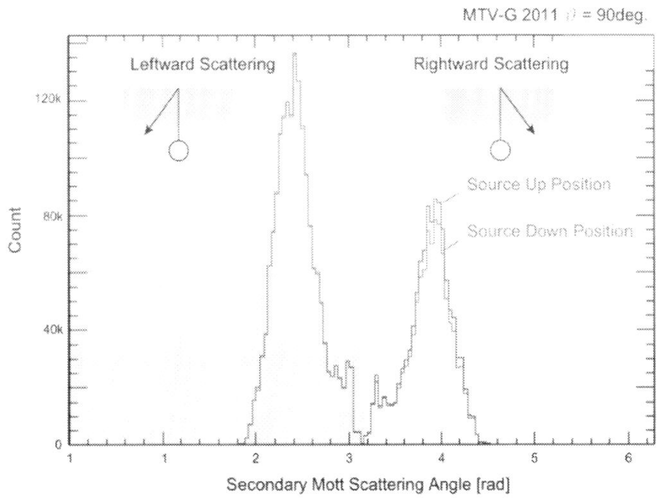

Figure 5. An example of a backscattering angular distribution showing a clear parity violating asymmetry.

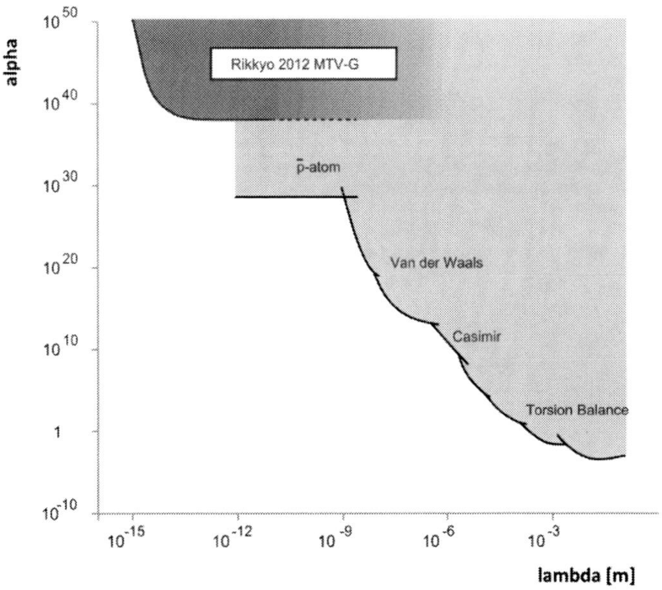

Figure 6. Constraint on the $\alpha - \lambda$ plot from the present result.

The obtained results are compared with possible Yukawa type interactions. In the Coulomb scattering, the electron spin precession is dominated by the electromagnetic Thomas precession, which exists even in zero magnetic fields. The contribution from the Thomas precession is estimated using a numerical simulation. After subtracting the Thomas precession contributions, the maximum allowed strength α is estimated supposing a classical geodetic precession formula. We set a possible constraint on the $\alpha - \lambda$ parameter space using the obtained results, as shown in Figure 6. In Figure 6, the experimental limit at the atomic scale is taken from an analysis of an anti-protonic atom [8]. Thus, the present study sets a new constraint at the shortest scale.

4. Discussion, Conclusions and Future plans

The present analysis supposes a classical geodetic precession expressed as

$$\vec{\Omega}_G = \frac{3}{2}\frac{GM}{r^3}\vec{r} \times \vec{v}, \tag{4}$$

which describes the trajectory of a spinning particle following a free fall motion in the gravitational field [5]. Here, M is the mass of the nuclei, r is the radius of the electron's orbit, $\vec{v}$ is the electron's velocity. The real situation is not a free fall, but it is dominated by the Coulomb potential. In addition, the phenomena are in a microscopic scale, therefore, it might not be appropriate to apply a classical treatment. Repeating the calculation of the present study by including a quantum gravitational treatment with a Coulomb field should be theoretically interesting and a challenging subject for theorists.

The results shown in this paper are based on a first stage experiment with many parameter ambiguities, such as de-polarization factor, precision estimation of electromagnetic Thomas precession etc. We are now going to build a next generation experiment using cylindrical drift chamber (CDC) shown in Figure 7, which may provide better results with increased precision.

Figure 7. The new setup and the image of the electron scattering

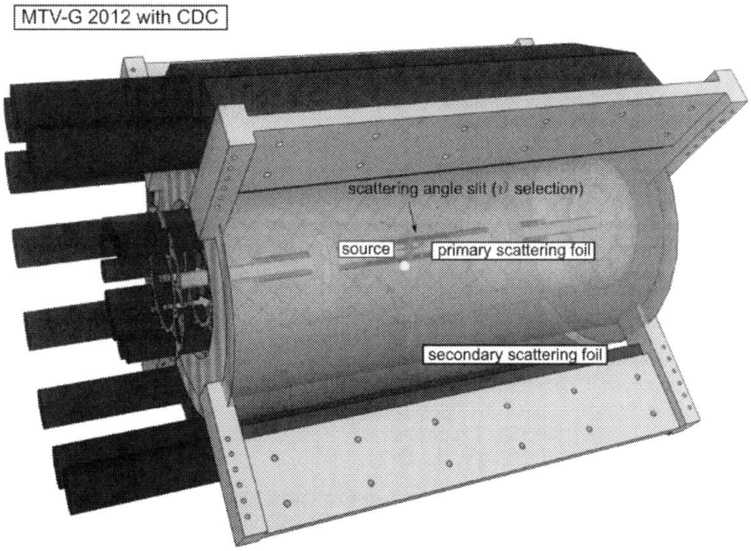

Figure 8. The new setup and the image of the electron scattering

Figure 7 and 8 show the experimental setup using the CDC. The radiation source is set at the center of the CDC, together with the primary scattering foil. The secondary Mott analyzing foil is set outside of the CDC, followed by stopping scintillation counters. By measuring the azimuthal angular dependence around the symmetry axis of the CDC setup, experimental reliability is significantly improved from the previous setup with the planer drift chamber shown in Figure 3.

In conclusion, we have performed the first MTV-G experiment aiming to probe a strong gravitational field around nuclei by utilizing geodetic precession in an electron-nuclear scattering phenomenon. As a result, we have succeeded to set a new constraint on the shortest length scale around 100 fm on the $\alpha - \lambda$ plot. Now, we are going to improve this experiment with the new CDC, and will take special care to reduce the systematic errors. The first test experiment using the CDC was already performed in 2012, and the full-scale data production is scheduled in 2013.

References

[1] N. Arkani-Hamed, S. Dimopoulos, G. Dvali, Phys. Lett. **B429**, 263 (1998).
[2] M. Hata *et al.*, J. Phys. **CS189**, 012019 (2009).
[3] ATLAS Collaboration, Phys. Lett. **B705**, 294 (2011).
[4] E. Fischbach and C.L. Talmadge, The search for non-Newtonian gravity, Springer Verlag, (1999).
[5] Hans Ohanian and Remo Ruffini, □ Gravitation and Spacetime, W.W. Norton & Co.
[6] C.W.F. Everitte *et al.*, Phys. Rev. Lett. **106** 221101 (2011).
[7] J. Murata *et al.*, J. Phys. **CS 312** 102011 (2011); J. Onishi *et al.*, J. Phys. **CS312** 102012 (2011).
[8] V.V. Nesvizhevsky, G. Pignol and K.V. Protasov, Phys. Rev. D **77** 034020 (2008).

Journal of Physics: Conference Series **453** (2013) 012019 doi:10.1088/1742-6596/453/1/012019

Experimental determination of gravitomagnetic effects by means of ring lasers

Angelo Tartaglia
Politecnico, corso Duca degli Abruzzi 24, 10129 Torino, Italy, and INFN
E-mail: angelo.tartaglia@polito.it

Abstract. A new experiment aimed to the detection of the gravito-magnetic Lense-Thirring effect at the surface of the Earth will be presented; the name of the experiment is GINGER. The proposed technique is based on the behavior of light beams in ring-lasers, also known as gyrolasers. A three-dimensional array of ringlasers will be attached to a rigid "monument"; each ring will have a different orientation in space. Within the space-time of a rotating mass the propagation of light is indeed anisotropic; part of the anisotropy is purely kinematical (Sagnac effect), part is due to the interaction between the gravito-electric field of the source and the kinematical motion of the observer (de Sitter effect), finally there is a contribution from the gravito-magnetic component of the Earth (gravito-magnetic frame dragging or Lense-Thirring effect). In a ring-laser a light beam traveling counterclockwise is superposed to another beam traveling in the opposite sense. The anisotropy in the propagation leads to standing waves with slightly different frequencies in the two directions; the final effect is a beat frequency proportional to the size of the instrument and its effective rotation rate in space, including the gravito-magnetic drag. Current laser techniques and the performances of the best existing ring-lasers allow at the moment a sensitivity within one order of magnitude of the required accuracy for the detection of gravito-magnetic effects, so that the objective of GINGER is in the range of feasibility and aims to improve the sensitivity of a couple of orders of magnitude with respect to present. The experiment will be underground, probably in the Gran Sasso National Laboratories in Italy, and is based on an international collaboration among four Italian groups, the Technische Universität München and the University of Canterbury in Christchurch (NZ).

1. Introduction

The theory of relativity, both special (SR) and general (GR), tells us that the propagation of light along closed space contours appears to be anisotropic either when the observer is rotating or when the gravitational field belongs to a rotating mass, or of course when both circumstances are present. The purely kinematical effect had initially also a classical description and is known as the Sagnac effect [1]; it is easy to measure by modern interferometric techniques or using ring-lasers. The other two effects, due to the gravitational field of a rotating mass, arise from the coupling of the gravito-electric part of the field with the rotation of the observer (geodetic or de Sitter or Schiff effect, according to the experimental circumstances) [2] and from the gravito-magnetic component of the field (Lense-Thirring effect) [3]. I shall call the two latter phenomena physical or GR effects.

The physical rotation effects have been measured so far by two experiments in space. In both cases the precession induced by the rotation of the Earth on the axis of a gyroscope has been used. By these means the Gravity Probe B experiment, stemmed from an idea proposed by

Schiff [4], verified the terrestrial geodetic effect with a 0.28% accuracy and the Lense-Thirring drag with a 19% accuracy [5]. Another measurement has been made by Ciufolini using the data from the laser ranging of the LAGEOS satellites, launched for a different purpose. From the precession of the orbits of the LAGEOS the Lense-Thirring effect has been confirmed at 10% accuracy [6]. Another experiment based on the laser ranging technique is presently under way: the LARES satellite has been launched on February 13th, 2012, on a dedicated mission and is now collecting data with the purpose of measuring the Lense-Thirring at a 1% accuracy [7]. These experiments are not easy and have the inconveniences of being in space.

What I am presenting here is a different opportunity to measure the GR effects of rotation resting on the surface of the planet Earth (actually under the surface, in an underground location). The proposal is to use light as a probe of the configuration and properties of space-time. The device to be used is a ring-laser (actually a three-dimensional array of ring-lasers). The name of the experiment is GINGER and it will be described in the following. The GINGER proposal is the result of a collaboration whose Principal Investigator is Angela Di Virgilio of the Pisa section of the Italian INFN and involving primarily groups from a number of Italian institutions (the universities of Florence, Naples, Padua, Pisa, Turin-Politecnico; various INFN sections) and from the German Technische Universität München. A consultancy link is also active with a group of the University of Canterbury in Christchurch (NZ). The detailed proposal is presented in [8].

2. Propagation of light in the field of a rotating mass

The line element of the space-time of a steadily rotating mass may in general be written as:

$$ds^2 = g_{00}dt^2 + g_{rr}dr^2 + g_{\theta\theta}d\theta^2 + g_{\phi\phi}d\phi^2 + 2g_{0\phi}dtd\phi \tag{1}$$

where polar coordinates in space are used, to evidence the symmetry. The $g_{\mu\nu}$'s are the elements of the metric tensor and each of them does not depend on t and ϕ due to the symmetry. The mixed term $g_{0\phi}$ accounts for the rotation of the source, i.e. the central mass, and is responsible for the so called gravito-magnetic effects; $g_{0\phi}$ cannot be eliminated by any global coordinate transformation. Had we used different coordinates, we would have had in general three elements g_{0i}, with the index i labeling the three space coordinates. The three g_{0i}'s may be read as the components of a three-vector $\vec{h}$, which can be interpreted as the vector-potential of a gravito-magnetic field $\vec{B}_g$, i.e., of the rotation depending part of the gravitational field [9]. In conventional three-dimensional notation it is:

$$\vec{B}_g = \vec{\nabla} \wedge \vec{h}. \tag{2}$$

Such a gravito-magnetic field has the configuration of a dipole. Let us then evaluate the time of flight (TOF) of a light beam constrained by some physical system to travel along a closed space path; the TOF is the proper time interval of a laboratory within which the experiment is performed. The result is obtained from eq. (1) considering a null spacely closed world-line and integrating along it. The only term sensitive to the rotation sense is the one containing $g_{0\phi}$ that linearly multiplies $d\phi$, which is odd in the angle. That term will be the only one left when subtracting the clockwise from the counter-clockwise travel time and it reads:

$$\delta T = T_+ - T_- = -2\sqrt{g_{00}} \oint \frac{g_{0\phi}}{g_{00}} d\phi. \tag{3}$$

3. Ring-lasers

The anisotropy pointed out in the previous section is the basis of the operation of a ring-laser. Consider the scheme shown on fig. 1. An active cavity produces two light beams traveling in opposite directions; four mirrors deviate the beams to form a closed square path in space. In order to have a ring the mirrors cannot be less than three but of course could be more.

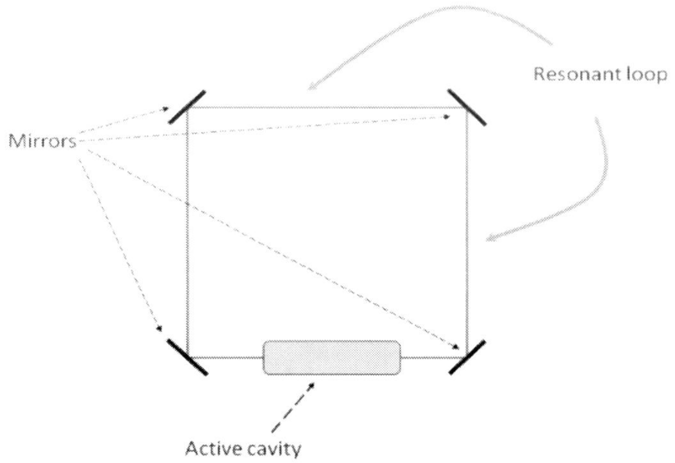

Figure 1. The schematic view of a square ring-laser is shown. The right- and left-handed beams exiting the active cavity are led to form a square resonant loop by four mirrors.

The loop forms in turn a resonant cavity. According to formula (3), there is a TOF difference between the two beams. Since we suppose to be in a stationary condition we have a pair of standing waves; each of them is formed by an integer number of wavelengths, but, due to the TOF difference (3), the clockwise and counter-clockwise wavelengths are slightly different from one another. In the lowest mode, the integer is the same for both beams and the superposition leads to a beat note whose frequency is proportional to the TOF difference [8]:

$$f_b = \frac{c^2}{\lambda P}\delta T \tag{4}$$

P is the length of the loop and λ is the fiducial wavelength of the laser. The beat frequency can be extracted and read at one of the corners of the ring.

In the case of an earth-bound laboratory, the relevant metric tensor elements can be approximated to the lowest significant order including the angular momentum of the Earth, obtaining:

$$\begin{aligned} g_{0\phi} &\simeq (2\frac{GJ_\oplus}{c^3 r} - r^2\frac{\omega}{c} - 2G\frac{M_\oplus}{c^2}\frac{r\Omega_\oplus}{c})sin^2\theta \\ g_{00} &\simeq 1 - 2\frac{GM_\oplus}{c^2 r} - \frac{\omega^2 r^2}{c^2}sin^2\theta. \end{aligned} \tag{5}$$

In (5) the label $_\oplus$ designates physical quantities belonging to the planet Earth: the total mass M, the angular momentum J and the angular velocity Ω; ω is the angular velocity of the apparatus and, in the following, we shall assume that $\omega = \Omega_\oplus$; θ is the colatitude of the laboratory and r is the distance from the center of the Earth.

Using the above approximation it is possible to write down the expected signal in the form of the beat frequency [8]:

$$\delta f = 4\frac{A}{\lambda P}[\overrightarrow{\Omega}_{\oplus} - 2\frac{GM_{\oplus}}{c^2R}\Omega_{\oplus} sin\theta\hat{u}_{\theta} + \frac{GJ_{\oplus}}{c^2R^3}(2\cos\theta\hat{u}_r + \sin\theta\hat{u}_{\theta})]\cdot\hat{u}_n \qquad (6)$$

A is now the area contoured by the light beams; R is the radius of the Earth; the unit vectors $\hat{u}_r$, $\hat{u}_{\theta}$ and $\hat{u}_n$ are respectively: radial, along the local meridian in the sense of increasing colatitude, perpendicular to the plane of the ring (provided it is in a plane). The whole quantity in front of the square bracket is called the scale factor: the bigger it is, the stronger is the signal from the apparatus.

Within the square brackets of formula (6), the three terms represent three real or effective angular velocities. The first is the rotation rate of the Earth and accounts for the kinematical Sagnac effect; the second term corresponds to the geodetic or de Sitter effect; the third and last term is the gravito-magnetic contribution and accounts for the Lense-Thirring effect. The two physical terms (as they are also called) in (6) are of the same order of magnitude and nine orders of magnitude smaller than the dominant Sagnac term.

3.1. Commercial gyrolasers

Ring-lasers are already in use as "gyrolasers" for the measurement of rotation rates, for instance in airplanes or in submarines. They are replacing the mechanical gyroscopes formerly used for the same purpose, whence the name of gyrolasers. Commercial gyrolasers are compact objects with sensitivities in the order of $\sim 10^{-7} rad/s/\sqrt{Hertz}$ obtained with appropriate values of the scale factor resulting from the use of multiply wound optical fibers. Often one single device is made of three gyrolasers aligned along three mutually perpendicular axes in order to sense the full angular velocity three-vector. Fig. (2) shows an example of a commercial gyrolaser.

Figure 2. A commercial optical fiber gyrolaser.

3.2. G-Pisa

The accuracy of commercial gyrolasers is not enough to perform scientific experiments and can hardly detect the diurnal rotation of the Earth. For scientific purposes, more refined and bigger instruments are required. An example is G-Pisa, initially developed to test the rotational stability of the VIRGO gravitational interferometer located at Cascina near Pisa in Italy. G-Pisa is a square ring with a 1.35 m long side; it is visible in figs. 3 and 4, both in its real aspect and in a schematic drawing.

The source of light is a He-Ne laser with an adjustable output power of a few tens of nW. The device is mounted on a thick granite table that can be held at various pitches from horizontal to vertical.

Figure 3. Picture of the G-Pisa ring-laser on its granite support in a tilted configuration.

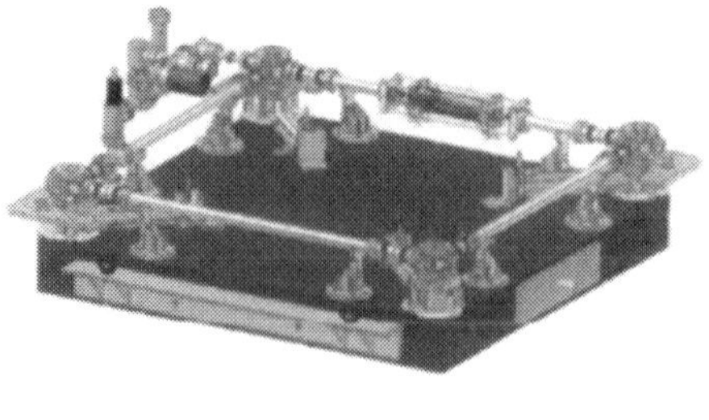

Figure 4. Schematic view of G-Pisa. The light beams are contained into tubes filled of low pressure He gas; the mirrors are mounted at the four corners of the device; the active laser cavity is visible; the extraction of the signal happens at the left upper corner.

The sensitivity of G-Pisa is somewhat in between 10^{-9} and 10^{-10} $rad/s/\sqrt{Hertz}$ [10]. It is not enough to sense the GR effects, but permits to measure, besides the rotation of the Earth, many interesting motions of the surface of the Earth at the laboratory, of geophysical origin. G-Pisa is destined to become a test-bed for technologies in preparation for the future GINGER instrument.

3.3. The Cashmere cavern instruments

During the years a number of ring-lasers for fundamental research have been built by a group of the University of Canterbury in Christchurch, NZ. The first one was C-I [11], others followed up to UG-2 whose interclosed area is 834 m^2 [12]. They were all located underground in the Cashmere cavern, near Christchurch (see fig. 5).

Figure 5. View of one of the vacuum pipes hosting the laser beam of a ring-laser in the Cashmere cavern near Christchurch, NZ.

Unfortunately the important gain in the scale factor obtained thanks to the size of the apparatus is overridden by the mechanical instabilities due to the length of the arms, so UG-2 did not permit measurements of the GR effects.

The laboratory in the Cashmere cavern suffered relevant damages from the severe earthquakes that shook the Christchurch area in September 2010 and February 2011, so that now it is not in operation. For the future, however, the group in the University of Canterbury is planning to build a new triangular ring-laser with 5 m sides, mounted on a rigid support, that promises very good performances.

3.4. G in Wettzell

The best ring-laser in the world is at the moment the G ring at the Geodätisches Observatorium Wettzell in Bavaria. It is a square ring, 4 m side, mounted on a monolithic block made of zerodur, a rigid ceramic material with a very small thermal expansion coefficient.The source of light is a He-Ne laser with 20 nW output power. The laboratory is under an artificial mound, in order to isolate it as far as possible from environmental disturbances; the apparatus is in turn kept within a chamber where pressure and temperature are stabilized; the zerodur table rests on a concrete pillar attached to the natural rock pavement under ground. Figs. 6 and 7 present a picture and a scheme of the instrument.

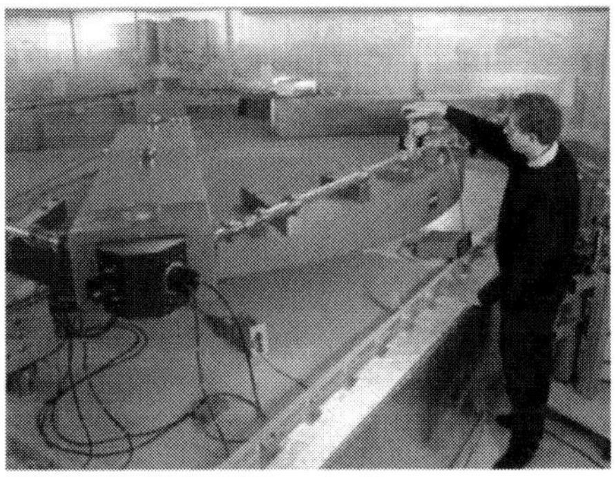

Figure 6. Picture of the G ring-laser located in Wettzell on its zerodur support; the apparatus is shown without the temperature and pressure controlled casing that covers it when in operation.

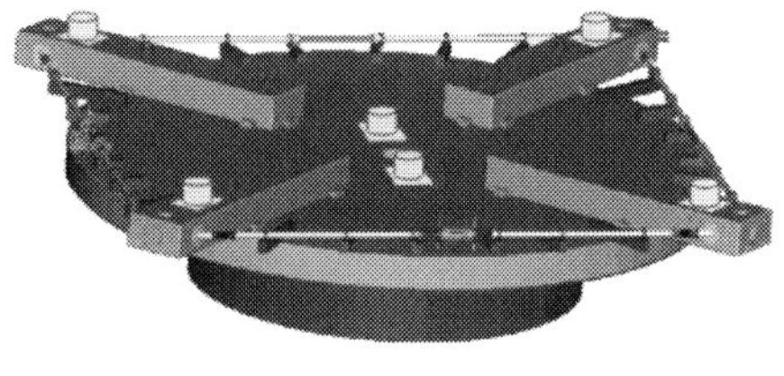

Figure 7. Schematic view of the G ring-laser. The device is mounted on a single block of zerodur, an extremely rigid and thermally stable ceramic material

The sensitivity of G arrives to $4.5 \times 10^{-12}\ rad/s/\sqrt{Hz}$ and is already close to the value required in order to reveal the GR effects of rotation. Fig. 8 shows an example of data taken by G during thirteen days. The peaks correspond to the diurnal polar motion of the terrestrial axis; each of them has a height of the order of 50 μHz or less, i.e. $\sim 10^{-8}$ times the Sagnac signal above which the peaks rise.

The accuracy of G is within one order of magnitude from the expected physical rotation terms. The instrument is an excellent sensor of any sort of geophysical rotations as well as of tiny pushes from environmental interactions [13].

4. GINGER

As we have seen, the G ring-laser in Wettzell is already close to the sensitivity needed for gravito-magnetic measurements so that an additional effort allowed by present laser technology can make an Earth experiment to reveal the Lense-Thirring effect feasible. To this purpose the collaboration presented in the introduction is proposing a new experiment relying on the expertise of all members: GINGER.

GINGER will not be a single ring-laser, but rather a three-dimensional array of rings. Each ring will be a square loop with a 6 m long side. The minimum number of rings is three, with the

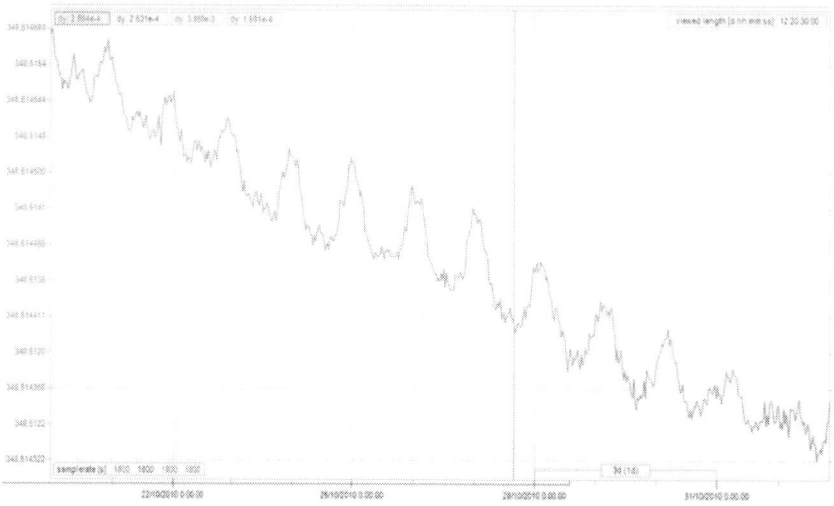

Figure 8. Signal from the G ring-laser during 13 days. Superposed to the Sagnac signal it can easily be seen the diurnal polar motion of the rotation axis of the Earth; the height of individual peaks is a bit less than 50 μHz.

three normals oriented along mutually perpendicular directions in space; in this way the three space components of the effective angular velocities would simultaneously be measured.

A possibility is to actually have six loops, coupled in pairs of equal orientation; the consequent redundancy would allow a better control of noise. Fig. 9 shows two possible configurations of GINGER.

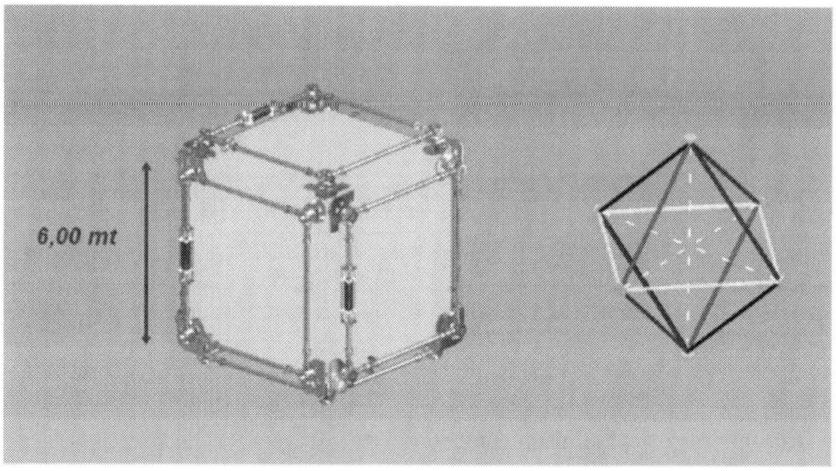

Figure 9. Two possible configurations for GINGER. On the left a cubic concrete monument carries six ring-lasers, perpendicular to each other in pairs; on the right an octahedral structure is presented with three mutually perpendicular square rings.

One is a cube whose faces support pairs of equal square loops; the other is an octahedron, carrying three square loops. The octahedral structure allows a better control of the geometry through the diagonals, along which three Fabry-Pérot interferometers would be placed. In all cases, the mirrors will be attached to a concrete "monument" and the stability of the configuration will be insured more through active control of the position of the mirrors than via passive rigidity of the support.

The design power of the light source is 200 nW and the quality factor of the cavity is $Q = 3 \times 10^{12}$.

The site for GINGER could probably be the underground Gran Sasso National Laboratory (LNGS) of the INFN in Italy where the appropriate technological facilities are already available and more than 1000 m of rock above the apparatus will be a most effective screen against surface noise.

The purpose is to measure the Lense-Thirring effect with a 1% accuracy after one year of integration.

5. Conclusion

We have seen how ring-lasers can be used for fundamental tests of GR on Earth, having attained, in recent years, an unprecedented accuracy. A synergy between three groups working on ring-lasers in the world has brought to the proposal of a dedicated new experiment named GINGER, whose technological improvements will indeed allow for the measurement of the general relativistic effects originated from the rotation of the mass of the Earth. Actually, if the accuracy will be good enough, it is also possible to arrive to put constraints on a couple of PPN parameters. In fact, the PPN form of three physical effective rotations that GINGER could reveal is:

$$\begin{aligned}
\overrightarrow{\Omega}_G &= -(1+\gamma)\Omega_\oplus \frac{GM}{c^2 R}\sin\theta \hat{u}_\theta \\
\overrightarrow{\Omega}_B &= -\frac{1+\gamma+\alpha_1/4}{2}\frac{G}{c^2R^3}(\overrightarrow{J}_\oplus - 3(\overrightarrow{J}_\oplus \cdot \hat{u}_r)\hat{u}_r) \\
\overrightarrow{\Omega}_W &= -\frac{\alpha_1}{4}\frac{GM}{c^2R^2}\hat{u}_r \wedge \overrightarrow{W}.
\end{aligned} \tag{7}$$

The γ and α_1 parameters would account for possible anomalous (from the view point of GR) dependencies of the curvature from the mass of the source and for possible preferred frame effects; the $\overrightarrow{W}$ appearing in formula (7) is a gravitational three-vector potential associated with the presence of a preferred frame. In GR, it is $\gamma = 1$ and $\alpha_1 = 0$.

GINGER has got a principle approbation from the Italian INFN. The seismic noise of the possible location in the LNGS has been checked at the beginning of 2011 by a German team; the G-Pisa ring will be moved, at the beginning of 2013, to the LNGS in order to characterize the place from the viewpoint of the rotational noise. Meanwhile, a new ring approximately the same size as G-Pisa will be built in order to serve as a test bed for mirrors stability and control.

Summing up, GINGER is on a track that will lead it to completion and we are confident that its performance will match up with the promises and expectations, paving the way for a number of future investigations using light as an intrinsically relativistic probe for the structure of space-time.

Acknowledgments
This presentation has been given on behalf of the whole GINGER collaboration mentioned in the introduction and, in particular, of the authors of ref. [8].

References

[1] Sagnac G 1913 *Comptes Rendus* **157** 708
[2] de Sitter W 1916 *Mon. Not. Roy. Astron. Soc.* **77** 155-184
[3] Thirring H 1918 *Physikalische Zeitschrift* **19** 33-46,
Lense J and Thirring H 1918 *Physikalische Zeitschrift* **19** 156-163,
Thirring H 1921 *Physikalische Zeitschrift* **22** 29-31
[4] Schiff L I 1960 *Phys. Rev. Lett.* **4** 215-217
[5] Everitt C W F *et al.* 2011 *Phys. Rev. Lett.* **106** 221101
[6] Ciufolini I, Paolozzi A, Pavlis E, Ries J, König R, Matzner R, Sindoni G and Neumayer H 2011 *Eur. Phys. J. Plus* **126** 72
[7] Ciufolini I, Paolozzi A, Pavlis E, Ries J, Gurzadyan V, König R, Matzner R, Penrose R and Sindoni G 2012 *Eur. Phys. J. Plus* **127** 133
[8] Bosi F *et al.* 2011 *Phys. Rev.* D **84** 122002
[9] Tartaglia A 2005 *General relativity and gravitational physics - Proceedings of the 16th SIGRAV conference (Vietri sul mare)* ed. G. Esposito, G. Lambiase, G. Marmo, G. Scarpetta and G. Vilasi (Melville: AIP conference proceedings) vol **751** pp 136-145,
Ruggiero M L and Tartaglia A 2002 *Nuovo Cimento* B **117** 743-767
[10] Belfi J *et al.* 2012 *J. Seismol.* **16** 757-766
[11] Stedman G E 1997 *Rep. Prog. Phys.* **60** 615-688
[12] Hurst R B, Stedman G E, Schreiber K U, Thirkettle R J, Graham R D, Rabeendran N and Wells J-P R 2009 *J. Appl. Phys.* **105** 113115
[13] Gebauer A, Schreiber K U, Klügel T, Schön N and Ulbrich U 2012 *Journal of Seismology* **16** 777-786

Geometrization of Lie and Noether symmetries with applications in Cosmology

Michael Tsamparlis

Department of Physics, Section of Astronomy, Astrophysics and Mechanics, University of Athens, Panepistemiopolis, Athens 157 83, Greece

E-mail: mtsampa@phys.uoa.gr

Abstract. We derive the Lie and the Noether conditions for the equations of motion of a dynamical system in a $n-$dimensional Riemannian space. We solve these conditions in the sense that we express the symmetry generating vectors in terms of the special projective and the homothetic vectors of the space. Therefore the Lie and the Noether symmetries for these equations are geometric symmetries or, equivalently, the geometry of the space is modulating the motion of dynamical systems in that space. We give two theorems which contain all the necessary conditions which allow one to determine the Lie and the Noether symmetries of a specific dynamical system in a given Riemannian space. We apply the theorems to various interesting situations covering Newtonian 2d and 3d systems as well as dynamical systems in cosmology.

Keywords: Lie Symmetries, Noether symmetries, Cosmology
Pacs - numbers:98.80.-k,04.20.-q, 45.20.D-,02.20Sv

1. Introduction

In a Riemannian space the affinely parameterized geodesics are determined uniquely by the metric. Therefore one should expect a close relation between the geodesics as a set of homogeneous ordinary differential equations (ODE) linear in the highest order term and quadratically non-linear in first order terms, and the metric as a second order symmetric tensor. A system of such ODEs is characterized (not fully) by its Lie symmetries and correspondingly a metric is characterized (again not fully) by its collineations. Therefore it is reasonable to expect that the Lie symmetries of the system of geodesic equation of a metric will be closely related with the collineations of the metric. That such a relation exists it is easy to see by the following simple example. Consider on the Euclidian plane a family of straight lines parallel to the $x-$axis. These curves can be considered either as the integral curves of the ODE $\frac{d^2y}{dx^2}=0$ or as the geodesics of the Euclidian metric dx^2+dy^2. Subsequently consider a symmetry operation defined by a reshuffling of these lines without preserving necessarily their parametrization. According to the first interpretation this symmetry operation is a Lie symmetry of the ODE $\frac{d^2y}{dx^2}=0$ and according to the second interpretation it is a (special) projective symmetry of the Euclidian two dimensional metric.

What has been said for a Riemannian space can be generalized to a space in which there is only a linear connection. In this case the geodesics are called autoparallels (or paths) and they comprise again a system of ODEs linear in the highest order term and quadratically non-linear in the first order terms. In this case one is looking for relations between the Lie symmetries of the autoparallels and the projective collineations of the connection.

A Lie point symmetry of an ordinary differential equation (ODE) is a point transformation in the space of variables which preserves the set of solutions of the ODE [1–3]. If we look at these solutions as curves in the space of variables, then we may equivalently consider a Lie point symmetry as a point transformation which preserves the set of the solution curves. Applying this observation to the geodesic curves in a Riemannian (affine) space, we infer that the Lie point symmetries of the geodesic equations in any Riemannian space are the automorphisms which preserve the set of these curves. However we know from Differential Geometry that the point transformations of a Riemannian (affine) space which preserve the set of geodesics are the projective transformations. Therefore it is reasonable to expect a correspondence between the Lie symmetries of the geodesic equations and the projective algebra of the space.

The equation of geodesics in an arbitrary coordinate frame is a second order ODE of the form

$$\ddot{x}^i + \Gamma^i_{jk}\dot{x}^j\dot{x}^k + F(x^i, \dot{x}^j) = 0 \tag{1}$$

where $F(x^i, \dot{x}^j)$ is an arbitrary function of its argument and the functions Γ^i_{jk} are the connection coefficients of the space. However equation (1) is also the equation of motion of a dynamical system moving in a Riemannian (affine) space under the action of a velocity dependent force. According to the above argument we expect that the Lie symmetries of the ODE (1) for a given function $F(x^i, \dot{x}^j)$ will be a subalgebra of the projective algebra of the space. This subalgebra is selected by means of certain constraint conditions which will involve geometric quantities of the space and the function $F(x^i, \dot{x}^j)$ [4–8].

The determination of the Lie point symmetries of a given system of ODEs consists of two steps (a) the determination of the conditions which the components of the Lie symmetry vector must satisfy and (b) the solution of the system of these conditions. Step (a) is formal and it is outlined in e.g. [1–3]. The second step is the key one and, for example, in higher dimensions where one has a large number of simultaneous equations the solution can be quite involved and perhaps prohibitive. However if one expresses the system of Lie symmetry conditions of (1) in terms of collineation (i.e. symmetry) conditions of the metric, then the determination of Lie symmetries is transferred to the geometric problem of determining the generators of a collineation group of the metric and, as it will be shown below, specifically the generators of the special projective group of the metric. In this field there is a vast amount of work from Differential Geometry waiting to be used. Indeed the projective symmetries are already known for many spaces or they can be determined by existing general theorems. For example the projective algebra and all its subalgebras are known for the important case of spaces of constant curvature [10] and in particular for the flat spaces. This implies that, for example, the Lie symmetries of *all* Newtonian dynamical systems as well as those of Special Relativity can be determined by simple differentiation from the known projective algebra of these spaces!

What has been said for the Lie point symmetries of (1) applies also to Noether point symmetries (provided (1) follows from a Lagrangian). The Noether symmetries are Lie point symmetries which satisfy the additional constraint

$$X^{[1]}L + L\frac{d\xi}{dt} = \frac{df}{dt}. \tag{2}$$

Noether symmetries form a closed subalgebra of the Lie point symmetries algebra. In accordance to the above this implies that the Noether symmetries will be related with a subalgebra of the projection algebra of the space where 'motion' occurs. As it will be shown this subalgebra is contained in the homothetic algebra of the space. As it is well known with each Noether point symmetry it is associated a conserved current (i.e. a Noether first integral). This leads us to the important conclusion that the (standard) conserved quantities of a dynamical system depend on the space it moves and the type of force $F(x^i, \dot{x}^j)$ which modulates the motion. In particular in 'free fall', that is when $F(x^i, \dot{x}^j) = 0$, the orbits are the geodesics and the *geometry* of the space is the sole factor which determines the conserved quantities of motion. This conclusion is by no means trivial and means that the space where motion occurs is not a pathetic carrier of motion but it is the major modulator in the evolution of a dynamical system. In other words there is a strong and deep relation between Geometry of the space and Physics in that space!

The above matters have been discussed extensively in a series of interesting papers by Aminova [11–13] who has given a partial answer. Furthermore in a recent work [14] the authors have considered the KVs of the metric and their relation to the Lie symmetries of the system of affinely parameterized geodesics in maximally symmetric spaces of low dimension. In the same paper a conjecture is made, which essentially says that the maximally symmetric spaces of non-vanishing curvature do not admit further Lie symmetries.

The purpose of the present work is to present recent results concerning the above approach. More specifically we do the following

a. We derive the general expression for the Lie symmetry conditions of (1) assuming a general (i.e. not necessarily symmetric) connection in terms of geometric quantities

b. For the case that the function F depends only on the coordinates, i.e. $F(x^i)$ we give two theorems which establish the exact relation between the projective/homothetic algebra of the space and the Lie/Noether symmetry algebra of (1).

c. We consider applications in Newtonian Physics, General Relativity and in Cosmology

The structure of the paper is as follows. In section 3 we determine the conditions for Lie symmetries in covariant form. We find that the major symmetry condition relates the Lie symmetries with the special projective symmetries of the connection. A similar result has been obtained by Prince and Crampin in [4] using the bundle formulation of second order ODEs. In section 4 we apply these conditions in the case of Riemannian spaces. We solve the symmetry conditions and in Theorem 1 we give the Lie symmetry vectors in terms of the collineations of the metric. In section 5 we apply Theorem 1 to the case of spaces of constant curvature of dimension n and compute the complete set of Lie symmetries of the system of affinely parameterized geodesic equations in these spaces. We distinguish the case of the flat space and the spaces of non-zero curvature and prove the validity of the conjecture made in [14]. In the remaining sections we consider applications in General Relativity and in Cosmology.

2. Collineations of Riemannian spaces

A collineation in a Riemannian space is a vector field $\mathbf{X}$ which satisfies an equation of the form

$$\mathcal{L}_X \mathbf{A} = \mathbf{B} \tag{3}$$

where $\mathcal{L}_X$ denotes Lie derivative, $\mathbf{A}$ is a geometric object (not necessarily a tensor) defined in terms of the metric and its derivatives (e.g. connection, Ricci tensor, curvature tensor etc.) and $\mathbf{B}$ is an arbitrary tensor with the same tensor indices as $\mathbf{A}$. The collineations in a Riemannian space have been classified by Katzin et al. [15]. In the following we use only certain collineations.

A conformal Killing vector (CKV) is defined by the relation

$$\mathcal{L}_X g_{ij} = 2\psi\left(x^k\right) g_{ij}. \tag{4}$$

If $\psi = 0$, $\mathbf{X}$ is called a Killing vector (KV), if ψ is a non-vanishing constant a homothetic vector (HV) and if $\psi_{;ij} = 0$, a special conformal Killing vector (SCKV). A CKV is called proper if it is not a KV, HV or a SCKV.

Table 1. The projective algebra of a Riemannian space

Collineation	A	B
Killing vector (KV)	g_{ij}	0
Homothetic vector (HV)	g_{ij}	$\psi g_{ij}, \psi_{,i} = 0$
Conformal Killing vector (CKV)	g_{ij}	$\psi g_{ij}, \psi_{,i} \neq 0$
Projective collineation (PC)	Γ^i_{jk}	$2\phi_{(,j}\delta^i_{k)}$, $\phi_{,i} \neq 0$
Special Projective collineation (SPC)	Γ^i_{jk}	$2\phi_{(,j}\delta^i_{k)}$, $\phi_{,i} \neq 0$ and $\phi_{,jk} = 0$

Table 2. The projective algebra of the Euclidian space

Collineation	Gradient	Non-gradient
Killing vectors (KV)	$\mathbf{S}_I = \delta^i_I \partial_i$	$\mathbf{X}_{IJ} = \delta^j_{[I}\delta^i_{J]} x_j \partial_i$
Homothetic vector (HV)	$\mathbf{H} = x^i \partial_i$	
Affine Collineation (AC)	$\mathbf{A}_{II} = x_I \delta^i_I \partial_i$	$\mathbf{A}_{IJ} = x_J \delta^i_I \partial_i$
Special Projective collineation (SPC)		$\mathbf{P}_I = S_I \mathbf{H}$.

Where the indices $I, J = 1, 2, \ldots, n$.

A projective collineation (PC) is defined by the equation

$$\mathcal{L}_X \Gamma^i_{jk} = 2\phi_{(,j}\delta^i_{k)}. \tag{5}$$

If $\phi = 0$ the PC is called an affine collineation (AC) and if $\phi_{;ij} = 0$ a special projective collineation (SPC). A proper PC is a PC which is not an AC, HV or KV or SPC. The PCs form a Lie algebra whose ACs, HV and KVs form subalgebras. It has been shown that if a metric admits a SCKV then also admits a SPC, a gradient HV and a gradient KV [16]. We summarize the above in the Table 1

In the following we shall need the symmetry algebra of spaces of constant curvature. In [10] it has been shown that the PCs of a space of constant non-vanishing curvature consist of proper PCs and KVs only and if the space is flat then the algebra of the PCs consists of KVs/HV/ACs and SPCs. For convenience we summarize these results in Table 2.

The Lie point symmetries of all Newtonian dynamical systems are amongst the vectors in the above table. Also the Noether point symmetries of all Newtonian dynamical systems (and more general all systems moving in a flat space) follow from the elements of the first two rows of Table 2 (apart form some differences in sign depending on the signature of the metric).

3. The Lie point symmetry conditions in an affine space

We consider the system of ODEs:

$$\ddot{x}^i + \Gamma^i_{jk}\dot{x}^j\dot{x}^k + \sum_{m=0}^{n} P^i_{j_1 \ldots j_m}\dot{x}^{j_1}\ldots\dot{x}^{j_m} = 0 \tag{6}$$

where Γ^i_{jk} are the connection coefficients of the space and $P^i_{j_1 \ldots j_m}(t, x^i)$ are smooth polynomials completely symmetric in the lower indices and derive the Lie point symmetry conditions in geometric form using the standard approach. Equation (6) is quite general and covers most of the standard cases autonomous and non autonomous and in particular equation (1). Furthermore because the Γ^i_{jk}'s are not assumed to be symmetric, the results are valid in a space with torsion. Obviously they hold in a Riemannian space where the connection coefficients are given in terms of the Christofell symbols.

The detailed calculation can be found in [8] and we shall not be repeat it here. In the following we summarize for convenience the results.

The terms $\dot{x}^{j_1}\ldots\dot{x}^{j_m}$ for $m \leq 4$ give the equations:

$$L_\eta P^i + 2\xi_{,t}\, P^i + \xi P^i{}_{,t} + \eta^i{}_{,tt} + \eta^j{}_{,t}\, P^i_{,j} = 0 \tag{7}$$

$$L_\eta P^i_j + \xi_{,t}\, P^i_j + \xi P^i_{j,t} + \left(\xi_{,k}\,\delta^i_j + 2\xi_{,j}\,\delta^i_k\right) P^k + 2\eta^i{}_{,t|j} - \xi_{,tt}\,\delta^i_k + 2\eta^k{}_{,t}\, P^i_{.jk} = 0 \tag{8}$$

$$L_\eta P^i_{jk} + L_\eta \Gamma^i_{jk} + \left(\xi_{,d}\,\delta^i_{(k} + \xi_{,(k}\,\delta^i_{|d|}\right) P^d_{.j)} + \xi P^i_{.kj,t} - 2\xi_{,t(j}\,\delta^i_{k)} + 3\eta^d{}_{,t}\, P^i_{.dkj} = 0 \tag{9}$$

$$L_\eta P^i_{.jkd} - \xi_{,t}\, P^i_{.jkd} + \xi_{,e}\,\delta^i_{(k} P^e_{.dj)} + \xi P^i_{.jkd,t} + 4\eta^e{}_{,t}\, P^i_{.jkde} - \xi_{(,j|k}\delta^i_{d)} = 0 \tag{10}$$

and the conditions due to the terms $\dot{x}^{j_1}\ldots\dot{x}^{j_m}$ for $m > 4$ are given by the following general formula:

$$\begin{aligned}&L_\eta P^i_{j_1\cdots j_m} + P^i_{j_1\cdots j_m}{}_{,t}\xi + (2-m)\,\xi_{,t} P^i_{j_1\cdots j_m} + \\ &+ \xi_{,r}\left(2-(m-1)\right) P^i_{j_1\cdots j_{m-1}}\delta^r_{j_m} + (m+1)\, P^i_{j_1\cdots j_{m+1}}\eta^{j_{m+1}}_{,t} + \xi_{,j} P^j_{j_1\cdots j_{m-1}}\delta^i_{j_m} = 0.\end{aligned} \tag{11}$$

We note the appearance of the term $L_\eta \Gamma^i_{jk}$ in these expressions.

Eqn (1) is obtained for $m = 0$, $P^i = F^i$ in which case the Lie symmetry conditions read:

$$L_\eta P^i + 2\xi_{,t}\, P^i + \xi P^i{}_{,t} + \eta^i{}_{,tt} = 0 \tag{12}$$

$$\left(\xi_{,k}\,\delta^i_j + 2\xi_{,j}\,\delta^i_k\right) P^k + 2\eta^i{}_{,t|j} - \xi_{,tt}\,\delta^i_k = 0 \tag{13}$$

$$L_\eta \Gamma^i_{jk} - 2\xi_{,t(j}\,\delta^i_{k)} = 0 \tag{14}$$

$$\xi_{(,j|k}\delta^i_{d)} = 0. \tag{15}$$

If $F^i = 0$ we obtain the Lie symmetry conditions for the geodesic equations (see [8], [9]) .

4. The autonomous dynamical system moving in a Riemannian space

We 'solve' the Lie symmetry conditions (12) - (15) for an autonomous dynamical system in the sense that we express them in terms of the collineations of the metric of the space where motion occurs.

Equation (15) means that $\xi_{,j}$ is a gradient KV of g_{ij}. This implies that the metric g_{ij} is decomposable. Equation (14) means that η^i is a projective collineation of the metric with projective function $\xi_{,t}$. The remaining two equations are constraint conditions, which relate the components ξ, n^i of the Lie symmetry vector with the vector $F^i(x^j)$. Equation (12) gives

$$\left(L_\eta g^{ij}\right) F_j + g^{ij} L_\eta F_j + 2\xi_{,t} g^{ij} F_j + \eta^i_{,tt} = 0. \tag{16}$$

This equation is an additional restriction for η^i because it relates it directly to the symmetries of the metric. Finally equation (13) gives

$$-\delta^i_j \xi_{,tt} + \left(\xi_{,j}\delta^i_k + 2\delta^i_j \xi_{,k}\right) F^k + 2\eta^i_{,tj} + 2\Gamma^i_{jk}\eta^k_{,t} = 0. \tag{17}$$

We conclude that the Lie symmetry equations are equations (16) ,(17) where $\xi(t,x)$ is a gradient KV of the metric and $\eta^i(t,x)$ is a special Projective collineation of the metric with projective function $\xi_{,t}$. We state this result in theorem 1 [8].

Theorem 1 *The Lie point symmetries of the system of equations of motion of an autonomous system under the action of the force $F^j(x^i)$ in a general Riemannian space with metric g_{ij}, namely*

$$\ddot{x}^i + \Gamma^i_{jk}\dot{x}^j\dot{x}^k = F^i \tag{18}$$

are given in terms of the generators Y^i of the special projective algebra of the metric g_{ij}.

If the force F^i is derivable from a potential $V(x^i)$, so that the equations of motion follow from the standard Lagrangian

$$L\left(x^j, \dot{x}^j\right) = \frac{1}{2}g_{ij}\dot{x}^i\dot{x}^j - V\left(x^j\right) \tag{19}$$

with Hamiltonian

$$E = \frac{1}{2}g_{ij}\dot{x}^i\dot{x}^j + V\left(x^j\right) \tag{20}$$

it can be shown that the Noether conditions are

$$V_{,k}\eta^k + V\xi_{,t} = -f_{,t} \tag{21}$$
$$\eta^i_{,t}g_{ij} - \xi_{,j}V = f_{,j} \tag{22}$$
$$L_\eta g_{ij} = 2\left(\frac{1}{2}\xi_{,t}\right)g_{ij} \tag{23}$$
$$\xi_{,k} = 0. \tag{24}$$

Last equation implies $\xi = \xi(t)$ and reduces the system as follows

$$L_\eta g_{ij} = 2\left(\frac{1}{2}\xi_{,t}\right)g_{ij} \tag{25}$$
$$V_{,k}\eta^k + V\xi_{,t} = -f_{,t} \tag{26}$$
$$\eta_{i,t} = f_{,i}. \tag{27}$$

Equation (25) implies that η^i is a conformal Killing vector of the metric provided $\xi_{,t} \neq 0$. Because g_{ij} is independent of t and $\xi = \xi(t)$ the η^i must be is a HV of the metric. This means that $\eta^i(t,x) = T(t)Y^i\left(x^j\right)$ where Y^i is a HV. If $\xi_{,t} = 0$ then η^i is a Killing vector of the metric. Equations (26), (27) are the constraint conditions, which the Noether symmetry and the potential must satisfy for the former to be admitted. The above lead to the following theorem [8].

Theorem 2 *The Noether point symmetries of the Lagrangian (19) are generated from the homothetic algebra of the metric g_{ij}.*

More specifically, concerning the Noether symmetries, we have the following.

All autonomous systems admit the Noether symmetry ∂_t whose Noether integral is the Hamiltonian E given in equation (20). For the rest of the Noether symmetries we have the following cases

Case I Noether point symmetries generated by the homothetic algebra.

The Noether symmetry vector and the Noether function $G\left(t, x^k\right)$ are

$$\mathbf{X} = 2\psi_Y t\partial_t + Y^i\partial_i\ ,\ G\left(t, x^k\right) = pt \tag{28}$$

where ψ_Y is the homothetic factor of Y^i ($\psi_Y = 0$ for a KV and 1 for the HV) and p is a constant, provided the potential satisfies the condition

$$\mathcal{L}_Y V + 2\psi_Y V + p = 0. \tag{29}$$

Case II Noether point symmetries generated by the gradient homothetic Lie algebra i.e. both KVs and the HV must be gradient.

In this case the Noether symmetry vector and the Noether function are

$$\mathbf{X} = 2\psi_Y \int T(t)\, dt\partial_t + T(t)\, H^i\partial_i \ , \ G\left(t, x^k\right) = T_{,t}H\left(x^k\right) \ + p\int T dt \tag{30}$$

where H^i is the gradient HV or a gradient KV, the function $T(t)$ is computed from the relation $T_{,tt} = mT$ where m is a constant and the potential satisfies the condition

$$\mathcal{L}_H V + 2\psi_Y V + mH + p = 0. \tag{31}$$

Concerning the Noether integrals we have the following result (not including the Hamiltonian)

Corollary 1 *The Noether integrals of Case I and Case II are respectively*

$$I_{C_I} = 2\psi_Y tE - g_{ij}Y^i\dot{x}^j + pt \tag{32}$$

$$I_{C_{II}} = 2\psi_Y \int T(t)\, dt\ E - g_{ij}H^{,i}\dot{x}^j + T_{,t}H + p\int T dt. \tag{33}$$

where E is the Hamiltonian given in (20).

We remark that theorems 1 and 2 do not apply to generalized symmetries [17, 18].

It would be of interest to examine if the above close relation of the Lie and the Noether symmetries of the second order ODEs of the form (6) with the collineations of the metric is possible to be carried over to some types of second order partial differential equations (PDEs). To this question it is not possible to give a global answer, due to the complexity of the study and the great variety of PDEs. However for the case of the generalized heat equation the problem has been solved (see [19]).

5. The Lie symmetries of geodesic equations in an Einstein space

Spaces of constant curvature are Einstein spaces whose curvature tensor is of the form $R_{abcd} = \frac{R}{n(n-1)}(g_{ac}g_{bd} - g_{ad}g_{bc})$ where R is a constant. In this section we generalize the results of the last section to proper Einstein spaces in which $R_{ab} = \frac{R}{n}g_{ab}$ but [1] $R_{abcd} \neq \frac{R}{n(n-1)}(g_{ac}g_{bd} - g_{ad}g_{bc})$.

Suppose X^a is a projective collineation with projection function $\phi(x^a)$, so that $\mathcal{L}_X\Gamma^a_{bc} = \phi_{,b}\delta^a_c + \phi_{,c}\delta^a_b$. For a proper Einstein space :

$$\mathcal{L}_X g_{ab} = \frac{n(1-n)}{R}\phi_{;ab}.$$

It follows that if X^a generates either an affine or a special projective collineation then $\phi_{;ab} = 0$ hence X^a reduces to a KV. This means that proper Einstein spaces do not admit homothetic vector, affine collineations or special projective collineations.

The above results and Theorem 1 lead to the following conclusion:

[1] Using the contracted Bianchi identity $(R^{ij} - \frac{1}{2}Rg^{ij})_{;j} = 0$ it follows that in an Einstein space of dimension $n > 2$ the curvature scalar R =constant.

Table 3. First family of 3d Newtonian dynamical systems admitting Lie symmetries

Lie symmetry	$\mathbf{F}(x_\mu, x_\nu, x_\sigma)$
$\frac{d}{2}t\partial_t + \partial_\mu$	$e^{-dx_\mu} f_{\mu,\nu,\sigma}(x_\nu, x_\sigma)$
$\frac{d}{2}t\partial_t + \partial_{\theta_{(\mu\nu)}}$	$e^{-d\theta_{(\mu\nu)}} f_{\mu,\nu,\sigma}(r_{(\mu\nu)}, x_\sigma)$
$\frac{d}{2}t\partial_t + R\partial_R$	$x_\mu^{1-d} f_{\mu,\nu,\sigma}\left(\frac{x_\nu}{x_\mu}, \frac{x_\sigma}{x_\mu}\right)$
$\frac{d}{2}t\partial_t + x_\mu\partial_\mu$	$x_\mu^{1-d} f_{\mu,\nu,\sigma}(x_\nu, x_\sigma)$
$\frac{d}{2}t\partial_t + x_\nu\partial_\mu$	$e^{-d\frac{x_\mu}{x_\nu}}\left[\frac{x_\mu}{x_\nu} f_\nu(x_\nu, x_\sigma) + f_\mu(x_\nu, x_\sigma)\right]\partial_\mu + f_\nu\partial_\nu + f_\sigma\partial_\sigma$

Theorem 3 *The Lie point symmetries of the geodesic equations in a proper Einstein space of curvature scalar R ($\neq 0$) are given by the vectors $\mathbf{X} = \xi(t,x)\partial_t + \eta^i(t,x)\partial_{x^i}$ where*

$$\xi(t,x) = Kt + L \quad (34)$$
$$\eta^i(t,x) = D^i(x) \quad (35)$$

where $D^i(x)$ is a KV of the metric and K, L are constants.

Theorem 3 extends and amends the conjecture of [14] to the more general case of Einstein spaces.

6. Applications in Newtonian Physics

One important problem to consider is to find all 2d and 3d dimensional autonomous Newtonian dynamical systems, that is, all dynamical systems moving in flat Euclidian space with equation of motion $\ddot{x}^i = F^i(x^j)$ where $i, j = 1, 2$ or $1, 2, 3$ which admit Lie and Noether symmetries and of course to determine the admitted symmetries. The answer to this question consists of finding all the forces $F^i(x^j)$ which result in this property.

The importance of this problem is twofold. Indeed the knowledge of the Lie symmetries makes possible the computation of the invariants of the dynamical system; also the knowledge of the Noether symmetries gives the Noether integrals (conserved currents) of the system. Both these data can be used to simplify the equations of motion and even lead to their analytic solution.

The two dimensional case has been considered originally by [20] and the 3d case by [21, 22]. Both cases have been considered anew following the geometric approach of the previous paragraphs. More specifically the 2d case has been considered in [23] and the 3d case in [24]. In both cases it has been shown that the existing results were incomplete.

It is to be noted that from the applications point of view the 2d case is important because it applies to the mini super space which is particularly useful in cosmology. Indeed in the standard cosmological model and much of its extensions the resulting dynamical system reduces to a two dimensional dynamical system in a flat 2d Lorentz space (see [25]). For convenience we collect the results of the 3d case in Tables 3 and 4. Concerning the determination of all 3d Newtonian dynamical systems, which admit Noether point symmetries and subsequently the ones which are integrable via Noether integrals we have the results of the Table 5 and Table 6.

Noether symmetries

In order to show the use of these tables are used we consider two applications; one concerning the motion of a Newtonian system on a 2d space and the second motion in a 3d space. Before we

Table 4. Second family of 3d Newtonian dynamical systems admitting Lie symmetries

Lie symmetry	$F_\mu(x_\mu, x_\nu, x_\sigma)$
$t\partial_\mu$	$f_{\mu,\nu,\sigma}(x_\nu, x_\sigma)$
$t^2\partial_t + tR\partial_R$	$\frac{1}{x_\mu^3} f_{\mu,\nu,\sigma}\left(\frac{x_\nu}{x_\mu}, \frac{x_\sigma}{x_\mu}\right)$
$e^{\pm t\sqrt{m}}\partial_\mu$	$-mx_\mu + f_{\mu,\nu,\sigma}(x_\nu, x_\sigma)$
$\frac{1}{\sqrt{m}} e^{\pm t\sqrt{m}}\partial_t + e^{\pm t\sqrt{m}} R\partial_R$	$-\frac{m}{4}x_\mu + \frac{1}{x_\mu^3} f_{\mu,\nu,\sigma}\left(\frac{x_\nu}{x_\mu}, \frac{x_\sigma}{x_\mu}\right)$

Table 5. First family of 3d Newtonian dynamical systems which admit Noether symmetries

Noether	$d = 0$	$d \neq 2$
$\frac{d}{2}t\partial_t + \partial_\mu$	$c_1 x_\mu + f(x_\nu, x_\sigma)$	$e^{-dx_\mu} f(x_\nu, x_\sigma)$
$\frac{d}{2}t\partial_t + \partial_{\theta_{(\mu\nu)}}$	$c_1\theta_{(\mu\nu)} + f\left(r_{(\mu\nu)}, x_\sigma\right)$	$e^{-d\theta_{(\mu\nu)}} f\left(r_{(\mu\nu)}, x_\sigma\right)$
$\frac{d}{2}t\partial_t + R\partial_R$	$x^2 f\left(\frac{x_\nu}{x_\mu}, \frac{x_\sigma}{x_\mu}\right)$	$x^{2-d} f\left(\frac{x_\nu}{x_\mu}, \frac{x_\sigma}{x_\mu}\right)$
$\frac{d}{2}t\partial_t + x_\mu\partial_\mu$	$c_1 x_\mu^2 + f(x_\nu, x_\sigma)$	$\nexists$
$\frac{d}{2}t\partial_t + x_\nu\partial_\mu$	$c_1 x_\mu + c_2\left(x_\mu^2 + x_\nu^2\right) + f(x_\sigma)$	$\nexists$

Table 6. Second family of 3d Newtonian dynamical systems which admit Noether symmetries

Noether	$V(x, y, z)$
$t\partial_\mu$	$c_1 x_\mu + f(x_\nu, x_\sigma)$
$t^2\partial_t + tR\partial_R$	$\frac{1}{x_\mu^2} f\left(\frac{x_\nu}{x_\mu}, \frac{x_\sigma}{x_\mu}\right)$
$e^{\pm t\sqrt{m}}\partial_\mu$	$-\frac{m}{2}x_\mu^2 + c_1 x_\mu + f(x_\nu, x_\sigma)$
$\frac{1}{\sqrt{m}} e^{\pm t\sqrt{m}}\partial_t + e^{\pm t\sqrt{m}} R\partial_R$	$-\frac{m}{8}R^2 + \frac{1}{x_\mu^2} f\left(\frac{x_\nu}{x_\mu}, \frac{x_\sigma}{x_\mu}\right)$

proceed we recall that in order a 3d Newtonian dynamical system to be integrable via Noether point symmetries it must admit at least 3 Noether first integrals.

6.1. Newtonian motion on the two dimensional sphere

The motion of a system moving on the two dimensional sphere under the potential function $V(\theta, \phi)$ has Lagrangian

$$L\left(\phi, \theta, \dot{\phi}, \dot{\theta}\right) = \frac{1}{2}\left(\dot{\phi}^2 + \mathrm{Sinn}^2\phi\ \dot{\theta}^2\right) - V(\theta, \phi) \tag{36}$$

Table 7. The potentials $V(\theta,\phi)$ for which the dynamical system (36) is integrable via Noether point symmetries

$V(\theta,\phi)$	Noether Integral
$F(\cos\theta\mathrm{Sinn}\phi)$	$I_{CK^1_{e,h}}$
$F(\sin\theta\mathrm{Sinn}\phi)$	$I_{CK^2_{e,h}}$
$F(\phi)$	$I_{CK^3_{e,h}}$
$F\left(\frac{1+\tan^2\theta}{\mathrm{Sinn}^2\phi\ (a-b\tan\theta)^2}\right)$	$aI_{CK^1_{e,h}}+bI_{CK^2_{e,h}}$
$F(a\cos\theta\mathrm{Sinn}\phi-K\ b\mathrm{Cosn}\phi)$	$aI_{CK^1_{e,h}}+bI_{CK^3_{e,h}}$
$F(a\sin\theta\mathrm{Sinn}\phi-K\ b\mathrm{Cosn}\phi)$	$aI_{CK^2_{e,h}}+bI_{CK^3_{e,h}}$
$F\left(\begin{array}{c}(a\cos\theta-b\sin\theta)\mathrm{Sinn}\phi+\\ -K\ c\mathrm{Cosn}\phi\end{array}\right)$	$aI_{CK^1_{e,h}}+bI_{CK^2_{e,h}}+cI_{CK^3_{e,h}}$

where

$$\mathrm{Sinn}\phi=\left\{\begin{array}{ll}\sin\phi & K=1\\ \sinh\phi & K=-1\end{array}\right.\quad \mathrm{Cosn}\phi=\left\{\begin{array}{ll}\cos\phi & K=1\\ \cosh\phi & K=-1.\end{array}\right.$$

The equations of motion are:

$$\ddot{\phi}-\mathrm{Sinn}\phi\ \mathrm{Cosn}\phi\ \dot{\theta}^2+V_{,\phi} = 0 \tag{37}$$

$$\ddot{\theta}+2\frac{\mathrm{Cosn}\phi}{\mathrm{Sinn}\phi}\dot{\theta}\dot{\phi}+\frac{1}{\mathrm{Sinn}^2\phi}V_{,\theta} = 0. \tag{38}$$

The problem to find all potential functions for which the system is integrable via Noether point symmetries is to find all potentials for which the system admits at least one extra Noether symmetry, in addition to the trivial one which is the energy (Hamiltonian)

$$E=\frac{1}{2}\left(\dot{\phi}^2+\mathrm{Sinn}^2\phi\ \dot{\theta}^2\right)+V(\theta,\phi).$$

In order to do that we consider the kinematic metric $g_{ij}=\frac{1}{2}\left(\dot{\phi}^2+\mathrm{Sinn}^2\phi\ \dot{\theta}^2\right)$ in the configuration space, which is defined by the kinematic part of the Lagrangian and apply theorem 2.

This has been done in [24] and the answer is given in Table 7. Where $I_{CK^{1,2,3}}$ are the Noether Integrals

$$\begin{aligned} I_{CK^3} &= \dot{\theta}\mathrm{Sinn}^2\phi.\ ,\quad I_{CK^1}=\dot{\phi}\sin\theta+\dot{\theta}\cos\theta\mathrm{Sinn}\phi\mathrm{Cosn}\phi \\ I_{CK^2} &= \dot{\phi}\cos\theta-\dot{\theta}\sin\theta\mathrm{Sinn}\phi\mathrm{Cosn}\phi \end{aligned}$$

Corollary 2 *A dynamical system with Lagrangian (36) has one, two or four Noether point symmetries hence Noether integrals.*

Proof: For the case of the free particle we have the maximum number of four Noether symmetries (the rotation group so(3) plus the ∂_t). In the case the potential is not constant the Noether symmetries are produced by the non-gradient KVs with Lie algebra $[X_A,X_B]=C^C_{AB}X_C$ where $C^3_{12}=C^2_{31}=C^1_{23}=1$ for $\varepsilon=1$ and $\bar{C}^3_{21}=\bar{C}^1_{23}=\bar{C}^2_{31}=1$ for $\varepsilon=-1$. Because *the Noether point symmetries form a Lie algebra and the Lie algebra of the KVs is semisimple the system will admit either none, one or three Noether symmetries generated from the KVs. The case of three is when $V(\theta,\phi) = V_0$ that is the case of geodesics, therefore the Noether point symmetries will be (including ∂_t) either one, two or four.*

6.2. Newtonian motion in 3d space

We consider the dynamical system defined by the equations of motion

$$\ddot{x}^{\mu} = -\frac{m}{4}x_{\mu} + \frac{1}{x_{\mu}^{3}} f_{\mu,\nu,\sigma}\left(\frac{x_{\nu}}{x_{\mu}}, \frac{x_{\sigma}}{x_{\mu}}\right)$$

This system is the well known Ermakov system which has a long history in the older and recent literature [26–30].

From Table 6 we find without any calculations that the system admits the Lie symmetry $\mathbf{X} = \frac{1}{\sqrt{m}} e^{\pm t\sqrt{m}} \partial_t + e^{\pm t\sqrt{m}} R\partial_R$.

7. Applications in cosmology

In this section we consider applications of Theorem 1 and Theorem 2 to dynamical systems occurring in Cosmology.

7.1. Lie and Noether symmetries of Bianchi class A homogeneous cosmologies with a scalar field

The Bianchi models in the ADM formalism are described by the metric

$$ds^2 = -N^2(t)dt^2 + g_{\mu\nu}\omega^{\mu} \otimes \omega^{\nu} \tag{39}$$

where $N(t)$ is the lapse function and $\{\omega^a\}$ is the canonical basis 1-forms which satisfy the Lie algebra $d\omega^i = C^i_{jk}\omega^j \wedge \omega^k$ where C^i_{jk} are the structure constants of the algebra.

The spatial metric $g_{\mu\nu}$ splits so that $g_{\mu\nu} = \exp(2\lambda)\exp(-2\beta)_{\mu\nu}$ where $\exp(2\lambda)$ is the scale factor of the universe and $\beta_{\mu\nu}$ is a 3×3 symmetric, traceless matrix, which can be written in a diagonal form with two independent quantities, known as the anisotropy parameters β_+, β_-, as follows:

$$\beta_{\mu\nu} = diag\left(\beta_+, -\frac{1}{2}\beta_+ + \frac{\sqrt{3}}{2}\beta_-, -\frac{1}{2}\beta_+ - \frac{\sqrt{3}}{2}\beta_-\right). \tag{40}$$

It has been shown that the dynamics of the class A Bianchi models with a scalar field is described by the Lagrangian

$$L = e^{3\lambda}\left[R^* + 6\lambda - \frac{3}{2}(\dot{\beta}_1^2 + \dot{\beta}_2^2) - \dot{\phi}^2 + V(\phi)\right] \tag{41}$$

where R^* is the Ricci scalar of the 3 dimensional spatial hypersurfaces given by the expression:

$$R^* = -\frac{1}{2}e^{-2\lambda}\left[N_1^2 e^{4\beta_1} + e^{-2\beta_1}\left(N_2 e^{\sqrt{3}\beta_2} - N_3 e^{-\sqrt{3}\beta_2}\right)^2 - 2N_1 e^{\beta_1}\left(N_2 e^{\sqrt{3}\beta_2} - N_3 e^{-\sqrt{3}\beta_2}\right)\right] + \frac{1}{2}N_1N_2N_3(1 + N_1N_2N_3).$$

The constants N_1, N_2, and N_3 are the components of the classification vector n^{μ} and $\beta_1 = -\frac{1}{2}\beta_+ + \frac{\sqrt{3}}{2}\beta_-$, $\beta_2 = -\frac{1}{2}\beta_+ - \frac{\sqrt{3}}{2}\beta_-$.

It is important to note that the curvature scalar R^* does not depend on the derivatives of the anisotropy parameters β_+, β_- , equivalently of β_1, β_2.

The Euler Lagrange equations due to the Lagrangian (41) are:

$$\ddot{\lambda}+\frac{3}{2}\dot{\lambda}^2+\frac{3}{8}(\dot{\beta}_1^2+\dot{\beta}_2^2)+\frac{1}{4}\dot{\phi}^2-\frac{1}{12}e^{-3\lambda}\frac{\partial}{\partial\lambda}\left(e^{3\lambda}R^*\right)-\frac{1}{2}V(\phi)=0$$
$$\ddot{\beta}_1+3\dot{\lambda}\dot{\beta}_1+\frac{1}{3}\frac{\partial R^*}{\partial\beta_1}=0$$
$$\ddot{\beta}_2+3\dot{\lambda}\dot{\beta}_2+\frac{1}{3}\frac{\partial R^*}{\partial\beta_2}=0$$
$$\ddot{\phi}+3\dot{\phi}\dot{\lambda}+\frac{\partial V}{\partial\phi}=0$$

where a dot over a symbol indicates derivative with respect to t.

We apply Theorem 1 and Theorem 2 in order to compute the Lie and the Noether symmetries of class A Bianchi models with a scalar field. Similar (but incomplete) works on that topic can be found in [31–33].

We consider the four dimensional Riemannian space with coordinates $x^i = (\lambda, \beta_1, \beta_2, \phi)$ and metric

$$ds^2 = e^{3\lambda}\left(12d\lambda^2 - 3d\beta_1^2 - 3d\beta_2^2 - 2d\phi^2\right). \tag{42}$$

It is easy to show that this metric is the conformally flat FRW spacetime whose special projective algebra consists of the non gradient KVs

$$Y^1 = \partial_{\beta_1},\ Y^2 = \partial_{\beta_2},\ Y^3 = \partial_\phi,\ Y^4 = \beta_2\partial_{\beta_1} - \beta_1\partial_{\beta_2}$$
$$Y^5 = \phi\partial_{\beta_1} - \frac{3}{2}\beta_1\partial_\phi,\ Y^6 = \phi\partial_{\beta_2} - \frac{3}{2}\beta_2\partial_\phi$$

and the gradient HV

$$H^i = \frac{2}{3}\partial_\lambda\ ,\ \psi = 1.$$

The Lagrangian is written $L = T - U(x^i)$ where $T = \frac{1}{2}g_{ij}\dot{x}^i\dot{x}^i$ is the geodesic Lagrangian, the potential function is

$$U(x^i) = -e^{3\lambda}\left(V(\phi) + R^*\right) \tag{43}$$

and we have used the fact that the curvature scalar does not depend on the derivatives of the coordinates β_1, β_2.

We apply Theorem 1 and Theorem 2 to determine the Lie and the Noether symmetries of the dynamical system with Lagrangian (41) in the following cases:

Case 1. Vacuum. In this case ϕ =constant.

Case 2. Zero potential $V(\phi) = 0$, $\dot{\phi} \neq 0$

Case 3. Constant Potential $V(\phi)$ =constant, $\dot{\phi} \neq 0$

Case 4. Arbitrary Potential $V(\phi)$, $\dot{\phi} \neq 0$.

The detailed calculations can be found in [8]. Below we give the results for Bianchi I in the Table 8, for Bianchi II in the Table 9, for Bianchi VI and VII in the Table 10 and for Bianchi VIII/IX spacetimes in the Table 11.

8. Conclusion

The Lie and the Noether symmetry vectors of the equations of motion in a general Riemannian space are not independent dynamical quantities but instead they are determined completely by the geometry of the space. Indeed Theorem 1 and Theorem 2 show that the Lie point

Journal of Physics: Conference Series **453** (2013) 012020 doi:10.1088/1742-6596/453/1/012020

Table 8. Bianchi I spacetime

Bianchi I	Noether Sym.	Lie Sym.
Vacuum	$\partial_t,\ Y^1,\ Y^2,\ Y^4$	$\partial_t,\ t\partial_t,\ Y^{1,2,4},\ H^i$
	$2t\partial_t + H^i,\ t^2\partial_t + tH^i$	$t^2\partial_t + tH^i$
Zero Pot.	$\partial_t, Y^{1-6}\ ,\ 2t\partial_t + H^i$	$\partial_t,\ t\partial_t,\ Y^{1-6}$
	$t^2\partial_t + tH^i$	$H^i,\ t^2\partial_t + tH^i$
Constant Pot.	$\partial_t,\ Y^{1-6}$	$\partial_t,\ Y^{1-6}, H^i$
	$\frac{1}{C}e^{\pm Ct}\partial_t \pm e^{\pm Ct}H^i$	$\frac{1}{C}e^{\pm Ct}\partial_t \pm e^{\pm Ct}H^i$
Arbitrary Pot.	$\partial_t,\ Y^{1,2,4}$	$\partial_t,\ Y^{1,2,4}\ , H^i$
Exponential Pot.	$\partial_t,\ Y^{1,2,4}$	$\partial_t,\ Y^{1,2,4}\ , H^i$
	$2t\partial_t + H^i + \frac{4}{d}Y^3$	$t\partial_t + \frac{2}{d}Y^3$

Table 9. Bianchi II spacetime

Bianchi II	Noether Sym.	Lie Sym.
Vacuum	$\partial_t,\ Y^2$	$\partial_t,\ Y^2$
	$6t\partial_t + 3H^i - 5Y^1$	$\frac{1}{3}t\partial_t + H^i,\ t\partial_t - Y^1$
Zero Pot.	$\partial_t,\ Y^2,\ Y^3,\ Y^6$	$\partial_t,\ Y^2,\ Y^3,\ Y^6$
	$6t\partial_t + 3H^i - 5Y^1$	$\frac{1}{3}t\partial_t + H^i,\ t\partial_t - Y^1$
Constant Pot.	$\partial_t,\ Y^2,\ Y^3,\ Y^6$	$\partial_t,\ Y^2,\ Y^3,\ Y^6$
		$3H^i + Y^1$
Arbitrary Pot.	$\partial_t,\ Y^2$	$\partial_t,\ Y^2,\ 3H^i + Y^1$
Exponential Pot.	$\partial_t,\ Y^2$	$\partial_t,\ Y^2,\ 3H^i + Y^1$
	$2t\partial_t + H^i - \frac{5}{3}Y^1 + \frac{4}{d}Y^3$	$t\partial_t + \frac{2}{d}Y^3$

Table 10. Bianchi VI/VII spacetimes

Bianchi VI_0 / VII_0	Noether Sym.	Lie Sym.
Vacuum	$\partial_t,\ \ 6t\partial_t + 3H^i +$	$\partial_t,\ H^i + \frac{1}{3}Y^1 + \frac{\sqrt{3}}{3}Y^2$
	$-\ 2Y^1 - 2\sqrt{3}Y^2$	$2t\partial_t - Y^1 - \sqrt{3}Y^2$
Zero Pot.	$\partial_t,\ Y^3\ , 6t\partial_t + 3H^i +$	$\partial_t,\ H^i + \frac{1}{3}Y^1 + \frac{\sqrt{3}}{3}Y^2$
	$-\ 2Y^1 - 2\sqrt{3}Y^2$	$Y^3, 2t\partial_t - Y^1 - \sqrt{3}Y^2$
Constant Pot.	$\partial_t,\ Y^3$	$\partial_t,\ Y^3,\ H^i + \frac{1}{3}Y^1 + \frac{\sqrt{3}}{3}Y^2$
Arbitrary Pot.	∂_t	$\partial_t,\ H^i + \frac{1}{3}Y^1 + \frac{\sqrt{3}}{3}Y^2$
Exponential Pot.	$\partial_t\ ,\ 6t\partial_t + 3H^i - 2Y^1 +$	$\partial_t,\ H^i + \frac{1}{3}Y^1 + \frac{\sqrt{3}}{3}Y^2$
	$-2\sqrt{3}Y^2 + \frac{6}{d}Y^3$	$t\partial_t + \frac{1}{d}Y^3$

Table 11. Bianchi VIII/IX spacetimes

Bianchi VIII	Noether Sym.	Lie Sym.
Vacuum	∂_t	$\partial_t,\ \frac{2}{3}t\partial_t + H^i$
Zero Pot.	$\partial_t,\ Y^3$	$\partial_t,\ Y^3,\ \frac{2}{3}t\partial_t + H^i$
Constant Pot.	$\partial_t,\ Y^3$	$\partial_t,\ Y^3$
Arbitrary Pot.	∂_t	∂_t
Bianchi IX	Noether Sym.	Lie Sym.
Vacuum	∂_t	∂_t
Zero Pot.	$\partial_t,\ Y^3$	$\partial_t,\ Y^3$
Constant Pot.	$\partial_t,\ Y^3$	$\partial_t,\ Y^3$
Arbitrary Pot.	∂_t	∂_t

symmetries are elements of the special projective algebra of the space and the Noether point symmetries elements of the homothetic algebra of the space. The selection of the particular vectors for a given 'force' modulating the motion of a particular system is done by means of certain compatibility conditions of differential nature which can easily be managed. Using the above theorems we have determined all 2d and 3d Newtonian dynamical systems which admit Lie and Noether point symmetries and have demonstrated how the results can be used to determine these symmetries in specific situations. Furthermore we have applied the theorems in the case of Cosmology and have determined the Lie and the Noether point symmetries of the Bianchi Class A homogeneous models. More applications of these theorems in cosmology can be found in [25, 34], where one is able to determine analytic solutions using the integration of field equations by means of Noether integrals.

Acknowledgments

This research was partially funded by the University of Athens Special Account of Research Grants no 10812.

References

[1] Bluman W G and Kumei S 1989 *Symmetries and Differential Equations* (Springer Verlag New York)
[2] Olver P J 1986 *Application of Lie groups to differential equations* (Springer Graduate texts in Mathematics, New York: Springer)
[3] Stephani H 1989 *Differential Equations: Their Solutions using Symmetry* (Cambridge University Press)
[4] Prince G E and Crampin M (1984) *Projective differential geometry and ... servation laws in general relativity. I Projective actions*,Gen Rel Grav. **16**, 921 –942.
[5] Aminova A V and Aminov N A 2006 Sbornic Mathematics **197** 951
[6] Aminova A V and Aminov N A 2010 Sbornic Mathematics **201** 631
[7] Tsamparlis M and Paliathanasis A (2010) *Lie and Noether symmetries of geodesic equations and collineations*, Gen. Relativ. Gravit. **42** 2957-2980; (Preprint arXiv:1101.5769)
[8] Tsamparlis M and Paliathanasis A (2011) *The geometric nature of Lie and Noether symmetries*; Gen. Relativ. Gravit. **43** 1861
[9] Tsamparlis M, Paliathanasis A (2010), *Lie symmetries of geodesic equations and projective collineations*, Nonlinear Dynamics **62**, 203
[10] Barnes, A. 1993, *Projective collineations in Einstein spaces*, Class. Quantum Grav., **10**, 1139 - 1145
[11] Aminova, A. V. 1994, *Automorphisms of geometric structures as symmetries of diferential equations* (in Russian), *Izv. - Vyssh. - Uchebn. - Zaved. - Mat.* **2**, 3 - 10

[12] Aminova, A. V. 1995, *Projective transformations and symmetries of differential equations*, Sbornik Mathematics **186** (12), 1711 - 1726

[13] Aminova, A. V. 2000, *Projective geometry of systems of differential equations: General conceptions*", Tensor, N.S., 65 - 86

[14] Feroze, T., Mahomed, F.M., Qadir, A. 2006, *The connection between isometries and symetries of geodesic equations of the underlying spaces*, Nonlinear Dynamics 65 - 74

[15] Katzin G H Levine J and Davis R W 1969 J. Math. Phys. **10** 617

[16] Hall G S and Roy I M 1997 Gen. Relativ. Gravit. **29** 827

[17] Sarlet W and Cantrijin F 1981 J. Phys. A: Math. Gen. **14** 479

[18] Kalotas T M and Wybourne B G 1982 J. Phys A: Math. Gen. **15** 2077

[19] Andronikos Paliathanasis, Michael Tsamparlis (2012) '*Lie point symmetries of a general class of PDEs: The heat equation*' J. Geom. Phys. 62 (2012) 2443–2456

[20] Sen T 1987 Phys. Lett. A **122** 327

[21] Damianou P A and Sophocleous C (1999) *Symmetries of Hamiltonian Systems with two degrees of freedom* J. Math. Phys. **40** 210

[22] Damianou P A and Sophocleous C (2004) *Noether symmetries for Lagrangians in 3 degrees of freedom*, Nonlinear Dynamics **36** 3

[23] Tsamparlis M. and Paliathanasis A (2011) *Two-dimensional dynamical systems which admit Lie and Noether symmetries* J. Phys. A: Math. and Theor. **44** 175202; (Preprint arxiv:1205.4114)

[24] Tsamparlis M. Paliathanasis A *Autonomous three dimensional Newtonian systems which admit Lie and Noether point symmetries*, J. Phys. A (2012) ; (Preprint arXiv:1111.0810)+

[25] Basilakos S, Tsamparlis M and Paliathanasis A (2011) *Using the Noether symmetry approach to probe the nature of dark energy* Phys. Rev. D **83,** 103512; (Preprint arXiv:1104.2980)

[26] Moyo S and Leach P G L 2002 J. Phys. A: Math and Gen **35** 5333

[27] Haas F and Goedert J 2001 Phys. Lett. A **279** 181

[28] Athorne C 1991 Phys. Lett. A 159 375

[29] Govinder K S Athorne C Leach P G L 1993 J. Phys. A: Math and Gen **26** 4035

[30] Govinder K S Leach P G L 1993 Quaestiones Mathematicae **16** 405

[31] Capozzielo S., Marmo G., Rubano C., Scudellaro P. (1997) *Noether Symmetries in Bianchi Universes* Int. J. Mod. Phys. D. **6**, 491-503

[32] Cotsakis S., Leach P.G.L., Pantazi H. (1998), *Symmetries of Homogeneous Cosmologies*, Grav. Cosm. **4**, 314 - 325

[33] Vakili B., Khosvravi N., Sepangi H R. (2007), *Bianchi spacetimes in noncommutative phase space*, **24**, 931 - 949

[34] Paliathanasis A, Tsamparlis M, Basilakos S (2011) *Constraints and analytical solutions of f (R) theories of gravity using Noether symmetries*, Phys. Rev. D **84** 123514; (Preprint arXiv:1111.4547)

Journal of Physics: Conference Series **453** (2013) 012021 doi:10.1088/1742-6596/453/1/012021

Decoupling of Field Equations in Einstein and Modified Gravity

Sergiu I. Vacaru

Alexandru Ioan Cuza University at Iaşi (Yassy), UAIC,
Alexandru Lapuşneanu street, nr. 14, Corpus R, office 323; Iaşi, Romania, 700057

E-mail: sergiu.vacaru@uaic.ro

Abstract. This paper is concerned with giving the proof that there is a general decoupling property of vacuum and nonvacuum gravitational field equations in Einstein gravity and $f(R,T)$–modifications. The constructions are possible in terms of geometric and physical objects adapted to certain nonholonomic 2+2 splitting with local fibred structure. This allows us to generate exact and/or parametric depending solutions with generic off–diagonal metrics and generalized, or Levi–Civita, connections. Different classes of modified spacetimes are determined by corresponding generating and integration functions depending, in general, on all space and time coordinates and may possess, or not, Killing symmetries. The initial data sets for the Cauchy problem and their global properties are analyzed. There are formulated the criteria of evolution with spacetime splitting and decoupling of fundamental field equations. Examples of exact solutions defining ellipsoid deformations of black hole metrics and solitonic configurations are provided.

1. Introduction

The issue of constructing exact and approximate solutions of (modified) gravitational and matter field equations is of interest in mathematical physics and for various applications in modern cosmology and astrophysics. Mathematically, the fundamental field equations in gravitational physics are defined by sophisticate systems of nonlinear partial differential equations (PDE) which are very difficult to be integrated and studied in general forms.

In this work, we address again and develop a geometric approach (the so–called anholonomic frame deformation method, AFDM, see [20, 21, 22]) to constructing exact solutions in gravity described by generic off–diagonal and nonlinear gravitational, gauge and scalar field interactions. Such a geometric method allows us to generate exact solutions when the coefficients of generic off–diagonal metrics and various classes of connections depend on all spacetime coordinates. Our goal is to provide new results on the decoupling property[1] of modified gravity equations and formal integration of such nonlinear systems of PDEs. Certain issues on the Cauchy problem and decoupling of gravitational equations and the initial data sets and nonholonomic evolution (with non–integrable constraints and/or with respect to anholonomic frames) will be analyzed. Examples of off–diagonal solutions for modified black holes/ellpsoids and solitonic configurations will be given.

[1] It is used also the term "separation" of equations, which should not be confused with separation of variables.

Journal of Physics: Conference Series **453** (2013) 012021 doi:10.1088/1742-6596/453/1/012021

Section 2 contains geometric preliminaries on modified gravity theories written in nonholonomic variables. In Section 3 we provide the main Theorems on decoupling the gravitational field equations for generic off–diagonal metrics with one Killing symmetry. We consider extensions to "non–Killing" configurations with coefficients depending on all set of four coordinates in section 4. Section 5 contains a study of the Cauchy problem in connection to the decoupling property of gravitational field equations. Explicit examples of generic off–diagonal exact solutions are studied in the next two sections. In Section 6, there are constructed nonholonomic vacuum deformations of black holes determined by off–diagonal gravitational interactions. Ellipsoid–solitonic configurations are analyzed in Section 7. The most important formulas and computations which are necessary to prove the decoupling property are given in Appendix Appendix A.

2. Einstein and Modified Gravity in Nonholonomic Variables

We provide geometric preliminaries on $f(\ ^{s}R,T)$–gravity models written in nonholonomic variables.[2] In this work, the scalar curvature $\ ^{s}R$ is constructed for an auxiliary connection $\mathbf{D}$ completely defined by the metric structure $\mathbf{g}$ and T is the trace of the energy–momentum tensor for matter fields. If $\mathbf{D}=\nabla$, where ∇ is the Levi–Civita connection, we get as particular cases the theories studied, for instance, in Ref. [11], when $\ ^{s}R=R$ is just the scalar curvature of a pseudo–Riemannian space. For $f(\ ^{s}R,T)=R$, such a modified gravity model transforms into the "standard" Einstein gravity theory.

2.1. Nonholonomic 2+2 splitting

In general relativity (GR), the curved spacetime $(V,\mathbf{g})$ is defined by a pseudo–Riemannian manifold V endowed with a Lorentzian metric $\mathbf{g}$ as a solution of the Einstein equations[3]. The space of classical physical events is modelled as a Lorentzian four dimensional, 4-d, manifold $\mathcal{V}$ (of necessary smooth class, Housdorf and paracompact one) when the symmetric 2–covariant tensor $\mathbf{g}=\{g_{\alpha\beta}\}$ defines in each point $u\in\mathcal{V}$ and nondegenerate bilinear form on the tangent space $T_u\mathcal{V}$, for instance, of signature $(+++-)$. The assumption that $T_u\mathcal{V}$ has its prototype (local fiber) the Minkowski space $\mathbb{R}^{3,1}$ leads at a causal character of positive/negative/null vectors on $\mathcal{V}$, i.e. for the module $\mathcal{X}$ of vectors fields $X,Y,...\in\mathcal{X}(\mathcal{V})$, which is similar to that in special relativity.

Let us denote by e_α and e^β a local frame and, respectively, its dual frame [we can consider orthonormal (co) bases], where Greek indices $\alpha,\beta,..$ may be abstract ones, or running values $1,2,3,4$. For a coordinate base $u=\{u^\alpha\}$ on a chart $U\subset\mathcal{V}$, we can write $e_\alpha=\partial_\alpha=\partial/\partial u^\beta$ and $e^\beta=du^\beta$ and, for instance, define the coefficients of a vector X and a metric $\mathbf{g}$, respectively, in the forms $X=X^\alpha e_\alpha$ and

$$\mathbf{g}=g_{\alpha\beta}(u)e^\alpha\otimes e^\beta, \tag{1}$$

where $g_{\alpha\beta}:=\mathbf{g}(e_\alpha,e_\beta)$.[4] We consider bases with non–integrable (equivalently, nonholonomic/anholonomic) $2+2$ splitting for conventional, horizontal, h, and vertical, v, decomposition, when for the tangent bundle $T\mathcal{V}:=\bigcup_u T_u\mathcal{V}$ a Whitney sum

$$\mathbf{N}:\ T\mathcal{V}=h\mathcal{V}\oplus v\mathcal{V} \tag{2}$$

is globally defined. Such a nonholonomic distribution is determined locally by its coefficients $N_i^a(u)$, when $\mathbf{N}=N_i^a(x,y)dx^i\otimes\partial/\partial y^a$, where $u^\alpha=(x^i,y^a)$ splits into h–coordinates, $x=(x^i)$,

[2] Such a formulation is necessary for proving the main results on decoupling and integrating in certain general forms the gravitational field equations, see sections 3 and 4.

[3] We assume that readers are familiar with basic concepts and results on mathematical relativity and methods of constructing exact solutions outlined, for instance, in above mentioned monographs and reviews.

[4] The summation rule on repeating low–up indices will be applied if the contrary is not stated.

and v–coordinates, $y = (y^a)$, with indices running, respectively, values $i, j, k, ... = 1, 2$ and $a, b, c, ... = 3, 4.$[5]

A spacetime $(V, \mathbf{g})$ can be equipped with a non–integrable fibred structure (2) and such a manifold is called nonholonomic (equivalently, N–anholonomic). We use "boldface" letters in order to emphasize that certain spaces and geometric objects/constructions are "N–adapted", i.e., adapted to a h–v–splitting. The geometric objects are called distinguished (in brief, d–objects, d–vectors, d–tensors etc). For instance, we write a d–vector as $\mathbf{X} = (hX, vX)$ for a nonholonomic Lorentz manifold/spacetime $(\mathbf{V}, \mathbf{g})$.

On a spacetime $(\mathbf{V}, \mathbf{g})$, we can perform/adapt the geometric constructions using "N–elongated" local bases (partial derivatives), $\mathbf{e}_\nu = (\mathbf{e}_i, e_a)$, and cobases (differentials), $\mathbf{e}^\mu = (e^i, \mathbf{e}^a)$, when

$$\mathbf{e}_i = \partial/\partial x^i - N_i^a(u)\partial/\partial y^a, \ e_a = \partial_a = \partial/\partial y^a, \tag{3}$$

$$\text{and } e^i = dx^i, \ \mathbf{e}^a = dy^a + N_i^a(u)dx^i. \tag{4}$$

Such (co) frame structures depend linearly on N–connection coefficients being, in general, nonholonomic. For instance, the basic vectors (3) satisfy certain nontrivial nonholonomy relations

$$[\mathbf{e}_\alpha, \mathbf{e}_\beta] = \mathbf{e}_\alpha \mathbf{e}_\beta - \mathbf{e}_\beta \mathbf{e}_\alpha = W_{\alpha\beta}^\gamma \mathbf{e}_\gamma, \tag{5}$$

with nontrivial anholonomy coefficients $W_{ia}^b = \partial_a N_i^b, W_{ji}^a = \Omega_{ij}^a = \mathbf{e}_j(N_i^a) - \mathbf{e}_i(N_j^a)$.

Any spacetime metric $\mathbf{g} = \{g_{\alpha\beta}\}$ (1), via frame/coordinate transforms can be represented equivalently in N–adapted form as a d–metric

$$\mathbf{g} = g_{ij}(x, y) \ e^i \otimes e^j + g_{ab}(x, y) \ \mathbf{e}^a \otimes \mathbf{e}^b, \tag{6}$$

or, with respect to a coordinate local cobasis $du^\alpha = (dx^i, dy^a)$, as an off–diagonal metric $\mathbf{g} = \underline{g}_{\alpha\beta} du^\alpha \otimes du^\beta$, where

$$\underline{g}_{\alpha\beta} = \begin{bmatrix} g_{ij} + N_i^a N_j^b g_{ab} & N_j^e g_{ae} \\ N_i^e g_{be} & g_{ab} \end{bmatrix}. \tag{7}$$

A metric $\mathbf{g}$ is generically off–diagonal, if (7) can not be diagonalized via coordinate transforms. Ansatzes of this type are used in Kaluza–Klein gravity when $N_i^a(x, y) = \Gamma_{bi}^a(x) y^a$ and y^a are "compactified" extra–dimensions coordinates, or in Finsler gravity theories, see details in [17, 22]. In this work, we restrict our considerations only to the 4–d gravity theories. The principle of general covariance in GR, allows us to consider any frame/coordinate transforms and write a spacetime metric $\mathbf{g}$, equivalently, in any of the above form (1), (7) and/or (6). The last mentioned parametrization will allow us to prove a very important property of decoupling of gravitational field equations with respect to N–adapted bases (3) and (4).

Via frame/coordinate transforms $e_\alpha = e_\alpha^{\alpha'}(x, y) e_{\alpha'}, \ \underline{g}_{\alpha\beta} = e^{\alpha'}_{\ \alpha} e^{\beta'}_{\ \beta} \ \underline{g}_{\alpha'\beta'}$, a metric $\mathbf{g}$ (6) can be written in a form with separation of v–coordinates and nontrivial vertical conformal transforms,

$$\begin{aligned} \mathbf{g} &= g_i dx^i \otimes dx^i + \omega^2 h_a \underline{h}_a \mathbf{e}^a \otimes \mathbf{e}^a, \\ \mathbf{e}^3 &= dy^3 + (w_i + \underline{w}_i) \, dx^i, \ \mathbf{e}^4 = dy^4 + (n_i + \underline{n}_i) \, dx^i, \end{aligned} \tag{8}$$

$$\begin{aligned} \text{were } g_i &= g_i(x^k), g_a = \omega^2(x^i, y^c) \ h_a(x^k, y^3) \underline{h}_a(x^k, y^4), \\ N_i^3 &= w_i(x^k, y^3) + \underline{w}_i(x^k, y^4), N_i^4 = n_i(x^k, y^3) + \underline{n}_i(x^k, y^4), \end{aligned} \tag{9}$$

[5] We note that the 2+2 splitting can be considered as an alternative to the well known 3+1 splitting. The first one is convenient, for instance, for constructing generic off–diagonal solutions and elaborating models of deformation and/or A–brane quantization of gravity, but the second one is more important for canonical/loop quantization etc.

are functions of necessary smooth class which will be defined in a form to generate solutions of gravitational field equations[6].

The aim of this section is to prove that the gravitational field equations in the vacuum cases and certain very general classes of matter field sources decouple for parameterizations of metrics in the form (9).

2.2. *Torsions and curvatures*

In a general case, a metric–affine manifold V is endowed with a metric structure $\mathbf{g}$ and an affine (linear) connection structure D (as a covariant differentiation operator). A linear connection gives us the possibility to compute the directional derivative $D_X Y$ of a vector field Y in the direction of X. It is characterized by three fundamental geometric objects

(i) the torsion field is (by definition) $\mathcal{T}(X,Y) := D_{\mathbf{X}}Y - D_{\mathbf{Y}}X - [X,Y]$;
(ii) the curvature field is $\mathcal{R}(X,Y) := D_{\mathbf{X}}D_{\mathbf{Y}} - D_{\mathbf{Y}}D_{\mathbf{X}} - D_{[\mathbf{X},\mathbf{Y}]}$;
(iii) the nonmetricity field is $\mathcal{Q}(X) := D_{\mathbf{X}}\mathbf{g}$.

Introducing $\mathbf{X} = \mathbf{e}_\alpha$ and $\mathbf{Y} = \mathbf{e}_\beta$, defined by (3), into above formulas, we compute the N–adapted coefficients $D = \{\Gamma^\gamma_{\ \alpha\beta}\}$ and corresponding fundamental geometric objects,

$$\begin{aligned}
\mathcal{T} &= \{T^\gamma_{\ \alpha\beta} = \left(T^i_{\ jk}, T^i_{\ ja}, T^a_{\ ji}, T^a_{\ bi}, T^a_{\ bc}\right)\};\\
\mathcal{R} &= \{R^\alpha_{\ \beta\gamma\delta} = \left(R^i_{\ hjk}, R^a_{\ bjk}, R^i_{\ hja}, R^c_{\ bja}, R^i_{\ hba}, R^c_{\ bea}\right)\};\ \mathcal{Q} = \{Q^\gamma_{\ \alpha\beta}\}.
\end{aligned}$$

Every (pseudo) Riemannian manifold $(V,\mathbf{g})$ is naturally equipped with a Levi–Civita connection $D = \nabla = \{\ {}^\nabla\Gamma^\gamma_{\ \alpha\beta}\}$ completely defined by $\mathbf{g} = \{g_{\alpha\beta}\}$ if and only if the metric compatibility, i.e., ${}^\nabla\mathcal{Q}(X) = \nabla_{\mathbf{X}}\mathbf{g} = 0$, and zero torsion, i.e., ${}^\nabla\mathcal{T} = 0$, conditions are satisfied. Hereafter, we shall write, for simplicity, ${}^\nabla\Gamma^\gamma_{\ \alpha\beta} = \Gamma^\gamma_{\ \alpha\beta}$. It should be emphasized that ∇ does not preserve under parallelism and general frame/coordinate transforms a N–splitting (2). Nevertheless, it is possible to construct a unique distortion relation

$$\nabla = \widehat{\mathbf{D}} + \widehat{\mathbf{Z}}, \tag{10}$$

where both linear connections ∇ and $\widehat{\mathbf{D}}$ (the second one can be considered as an auxiliary linear connection, which in literature is called the canonical distinguished connection; in brief, d–connection) and the distortion tensor $\widehat{\mathbf{Z}}$, i.e. all values in the above formula, are completely defined by $\mathbf{g} = \{g_{\alpha\beta}\}$ for a prescribed $\mathbf{N} = \{N^a_i\}$, see details in [20, 21, 22].

Theorem 2.1 *With respect to N–adapted frames (3) and (4), the coefficient of distortion relation (10) are computed*

$$\Gamma^\gamma_{\ \alpha\beta} = \widehat{\mathbf{\Gamma}}^\gamma_{\ \alpha\beta} + \widehat{\mathbf{Z}}^\gamma_{\ \alpha\beta}, \tag{11}$$

where the canonical d–connection $\widehat{\mathbf{D}} = \{\ \widehat{\mathbf{\Gamma}}^\gamma_{\ \alpha\beta} = (\widehat{L}^i_{jk}, \widehat{L}^a_{bk}, \widehat{C}^i_{jc}, \widehat{C}^a_{bc})\}$ *is defined by coefficients*

$$\begin{aligned}
\widehat{L}^i_{jk} &= \frac{1}{2}g^{ir}\left(\mathbf{e}_k g_{jr} + \mathbf{e}_j g_{kr} - \mathbf{e}_r g_{jk}\right), \widehat{L}^a_{bk} = e_b(N^a_k) + \frac{1}{2}g^{ac}\left(\mathbf{e}_k g_{bc} - g_{dc}\ e_b N^d_k - g_{db}\ e_c N^d_k\right),\\
\widehat{C}^i_{jc} &= \frac{1}{2}g^{ik} e_c g_{jk},\ \widehat{C}^a_{bc} = \frac{1}{2}g^{ad}\left(e_c g_{bd} + e_b g_{cd} - e_d g_{bc}\right),
\end{aligned} \tag{12}$$

[6] There is not summation on repeating "low" indices a in formulas (9) but such a summation is considered for crossing "up–low" indices i and a in (8). We shall underline a function if it positively depends on y^4 but not on y^3 and write, for instance, $\underline{n}_i(x^k, y^4)$.

and the distortion tensor $\widehat{\mathbf{Z}}^{\gamma}{}_{\alpha\beta}$ *is*

$$
\begin{aligned}
Z^a_{jk} &= -\widehat{C}^i_{jb} g_{ik} g^{ab} - \frac{1}{2}\Omega^a_{jk},\ Z^i_{bk} = \frac{1}{2}\Omega^c_{jk} g_{cb} g^{ji} - \Xi^{ih}_{jk}\ \widehat{C}^j_{hb},\ Z^a_{bk} = {}^{+}\Xi^{ab}_{cd}\ \widehat{T}^c_{kb},\ Z^i_{jk} = 0, \qquad (13)\\
Z^i_{kb} &= \frac{1}{2}\Omega^a_{jk} g_{cb} g^{ji} + \Xi^{ih}_{jk}\ \widehat{C}^j_{hb},\ Z^a_{jb} = -\ {}^{-}\Xi^{ad}_{cb}\ \widehat{T}^c_{jd},\ Z^a_{bc} = 0,\ Z^i_{ab} = -\frac{g^{ij}}{2}\left[\widehat{T}^c_{ja} g_{cb} + \widehat{T}^c_{jb} g_{ca}\right],
\end{aligned}
$$

for $\Xi^{ih}_{jk} = \frac{1}{2}(\delta^i_j \delta^h_k - g_{jk} g^{ih})$ *and* ${}^{\pm}\Xi^{ab}_{cd} = \frac{1}{2}(\delta^a_c \delta^b_d + g_{cd} g^{ab})$*. The nontrivial coefficients* Ω^a_{jk} *and* $\widehat{\mathbf{T}}^{\gamma}{}_{\alpha\beta}$ *are given, respectively, by* $W^{\alpha}_{\beta\gamma}$ *as formulas (5) and, see below, (14).*

Proof. It follows from a straightforward verification in N–adapted frames that the sums of coefficients (12) and (13) result in the coefficients of the Levi–Civita connection $\Gamma^{\gamma}{}_{\alpha\beta}$ for a general metric parametrized as a d–metric $\mathbf{g} = [g_{ij}, g_{ab}]$ (6). □

All geometric constructions and physical theories derived for geometric data $(\mathbf{g}, \nabla)$ can be equivalently modeled by geometric data $\left(\mathbf{g}, \mathbf{N}, \widehat{\mathbf{D}}\right)$ because of unique distortion relation (10).

Theorem 2.2 *The nonholonomically induced torsion* $\widehat{\mathcal{T}} = \{\widehat{\mathbf{T}}^{\gamma}{}_{\alpha\beta}\}$ *of* $\widehat{\mathbf{D}}$ *is determined in a unique form by the metric compatibility condition,* $\widehat{\mathbf{D}}\mathbf{g} = 0$*, and zero horizontal and vertical torsion coefficients,* $\widehat{T}^i{}_{jk} = 0$ *and* $\widehat{T}^a{}_{bc} = 0$*, but with nontrivial h–v– coefficients*

$$\widehat{T}^i{}_{jk} = \widehat{L}^i_{jk} - \widehat{L}^i_{kj}, \widehat{T}^i{}_{ja} = \widehat{C}^i_{jb}, \widehat{T}^a{}_{ji} = -\Omega^a{}_{ji}, \widehat{T}^c_{aj} = \widehat{L}^c_{aj} - e_a(N^c_j), \widehat{T}^a{}_{bc} = \widehat{C}^a_{bc} - \widehat{C}^a_{cb}. \qquad (14)$$

Proof. The coefficients (14) are computed by introducing $D = \widehat{\mathbf{D}}$, with coefficients (12), and $X = \mathbf{e}_\alpha, Y = \mathbf{e}_\beta$ (for N–adapted frames (3)) into standard formula for torsion, $\mathcal{T}(X,Y) := D_{\mathbf{X}}Y - D_{\mathbf{Y}}X - [X,Y]$. □

In a similar form, introducing $\widehat{\mathbf{D}}$ and $X = \mathbf{e}_\alpha, Y = \mathbf{e}_\beta, Z = \mathbf{e}_\gamma$ into $\mathcal{R}(X,Y) := D_{\mathbf{X}}D_{\mathbf{Y}} - D_{\mathbf{Y}}D_{\mathbf{X}} - D_{[\mathbf{X},\mathbf{Y}]}$, we prove

Theorem 2.3 *The curvature* $\widehat{\mathcal{R}} = \{\widehat{\mathbf{R}}^{\alpha}{}_{\beta\gamma\delta}\}$ *of* $\widehat{\mathbf{D}}$ *is characterized by N–adapted coefficients*

$$
\begin{aligned}
\widehat{R}^i{}_{hjk} &= e_k\widehat{L}^i{}_{hj} - e_j\widehat{L}^i{}_{hk} + \widehat{L}^m{}_{hj}\widehat{L}^i{}_{mk} - \widehat{L}^m{}_{hk}\widehat{L}^i{}_{mj} - \widehat{C}^i{}_{ha}\Omega^a{}_{kj},\\
\widehat{R}^a{}_{bjk} &= e_k\widehat{L}^a{}_{bj} - e_j\widehat{L}^a{}_{bk} + \widehat{L}^c{}_{bj}\widehat{L}^a{}_{ck} - \widehat{L}^c{}_{bk}\widehat{L}^a{}_{cj} - \widehat{C}^a{}_{bc}\Omega^c{}_{kj}, \qquad (15)\\
\widehat{R}^i{}_{jka} &= e_a\widehat{L}^i{}_{jk} - \widehat{D}_k\widehat{C}^i{}_{ja} + \widehat{C}^i{}_{jb}\widehat{T}^b{}_{ka}, \widehat{R}^c{}_{bka} = e_a\widehat{L}^c{}_{bk} - D_k\widehat{C}^c{}_{ba} + \widehat{C}^c{}_{bd}\widehat{T}^c{}_{ka},\\
\widehat{R}^i{}_{jbc} &= e_c\widehat{C}^i{}_{jb} - e_b\widehat{C}^i{}_{jc} + \widehat{C}^h{}_{jb}\widehat{C}^i{}_{hc} - \widehat{C}^h{}_{jc}\widehat{C}^i{}_{hb},\ \widehat{R}^a{}_{bcd} = e_d\widehat{C}^a{}_{bc} - e_c\widehat{C}^a{}_{bd} + \widehat{C}^e{}_{bc}\widehat{C}^a{}_{ed} - \widehat{C}^e{}_{bd}\widehat{C}^a{}_{ec}.
\end{aligned}
$$

We can re–define the differential geometry of a (pseudo) Riemannian space $\mathbf{V}$ in nonholonomic form in terms of geometric data $(\mathbf{g}, \widehat{\mathbf{D}})$, which is equivalent to the formulation with $(\mathbf{g}, \nabla)$.

Corollary 2.1 *The Ricci tensor* $\widehat{\mathbf{R}}_{\alpha\beta} := \widehat{\mathbf{R}}^{\gamma}{}_{\alpha\beta\gamma}$ *(19) of* $\widehat{\mathbf{D}}$ *is characterized by coefficients*

$$\widehat{\mathbf{R}}_{\alpha\beta} = \{\widehat{R}_{ij} := \widehat{R}^k{}_{ijk},\ \widehat{R}_{ia} := -\widehat{R}^k{}_{ika},\ \widehat{R}_{ai} := \widehat{R}^b{}_{aib},\ \widehat{R}_{ab} := \widehat{R}^c{}_{abc}\}. \qquad (16)$$

Proof. The formulas for h–v–components (16) are obtained by contracting, respectively, the coefficients (15). Using $\widehat{\mathbf{D}}$ (12), we express such formulas in terms of partial derivatives of coefficients of metric $\mathbf{g}$ (1) and any equivalent parametrization in the form (6), or (7). □

The scalar curvature ${}^s\widehat{R}$ of $\widehat{\mathbf{D}}$ is by definition

$${}^s\widehat{R} := \mathbf{g}^{\alpha\beta}\widehat{\mathbf{R}}_{\alpha\beta} = g^{ij}\widehat{R}_{ij} + g^{ab}\widehat{R}_{ab}. \qquad (17)$$

In order to elaborate models of gravity theories for ∇ and/or $\widehat{\mathbf{D}}$, we have to consider the corresponding Ricci tensors,

$$Ric = \{R_{\ \beta\gamma} := R^{\alpha}_{\ \beta\gamma\alpha}\}, \text{ for } \nabla = \{\Gamma^{\gamma}_{\ \alpha\beta}\}, \quad (18)$$

$$\text{and } \widehat{Ric} = \{\widehat{\mathbf{R}}_{\ \beta\gamma} := \widehat{\mathbf{R}}^{\alpha}_{\ \beta\gamma\alpha}\}, \text{ for } \widehat{\mathbf{D}} = \{\widehat{\mathbf{\Gamma}}^{\gamma}_{\ \alpha\beta}\}. \quad (19)$$

For instance, using (16) and (17), we can compute the Einstein tensor $\widehat{\mathbf{E}}_{\alpha\beta}$ of $\widehat{\mathbf{D}}$,

$$\widehat{\mathbf{E}}_{\alpha\beta} \doteqdot \widehat{\mathbf{R}}_{\alpha\beta} - \frac{1}{2}\mathbf{g}_{\alpha\beta}\ {}^{s}\widehat{R}. \quad (20)$$

In general, this tensor is different from a similar one constructed with (18) for the Levi–Civita connection ∇.

Proposition 2.1 *The N–adapted coefficients $\widehat{\mathbf{\Gamma}}^{\gamma}_{\ \alpha\beta}$ of $\widehat{\mathbf{D}}$ are identic to the coefficients $\Gamma^{\gamma}_{\ \alpha\beta}$ of ∇, both sets being computed with respect to N–adapted frames (3) and (4), if and only if the conditions $\widehat{L}^{c}_{aj} = e_a(N^c_j), \widehat{C}^{i}_{jb} = 0$ and $\Omega^{a}_{\ ji} = 0$ are satisfied.*

Proof. If the conditions of the Proposition, i.e., constraints (27), are satisfied, all N–adapted coefficients of the torsion $\widehat{\mathbf{T}}^{\gamma}_{\ \alpha\beta}$ (14) are zero. In such a case, the distortion tensor $\widehat{\mathbf{Z}}^{\gamma}_{\ \alpha\beta}$ (13) is also zero. Following formula (11), we get $\Gamma^{\gamma}_{\ \alpha\beta} = \widehat{\mathbf{\Gamma}}^{\gamma}_{\ \alpha\beta}$. Inversely, if the last equalities of coefficients are satisfied for a chosen splitting (2), we get trivial torsions and distortions of ∇. We emphasize that $\widehat{\mathbf{D}} \neq \nabla$ because such connections have different transformation rules under frame/coordinate transforms. Nevertheless, if $\Gamma^{\gamma}_{\ \alpha\beta} = \widehat{\mathbf{\Gamma}}^{\gamma}_{\ \alpha\beta}$ in a N–adapted frame of reference, we get corresponding equalities for the Riemann and Ricci tensors etc. This means that the N–coefficients are such way fixed via frame transforms that the nonholonomic distribution became integrable even, in general, the frames (3) and (4) are nonholonomic (because not all anholonomy coefficients are not obligatory zero, for instance, $w^{b}_{ia} = \partial_a N^b_i$ may be nontrivial. □

2.3. Field equations in nonholonomically modified gravity

We study modified gravity theories derived for the action

$$S = \frac{1}{16\pi}\int \delta u^4 \sqrt{|\mathbf{g}_{\alpha\beta}|}\left(f({}^{s}\widehat{R}, T) + {}^{m}L\right) \quad (21)$$

where ${}^{m}L$ is the matter Lagrangian density for which the stress–energy tensor of matter is defined via variation on inverse metric tensor as $\mathbf{T}_{\alpha\beta} = -\frac{2}{\sqrt{|\mathbf{g}_{\mu\nu}|}}\frac{\delta(\sqrt{|\mathbf{g}_{\mu\nu}|}\ {}^{m}L)}{\delta \mathbf{g}^{\alpha\beta}}$, trace $T := \mathbf{g}^{\alpha\beta}\mathbf{T}_{\alpha\beta}$, and $f({}^{s}\widehat{R}, T)$ is an arbitrary functional on ${}^{s}\widehat{R}$ (17) and T. For simplicity, we assume that the stress–energy tensor of the matter is given by

$$\mathbf{T}_{\alpha\beta} = (\rho + p)\mathbf{v}_{\alpha}\mathbf{v}_{\beta} - p\mathbf{g}_{\alpha\beta}, \quad (22)$$

where in the approximation of perfect fluid matter ρ is the energy density, p is the pressure and the four–velocity $\mathbf{v}_{\alpha}$ is subjected to the conditions $\mathbf{v}_{\alpha}\mathbf{v}^{\alpha} = 1$ and $\mathbf{v}^{\alpha}\widehat{\mathbf{D}}_{\beta}\mathbf{v}_{\alpha} = 0$, for ${}^{m}L = -p$ in a corresponding local frame. We also consider approximations for type

$$f({}^{s}\widehat{R}, T) = {}^{1}f({}^{s}\widehat{R}) + {}^{2}f(T) \quad (23)$$

and denote by ${}^{1}F({}^{s}\widehat{R}) := \partial\ {}^{1}f({}^{s}\widehat{R})/\partial\ {}^{s}\widehat{R}$ and ${}^{2}F(T) := \partial\ {}^{2}f(T)/\partial\ T$.

Theorem 2.4 *The gravitational field equations for a modified gravity model (21) with f–functional (23) and perfect fluid stress–energy tensor (22) can be re–written equivalently using the canonical d–connection* $\widehat{\mathbf{D}}$,

$$\widehat{\mathbf{R}}_{\ \beta\delta} - \frac{1}{2}\mathbf{g}_{\beta\delta}\ {}^{s}R = \mathbf{\Upsilon}_{\beta\delta}, \qquad (24)$$

where the source d–tensor $\mathbf{\Upsilon}_{\beta\delta}$ *is such way constructed that* $\mathbf{\Upsilon}_{\beta\delta} \to 8\pi G T_{\beta\delta}$ *for* $\widehat{\mathbf{D}} \to \nabla$, *where* $T_{\beta\delta}$ *is the energy–momentum tensor in GR with coupling gravitational constant G. In explicit form,*

$$\mathbf{\Upsilon}_{\beta\delta} = {}^{ef}\eta\ G\ \mathbf{T}_{\beta\delta} + {}^{ef}\mathbf{T}_{\beta\delta},$$

where the effective polarization of cosmological constant is computed ${}^{ef}\eta = [1 + {}^{2}F/8\pi]/\ {}^{1}F$ *and the f–modification of the energy–momentum tensor is computed as an additional effective source* ${}^{ef}\mathbf{T}_{\beta\delta} = [\frac{1}{2}(\ {}^{1}f - {}^{1}F\ {}^{s}\widehat{R} + 2p\ {}^{2}F + {}^{2}f)\mathbf{g}_{\beta\delta} - (\mathbf{g}_{\beta\delta}\ \widehat{\mathbf{D}}_{\alpha}\widehat{\mathbf{D}}^{\alpha} - \widehat{\mathbf{D}}_{\beta}\widehat{\mathbf{D}}_{\delta})\ {}^{1}F]/\ {}^{1}F$.

Proof. By varying the action S (21) with respect to $\mathbf{g}^{\alpha\beta}$ and following a N–adapted covariand differential calculus with respect bases (3) and (4) (see similar results for the Levi–Civita connection in [11]) we obtain the gravitational field equations (24) and respective effective sources. □

We consider matter field sources in (24) which can be diagonalized with respect to N–adapted frames,

$$\mathbf{\Upsilon}^{\beta}{}_{\delta} = diag[\mathbf{\Upsilon}_{\alpha} : \mathbf{\Upsilon}^{1}{}_{1} = \mathbf{\Upsilon}^{2}{}_{2} = \Upsilon(x^{k}, y^{3}) + \underline{\Upsilon}(x^{k}, y^{4}); \mathbf{\Upsilon}^{3}{}_{3} = \mathbf{\Upsilon}^{4}{}_{4} = {}^{v}\Upsilon(x^{k})]. \qquad (25)$$

Such a formal diagonalization can be performed via corresponding frame/ coordinate transforms for very general distributions of matter fields. Such effective sources can be considered as nonholonomic constraints on the Ricci tensor (see Theorem 3.1) computed for certain general assumptions on modified off–diagonal gravitational interactions.

Corollary 2.2 *The gravitational field equations (24) transform into the Einstein equations in GR, in "standard" form for* ∇,

$$E_{\beta\delta} = R_{\beta\delta} - \frac{1}{2}\mathbf{g}_{\beta\delta}\ R = \varkappa T_{\beta\delta}, \qquad (26)$$

where $R := \mathbf{g}^{\beta\delta}R_{\ \beta\delta}$, *if* ${}^{2}f = 0$, ${}^{1}f(\ {}^{s}\widehat{R}) = R$, *for the same N–adapted coefficients for* $\widehat{\mathbf{D}}$ *and* ∇ *if*

$$\widehat{L}^{c}_{aj} = e_{a}(N^{c}_{j}),\ \widehat{C}^{i}_{jb} = 0,\ \Omega^{a}{}_{ji} = 0. \qquad (27)$$

Proof. In general, the systems of PDEs (24) and (26) are very different. But if the constraints (27) are imposed additionally on $\widehat{\mathbf{D}}$, we satisfy the conditions of Proposition 2.1, when $\Gamma^{\gamma}{}_{\alpha\beta} = \widehat{\mathbf{\Gamma}}^{\gamma}{}_{\alpha\beta}$ results in $R_{\beta\delta} = \widehat{\mathbf{R}}_{\ \beta\delta}$ and $E_{\alpha\beta} = \widehat{\mathbf{E}}_{\alpha\beta}$. □

3. Decoupling of Gravitational Field Equations

The Einstein equations and their generalizations to various models of commutative and/or noncommutative gravity (for instance, in Finsler variables) possess a very important property that they decouple with respect to certain N–adapted frames of reference (3) and (4) and for the canonical d–connection $\widehat{\mathbf{D}}$. The aim of this section is to show how the anholonomic frame deformation method [20, 21, 22], AFDM, can be applied to decouple the fundamental field equations (24) for modified gravity.

3.1. Off–diagonal spacetimes with Killing symmetry

We use the ansatz (8) when $\omega = 1, \underline{h}_3 = 1, \underline{w}_i = 0$ and $\underline{n}_i = 0$ in data (9) and $\underline{\Upsilon} = 0$ for (25). Such a generic off–diagonal metric does not depend on variable y^4, i.e., $\partial/\partial y^4$ is a Killing vector, if $\underline{h}_4 = 1$. Nevertheless, the decoupling property can proven for the same assumptions but arbitrary $\underline{h}_4(x^k, y^4)$ with nontrivial dependence on y^4. We call this class of metrics to be with weak Killing symmetry because they result in systems of PDEs (24) as for the Killing case but there are differences in (27) if $\underline{h}_4 \neq 1$. It will be convenient to use brief denotations for partial derivatives, $a^\bullet = \partial a/\partial x^1, a' = \partial a/\partial x^2, a^* = \partial a/\partial y^3, a^\circ = \partial a/\partial y^4$. The equations will be written with respect to N–adapted frames of type (3) and (4).

Theorem 3.1 *The gravitational field equations (24) with possible constraints (27) for a metric* $\mathbf{g}$ *(9) with* $\omega = \underline{h}_3 = 1$ *and* $\underline{w}_i = \underline{n}_i = 0$ *and* $\underline{\Upsilon} = 0$ *in matter source* $\mathbf{\Upsilon}^{\beta}{}_{\delta}$ *(25) are equivalent, respectively, to*

$$\widehat{R}^1_1 = \widehat{R}^2_2 = \frac{-1}{2g_1g_2}[g_2^{\bullet\bullet} - \frac{g_1^\bullet g_2^\bullet}{2g_1} - \frac{(g_2^\bullet)^2}{2g_2} + g_1'' - \frac{g_1' g_2'}{2g_2} - \frac{(g_1')^2}{2g_1}] = -\,{}^v\Upsilon, \qquad (28)$$

$$\widehat{R}^3_3 = \widehat{R}^4_4 = -\frac{1}{2h_3h_4}[h_4^{**} - \frac{(h_4^*)^2}{2h_4} - \frac{h_3^* h_4^*}{2h_3}] = -\Upsilon, \qquad (29)$$

$$\widehat{R}_{3k} = \frac{w_k}{2h_4}[h_4^{**} - \frac{(h_4^*)^2}{2h_4} - \frac{h_3^* h_4^*}{2h_3}] + \frac{h_4^*}{4h_4}\left(\frac{\partial_k h_3}{h_3} + \frac{\partial_k h_4}{h_4}\right) - \frac{\partial_k h_4^*}{2h_4} = 0, \qquad (30)$$

$$\widehat{R}_{4k} = \frac{h_4}{2h_3}n_k^{**} + \left(\frac{h_4}{h_3}h_3^* - \frac{3}{2}h_4^*\right)\frac{n_k^*}{2h_3} = 0, \qquad (31)$$

$$\begin{aligned} \textit{and}\ w_i^* &= (\partial_i - w_i\partial_3)\ln|h_4|, \partial_k w_i = \partial_i w_k, \\ n_k \underline{h}_4^\circ &= \partial_k \underline{h}_4, n_i^* = 0, \partial_i n_k = \partial_k n_i. \end{aligned} \qquad (32)$$

Proof. See Appendix A. □

Let us discuss the decoupling (splitting) property of the modified and Einstein equations with respect to certain classes of N–adapted frames which is contained in the system of PDEs (28)–(32). For instance, the first equation is for a 2–d metric which always can be diagonalized, $[g_1, g_2]$, and/or made to be conformally flat. Prescribing a function g_1 and source ${}^v\Upsilon$, we can find g_2, or inversely. The equation (29) contains only the first and second derivatives on $\partial/\partial y^3$ and relates two functions h_3 and h_4. Prescribing one of such functions and source Υ, we can define the second one taking, respectively, one or two derivations on y^3. The equation (30) is a linear algebraic system for w_k if the coefficients h_a have been already defined as a solution of (29). Nevertheless, we have to solve a system of first order PDEs on x^k and y^3 in order to find w_k resulting in zero torsion conditions (32). Such conditions do not allow a complete decoupling because the first equations relate w_i to $H = \ln|h_4|$ via corresponding first order PDEs. Nevertheless, it is possible, for instance, to integrate such solutions for any prescribed H (see more details below, in Remark 4.1). The fourth equation (31) becomes trivial for any $n_i^* = 0$ if we want to satisfy completely such zero torsion conditions[7]. A nontrivial function $\underline{h}_4$ is explicitly present in the conditions (32). If such restrictions are satisfied, this allows us to eliminate $\underline{h}_4$ from (30).

We conclude that the modified equations for metrics with one Killing symmetry can be in such way parametrized with respect to N–adapted frames that they decouple and separate into "quite simple" PDEs for h–components, g_i, and then for v–components, h_a. The N–connection

[7] Nontrivial solutions and nonzero torsion configurations present interest in modified theories of gravity and brane physics, see such examples in Ref. [24].

coefficients also separate and can be defined from corresponding algebraic and/or first order PDE. The "zero torsion" conditions impose certain additional constraints (as some simple first order PDE with possible separation of variables) on N–coefficients and coefficients of v–metric.

In a similar form, we can decouple the modified gravitational field equations for spacetimes with one Killing symmetry on $\partial/\partial y_4$.

Corollary 3.1 *The equations (24) and (27) for a metric* $\mathbf{g}$ *(9) with* $\omega = h_4 = 1$ *and* $w_i = n_i = 0$ *and* $\boldsymbol{\Upsilon} = 0$ *in matter source* ${\boldsymbol{\Upsilon}^{\beta}}_{\delta}$ *(25) are equivalent, respectively, to*

$$
\begin{aligned}
\widehat{R}_1^1 &= \widehat{R}_2^2 = \frac{-1}{2g_1g_2}[g_2^{\bullet\bullet} - \frac{g_1^{\bullet}g_2^{\bullet}}{2g_1} - \frac{(g_2^{\bullet})^2}{2g_2} + g_1'' - \frac{g_1'g_2'}{2g_2} - \frac{(g_1')^2}{2g_1}] = -\ {}^{v}\Upsilon, && (33)\\
\widehat{R}_3^3 &= \widehat{R}_4^4 = -\frac{1}{2\underline{h}_3\underline{h}_4}[\underline{h}_3^{\circ\circ} - \frac{(\underline{h}_3^{\circ})^2}{2\underline{h}_3} - \frac{\underline{h}_3^{\circ}\underline{h}_4^{\circ}}{2\underline{h}_4}] = -\underline{\Upsilon}, && (34)\\
\widehat{R}_{3k} &= +\frac{\underline{h}_3}{2\underline{h}_4}\underline{w}_k^{\circ\circ} + \left(\frac{\underline{h}_3}{\underline{h}_4}\underline{h}_4^{\circ} - \frac{3}{2}\underline{h}_3^{\circ}\right)\frac{\underline{h}_k^{\circ}}{2\underline{h}_4} = 0, && (35)\\
\widehat{R}_{4k} &= \frac{\underline{n}_k}{2\underline{h}_3}[\underline{h}_3^{\circ\circ} - \frac{(\underline{h}_3^{\circ})^2}{2\underline{h}_3} - \frac{\underline{h}_3^{\circ}\underline{h}_4^{\circ}}{2\underline{h}_4}] + \frac{\underline{h}_3^{\circ}}{4\underline{h}_3}\left(\frac{\partial_k\underline{h}_3}{\underline{h}_3} + \frac{\partial_k\underline{h}_4}{\underline{h}_4}\right) - \frac{\partial_k h_3^{\circ}}{2h_3} = 0, && (36)
\end{aligned}
$$

$$
\begin{aligned}
\textit{and}\ \ \underline{n}_i^{\circ} &= (\partial_i - \underline{n}_i\partial_4)\ln|\underline{h}_3|, (\partial_k - \underline{n}_k\partial_4)\underline{n}_i = (\partial_i - \underline{n}_i\partial_4)\underline{n}_k, && (37)\\
\underline{w}_k h_3^{*} &= \partial_k h_3, \underline{w}_i^{\circ} = 0, \partial_i\underline{w}_k = \partial_k\underline{w}_i.
\end{aligned}
$$

Proof. It is similar to that for Theorem 3.1 provided in Appendix Appendix A. We do not repeat such computations. □

Using the above Theorem and Corollary, we can state:

Conclusion 3.1 *The nonlinear systems of PDEs corresponding to (modified) gravitational equations (24) and (27) for metrics* $\mathbf{g}$ *(9) with Killing symmetry on* $\partial/\partial y_4$*, when* $\omega = \underline{h}_3 = 1$ *and* $\underline{w}_i = \underline{n}_i = 0$ *and* $\underline{\boldsymbol{\Upsilon}} = 0$ *in matter source* ${\boldsymbol{\Upsilon}^{\beta}}_{\delta}$ *(25), can be transformed into respective systems of PDEs for data with Killing symmetry on* $\partial/\partial y_3$*, when* $\omega = h_4 = 1$ *and* $w_i = n_i = 0$ *and* $\boldsymbol{\Upsilon} = 0$*, if* $h_3(x^i, y^3) \rightarrow \underline{h}_4(x^i, y^4)$*,* $h_4(x^i, y^3) \rightarrow \underline{h}_3(x^i, y^4)$*,* $w_k(x^i, y^3) \rightarrow \underline{n}_k(x^i, y^4)$ *and* $n_k(x^i, y^3) \rightarrow \underline{w}_k(x^i, y^4)$*.*

The above presented method of nonholonomic deformations can be used for decoupling gravitational field equations if some metrics do not possess, in general, any Killing symmetries. The generic nonlinear character of such systems of PDEs does not allow us to use a principle of superposition of solutions. Nevertheless, certain classes of conformal transforms for the v–components of d–metrics and nonholonomic constraints of integral varieties give us the possibility to extend the anholonomic deformation method to "non–Killing" vacuum and nonvacuum gravitational interactions. In the next two subsections, we analyze two possibilities to decouple the effective Einstein equations for metrics with coefficients depending on all spacetime coordinates.

3.2. Preserving decoupling under v–conformal transforms

This property is stated by

Lemma 3.1 *The gravitational field equations (24) for geometric data (A.1), i.e., the system of PDEs (29)–(31), do not change under a "vertical" conformal transform with nontrivial* $\omega(x^k, y^a)$ *to a d–metric (9) if the following conditions are satisfied,*

$$\partial_k\omega - w_i\omega^{*} - n_i\omega^{\circ} = 0 \ \textit{and}\ \ \widehat{T}^a_{kb} = 0. \qquad (38)$$

Proof. It follows from straightforward computations when coefficients $g_i(x^k), g_3 = h_3(x^k, y^3), g_4 = h_4(x^k, y^3)\underline{h}_4(x^k, y^4), N_i^3 = w_i(x^k, y^3), N_i^4 = n_i(x^k, y^3)$ are generalized to a nontrivial $\omega(x^k, y^a)$ with ${}^\omega g_3 = \omega^2 h_3$ and ${}^\omega g_4 = \omega^2 h_4 \underline{h}_4$. Using, respectively, formulas (11), (12), (15) and (16), we get distortion relations for the Ricci d–tensors (19), ${}^\omega\widehat{R}^a{}_b = \widehat{R}^a{}_b + {}^\omega\widehat{Z}^a{}_b$ and ${}^\omega\widehat{R}_{bi} = \widehat{R}_{bi} = 0$, where $\widehat{R}^a{}_b$ and $\widehat{R}_{bi}$ are those computed for $\omega = 1$, i.e., (29)–(31). The values ${}^\omega\widehat{R}^a{}_b$ and ${}^\omega\widehat{Z}^a{}_b$ are defined by a nontrivial ω and computed using the same formulas. We do not provide certain similar details from Refs. [21, 22] for $\underline{h}_4 = 1$ because a nontrivial $\underline{h}_4$ does not modify substantially the proof that ${}^\omega\widehat{Z}^a{}_b = 0$ if the conditions (38) are satisfied. □

Using the Theorem 3.1, Corollary 3.1, Conclusion 3.1, Lemma 3.1, we prove

Theorem 3.2 *A d–metric*

$$\begin{aligned} \mathbf{g} &= g_i(x^k)dx^i \otimes dx^i + \omega^2(x^k, y^a)\left(h_3 \mathbf{e}^3 \otimes \mathbf{e}^3 + h_4\ \underline{h}_4 \mathbf{e}^4 \otimes \mathbf{e}^4\right), \\ \mathbf{e}^3 &= dy^3 + w_i(x^k, y^3)dx^i,\ \mathbf{e}^4 = dy^4 + n_i(x^k)dx^i, \end{aligned} \quad (39)$$

satisfying the PDEs (28)– (32) and $\partial_k\omega - w_i\omega^* - n_i\omega^\circ = 0$, *or a d–metric*

$$\begin{aligned} \mathbf{g} &= g_i(x^k)dx^i \otimes dx^i + \omega^2(x^k, y^a)\left(h_3\underline{h}_3 \mathbf{e}^3 \otimes \mathbf{e}^3 + \underline{h}_4 \mathbf{e}^4 \otimes \mathbf{e}^4\right), \\ \mathbf{e}^3 &= dy^3 + \underline{w}_i(x^k)dx^i,\ \mathbf{e}^4 = dy^4 + \underline{n}_i(x^k, y^4)dx^i, \end{aligned} \quad (40)$$

satisfying the PDEs (33)– (37) and $\partial_k\omega - \underline{w}_i\omega^* - \underline{n}_i\omega^\circ = 0$, *define, in general, two different classes of generic off–diagonal solutions of Einstein equations (24) and (27) with respective sources of type (25).*

Both ansatzes of type (39) and (40) consist particular cases of parametrizations of metrics in the form (8). Via frame/coordinate transform into a finite region of a point ${}^0u \in \mathbf{V}$ any spacetime metric in GR and f–modifications can be represented in an above mentioned d–metric form. If only one of coordinates y^a is timelike, the solutions of type (39) and (40) can not be transformed mutually via nonholonomic frame deformations preserving causality.

3.3. Decoupling with effective linearization of the Ricci tensors

The explicit form of field equations for vacuum and nonvacuum gravitational interactions depends on the type of frames and coordinate systems we consider for decoupling such PDEs. We can split such systems for more general parameterizations (than ansatzes (39) and (40)) in a form (8) with nontrivial ω and N–coefficients in (9). This is possible in any open region $U \subset \mathbf{V}$ where for computing the N–adapted coefficients of the Riemann and Ricci d–tensors, see formulas (15) and (16), we can neglect contributions from quadratic terms of type $\widehat{\Gamma}\cdot\widehat{\Gamma}$ but preserve values of type $\partial_\mu\widehat{\Gamma}$. For such constructions, we have to introduce a class of N–adapted normal coordinates when $\widehat{\Gamma}(u_0) = 0$ for points u_0, for instance, belonging to a line on U. Such conditions can be satisfied for decompositions of metrics and connections on a small parameter like it is explained in details in Ref. [20] (see decompositions on a small eccentricity parameter ε in Section 5). Other possibilities can be found if we impose nonholonomic constraints, for instance, of type $h_4^* = 0$ but for nonzero $h_4(x^k, y^3)$ and/or $h_4^{**}(x^k, y^3)$; such constraints can be solved in non–explicit form and define a corresponding subclass of N–adapted frames. Considering further nonholonomic deformations with a general decoupling with respect to a “convenient” system of reference/coordinates, we can deform the equations and solutions to configurations when terms of type $\widehat{\Gamma}\cdot\widehat{\Gamma}$ become important.

Theorem 3.3 ("Non–quadratic" decoupling) *The effective Einstein equations in modified gravity (for instance, in the form (24) and (27)), via nonholonomic frame deformations to a metric* $\mathbf{g}$ *(9) and matter source* $\mathbf{\Upsilon}^{\beta}{}_{\delta}$ *(25), when contributions from terms of type* $\widehat{\Gamma}\cdot\widehat{\Gamma}$ *are considered small for an open region* $U \subset \mathbf{V}$*, can be transformed equivalently into a system of PDEs with h–v–decoupling*

$$\widehat{R}^1_1 = \widehat{R}^2_2 = \frac{-1}{2g_1g_2}[g_2^{\bullet\bullet} - \frac{g_1^{\bullet}g_2^{\bullet}}{2g_1} - \frac{(g_2^{\bullet})^2}{2g_2} + g_1^{\prime\prime} - \frac{g_1^{\prime}g_2^{\prime}}{2g_2} - \frac{(g_1^{\prime})^2}{2g_1}] = -\,{}^v\Upsilon, \quad (41)$$

$$\widehat{R}^3_3 = \widehat{R}^4_4 = -\frac{1}{2h_3h_4}[h_4^{**} - \frac{(h_4^*)^2}{2h_4} - \frac{h_3^*h_4^*}{2h_3}] - \frac{1}{2\underline{h}_3\underline{h}_4}[\underline{h}_3^{\circ\circ} - \frac{(\underline{h}_3^{\circ})^2}{2\underline{h}_3} - \frac{\underline{h}_3^{\circ}\underline{h}_4^{\circ}}{2\underline{h}_4}] = -\Upsilon - \underline{\Upsilon}, \quad (42)$$

$$\widehat{R}_{3k} = \frac{w_k}{2h_4}[h_4^{**} - \frac{(h_4^*)^2}{2h_4} - \frac{h_3^*h_4^*}{2h_3}] + \frac{h_4^*}{4h_4}\left(\frac{\partial_k h_3}{h_3} + \frac{\partial_k h_4}{h_4}\right) - \frac{\partial_k h_4^*}{2h_4} + \frac{\underline{h}_3}{2\underline{h}_4}\underline{n}_k^{\circ\circ} + \left(\frac{\underline{h}_3}{\underline{h}_4}\underline{h}_4^{\circ} - \frac{3}{2}\underline{h}_3^{\circ}\right)\frac{\underline{n}_k^{\circ}}{2\underline{h}_4} = 0, \quad (43)$$

$$\widehat{R}_{4k} = \frac{\underline{w}_k}{2\underline{h}_3}[\underline{h}_3^{\circ\circ} - \frac{(\underline{h}_3^{\circ})^2}{2\underline{h}_3} - \frac{\underline{h}_3^{\circ}\underline{h}_4^{\circ}}{2\underline{h}_4}] + \frac{\underline{h}_3^{\circ}}{4\underline{h}_3}\left(\frac{\partial_k \underline{h}_3}{\underline{h}_3} + \frac{\partial_k \underline{h}_4}{\underline{h}_4}\right) - \frac{\partial_k \underline{h}_3^{\circ}}{2\underline{h}_3} + \frac{h_4}{2h_3}n_k^{**} + \left(\frac{h_4}{h_3}h_3^* - \frac{3}{2}h_4^*\right)\frac{n_k^*}{2h_3} = 0, \quad (44)$$

$$w_i^* = (\partial_i - w_i\partial_3)\ln|h_4|, (\partial_k - w_k\partial_3)w_i = (\partial_i - w_i\partial_3)w_k, n_i^* = 0, \partial_i n_k = \partial_k n_i,$$
$$\underline{w}_i^{\circ} = 0,\ \partial_i\underline{w}_k = \partial_k\underline{w}_i, \underline{n}_i^{\circ} = (\partial_i - \underline{n}_i\partial_4)\ln|\underline{h}_3|, (\partial_k - \underline{n}_k\partial_4)\underline{n}_i = (\partial_i - \underline{n}_i\partial_4)\underline{n}_k,$$
$$\mathbf{e}_k\omega = \partial_k\omega - (w_i + \underline{w}_i)\,\omega^* - (n_i + \underline{n}_i)\,\omega^{\circ} = 0. \quad (45)$$

Proof. It is a consequence of Conclusion 3.1 and Theorems 3.1 and 3.2 for superpositions of ansatzes (39) and (40) resulting into (8). If we repeat the computations from Appendix Appendix A for geometric data (9) considering that contributions of type $\widehat{\Gamma}\cdot\widehat{\Gamma}$ are small, we can see that the equations (42)–(44) are derived to be, respectively, equivalent to sums of (29)–(31) and (34)–(36). The torsionless conditions (45) consist a sum of similar conditions (32) and (37). □

In general, the solutions defined by a system (41)–(45) can not be transformed into solutions parameterized by an ansatz (39) and/or (40). As we shall prove in Section 4, the general solutions of the such systems of PDEs are determined by corresponding sets of generating and integration functions. A solution for (41)–(45) contains a larger set of $h-v$–generating functions than those with some N–coefficients stated to be zero.

4. Generic Off–Diagonal Solutions

The goal of this section is to show how decoupling effective Einstein equations we can integrate such PDEs in very general forms depending on the properties of the coefficients of an ansatz for metrics. Some similar theorems for 4–d, 5–d and higher dimension modifications of GR were proven in Refs [21, 22, 20]. In this work, we generalize those results for f–modifications.

4.1. Generating solutions with weak one Killing symmetry

We prove that the modified gravitational equations encoding gravitational and generic off–diagonal nonlinear interactions and satisfying the conditions of Theorem 3.1 can be integrated in general forms for $h_a^* \neq 0$ and certain special cases with zero and non–zero sources (25). In general, such generic off–diagonal metrics are determined by generating functions depending on three/four coordinates. The bulk of known exact solutions with diagonalizable metrics and coefficients depending on two coordinates (in certain special frames of references) can be included as special cases for more general nonholonomic configurations.

4.1.1. (Non) vacuum metrics with $h_a^* \neq 0$ For ansatz (8) with data $\omega = 1, \underline{h}_3 = 1, \underline{w}_i = 0$ and $\underline{n}_i = 0$ for (9), when $h_a^* \neq 0$, and the condition that the source

$$\Upsilon^\beta{}_\delta = diag[\Upsilon_\alpha : \Upsilon^1{}_1 = \Upsilon^2{}_2 = \Upsilon(x^k, y^3); \Upsilon^3{}_3 = \Upsilon^4{}_4 = {}^v\Upsilon(x^k)], \quad (46)$$

is not zero, the solutions of gravitational field equations can be constructed as follows

Theorem 4.1 *The system (28)–(31) with source (46) can be integrated in general forms by*

$$\begin{aligned} \mathbf{g} &= \epsilon_i e^{\psi(x^k)} dx^i \otimes dx^i + h_3(x^k, y^3)\mathbf{e}^3 \otimes \mathbf{e}^3 + h_4(x^k, y^3)\underline{h}_4(x^k, y^4)\mathbf{e}^4 \otimes \mathbf{e}^4, \\ \mathbf{e}^3 &= dy^3 + w_i(x^k, y^3)dx^i, \ \mathbf{e}^4 = dy^4 + n_i(x^k)dx^i, \quad (47) \end{aligned}$$

with coefficients determined by generating functions $\psi(x^k), \phi(x^k, y^3)$, $\phi^* \neq 0$, $n_i(x^k)$ and $\underline{h}_4(x^k, y^4)$, and integration functions ${}^0\phi(x^k)$ following recurrent formulas and conditions

$$\begin{aligned} \epsilon_1 \psi^{\bullet\bullet} + \epsilon_2 \psi'' &= 2 \ {}^v\Upsilon; \quad (48) \\ h_4 &= \pm\frac{1}{4}\int |\Upsilon|^{-1}\left(e^{2\phi}\right)^* dy^3, \ or \ = \pm\frac{1}{4\Lambda}e^{2[\phi - {}^0\phi]}, \ if\ \Upsilon = \Lambda = const; \quad (49) \\ h_3 &= \pm\left[\left(\sqrt{|h_4|}\right)^*\right]^2 e^{-2\phi} = \frac{\phi^*}{2|\Upsilon|}\left(\ln|h_4|\right)^* \ or \ = \pm\frac{(\phi^*)^2}{4\Lambda}, \ if\ \Upsilon = \Lambda = const; \\ w_i &= -\partial_i\phi/\phi^*, \quad (50) \end{aligned}$$

where constraints

$$w_i^* = (\partial_i - w_i\partial_3)\ln|h_4|, \partial_k w_i = \partial_i w_k, \ n_k \underline{h}_4^\circ = \partial_k \underline{h}_4, \partial_i n_k = \partial_k n_i. \quad (51)$$

must be imposed in order to satisfy the zero torsion conditions (32); we should take respective values $\epsilon_i = \pm 1$ and $\pm$ in (49) and/or (50) if we want to fix a necessary spacetime signature.

Proof. We sketch a proof which transforms into similar ones in [21, 22, 20] if $\underline{h}_4 = 1$.

- A horizontal metric $g_i(x^2)$ is for 2–d and can be always represented in a conformally flat form $\epsilon_i e^{\psi(x^k)} dx^i \otimes dx^i$. For such an h–metric, the equation (28) is a 2-d Laplace/wave equation (48) which can be solved exactly if a source ${}^v\Upsilon(x^k) - 4s^2$ is prescribed from certain physical conditions.
- If $h_4^* \neq 0$, we can define nontrivial functions

$$\phi = \ln\left|\frac{h_4^*}{\sqrt{|h_3 h_4|}}\right|, \ \gamma := \left(\ln\frac{|h_4|^{3/2}}{|h_3|}\right)^*, \quad \alpha_i = h_4^*\partial_i\phi, \ \beta = h_4^*\phi^* \quad (52)$$

for a function $\phi(x^k, y^3)$. If $\phi^* \neq 0$, we can write, respectively, the equations (29)–(31) in the forms,

$$\begin{aligned} \phi^* h_4^* &= 2h_3 h_4 \Upsilon. \quad (53) \\ \beta w_i + \alpha_i &= 0, \quad (54) \\ n_i^{**} + \gamma n_i^* &= 0. \end{aligned}$$

For the last equation, we must take any trivial solution given by functions $n_i(x^k)$ satisfying the conditions $\partial_i n_j = \partial_j n_i$ in order to solve the constraints (32). Using coefficients (52) with $\alpha_i \neq 0$ and $\beta \neq 0$, we can always express w_i via derivatives of ϕ, i.e., in the form (50).

We can choose any $\phi(x^k, y^3)$ with $\phi^* \neq 0$ as a generating function and express h_4 and, after that, h_3 as some integrals/derivatives of functions depending on ϕ and source $\Upsilon(x^k, y^3)$, see corresponding formulas (49) and (50). The integrals can be computed in a general explicit form if $\Upsilon(x^k, y^3) = \Lambda = const$, when possible matter field interactions and f–modifications are approximated by an energy–momentum tensor as a cosmological constant,

$$\mathbf{\Upsilon}^{\beta}_{\ \delta} = \delta^{\beta}_{\ \delta}\Upsilon = \delta^{\beta}_{\ \delta}\Lambda. \tag{55}$$

- A possible dependence on y^4 is present in function $\underline{h}_4$ which must satisfy conditions of type (A.6) in order to be compatible with (32). It is not possible to write, in explicit form, the solutions for the zero torsion condition if the source $\mathbf{\Upsilon}^{\beta}_{\ \delta}$ is parameterized by arbitrary functions. Nevertheless, if $\mathbf{\Upsilon}^{\beta}_{\ \delta}$ is of type (55), we get $h_4 \sim e^{2\phi}$ and $w_i \sim \partial_i\phi/\phi^*$ positively solve the constraints

$$w_i^* = (\partial_i - w_i\partial_3)\ln|h_4| \text{ and } \partial_i w_j = \partial_j w_i, \tag{56}$$

transformed into $\phi^{**}\partial_i\phi - \phi^*(\partial_i\phi)^* = 0$. By straightforward computation, we can check that (32) are satisfied by (51) when $n_i^* = 0$ and w_i is determined by (50). □

The solutions constructed in Theorem 4.1, and those which can be derived following Corollary 3.1 are very general ones and contain as particular cases (perhaps) all known exact solutions for (non) holonomic effective Einstein spaces with Killing symmetries. They also can be generalized to include arbitrary finite sets of parameters as it is proven in Ref. [20].

Remark 4.1 ***(–Example)*** *Introducing $H = -\ln|h_4|$, the system of equations (53), (54) and (56) can be written in the form*

$$\begin{aligned} h_3 &= -\phi^*\ H^*/2\Upsilon_2, && (57)\\ \phi^* w_i + \partial_i\phi &= 0, && (58)\\ w_i^* + \partial_i H - w_i H^* &= 0. && (59) \end{aligned}$$

Considering $H(x^k, y^3)$ as a generating function, we can integrate (59) in explicit form using parameterizations $w_i = {}_1w_i(x^k, y^3) \times {}_2w_i(x^k, y^3)$ (in this formula, we do not consider summation on i). We choose $\phi(x^k, y^3)$ as a nontrivial solution of the system of first order PDEs (58) for any found w_i. So, we can define h_3 for any prescribed source Υ_2 in (57). We conclude that the LC–conditions (32) (relating w_i to h_4, which does not allow a "complete" decoupling) can be solved also in very general forms by further fixing of parameterizations, classes of functions and boundary conditions for H, ϕ and Υ_2.

Nevertheless, we note that an explicit example of exact solutions can be derived by taking derivative on $*$ for (58) and using a subclass of functions ϕ when $(\partial_i\phi)^* = \partial_i(\phi^*)$. We obtain $w_i\frac{\phi^{**}}{\phi^*} + w_i^* + \frac{\partial_i\phi^*}{\phi^*} = 0$. This system is equivalent/compatible to (59) if, for instance, $\Phi := \ln|\phi^*|$, $H^* = -\Phi^*$, $\partial_i H = \partial_i\Phi$. Such equations are satisfied by any functions $H({}^-\varsigma)$ and $\Phi({}^+\varsigma)$, where ${}^{\pm}\varsigma := \pm v - x^1 - x^2$.

4.1.2. Effective modified vacuum gravitational configurations We can consider a subclass of generic off–diagonal modified gravitation interactions which can be encoded as effective Einstein manifolds. In general, such classes of solutions do not have a smooth limit from non-vacuum to vacuum models.

Corollary 4.1 *The effective vacuum solutions for the modified gravitational equations with ansatz for metrics of type (47) with vanishing source (46) are parametrized in the form*

$$\begin{aligned} \mathbf{g} &= \epsilon_i e^{\psi(x^k)} dx^i \otimes dx^i + h_3(x^k, y^3)\mathbf{e}^3 \otimes \mathbf{e}^3 + h_4(x^k, y^3)\underline{h}_4(x^k, y^4)\mathbf{e}^4 \otimes \mathbf{e}^4, \\ \mathbf{e}^3 &= dy^3 + w_i(x^k, y^3)dx^i, \ \mathbf{e}^4 = dy^4 + n_i(x^k)dx^i, \end{aligned} \tag{60}$$

where coefficients are defined by solutions of the system

$$\ddot{\psi} + \psi'' = 0, \tag{61}$$

$$\phi^* \, h_4^* = 0, \tag{62}$$

$$\beta w_i + \alpha_i = 0, \tag{63}$$

and with coefficients computed following formulas (52) for nonzero ϕ^* and h_4^* with possible further zero limit; such coefficients and h_3 and $\underline{h}_4$ are subjected, additionally, to the zero–torsion conditions (51).

Proof. Considering the system of equations (28)– (31) with zero right sides, we obtain, respectively, the equations (61)– (63). For positive signatures on h–subspace and equation (61), we can take $\psi = 0$, or consider a trivial 2-d wave equation if one of coordinates x^k is timelike. There are two possibilities to satisfy the condition (62). The first one is to consider that $h_4 = h_4(x^k)$, i.e., h_4^* which states that the equation (62) has solutions with zero source for arbitrary function $h_3(x^k, y^3)$ and arbitrary N–coefficients $w_i(x^k, y^3)$ as follows from (52). For such vacuum configurations, the functions h_3 and w_i can be taken as generation ones which should be constrained only by the conditions (51). Equations of type (56) constrain substantially the class of admissible w_i if h_4 depends only on x^k. Nevertheless, h_3 can be an arbitrary one generating solutions which can be extended for nontrivial sources $\underline{\Upsilon}$ and systems (33)– (37) and/or (41)– (45).

A different class of solutions can be generated if we state, after corresponding coordinate transforms, $\phi = \ln \left| h_4^* / \sqrt{|h_3 h_4|} \right| = {}^0\phi = const$, $\phi^* = 0$. For such configurations, we can consider $h_4^* \neq 0$, and solve (62) as

$$\sqrt{|h_3|} = {}^0h(\sqrt{|h_4|})^*, \tag{64}$$

for ${}^0h = const \neq 0$. Such v–metrics are generated by any $f(x^i, y^3)$, $f^* \neq 0$, when

$$h_4 = -f^2\left(x^i, y^3\right) \text{ and } h_3 = ({}^0h)^2 \left[f^*\left(x^i, y^3\right)\right]^2, \tag{65}$$

where the signs are such way fixed that for $N_i^a \to 0$ we obtain diagonal metrics with signature $(+,+,+,-)$. The coefficients $\alpha_i = \beta = 0$ in (63) and $w_i(x^k, y^3)$ can be any functions solving (51). This is equivalent to

$$\begin{aligned} w_i^* &= 2\partial_i \ln|f| - 2w_i(\ln|f|)^*, \\ \partial_k w_i - \partial_i w_k &= 2(w_k\partial_i - w_i\partial_k)\ln|f|, \end{aligned} \tag{66}$$

for any $n_i(x^k)$ when $\partial_i n_k = \partial_k n_i$. Constraints of type $n_k \underline{h}_4^{\circ} = \partial_k \underline{h}_4$ (A.6) have to be imposed for a nontrivial multiple $\underline{h}_4$. □

Using Corollary 3.1, the ansatz (60) can be "dualized" to generate effective vacuum solutions with weak Killing symmetry on $\partial/\partial y^3$. Finally, we note that the signature of the generic off–diagonal metrics generated in this subsection depend on the fact which coordinate x^1, x^2, y^3 or y^4 is chosen to be a timelike one.

4.2. Non–Killing effective Einstein configurations

The Theorem 3.2 can be applied for constructing non–vacuum and effective vacuum solutions of the modified gravitational equations depending on all coordinates without explicit Killing symmetries.

4.2.1. Non–vacuum off–diagonal solutions We can generate such effective Einstein manifolds as follows

Corollary 4.2 *An ansatz of type (39) with d–metric*

$$\begin{aligned} \mathbf{g} &= \epsilon_i e^{\psi} dx^i \otimes dx^i + \omega^2[\pm \frac{(\phi^*)^2}{4\Lambda}\mathbf{e}^3 \otimes \mathbf{e}^3 \pm \frac{1}{4\Lambda} e^{2[\phi - {}^0\phi]} \underline{h}_4 \mathbf{e}^4 \otimes \mathbf{e}^4], \\ \mathbf{e}^3 &= dy^3 - (\partial_i \phi/\phi^*)dx^i, \ \mathbf{e}^4 = dy^4 + n_i(x^k)dx^i, \end{aligned}$$

where the coefficients are subjected to conditions (48)–(51) and $\partial_k\omega + (\partial_i\phi/\phi^*)\omega^* - n_i\omega^\circ = 0$ *defines solutions of the Einstein equations* $R_{\alpha\beta} = \Lambda g_{\alpha\beta}$ *with nonholonomic interactions and modifications encoded effectively into the vacuum structure of GR with nontrivial cosmological constant,* $\Lambda \neq 0$.

Proof. It is a consequence of Theorem 3.2 and Corolarry 4.1. □

In a similar form, we can generate solutions of type (40) when the conformal factor is a solution of $\partial_k\omega - \underline{w}_i(x^k)\omega^* + (\partial_i\underline{\phi}/\underline{\phi}^\circ)\omega^\circ = 0$ with respective "dual" generating functions ω and $\underline{\phi}$ when the data (48)–(51) are re–defined for solutions with weak Killing symmetry on $\partial/\partial y^3$.

4.2.2. Effective vacuum off–diagonal solutions Vacuum Einstein spaces encoding nonholonomic interactions and f–modifications can be constructed using

Corollary 4.3 *An ansatz of type (39) with d–metric*

$$\begin{aligned} \mathbf{g} &= \epsilon_i e^{\psi(x^k)} dx^i \otimes dx^i + \omega^2(x^k, y^a)[({}^0h)^2 \ [f^*(x^i, y^3)]^2 \mathbf{e}^3 \otimes \mathbf{e}^3 - f^2(x^i, y^3)\underline{h}_4(x^k, y^4)\mathbf{e}^4 \otimes \mathbf{e}^4], \\ \mathbf{e}^3 &= dy^3 + w_i(x^k, y^3)dx^i, \ \mathbf{e}^4 = dy^4 + n_i(x^k)dx^i, \end{aligned}$$

where the coefficients are subjected to conditions (64)–(66), (51) and $\partial_k\omega - w_i\omega^* - n_i\omega^\circ = 0$ *defines generic off–diagonal solutions of* $R_{\alpha\beta} = 0$.

Proof. It is a consequence of Theorem 3.2 and Corollary 4.1. □

Solutions of type (40) can be defined if the conformal factor is a solution of $\partial_k\omega - \underline{w}_i(x^k)\omega^* - \underline{n}_i(x^k, y^4)\omega^\circ = 0$ with respective "dual" generating functions $\omega(x^k, y^a)$ and $\underline{\phi}(x^k, y^4)$ when the data (64)–(66) and (51) are re–defined for ansatz with weak Killing symmetry on $\partial/\partial y^3$.

Summarizing the results of this section, we state

Claim 4.1 *All generic off–diagonal non–vacuum and effective vacuum solutions of the modified gravitational equations determined, respectively, by Theorem 4.1 and Corollaries 4.1, 4.2 and 4.3 can be generalized to metrics of type (8). For such constructions, there are used nonlinear superposition of metrics and their "duals" on v–coordinates in order to define non–Killing solutions of respective systems of PDEs from Theorems 3.2 and/or 3.3*

The statements of this Claim are formulated following our experience on constructing generic off–diagonal vacuum and non–vacuum solutions. For such nonlinear systems, it is not possible to formulate certain general uniqueness and exhaustive criteria. Sure, not all solutions of modified gravitational equations can be constructed in such forms, or related to any sets of prescribed solutions via "nondegernerate" nonholonomic deformations. The length of this paper does not

Journal of Physics: Conference Series **453** (2013) 012021 doi:10.1088/1742-6596/453/1/012021

allow us to present all technical details and general formulas for coefficients for ansatzes for d–metrics[8] which can be constructed following this Claim. Explicit examples supporting our approach are given in sections 5 and 6.

5. The Cauchy Problem and Decoupling

The gravitational interactions in modified gravity studied in this work are described by off–diagonal solutions of

$$R_{\alpha\beta} = \Lambda g_{\alpha\beta}, \tag{67}$$

which can be found in very general forms with respect to N–adapted frames for certain nonintegrable spacetime $2+2$ splitting of type $\mathbf{N}$ (2). An effective cosmological constant Λ encodes a gravitational "vacuum" cosmological constant Λ in GR and f–modifications. Pseudo–Riemanian manifolds with metrics $g_{\alpha\beta}$ adapted to chosen non–integrable distribution with $2+2$ splitting and satisfying (67) are called *nonholonomic effective Einstein manifolds*. Hereafter, we shall refer to such systems of PDEs as *nonholonomic effective vacuum spacetimes*, regardless of whether, or not, an (effective) cosmological constant vanishes or can be "polarized" by gravitational and/or matter field interactions into some N–adapted diagonal sources admitting formal integration of gravitational field equations.

The equations (67), and their N–adapted equivalents (24)–(27), constitute a second–order system quasi–linear PDEs for the coefficients of spacetime metric $\mathbf{g} = \{g_{\alpha\beta}\}$. This means that given a manifold $\mathbf{V}$ of necessary smooth class such a quasi–linear system is linear in the second derivatives of the metric and quadratic in the first derivatives $\partial_\gamma g_{\alpha\beta}$ (the coefficients of such PDEs are rational functions of $g_{\alpha\beta}$). To be able to decouple and formally integrate such systems is necessary to consider special classes of nonholonomic frames and constraints. For GR, this type of equations do not fall in any of the standard cases of hyperbolic, parabolic, or elliptic systems which typically lead to unique solutions. It is important to formulate criteria when such general solutions would be unique ones with a topology and differential structure determined by some initial data. How the diffeomorphism, or coordinate, invariance and arbitrary frame transforms (principle of relativity) would be taken into account for N–splitting?

In mathematical relativity [5, 14, 8, 7, 9], it was proven a fundamental result (due to Choquet–Bruhat, 1952) that there exists a set of hyperbolic equations underlying (67). The goal of this section is to study the evolution (Cauchy) problem for the system (24)–(27) in N–adapted form and preserving the decoupling property.

5.1. The local N–adapted evolution problem

In the evolutionary approach, the topology of spacetime manifold is chosen in the form $\mathbf{V} = \mathbb{R} \times {}^3V$, where ${}^3V$ is a 3–d manifold carrying initial data. It should be noted here that there is no a priori known natural time–coordinate even we may fix a signature for metric and chose certain coordinates to be time or space like ones.

Definition 5.1 *A set of coordinates $\{\widehat{u}^\mu = (\widehat{x}^i, \widehat{y}^a)\}$ is canonically N–harmonic, i.e., it is both harmonic and adapted to a splitting* $\mathbf{N}$ *(2), if each of the functions $\widehat{u}^\mu$ satisfies the wave equation*

$$\widehat{\Box}\widehat{u}^\mu = 0, \tag{68}$$

where the canonical d'Alembert operator $\widehat{\Box} := \widehat{\mathbf{D}}_\alpha\, \widehat{\mathbf{D}}^\alpha$ acts on a scalar $f(x,y)$ in the form

$$\widehat{\Box} f := (\sqrt{|\mathbf{g}_{\alpha\beta}|})^{-1}\mathbf{e}_\mu \left(\sqrt{|\mathbf{g}_{\alpha\beta}|}\mathbf{g}^{\mu\nu}\mathbf{e}_\nu f\right) = (\sqrt{|g_{kl}|})^{-1}\mathbf{e}_i \left(\sqrt{|g_{kl}|}g^{ij}\mathbf{e}_j f\right) + (\sqrt{|g_{cd}|})^{-1}\partial_a \left(\sqrt{|g_{kl}|}g^{ab}e_b f\right),$$

[8] Such formulas are, for instance, of type (48)–(51), with functions $h_3(.,y^3)$ and $h_4(.,y^3)$ and further "dualization" $h_3 \to \underline{h}_4(.,y^4)$ and $h_4 \to \underline{h}_3(.,y^4)$ with corresponding re–definition of N–connection coefficients.

for a d–metric $\mathbf{g}_{\alpha\beta} = (g_{ij}, g_{ab})$ *(6) defined with respect to N–adapted (co) frames (3)–(4) and canonical d–connection* $\widehat{\mathbf{D}}_{\alpha}$ *(12).*

We can say that such coordinates $\widehat{u}^{\mu} = (\widehat{x}^{i}, \widehat{y}^{a})$ are N–adapted wave–coordinates.

Lemma 5.1 *In canonical N–harmonic coordinates, the effective Einstein equations (67) re–defined in canonical d–connection variables (24) can be written, equivalently,*

$$\widehat{\mathbf{E}}^{\alpha\beta} = \widehat{\Box}\mathbf{g}^{\alpha\beta} - \mathbf{g}^{\tau\theta}\left[(\mathbf{g}^{\alpha\mu}\widehat{\mathbf{\Gamma}}^{\beta}{}_{\mu\nu} + \mathbf{g}^{\alpha\mu}\widehat{\mathbf{\Gamma}}^{\beta}{}_{\mu\nu})\widehat{\mathbf{\Gamma}}^{\nu}{}_{\tau\theta} + 2\mathbf{g}^{\gamma\mu}\widehat{\mathbf{\Gamma}}^{\alpha}{}_{\mu\theta}\widehat{\mathbf{\Gamma}}^{\beta}{}_{\tau\gamma}\right] - 2\Lambda\mathbf{g}^{\alpha\beta} = 0, \qquad (69)$$

i.e., such PDEs, for $\mathbf{g}^{\alpha\beta}$ *(using algebraic transforms, for* $\mathbf{g}_{\alpha\beta}$*) form a system of second–order quasi–linear N–adapted wave–type equations.*

Proof. It is a standard computation with respect to N–adapted frames by using formulas (12), (20) and (16). If the zero torsion conditions (27) are imposed, we get well known results from GR but (in our case) adapted to 2 + 2 non–integrable splitting. □

This Lemma allows us to apply the standard theory of hyperbolic PDEs (see, for instance, [10]). Let us denote by H^{k}_{loc} the Sobolev spaces of functions which are in $L^2(K)$ for any compact set K when their distributional derivatives are considered up to an integer order k also in $L^2(K)$. We shall also use N–adapted wave coordinates with additional formal $3+1$ splitting, for instance, in a form $\widehat{u}^{\mu} = ({}^{t}\widehat{u}, \widehat{u}^{\bar{i}})$, where ${}^{t}\widehat{u}$ is used for the timelike coordinate and $\widehat{u}^{\bar{i}}$ are for 3 spacelike coordinates. "Hats" can be eliminated if such a splitting is considered for arbitrary local coordinates. Standard results from the theory of PDEs give rise to this

Theorem 5.1 *The field equations (24) for nonholonomic effective Einstein manifolds have a unique solution* $\mathbf{g}^{\alpha\beta}$ *defined by PDEs (69) stated on an open neighborhood* $\mathcal{U} \subset \mathbb{R} \times \mathbb{R}^3$ *of* $\mathcal{O} \subset \{0\} \times \mathbb{R}^3$ *with any initial data*

$$\mathbf{g}^{\alpha\beta}(0, \widehat{u}^{\bar{i}}) \in H^{k+1}_{loc} \quad and \quad \frac{\partial \mathbf{g}^{\alpha\beta}(0, \widehat{u}^{\bar{i}})}{\partial({}^{t}\widehat{u})} \in H^{k+1}_{loc}, k > 3/2. \qquad (70)$$

The set $\mathcal{U}$ *can be chosen in a form that* $(\mathcal{U}, \mathbf{g}^{\alpha\beta})$ *is globally hyperbolic with Cauchy surface* $\mathcal{O}$[9].

There is no reason that a solution constructed using the anholonomic deformation method as we considered in section 3 will satisfy the wave conditions (68), even if we give certain initial data for equation (69). In order to establish an hyperbolic (and evolutionary) form of such nonholonomic vacuum gravitational equations we should re–define the Choquet–Bruhat problem (see details and references in [5, 14, 8, 18, 9]) with respect to N–adapted frames. Using a 3+1 splitting of N–adapted coordinates, we wirte $\mathbf{g}^{\alpha\beta} = (\mathbf{g}^{t\beta}, \mathbf{g}^{\bar{i}\beta})$ and $\mathbf{e}_{\alpha} = (\mathbf{e}_{t}, \mathbf{e}_{\bar{i}})$ and consider the d–vector field $n^{\mu}(x,y)$ of unit timelike normals to the hypersurface $\{{}^{t}\widehat{u} = 0\}$[10]. There is no loss of generality if we assume that on such a hypersurface we have

$$\mathbf{g}^{tt} = -1 \text{ and } \mathbf{g}^{t\bar{i}} = 0. \qquad (71)$$

We can state such conditions via additional N–adapted frame/coordinate transform for any d–metric (6) with prescribed signature. It is also possible to re–define the generating functions for

[9] Choosing corresponding classes of nonholonomic distributions **N** (2), we can relax the conditions of differentiability as in Refs. [15, 19] (we omit such constructions in this work).

[10] We do not use labels for coordinates like $0,1,2,3$ because the decoupling property of the Einstein equations and general solutions can be proven for arbitrary signature, for instance $(-+++)$, $(++-+)$, etc. and for any set of local coordinates u^{α} with $\alpha = 1,2,3,4$.

any class of off–diagonal solutions with decoupling of (modified) Einstein equations in order to satisfy (71).

Another necessary condition for the vanishing of a $\widehat{\square}u^{\mu}$ is that this value is stated zero at ${}^{t}\widehat{u}=0$. This allows us to express in N–adapted form the timelike derivatives through N–elongated space type ones,

$$\mathbf{e}_t\left(\sqrt{|\mathbf{g}_{\alpha\beta}|}\mathbf{g}^{t\overline{i}}\right)=-\mathbf{e}_{\overline{i}}\left(\sqrt{|\mathbf{g}_{\alpha\beta}|}\mathbf{g}^{t\overline{i}}\right) \quad (72)$$

with N–elongated operators. So, the initial data from Theorem 5.1 can not be fixed in arbitrary form if we want to establish a hyperbolic (evolutionary) character for nonholonomic vacuum Einstein equations, i.e., to satisfy both systems (69) and (68). Really, the last system of fist order PDEs allows us to compute the time–like derivatives $\partial \mathbf{g}^{t\overline{i}}(0,\widehat{u}^{\overline{i}})/\partial\, {}^{t}\widehat{u}\mid_{\{\ {}^{t}\widehat{u}=0\}}$if $\mathbf{g}_{\overline{i}\overline{j}}\mid_{\{\ {}^{t}\widehat{u}=0\}}$ and $\partial \mathbf{g}_{\overline{i}\overline{j}}/\partial\, {}^{t}\widehat{u}\mid_{\{\ {}^{t}\widehat{u}=0\}}$have been defined. We conclude that the essential data for formulating a N–adapted evolution problem should be formulated for a 3–d space d–metric

$${}^{[3]}\mathbf{g}:=\mathbf{g}_{\overline{i}\overline{j}}(x,y)\mathbf{e}^{\overline{i}}\otimes\mathbf{e}^{\overline{j}}, \quad (73)$$

where $\mathbf{e}^{\overline{j}}$ are N–elongated differentials of type (4), and its N–elongated (3) time–like derivatives.

Using the Theorem 5.1 and above presented considerations, we get the proof of the following theorem

Theorem 5.2 *If the initial data (70) satisfy the conditions (71) and (72), and the so–called effective Einstein N–adapted constraint equations,*

$$\left(\widehat{\mathbf{E}}_{\alpha\beta}+\Lambda\mathbf{g}_{\alpha\beta}\right)n^{\alpha}=0, \quad (74)$$

are computed for zero distortion in (10), then the d–metric stated by Theorem 5.1 defines solutions of the nonholonomic vacuum equations (67) and/or (24)–(27).

This theorem gives us the possibility to state the Cauchy data for decoupled modified gravity systems and their generic off–diagonal solutions in order to generate N–adapted evolutions.

5.2. On initial data sets and global nonholonomic evolution

We adopt this convention for spacetime nonholonomic manifolds $\mathbf{V}=\mathbb{R}\times\ {}^{3}V$, were ${}^{3}V$ is a N–adapted 3-d manifold, when a Whitney sum $T\ {}^{3}V=h\ {}^{3}V\oplus v\ {}^{3}V$ is stated by a spacelike nonholonomic distribution ${}^{3}\mathbf{N}$ of type (2) and there is an embedding $e:\ {}^{3}V\rightarrow\mathbf{V}$.

Using the d–metric ${}^{[3]}\mathbf{g}$ (73), we can define the second fundamental form $\mathbf{K}$ of a spacelike hypersurface ${}^{3}V$ in $\mathbf{V}$, $\widehat{\mathbf{K}}(\mathbf{X},\mathbf{Y}):=\mathbf{g}(\widehat{\mathbf{D}}_{\mathbf{X}}\mathbf{n},\mathbf{Y}),\forall\mathbf{X}\in T\ {}^{3}V$. The unity d–vector $\mathbf{n}=n^{\alpha}\mathbf{e}_{\alpha}=n_{\overline{i}}\mathbf{e}^{\overline{i}}=n_t\mathbf{e}^{t}=\left(\sqrt{|g^{tt}|}\right)^{-1}\mathbf{e}^{t}$, with normalization $\mathbf{g}(\mathbf{n},\mathbf{n})=\mathbf{g}^{\alpha\beta}n_{\alpha}n_{\beta}=\mathbf{g}^{tt}(n_t)^2=-1$, is time–future and normal to ${}^{3}V$. The value $\widehat{\mathbf{K}}=\{\widehat{K}_{\overline{i}\overline{j}}\}$ is the extrinsic canonical curvature d–tensor of ${}^{3}V$. Imposing the zero torsion conditions (27), $\widehat{\mathbf{K}}\rightarrow\{K_{\overline{i}\overline{j}}\}$, where $K_{\overline{i}\overline{j}}=-\frac{1}{2}\mathbf{g}^{t\alpha}\left(\mathbf{e}_{\overline{j}}\mathbf{g}_{\alpha\overline{i}}+\mathbf{e}_{\overline{i}}\mathbf{g}_{\alpha\overline{j}}-\mathbf{e}_{\alpha}\mathbf{g}_{\overline{i}\overline{j}}\right)n_t$ are components computed in standard form using the Levi–Civita connection, but with respect to N–adapted frames. We can invert the last formula and write $\partial_t\mathbf{g}_{\overline{i}\overline{j}}=2(\mathbf{g}^{tt}n_t)^{-1}K_{\overline{i}\overline{j}}+$ {terms determined by $\mathbf{g}_{\alpha\beta}$ and their space–derivatives}. Such formulas show that $K_{\overline{i}\overline{j}}$ and $\partial_t\mathbf{g}_{\overline{i}\overline{j}}$ are geometric counterparts on hypersurfaces $\{\ {}^{t}\widehat{u}=0\}$.

For the canonical d–connection $\widehat{\mathbf{D}}_{\alpha}=\left(\widehat{D}_t,\widehat{\mathbf{D}}_{\overline{i}}\right)$ and d–metric ${}^{[3]}\mathbf{g}$ induced on a spacelike hypersurface in a Lorentzian nonholonomic manifold $\mathbf{V}$, we can derive a N–adapted variant of Gauss–Codazzi equations

$${}^{[3]}\widehat{R}^{\overline{i}}{}_{\overline{j}\overline{k}\overline{l}}=\widehat{\mathbf{R}}^{\overline{i}}{}_{\overline{j}\overline{k}\overline{l}}+\widehat{K}^{\overline{i}}{}_{\overline{l}}\widehat{K}_{\overline{j}\overline{k}}-\widehat{K}^{\overline{i}}{}_{\overline{k}}\widehat{K}_{\overline{j}\overline{l}},\ \widehat{\mathbf{D}}_{\overline{i}}\widehat{K}_{\overline{j}\overline{k}}-\widehat{\mathbf{D}}_{\overline{j}}\widehat{K}_{\overline{i}\overline{k}}=\widehat{\mathbf{R}}^{\overline{l}}{}_{\overline{j}\overline{k}\overline{l}}n^{\overline{l}}.$$

In these formulas, ${}^{[3]}\widehat{R}^{\bar{i}}{}_{\bar{j}\bar{k}\bar{l}}$ is the canonical curvature d–tensor of ${}^{[3]}\mathbf{g}$, $\widehat{\mathbf{R}}^{\bar{i}}{}_{\bar{j}\bar{k}\bar{l}}$ is computed following formulas (15) as the spacetime canonical d–curvature tensor, $\mathbf{n}$ is the timelike normal to hypersurface ${}^{3}V$ when the local N–adapted coordinate system is such way chosen that d–vectors $\mathbf{e}_{\bar{j}}$ are tangent to ${}^{3}V$. Contracting indices, introducing divergence operator $\widehat{div}$ determined by N–elongated partial derivatives (3), trace operator tr and absolute differential d, we derive from above equations and (24) the following system of equations

$$\begin{aligned}
\widehat{div}\widehat{K} - d(tr\widehat{K}) &= 8\pi\widehat{J}, \text{ momentum constraint}; \qquad (75)\\
{}^{s}_{[3]}\widehat{R} - 2\Lambda - {}^{[3]}\mathbf{g}^{\bar{l}\bar{j}}\widehat{K}^{\bar{i}}{}_{\bar{l}}\widehat{K}_{\bar{j}\bar{i}} + (tr\widehat{K})^{2} &= 16\pi\widehat{\rho}, \text{ Hamiltonian constraint};\\
\mathcal{C}(\widehat{\mathcal{F}}, {}^{[3]}\mathbf{g}) &= 0, \text{ Einstein constraint eqs};
\end{aligned}$$

where ${}^{s}_{[3]}\widehat{R}$ is computed as the scalar (17) but for ${}^{[3]}\mathbf{g}$. In a general context, we consider that ${}^{3}V$ is embedded in a nonholonomic spacetime $\mathbf{V}$ with induced data $\left({}^{3}V, {}^{[3]}\mathbf{g}, \widehat{\mathbf{K}}, \widehat{\mathcal{F}}\right)$, we have $\widehat{J} := -(\mathbf{n}, \cdot)$ and $\widehat{\rho} := (\mathbf{n}, \mathbf{n})$. The term $\mathcal{C}(\widehat{\mathcal{F}}, {}^{[3]}\mathbf{g})$ denotes the set of additional constraints resulting from the non–gravitational part, including nonholonomic distributions $\mathbf{N}$ (2). If, in such a set, there are included the zero torsion conditions (27), we can omit "hats" on geometric/physical objects if they are written in "not" N–adapted frames of reference. The equations (75) form an undetermined system of PDEs. For 3-d, there are locally four equations for twelve unknown values given by the components of d–tensors ${}^{[3]}\mathbf{g}$ and $\widehat{\mathbf{K}}$. Using the conformal method with the Levi–Civita connection, see details and references in Ref. [6], we can study the existence and uniqueness of solutions to such systems.

The above considerations motivate the following definition

Definition 5.2 *A canonical vacuum initial N–adapted data set is defined by a triple* $\left({}^{3}V, {}^{[3]}\mathbf{g}, \widehat{\mathbf{K}}\right)$, *where* $({}^{[3]}\mathbf{g}, \widehat{\mathbf{K}})$ *are defined as a solution of (75).*

If the conditions (27) are imposed, the data $({}^{[3]}\mathbf{g}, \widehat{\mathbf{K}})$ are equivalent to similar ones $({}^{[3]}\mathbf{g}, \mathbf{K})$ with $\mathbf{K}$ computed for the Levi–Civita connection. Covering ${}^{3}V$ by coordinate neighborhoods $\mathcal{O}_{u}$ of points $u \in {}^{3}V$, we can use Theorem 5.1 to construct globally hyperbolic N–adapted developments $(\mathcal{U}_{u}, \mathbf{g}_{u})$ of an initial data set $\left(\mathcal{U}_{u} \subset {}^{3}V, {}^{[3]}\mathbf{g}, \widehat{\mathbf{K}}\right)$ as in the above definition. The d–metrics generated in such forms will coincide after performing suitable N–adapted frame/coordinated transforms on charts covering such a spacetime wherever such solutions are defined. We can patch all data $(\mathcal{U}_{u}, \mathbf{g}_{u})$ together to a globally hyperbolic Lorentzian nonholonomic spacetime containing a Cauchy surface ${}^{3}V$. We prove that

Theorem 5.3 *Any N–adapted initial data* $({}^{3}V, {}^{[3]}\mathbf{g}, \widehat{\mathbf{K}})$ *of differentiability class* $H^{s+1} \times H^{s}, s > 3/2$, *admits a globally hyperbolic, N–adapted and unique development (in the sense of Theorem 5.1 and of the above considered assumptions and proofs).*

6. Anholonomic Modifications of Black Holes

6.1. (Non) holonomic non–Abelian effective vacuum spaces

This class of effective vacuum solutions are generated not just as a simple limit $\Lambda \to 0$, for instance, for a class of solutions (47) with coefficients (48)–(50). We have to construct off–diagonal solutions of the effective Einstein equations for the canonical d–connection taking the vacuum equations $\widehat{\mathbf{R}}_{\alpha\beta} = 0$ and an ansatz $\mathbf{g}$ with coefficients satisfying the conditions

$$\begin{aligned}
&\epsilon_1 \psi^{\bullet\bullet}(r,\theta) + \epsilon_2 \psi''(r,\theta) = 0; && (76)\\
h_3 &= \pm e^{-2\ {}^0\phi} \frac{(h_4^*)^2}{h_4} \text{ for a given } h_4(r,\theta,\varphi),\ \phi(r,\theta,\varphi) = {}^0\phi = const; \\
w_i &= w_i(r,\theta,\varphi), \text{ for any such functions if } \lambda = 0; \\
n_i &= \begin{cases} {}^1n_i(r,\theta) + {}^2n_i(r,\theta)\int (h_4^*)^2 |h_4|^{-5/2} dv, & \text{if } n_i^* \neq 0; \\ {}^1n_i(r,\theta), & \text{if } n_i^* = 0. \end{cases}
\end{aligned}$$

Vacuum solutions of the effective Einstein equations for the Levi–Civita connection, i.e., of $R_{\alpha\beta} = 0$, are generated if we impose additional constraints on coefficients of d–metric, for $e^{-2\ {}^0\phi} = 1$, as solutions of (51),

$$\begin{aligned}
h_3 &= \pm 4 \left[\left(\sqrt{|h_4|} \right)^* \right]^2, \quad h_4^* \neq 0; && (77)\\
w_1 w_2 \left(\ln \left| \frac{w_1}{w_2} \right| \right)^* &= w_2^\bullet - w_1', \ w_i^* \neq 0; \ w_2^\bullet - w_1' = 0, \ w_i^* = 0; \\
{}^1n_1'(r,\theta) - {}^1n_2^\bullet(r,\theta) &= 0, \ n_i^* = 0. && (78)
\end{aligned}$$

The constructed class of vacuum solutions with coefficients subjected to conditions (76)–(78) is of type (60) for (61)–(63). Such metrics consist a particular case of vacuum ansatz defined by Corollary 4.3 with $\underline{h}_4 = 1$ and $\omega = 1$.

Here we note that former analytic and numeric programs (for instance, standard ones with Maple/ Mathematica) for constructing solutions in gravity theories can not be directly applied for alternative verifications of our solutions. Those approaches do not encode the nonholonomic constraints which we use for constructing integral varieties. Nevertheless, it is possible to check, in general, the analytic form, see all details summarized in Refs. [20, 21, 22] (and formulas from Appendix Appendix A), that the vacuum effective Einstein equations (41)–(32) with zero effective sources, ${}^v\Upsilon = 0$ and $\Upsilon = 0$, can be solved by using the above presented off–diagonal ansatz for metrics.

6.2. f–deformations of the Schwarzschild metric

We can consider a "prime" metric which, in general, is not a solution of (modified) Einstein equations,

$$ {}^\varepsilon\mathbf{g} = -d\xi \otimes d\xi - r^2(\xi)\, d\vartheta \otimes d\vartheta - r^2(\xi) \sin^2\vartheta\, d\varphi \otimes d\varphi + \varpi^2(\xi)\, dt \otimes dt. \qquad (79)$$

We shall deform nonholonomically this metric into a "target" off–diagonal one which will be a solution of the vacuum Einstein equations. The nontrivial metric coefficients in (79) are stated in the form

$$\check{g}_1 = -1, \ \check{g}_2 = -r^2(\xi), \ \check{h}_3 = -r^2(\xi)\sin^2\vartheta, \ \check{h}_4 = \varpi^2(\xi), \qquad (80)$$

for local coordinates $x^1 = \xi, x^2 = \vartheta, y^3 = \varphi, y^4 = t$, where $\xi = \int dr \left| 1 - \frac{2\mu_0}{r} + \frac{\varepsilon}{r^2} \right|^{1/2}$ and $\varpi^2(r) = 1 - \frac{2\mu_0}{r} + \frac{\varepsilon}{r^2}$. If we put $\varepsilon = 0$ with μ_0 considered as a point mass, the metric ${}^\varepsilon\mathbf{g}$ (79) determines the Schwarzschild solution. For simplicity, we analyze only the case of "pure" gravitational vacuum solutions, not considering a more general construction when $\varepsilon = e^2$ can be related to the electric charge for the Reissner–Nordström metric. In our approach, ε is a small parameter (eccentricity) defining a small deformation of a circle into an ellipse.

We generate exact solutions of the system (61)–(63) with effective $\Lambda = 0$ via nonholonomic deformations ${}^{\varepsilon}\mathbf{g} \rightarrow {}^{\varepsilon}_{\eta}\mathbf{g}$, when $g_i = \eta_i \check{g}_i$ and $h_a = \eta_a \check{h}_a$ and w_i, n_i define a target metric

$$\begin{aligned} {}^{\varepsilon}_{\eta}\mathbf{g} &= \eta_1(\xi)d\xi \otimes d\xi + \eta_2(\xi)r^2(\xi)\, d\vartheta \otimes d\vartheta + \eta_3(\xi,\vartheta,\varphi)r^2(\xi)\sin^2\vartheta\ \delta\varphi \otimes \delta\varphi - \eta_4(\xi,\vartheta,\varphi)\varpi^2(\xi)\ \delta t \otimes \delta t, \\ \delta\varphi &= d\varphi + w_1(\xi,\vartheta,\varphi)d\xi + w_2(\xi,\vartheta,\varphi)d\vartheta,\ \delta t = dt + n_1(\xi,\vartheta)d\xi + n_2(\xi,\vartheta)d\vartheta. \end{aligned} \quad (81)$$

The gravitational field equations for zero source relate the coefficients of the vertical metric and polarization functions,

$$h_3 = h_0^2(b^*)^2 = \eta_3(\xi,\vartheta,\varphi)r^2(\xi)\sin^2\vartheta,\ h_4 = -b^2 = -\eta_4(\xi,\vartheta,\varphi)\varpi^2(\xi), \quad (82)$$

for $|\eta_3| = (h_0)^2|\check{h}_4/\check{h}_3|[(\sqrt{|\eta_4|})^*]^2$. In these formulas, we have to choose $h_0 = const$ ($h_0 = 2$ in order to satisfy the first condition (78)), where $\check{h}_a$ are stated by the Schwarzschild solution for the chosen system of coordinates and η_4 can be any function satisfying the condition $\eta_4^* \neq 0$. We generate a class of solutions for any function $b(\xi,\vartheta,\varphi)$ with $b^* \neq 0$. For different purposes, it is more convenient to work directly with η_4, for $\eta_4^* \neq 0$, and/or h_4, for $h_4^* \neq 0$. The gravitational polarizations η_1 and η_2, when $\eta_1 = \eta_2 r^2 = e^{\psi(\xi,\vartheta)}$, are found from (41) with zero source, written in the form $\psi^{\bullet\bullet} + \psi'' = 0$.

Introducing the defined values of the coefficients in the ansatz (81), we find a class of exact off–diagonal vacuum solutions of the Einstein equations defining stationary nonholonomic deformations of the Schwarzschild metric,

$$\begin{aligned} {}^{\varepsilon}\mathbf{g} &= -e^{\psi}\,(d\xi \otimes d\xi + \ d\vartheta \otimes d\vartheta) - 4\left[(\sqrt{|\eta_4|})^*\right]^2 \varpi^2\ \delta\varphi \otimes\ \delta\varphi + \eta_4\varpi^2\ \delta t \otimes \delta t, \\ \delta\varphi &= d\varphi + w_1 d\xi + w_2 d\vartheta,\ \delta t = dt + {}^1n_1 d\xi + {}^1n_2 d\vartheta. \end{aligned} \quad (83)$$

The N–connection coefficients $w_i(\xi,\vartheta,\varphi)$ and ${}^1n_i(\xi,\vartheta)$ must satisfy the conditions (78) in order to get vacuum metrics in GR.

It should be emphasized here that, in general, the bulk of solutions from the set of target metrics do not define black holes and do not describe obvious physical situations. They preserve the singular character of the coefficient ϖ^2 vanishing on the horizon of a Schwarzschild black hole if we take only smooth integration functions for some small deformation parameters ε.

6.3. Linear parametric polarizations induced by f–modifications

We may select some locally anisotropic configurations with possible physical interpretation of gravitational vacuum configurations with spherical and/or rotoid (ellipsoid) symmetry if it is considered a generating function

$$b^2 = q(\xi,\vartheta,\varphi) + \varepsilon\varrho(\xi,\vartheta,\varphi). \quad (84)$$

For simplicity, we restrict our analysis only to linear decompositions on a small parameter ε, with $0 < \varepsilon << 1$.

Using (84), we compute $(b^*)^2 = [(\sqrt{|q|})^*]^2\ [1 + \varepsilon\frac{1}{(\sqrt{|q|})^*}(\varrho/\sqrt{|q|})^*]$ and the vertical coefficients of d–metric (83), i.e., h_3 and h_4 (and corresponding polarizations η_3 and η_4), see formulas (82)[11]. We model rotoid configurations if we choose

$$q = 1 - \frac{2\mu(\xi,\vartheta,\varphi)}{r} \text{ and } \varrho = \frac{q_0(r)}{4\mu^2}\sin(\omega_0\varphi + \varphi_0), \quad (85)$$

[11] Nonholonomic deformations of the Schwarzschild solution (not depending on ε) can be generated if we consider $\varepsilon = 0$ and $b^2 = q$ and $(b^*)^2 = \left[(\sqrt{|q|})^*\right]^2$.

for $\mu(\xi,\vartheta,\varphi)=\mu_0+\varepsilon\mu_1(\xi,\vartheta,\varphi)$ (supposing that the mass is locally anisotropically polarized) with certain constants μ,ω_0 and φ_0 and arbitrary functions/ polarizations $\mu_1(\xi,\vartheta,\varphi)$ and $q_0(r)$ to be determined from some boundary conditions, with ε being the eccentricity.[12]. This condition defines a small deformation of the Schwarzschild spherical horizon into an ellipsoidal one (rotoid configuration with eccentricity ε).

The resulting off–diagonal solution with rotoid type symmetry is

$$\begin{aligned} {}^{rot}\mathbf{g} &= -e^{\psi}\,(d\xi\otimes d\xi+\ d\vartheta\otimes d\vartheta)+(q+\varepsilon\varrho)\ \delta t\otimes\delta t-4[(\sqrt{|q|})^*]^2\ [1+\varepsilon\frac{1}{(\sqrt{|q|})^*}(\varrho/\sqrt{|q|})^*]\ \delta\varphi\otimes\ \delta\varphi, \\ \delta\varphi &= d\varphi+w_1d\xi+w_2d\vartheta,\ \delta t=dt+\ {}^1n_1d\xi+\ {}^1n_2d\vartheta. \end{aligned} \tag{86}$$

The functions $q(\xi,\vartheta,\varphi)$ and $\varrho(\xi,\vartheta,\varphi)$ are given by formulas (85) and the N–connection coefficients $w_i(\xi,\vartheta,\varphi)$ and $n_i=\ {}^1n_i(\xi,\vartheta)$ are subjected to conditions of type (78),

$$\begin{aligned} w_1w_2\left(\ln|\frac{w_1}{w_2}|\right)^* &= w_2^{\bullet}-w_1',\quad w_i^*\neq 0; \\ \text{or}\ \ w_2^{\bullet}-w_1' &= 0,\quad w_i^*=0;\ {}^1n_1'(\xi,\vartheta)-\ {}^1n_2^{\bullet}(\xi,\vartheta)=0 \end{aligned} \tag{87}$$

and $\psi(\xi,\vartheta)$ being any function for which $\psi^{\bullet\bullet}+\psi''=0$.

7. Ellipsoidal f–Configurations and Solitons

We can consider nonholonomic deformations in modified gravity for arbitrary signs of the cosmological constant $\Lambda\neq 0$ containing contributions from nonholonomic f–modifications. Such classes of solutions can be constructed in general form for a system (41)–(44) and (51) with coefficients of metric of type (47). Such metrics consist a particular case of non–vacuum ansatz defined by Corollary 4.2 with $\underline{h}_4=1$ and $\omega=1$.

7.1. *Nonholonomic rotoid deformations*

Let us consider a diagonal metric of type

$$ {}^{\varepsilon}_{\lambda}\mathbf{g}=d\xi\otimes d\xi+r^2(\xi)\ d\theta\otimes d\theta+r^2(\xi)\sin^2\theta\ d\varphi\otimes d\varphi+\ {}_{\lambda}\varpi^2(\xi)\ dt\otimes\ dt, \tag{88}$$

where nontrivial metric coefficients are parametrized in the form $\check{g}_1=1,\ \check{g}_2=r^2(\xi),\ \check{h}_3=r^2(\xi)\sin^2\vartheta,\ \check{h}_4=\ {}_{\lambda}\varpi^2(\xi)$, for local coordinates $x^1=\xi,x^2=\vartheta,y^3=\varphi,y^4=t$, with $\xi=\int dr/|q(r)|^{\frac{1}{2}}$, and ${}_{\lambda}\varpi^2(r)=-\sigma^2(r)q(r)$, for $q(r)=1-2m(r)/r-\Lambda r^2/3$.

The ansatz for such classes of solutions is chosen to be of the form

$$\begin{aligned} {}^{\lambda}\mathring{\mathbf{g}} &= e^{\underline{\phi}(\xi,\theta)}\ (d\xi\otimes d\xi+\ d\theta\otimes d\theta)+h_3(\xi,\theta,\varphi)\ \delta\varphi\otimes\delta\varphi+h_4(\xi,\theta,\varphi)\ \delta t\otimes\ \delta t, \\ \delta\varphi &= d\varphi+w_1\,(\xi,\theta,\varphi)\,d\xi+w_2\,(\xi,\theta,\varphi)\,d\theta,\ \delta t=dt+n_1\,(\xi,\theta,\varphi)\,d\xi+n_2\,(\xi,\theta,\varphi)\,d\theta, \end{aligned}$$

for $h_3=-h_0^2(b^*)^2=\eta_3(\xi,\theta,\varphi)r^2(\xi)\sin^2\vartheta,\ h_4=b^2=\eta_4(\xi,\theta,\varphi)\ {}_{\lambda}\varpi^2(\xi)$. The coefficients of this metric determine exact solutions if

$$\underline{\phi}^{\bullet\bullet}(\xi,\theta)+\underline{\phi}''(\xi,\theta)=2\Lambda; \tag{89}$$

$$\begin{aligned} h_3 &= \pm\frac{(\phi^*)^2}{4\Lambda}e^{-2\ {}^0\phi(\xi,\theta)},\ h_4=\mp\frac{1}{4\ \Lambda}e^{2(\phi-\ {}^0\phi(\xi,\theta))};\ w_i=-\partial_i\phi/\phi^*; \\ n_i &= {}^1n_i(\xi,\theta)+\ {}^2n_i(\xi,\theta)\int(\phi^*)^2\,e^{-2(\phi-\ {}^0\phi(\xi,\vartheta))}d\varphi, \\ &= \left\{\begin{array}{ll} {}^1n_i(\xi,\theta)+\ {}^2n_i(\xi,\theta)\int e^{-4\phi}\frac{(h_4^*)^2}{h_4}d\varphi, & \text{if}\quad n_i^*\neq 0; \\ {}^1n_i(\xi,\theta), & \text{if}\quad n_i^*=0; \end{array}\right. \end{aligned}$$

[12] We may treat ε as an eccentricity imposing the condition that the coefficient $h_4=b^2=\eta_4(\xi,\vartheta,\varphi)\varpi^2(\xi)$ becomes zero for data (85), if $r_+\simeq 2\mu_0/\left(1+\varepsilon\frac{q_0(r)}{4\mu^2}\sin(\omega_0\varphi+\varphi_0)\right)$.

Journal of Physics: Conference Series **453** (2013) 012021 doi:10.1088/1742-6596/453/1/012021

for any nonzero h_a and h_a^* and (integrating) functions ${}^1n_i(\xi,\theta)$, ${}^2n_i(\xi,\theta)$, generating function $\phi(\xi,\theta,\varphi)$, and ${}^0\phi(\xi,\theta)$ to be determined from certain boundary conditions for a fixed system of coordinates.

For nonholonomic effective ellipsoid de Sitter configurations, we parameterize

$$\begin{aligned}
{}_{\lambda}^{rot}\mathbf{g} &= -e^{\underline{\phi}(\xi,\theta)}(d\xi \otimes d\xi + d\theta \otimes d\theta) + (\underline{q} + \varepsilon \underline{\varrho})\ \delta t \otimes \delta t \\
&\quad -h_0^2 \left[(\sqrt{|\underline{q}|})^*\right]^2 [1 + \varepsilon \frac{1}{(\sqrt{|\underline{q}|})^*}(\underline{\varrho}/\sqrt{|\underline{q}|})^*]\ \delta\varphi \otimes \delta\varphi, \\
\delta\varphi &= d\varphi + w_1 d\xi + w_2 d\vartheta,\ \delta t = dt + n_1 d\xi + n_2 d\vartheta,
\end{aligned} \tag{90}$$

where $\underline{q} = 1 - \frac{2\ {}^1\underline{\mu}(r,\theta,\varphi)}{r}$, $\underline{\varrho} = \frac{\underline{q}_0(r)}{4\underline{\mu}_0^2}\sin(\omega_0\varphi + \varphi_0)$, are chosen to generate an anisotropic rotoid configuration for the smaller "horizon" (when $h_4 = 0$), $r_+ \simeq 2\ {}^1\underline{\mu}/\left(1 + \varepsilon \frac{\underline{q}_0(r)}{4\underline{\mu}_0^2}\sin(\omega_0\varphi + \varphi_0)\right)$, for a corresponding $\underline{q}_0(r)$.

We have to impose the condition that the coefficients of the above d–metric induce a zero torsion in order to generate solutions of the Einstein equations for the Levi–Civita connection. Using formula (87), for $\phi^* \neq 0$, we obtain that $\phi(r,\varphi,\theta) = \ln|h_4^*/\sqrt{|h_3 h_4|}|$ must be any function defined in non–explicit form from equation $2e^{2\phi}\phi = \Lambda$. The set of constraints for the N–connection coefficients is solved, if the integration functions in (89) are chosen in a form when $w_1 w_2 \left(\ln|\frac{w_1}{w_2}|\right)^* = w_2^\bullet - w_1'$ for $w_i^* \neq 0$; $w_2^\bullet - w_1' = 0$ for $w_i^* = 0$; and take $n_i = {}^1n_i(x^k)$ for ${}^1n_1'(x^k) - {}^1n_2^\bullet(x^k) = 0$.

In a particular case, in the limit $\varepsilon \to 0$, we get a subclass of solutions of type (90) with spherical symmetry but with generic off–diagonal coefficients induced by the N–connection coefficients. This class of spacetimes depend on cosmological constants polarized nonholonomically by f–modifications. We can extract from such configurations the Schwarzschild solution, if we select a set of functions with the properties $\phi \to const, w_i \to 0, n_i \to 0$ and $h_4 \to \varpi^2$.

7.2. Effective vacuum solitonic configurations

It is possible to construct off–diagonal vacuum spacetimes generated by 3–d solitons as examples of generic off–diagonal solutions with nontrivial vertical conformal factor ω. We consider that there are satisfied the conditions of Corollary 4.3 with $\underline{h}_4 = 1$ for effective vacuum solutions (such configurations may encode f–modifications) and the Cauchy problem is stated as in Section 3.

7.2.1. Solutions with solitonic factor $\omega(x^1,y^3,t)$ We take $\omega = \eta(x^1,y^3,t)$, when $y^4 = t$ is a timelike coordinate, as a solution of KdP equation [13]

$$\pm\eta^{**} + (\partial_t\eta + \eta\ \eta^\bullet + \epsilon\eta^{\bullet\bullet\bullet})^\bullet = 0, \tag{91}$$

with dispersion ϵ and possible dependencies on a set of parameters θ. It is supposed that, in the dispersionless limit $\epsilon \to 0$, the solutions are independent on y^3 and determined by Burgers' equation $\partial_t\eta + \eta\ \eta^\bullet = 0$. For such 3–d solitonic configurations, the conditions (51) are written in the form

$$\mathbf{e}_1\eta = \eta^\bullet + w_1(x^i,y^3)\eta^* + n_1(x^i)\partial_t\eta = 0.$$

For $\eta' = 0$, we can impose the condition $w_2 = 0$ and $n_2 = 0$.

Such vacuum solitonic metrics can be parametrized in the form

$$\begin{aligned}\mathbf{g} &= e^{\psi(x^k)}(dx^1 \otimes dx^1 + dx^2 \otimes dx^2) + \left[\eta(x^1, y^3, t)\right]^2 h_a(x^1, y^3)\ \mathbf{e}^a \otimes \mathbf{e}^a,\\ \mathbf{e}^3 &= dy^3 + w_1(x^k, y^3)dx^1,\ \mathbf{e}^4 = dy^4 + n_1(x^k)dx^1.\end{aligned}$$

This class of metrics does not have (in general) Killing symmetries but may possess symmetries determined by solitonic solutions of (91). Alternatively, we can consider that η is a solution of any three dimensional solitonic and/ or other nonlinear wave equations; in a similar manner, we can generate solutions for $\omega = \eta(x^2, y^3, t)$.

7.2.2. Solitonic metrics with factor $\omega(x^i, t)$ There are effective vacuum metrics when the solitonic dynamics does not depend on anisotropic coordinate y^3. In this case, $\omega = \widehat{\eta}(x^k, t)$ is a solution of KdP equation

$$\pm\widehat{\eta}^{\bullet\bullet} + (\partial_t\widehat{\eta} + \widehat{\eta}\ \widehat{\eta}' + \epsilon\widehat{\eta}''')' = 0. \tag{92}$$

In the dispersionless limit $\epsilon \to 0$, the solutions are independent on x^1 and determined by Burgers' equation $\partial_t\widehat{\eta} + \widehat{\eta}\ \widehat{\eta}' = 0$.

This class of vacuum solitonic modified gravity configurations is given by

$$\begin{aligned}{}_2\mathbf{g} &= e^{\psi(x^k)}(dx^1 \otimes dx^1 + dx^2 \otimes dx^2) + \left[\widehat{\eta}(x^k, t)\right]^2 h_a(x^k, y^3)\ \mathbf{e}^a \otimes \mathbf{e}^a,\\ \mathbf{e}^3 &= dy^3 + w_1(x^k, y^3)dx^1,\ \mathbf{e}^4 = dy^4 + n_1(x^k)dx^1;\end{aligned}$$

the conditions (51) are $\mathbf{e}_1\widehat{\eta} = \widehat{\eta}^{\bullet} + n_1(x^i)\partial_t\widehat{\eta} = 0,\ \mathbf{e}_2\widehat{\eta} = \widehat{\eta}' + n_2(x^i)\partial_t\widehat{\eta} = 0.$

It is possible to derive an infinite number of vacuum gravitational 2-d and 3-d configurations characterized by corresponding solitonic hierarchies and bi–Hamilton structures, for instance, related to different KdP equations (92) with possible mixtures with solutions for 2-d and 3-d sine–Gordon equations etc, see details in Ref. [23].

Acknowledgments: The work is partially supported by the Program IDEI, PN-II-ID-PCE-2011-3-0256. Author is grateful to T. Grammenos, A. P. Kouretisis, N. E. Mavromatos, P. C. Stavrinos and E. C. Vagenas for important discussions, kind support and hospitality.

Appendix A. Proof of Theorem 3.1

For $\omega = 1$ and $\underline{h}_a = const$, such proofs can be obtained by straightforward computations [20]. The approach was extended for $\omega \neq 1$ and higher dimensions in [21, 22]. In this section, we sketch a proof for ansatz (8) with nontrivial $\underline{h}_4$ depending on variable y^4 when $\omega = 1$ in data (9). At the next step, the formulas will be completed for nontrivial values $\omega \neq 1$.

If $\widehat{R}^1{}_1 = \widehat{R}^2{}_2$ and $\widehat{R}^3{}_3 = \widehat{R}^4{}_4$, the effective Einstein equations (24) for $\widehat{\mathbf{D}}$ and data (A.1) (see below) can be written for any source (25) in the form

$$\widehat{E}^1{}_1 = \widehat{E}^2{}_2 = -\widehat{R}^3{}_3 = \boldsymbol{\Upsilon}(x^k, y^3) + \underline{\boldsymbol{\Upsilon}}(x^k, y^3, y^4),\ \widehat{E}^3{}_3 = \widehat{E}^4{}_4 = -\widehat{R}^1{}_1 = {}^v\boldsymbol{\Upsilon}(x^k).$$

The geometric data for the conditions of Theorem 3.1 are $g_i = g_i(x^k)$ and

$$g_3 = h_3(x^k, y^3), g_4 = h_4(x^k, y^3)\underline{h}_4(x^k, y^4), N_i^3 = w_i(x^k, y^3), N_i^4 = n_i(x^k, y^3), \tag{A.1}$$

for $\underline{h}_3 = 1$ and local coordinates $u^\alpha = (x^i, y^a) = (x^1, x^2, y^3, y^4)$. For such values, we shall compute respectively the coefficients of $\Omega^{\ a}_{\alpha\beta}$, canonical d–connection $\widehat{\mathbf{\Gamma}}^\gamma{}_{\alpha\beta}$ (12), d–torsion $\widehat{\mathbf{T}}^\gamma{}_{\alpha\beta}$ (14), necessary coefficients of d–curvature $\widehat{\mathbf{R}}^\tau{}_{\alpha\beta\gamma}$ (15) with respective contractions for $\widehat{\mathbf{R}}_{\alpha\beta} := \widehat{\mathbf{R}}^\gamma{}_{\alpha\beta\gamma}$ (16) and resulting ${}^s\widehat{R}$ (17) and $\widehat{\mathbf{E}}_{\alpha\beta}$(20). Finally, we shall state the conditions (27) when general coefficients (A.1) are considered for d–metrics.

A.1. Coefficients of the canonical d–connection

There are horizontal nontrivial coefficients of $\widehat{\mathbf{\Gamma}}^{\gamma}{}_{\alpha\beta}$ (12),

$$\begin{aligned}\widehat{L}^{i}_{jk} &= \frac{1}{2}g^{i1}\left(\mathbf{e}_k g_{j1} + \mathbf{e}_j g_{k1} - \mathbf{e}_1 g_{jk}\right) + \frac{1}{2}g^{i2}\left(\mathbf{e}_k g_{j2} + \mathbf{e}_j g_{k2} - \mathbf{e}_2 g_{jk}\right),\\ \text{i.e., } \widehat{L}^{1}_{jk} &= \frac{1}{2g_1}\left(\partial_k g_{j1} + \partial_j g_{k1} - g^{\bullet}_{jk}\right), \widehat{L}^{2}_{jk} = \frac{1}{2g_2}\left(\partial_k g_{j2} + \partial_j g_{k2} - g'_{jk}\right).\end{aligned}$$

The h–v–components $\widehat{L}^{a}_{bk}$ are computed following formulas

$$\begin{aligned}\widehat{L}^{3}_{bk} &= e_b(N^3_k) + \frac{1}{2g_3}[\mathbf{e}_k g_{b3} - g_3\ e_b N^3_k - g_{3b}(N^3_k)^{*} - g_{4b}(N^4_k)^{\circ}] = e_b(N^3_k)\\ &+\frac{1}{2g_3}[\partial_k g_{b3} - N^3_k g^{*}_{b3} - N^4_k g^{\circ}_{b3} - g_3\ e_b N^3_k - g_{3b}(N^3_k)^{*} - g_{4b}(N^4_k)^{*}]\\ &= e_b(w_k + \underline{w}_k) + \frac{1}{2g_3}[\partial_k g_{b3} - (w_k + \underline{w}_k)g^{*}_{b3} - (n_k + \underline{n}_k)g^{\circ}_{b3} - g_3\ e_b(w_k + \underline{w}_k) - g_{3b}w_k{}^{*} - g_{4b}n_k{}^{*}],\end{aligned}$$

$$\begin{aligned}\widehat{L}^{4}_{bk} &= e_b(N^4_k) + \frac{1}{2g_4}[\mathbf{e}_k g_{b4} - g_4\ e_b N^4_k - g_{3b}(N^3_k)^{\circ} - g_{4b}(N^4_k)^{\circ}] = e_b(N^4_k)\\ &+\frac{1}{2g_4}[\partial_k g_{b4} - N^3_k g^{*}_{b4} - N^4_k g^{\circ}_{b4} - g_4\ e_b N^4_k - g_{3b}(N^3_k)^{\circ} - g_{4b}(N^4_k)^{\circ}]\\ &= e_b(n_k + \underline{n}_k) + \frac{1}{2g_4}[\partial_k g_{b4} - (w_k + \underline{w}_k)g^{*}_{b4} - (n_k + \underline{n}_k)g^{\circ}_{b4} - g_4\ e_b(n_k + \underline{n}_k) - g_{3b}\underline{w}^{\circ}_k - g_{4b}\underline{n}^{\circ}_k].\end{aligned}$$

In explicit form, we obtain these nontrivial values

$$\begin{aligned}\widehat{L}^{3}_{3k} &= w^{*}_k + \frac{1}{2g_3}[\partial_k g_3 - w_k g^{*}_3 - n_k g^{\circ}_3 - g_3\ w^{*}_k - g_3 w_k{}^{*}] = \frac{1}{2g_3}[\partial_k g_3 - w_k g^{*}_3] = \frac{\partial_k h_3}{2h_3} - w_k\frac{h^{*}_3}{2h_3},\\ \widehat{L}^{3}_{4k} &= \frac{1}{2g_3}[-g_4 n_k{}^{*}] = -\frac{h_4\underline{h}_4}{2h_3}n_k{}^{*},\ \widehat{L}^{4}_{3k} = n^{*}_k + \frac{1}{2g_4}[-g_4 n^{*}_k] = \frac{1}{2}n^{*}_k,\\ \widehat{L}^{4}_{4k} &= \frac{1}{2g_4}[\partial_k g_4 - w_k g^{*}_4 - n_k g^{\circ}_4] = \frac{\partial_k(h_4\underline{h}_4)}{2h_4\underline{h}_4} - w_k\frac{h^{*}_4}{2h_4} - n_k\frac{\underline{h}^{\circ}_4}{2\underline{h}_4}.\end{aligned}$$

For the set of h–v C–coefficients, we get $\widehat{C}^{i}_{jc} = \frac{1}{2}g^{ik}\frac{\partial g_{jk}}{\partial y^c} = 0$. The v–components of C–coefficients are computed using the following formulas

$$\widehat{C}^{3}_{bc} = \frac{1}{2g_3}\left(e_c g_{b3} + e_c g_{c3} - g^{*}_{bc}\right),\ \widehat{C}^{4}_{bc} = \frac{1}{2g_4}\left(e_c g_{b4} + e_b g_{c4} - g^{\circ}_{bc}\right),$$

i.e., $\widehat{C}^{3}_{33} = \frac{g^{*}_3}{2g_3} = \frac{h^{*}_3}{2h_3}$, $\widehat{C}^{3}_{34} = \frac{g^{\circ}_3}{2g_3} = 0$, $\widehat{C}^{3}_{44} = -\frac{g^{*}_4}{2g_3} = -\frac{h^{*}_4\underline{h}_4}{h_3}, \widehat{C}^{4}_{33} = -\frac{g^{\circ}_3}{2g_4} = 0$, $\widehat{C}^{4}_{34} = \frac{g^{*}_4}{2g_4} = \frac{h^{*}_4}{2h_4}, \widehat{C}^{4}_{44} = \frac{g^{\circ}_4}{2g_4} = \frac{\underline{h}^{\circ}_4}{2\underline{h}_4}$.

Putting together the above formulas, we find all nontrivial coefficients,

$$\begin{aligned}\widehat{L}^{1}_{11} &= \frac{g^{\bullet}_1}{2g_1},\ \widehat{L}^{1}_{12} = \frac{g'_1}{2g_1}, \widehat{L}^{1}_{22} = -\frac{g^{\bullet}_2}{2g_1},\ \widehat{L}^{2}_{11} = \frac{-g'_1}{2g_2},\ \widehat{L}^{2}_{12} = \frac{g^{\bullet}_2}{2g_2},\ \widehat{L}^{2}_{22} = \frac{g'_2}{2g_2},\\ \widehat{L}^{4}_{4k} &= \frac{\partial_k(h_4\underline{h}_4)}{2h_4\underline{h}_4} - \frac{w_k h^{*}_4}{2h_4} - (n_k + \underline{n}_k)\frac{\underline{h}^{\circ}_4}{2\underline{h}_4}, \widehat{L}^{3}_{3k} = \frac{\partial_k h_3}{2h_3} - \frac{w_k h^{*}_3}{2h_3}, \widehat{L}^{3}_{4k} = \frac{h_4\underline{h}_4}{-2h_3}n^{*}_k,\\ \widehat{L}^{4}_{3k} &= \frac{1}{2}n^{*}_k, \widehat{C}^{3}_{33} = \frac{h^{*}_3}{2h_3}, \widehat{C}^{3}_{44} = -\frac{h^{*}_4\underline{h}_4}{h_3},\ \widehat{C}^{4}_{33} = -\frac{h_3\underline{h}^{\circ}_3}{h_4\underline{h}_4},\ \widehat{C}^{4}_{34} = \frac{h^{*}_4}{2h_4}, \widehat{C}^{4}_{44} = \frac{\underline{h}^{\circ}_4}{2\underline{h}_4}.\end{aligned} \tag{A.2}$$

We shall need also the values

$$\widehat{C}_3 = \widehat{C}^{3}_{33} + \widehat{C}^{4}_{34} = \frac{h^{*}_3}{2h_3} + \frac{h^{*}_4}{2h_4}, \widehat{C}_4 = \widehat{C}^{3}_{43} + \widehat{C}^{4}_{44} = \frac{\underline{h}^{\circ}_4}{2\underline{h}_4}. \tag{A.3}$$

A.2. Coefficients for torsion of $\widehat{\mathbf{D}}$

Using data (A.1) for $\underline{w}_i = \underline{n}_i = 0$, the coefficients $\Omega^a_{ij} = \mathbf{e}_j(N^a_i) - \mathbf{e}_i(N^a_j)$, are computed

$$\begin{aligned}
\Omega^a_{ij} &= \partial_j(N^a_i) - \partial_i(N^a_j) - N^b_i \partial_b N^a_j + N^b_j \partial_b N^a_i \\
&= \partial_j(N^a_i) - \partial_i(N^a_j) - N^3_i (N^a_j)^* - N^4_i (N^a_j)^\circ + N^3_j (N^a_i)^* + N^4_j (N^a_i)^\circ \\
&= \partial_j(N^a_i) - \partial_i(N^a_j) - w_i (N^a_j)^* + w_j (N^a_i)^*,
\end{aligned}$$

and the nontrivial values for coefficients are

$$\begin{aligned}
\Omega^3_{12} &= -\Omega^3_{21} = \partial_2 w_1 - \partial_1 w_2 - w_1 w_2^* + w_2 w_1^* = w_1' - w_2^\bullet - w_1 w_2{}^* + w_2 w_1^*; \\
\Omega^4_{12} &= -\Omega^4_{21} = \partial_2 n_1 - \partial_1 n_2 - w_1 n_2^* + w_2 n_1^* = n_1' - n_2^\bullet - w_1 n_2^* + w_2 n_1^*. \qquad (A.4)
\end{aligned}$$

The nontrivial coefficients of d–torsion (14) are $\widehat{T}^a{}_{ji} = -\Omega^a{}_{ji}$ (A.4) and $\widehat{T}^c_{aj} = \widehat{L}^c_{aj} - e_a(N^c_j)$. We find for other types of coefficients that

$$\widehat{T}^i{}_{jk} = \widehat{L}^i_{jk} - \widehat{L}^i_{kj} = 0, \ \widehat{T}^i{}_{ja} = \widehat{C}^i_{jb} = 0, \ \widehat{T}^a{}_{bc} = \widehat{C}^a_{bc} - \widehat{C}^a_{cb} = 0.$$

We have such nontrivial N–adapted coefficients of d–torsion

$$\begin{aligned}
\widehat{T}^3_{3k} &= \widehat{L}^3_{3k} - e_3(N^3_k) = \frac{\partial_k h_3}{2h_3} - w_k \frac{h_3^*}{2h_3} - w_k^*, \\
\widehat{T}^3_{4k} &= \widehat{L}^3_{4k} - e_4(N^3_k) = -\frac{h_4 \underline{h}_4}{2h_3} n_k^*, \ \widehat{T}^4_{3k} = \widehat{L}^4_{3k} - e_3(N^4_k) = \frac{1}{2} n_k^* - n_k^* = -\frac{1}{2} n_k^*, \\
\widehat{T}^4_{4k} &= \widehat{L}^4_{4k} - e_4(N^4_k) = \frac{\partial_k (h_4 \underline{h}_4)}{2h_4 \underline{h}_4} - w_k \frac{h_4^*}{2h_4} - n_k \frac{\underline{h}_4^\circ}{2\underline{h}_4}, \\
-\widehat{T}^3_{12} &= w_1' - w_2^\bullet - w_1 w_2^* + w_2 w_1^*, \ -\widehat{T}^4_{12} = n_1' - n_2^\bullet - w_1 n_2^* + w_2 n_1^*. \qquad (A.5)
\end{aligned}$$

If all coefficients (A.5) are zero, then $\Gamma^\gamma{}_{\alpha\beta} = \widehat{\mathbf{\Gamma}}^\gamma{}_{\alpha\beta}$.

A.3. Calculation of the Ricci tensor

Let us compute the values $\widehat{R}_{ij} = \widehat{R}^k{}_{ijk}$ from (19) using (15),

$$\widehat{R}^i{}_{hjk} = \mathbf{e}_k \widehat{L}^i{}_{.hj} - \mathbf{e}_j \widehat{L}^i_{hk} + \widehat{L}^m_{hj} \widehat{L}^i_{mk} - \widehat{L}^m_{hk} \widehat{L}^i_{mj} - \widehat{C}^i_{ha} \Omega^a_{jk} = \partial_k \widehat{L}^i{}_{.hj} - \partial_j \widehat{L}^i_{hk} + \widehat{L}^m_{hj} \widehat{L}^i_{mk} - \widehat{L}^m_{hk} \widehat{L}^i_{mj},$$

where $\widehat{C}^i{}_{ha} = 0$ and $\mathbf{e}_k \widehat{L}^i_{hj} = \partial_k \widehat{L}^i_{hj} + N^a_k \partial_a \widehat{L}^i_{hj} = \partial_k \widehat{L}^i_{hj} + w_k \left(\widehat{L}^i_{hj}\right)^* + n_k \left(\widehat{L}^i_{hj}\right)^\circ = \partial_k \widehat{L}^i_{hj}$. Taking derivatives of (A.2), we obtain

$$\begin{aligned}
\partial_1 \widehat{L}^1{}_{11} &= \left(\frac{g_1^\bullet}{2g_1}\right)^\bullet = \frac{g_1^{\bullet\bullet}}{2g_1} - \frac{(g_1^\bullet)^2}{2\,(g_1)^2}, \ \partial_1 \widehat{L}^1{}_{12} = \left(\frac{g_1'}{2g_1}\right)^\bullet = \frac{g_1'^\bullet}{2g_1} - \frac{g_1^\bullet g_1'}{2\,(g_1)^2}, \\
\partial_1 \widehat{L}^1{}_{22} &= \left(-\frac{g_2^\bullet}{2g_1}\right)^\bullet = -\frac{g_2^{\bullet\bullet}}{2g_1} + \frac{g_1^\bullet g_2^\bullet}{2\,(g_1)^2}, \ \partial_1 \widehat{L}^2{}_{11} = \left(-\frac{g_1'}{2g_2}\right)^\bullet = -\frac{g_1'^\bullet}{2g_2} + \frac{g_1^\bullet g_2'}{2\,(g_2)^2}, \\
\partial_1 \widehat{L}^2{}_{12} &= \left(\frac{g_2^\bullet}{2g_2}\right)^\bullet = \frac{g_2^{\bullet\bullet}}{2g_2} - \frac{(g_2^\bullet)^2}{2\,(g_2)^2}, \ \partial_1 \widehat{L}^2{}_{22} = \left(\frac{g_2'}{2g_2}\right)^\bullet = \frac{g_2'^\bullet}{2g_2} - \frac{g_2^\bullet g_2'}{2\,(g_2)^2},
\end{aligned}$$

$$\begin{aligned}
\partial_2 \widehat{L}^1{}_{11} &= \left(\frac{g_1^\bullet}{2g_1}\right)' = \frac{g_1^{\bullet\prime}}{2g_1} - \frac{g_1^\bullet g_1'}{2\,(g_1)^2}, \ \partial_2 \widehat{L}^1{}_{12} = \left(\frac{g_1'}{2g_1}\right)' = \frac{g_1''}{2g_1} - \frac{(g_1')^2}{2\,(g_1)^2}, \\
\partial_2 \widehat{L}^1{}_{22} &= \left(-\frac{g_2^\bullet}{2g_1}\right)' = -\frac{g_2^{\bullet\prime}}{2g_1} + \frac{g_2^\bullet g_1'}{2\,(g_1)^2}, \ \partial_2 \widehat{L}^2{}_{11} = \left(-\frac{g_1'}{2g_2}\right)' = -\frac{g_1''}{2g_2} + \frac{g_1^\bullet g_1'}{2\,(g_2)^2}, \\
\partial_2 \widehat{L}^2{}_{12} &= \left(\frac{g_2^\bullet}{2g_2}\right)' = \frac{g_2^{\bullet\prime}}{2g_2} - \frac{g_2^\bullet g_2'}{2\,(g_2)^2}, \partial_2 \widehat{L}^2{}_{22} = \left(\frac{g_2'}{2g_2}\right)' = \frac{g_2''}{2g_2} - \frac{(g_2')^2}{2\,(g_2)^2}.
\end{aligned}$$

For these values, there are only two nontrivial components,

$$\begin{aligned}
\widehat{R}^{1}{}_{212} &= \frac{g_2^{\bullet\bullet}}{2g_1} - \frac{g_1^{\bullet}g_2^{\bullet}}{4\left(g_1\right)^2} - \frac{\left(g_2^{\bullet}\right)^2}{4g_1g_2} + \frac{g_1''}{2g_1} - \frac{g_1'g_2'}{4g_1g_2} - \frac{\left(g_1'\right)^2}{4\left(g_1\right)^2}, \\
\widehat{R}^{2}{}_{112} &= -\frac{g_2^{\bullet\bullet}}{2g_2} + \frac{g_1^{\bullet}g_2^{\bullet}}{4g_1g_2} + \frac{\left(g_2^{\bullet}\right)^2}{4(g_2)^2} - \frac{g_1''}{2g_2} + \frac{g_1'g_2'}{4(g_2)^2} + \frac{\left(g_1'\right)^2}{4g_1g_2}.
\end{aligned}$$

Considering $\widehat{R}_{11} = -\widehat{R}^{2}{}_{112}$ and $\widehat{R}_{22} = \widehat{R}^{1}{}_{212}$, when $g^i = 1/g_i$, we find

$$\widehat{R}^1_1 = \widehat{R}^2_2 = -\frac{1}{2g_1g_2}[g_2^{\bullet\bullet} - \frac{g_1^{\bullet}g_2^{\bullet}}{2g_1} - \frac{\left(g_2^{\bullet}\right)^2}{2g_2} + g_1'' - \frac{g_1'g_2'}{2g_2} - \frac{\left(g_1'\right)^2}{2g_1}],$$

which can be found in equations (42).

The next step is to derive the equations (43). We consider the third formula in (15),

$$\widehat{R}^{c}{}_{bka} = \frac{\partial \widehat{L}^c_{bk}}{\partial y^a} - (\frac{\partial \widehat{C}^c_{ba}}{\partial x^k} + \widehat{L}^c_{dk}\widehat{C}^d_{ba} - \widehat{L}^d_{bk}\widehat{C}^c_{da} - \widehat{L}^d_{ak}\widehat{C}^c_{bd}) + \widehat{C}^c_{bd}\widehat{T}^d_{ka} = \frac{\partial \widehat{L}^c_{bk}}{\partial y^a} - \widehat{C}^{c}{}_{ba|k} + \widehat{C}^{c}{}_{bd}\widehat{T}^{d}{}_{ka}.$$

Contracting indices, we get $\widehat{R}_{bk} = \widehat{R}^{a}{}_{bka} = \frac{\partial L^a_{bk}}{\partial y^a} - \widehat{C}^a_{ba|k} + \widehat{C}^a_{bd}\widehat{T}^d_{ka}$. For $\widehat{C}_b := \widehat{C}^c_{ba}$, we write

$$\widehat{C}_{b|k} = \mathbf{e}_k\widehat{C}_b - \widehat{L}^{d}{}_{bk}\widehat{C}_d = \partial_k\widehat{C}_b - N^e_k\partial_e\widehat{C}_b - \widehat{L}^{d}{}_{bk}\widehat{C}_d = \partial_k\widehat{C}_b - w_k\widehat{C}^*_b - n_k\widehat{C}^{\circ}_b - \widehat{L}^{d}{}_{bk}\widehat{C}_d.$$

We split conventionally $\widehat{R}_{bk} = {}_{[1]}R_{bk} + {}_{[2]}R_{bk} + {}_{[3]}R_{bk}$, where

$$\begin{aligned}
{}_{[1]}R_{bk} &= \left(\widehat{L}^3_{bk}\right)^* + \left(\widehat{L}^4_{bk}\right)^{\circ},\ {}_{[2]}R_{bk} = -\partial_k\widehat{C}_b + w_k\widehat{C}^*_b + n_k\widehat{C}^{\circ}_b + \widehat{L}^{d}{}_{bk}\widehat{C}_d, \\
{}_{[3]}R_{bk} &= \widehat{C}^a_{bd}\widehat{T}^d_{ka} = \widehat{C}^3_{b3}\widehat{T}^3_{k3} + \widehat{C}^3_{b4}\widehat{T}^4_{k3} + \widehat{C}^4_{b3}\widehat{T}^3_{k4} + \widehat{C}^4_{b4}\widehat{T}^4_{k4}.
\end{aligned}$$

Using formulas (A.2), (A.5) and (A.3), we compute

$$\begin{aligned}
{}_{[1]}R_{3k} &= \left(\widehat{L}^3_{3k}\right)^* + \left(\widehat{L}^4_{3k}\right)^{\circ} = \left(\frac{\partial_k h_3}{2h_3} - w_k\frac{h_3^*}{2h_3}\right)^* = -w_k^*\frac{h_3^*}{2h_3} - w_k\left(\frac{h_3^*}{2h_3}\right)^* + \frac{1}{2}\left(\frac{\partial_k h_3}{h_3}\right)^*, \\
{}_{[2]}R_{3k} &= -\partial_k\widehat{C}_3 + w_k\widehat{C}^*_3 + n_k\widehat{C}^{\circ}_3 + \widehat{L}^{3}{}_{3k}\widehat{C}_3 + \widehat{L}^{4}{}_{3k}\widehat{C}_4 = -\partial_k(\frac{h_3^*}{2h_3} + \frac{h_4^*}{2h_4}) + \\
&\quad w_k(\frac{h_3^*}{2h_3} + \frac{h_4^*}{2h_4})^* + (\frac{\partial_k h_3}{2h_3} - w_k\frac{h_3^*}{2h_3})(\frac{h_3^*}{2h_3} + \frac{h_4^*}{2h_4}) + \frac{1}{2}n_k^*\frac{\underline{h}_4^{\circ}}{2\underline{h}_4} \\
&= w_k[\frac{h_3^{**}}{2h_3} - \frac{3}{4}\frac{(h_3^*)^2}{(h_3)^2} + \frac{h_4^{**}}{2h_4} - \frac{1}{2}\frac{(h_4^*)^2}{(h_4)^2} - \frac{1}{4}\frac{h_3^*}{h_3}\frac{h_4^*}{h_4}] + n_k^*\frac{\underline{h}_4^{\circ}}{4\underline{h}_4} \\
&\quad + \left(\frac{\partial_k h_3}{2h_3} + \frac{\partial_k\underline{h}_3}{2\underline{h}_3}\right)(\frac{h_3^*}{2h_3} + \frac{h_4^*}{2h_4}) - \frac{1}{2}\partial_k(\frac{h_3^*}{h_3} + \frac{h_4^*}{h_4}), \\
{}_{[3]}R_{3k} &= \widehat{C}^3_{33}\widehat{T}^3_{k3} + \widehat{C}^3_{34}\widehat{T}^4_{k3} + \widehat{C}^4_{33}\widehat{T}^3_{k4} + \widehat{C}^4_{34}\widehat{T}^4_{k4} \\
&= -\frac{h_3^*}{2h_3}\left[\frac{\partial_k h_3}{2h_3} - w_k\frac{h_3^*}{2h_3} - w_k^*\right] - \frac{h_4^*}{2h_4}\left[\frac{\partial_k(h_4\underline{h}_4)}{2h_4\underline{h}_4} - w_k\frac{h_4^*}{2h_4} - n_k\frac{\underline{h}_4^{\circ}}{2\underline{h}_4}\right] \\
&= w_k^*\frac{h_3^*}{2h_3} + w_k\left(\frac{(h_3^*)^2}{4(h_3)^2} + \frac{(h_4^*)^2}{4(h_4)^2}\right)n_k\frac{h_4^*}{2h_4}\frac{\underline{h}_4^{\circ}}{2\underline{h}_4} - \frac{h_3^*}{2h_3}\frac{\partial_k h_3}{2h_3} - \frac{h_4^*}{2h_4}\left(\frac{\partial_k h_4}{2h_4} + \frac{\partial_k\underline{h}_4}{2\underline{h}_4}\right).
\end{aligned}$$

Summarizing, we get

$$\begin{aligned}
\widehat{R}_{3k} &= w_k\left[\frac{h_4^{**}}{2h_4} - \frac{1}{4}\frac{(h_4^*)^2}{(h_4)^2} - \frac{1}{4}\frac{h_3^*}{h_3}\frac{h_4^*}{h_4}\right] + n_k^*\frac{\underline{h}_4^{\circ}}{4\underline{h}_4} + n_k\frac{h_4^*}{2h_4}\frac{\underline{h}_4^{\circ}}{2\underline{h}_4} \\
&\quad + \frac{h_4^*}{2h_4}\frac{\partial_k h_3}{2h_3} - \frac{1}{2}\frac{\partial_k h_4^*}{h_4} + \frac{1}{4}\frac{h_4^*\partial_k h_4}{(h_4)^2} - \frac{h_4^*}{2h_4}\frac{\partial_k\underline{h}_4}{2\underline{h}_4}.
\end{aligned}$$

Journal of Physics: Conference Series **453** (2013) 012021 doi:10.1088/1742-6596/453/1/012021

which is equivalent to (43) if the conditions $n_k \underline{h}_4^\circ = \partial_k \underline{h}_4$, see below formula (A.6), are satisfied.

In a similar way, we compute $\widehat{R}_{4k} = {}_{[1]}R_{4k} + {}_{[2]}R_{4k} + {}_{[3]}R_{4k}$, where

$$
{}_{[1]}R_{4k} = \left(\widehat{L}^3_{4k}\right)^* + \left(\widehat{L}^4_{4k}\right)^\circ, \quad {}_{[2]}R_{4k} = -\partial_k \widehat{C}_4 + w_k \widehat{C}_4^* + n_k \widehat{C}_4^\circ + \widehat{L}^3{}_{4k}\widehat{C}_3
$$

$$
+\widehat{L}^4{}_{4k}\widehat{C}_4, \quad {}_{[3]}R_{4k} = \widehat{C}^a_{4d}\widehat{T}^d_{ka} = \widehat{C}^3_{43}\widehat{T}^3_{k3} + \widehat{C}^3_{44}\widehat{T}^4_{k3} + \widehat{C}^4_{43}\widehat{T}^3_{k4} + \widehat{C}^4_{44}\widehat{T}^4_{k4}.
$$

We get

$$
\begin{aligned}
{}_{[1]}R_{4k} &= -n_k^{**}\frac{h_4}{2h_3}\underline{h}_4 + n_k^*\left(-\frac{h_4^*}{2h_3} + \frac{h_4^* h_3^*}{2(h_3)^2}\right)\underline{h}_4 + \frac{\partial_k \underline{h}_4^\circ}{2\underline{h}_4} - \frac{\underline{h}_4^\circ \partial_k \underline{h}_4}{2(\underline{h}_4)^2}, \\
{}_{[2]}R_{4k} &= w_k\left(-\frac{h_4^*}{2h_4}\frac{\underline{h}_4^\circ}{2\underline{h}_4}\right) - n_k^*\frac{h_4\underline{h}_4}{2h_3\underline{h}_3}\left(\frac{h_3^*}{2h_3} + \frac{h_4^*}{2h_4}\right) + n_k[\left(\frac{\underline{h}_4^\circ}{2\underline{h}_4}\right)^\circ - \\
&\quad \frac{\underline{h}_4^\circ}{2\underline{h}_4}\frac{\underline{h}_4^\circ}{2\underline{h}_4}] + \frac{\partial_k h_4}{2h_4}\frac{\underline{h}_4^\circ}{2\underline{h}_4} + \frac{\partial_k \underline{h}_4}{2\underline{h}_4}\frac{\underline{h}_4^\circ}{2\underline{h}_4} - \frac{\partial_k \underline{h}_4^\circ}{2\underline{h}_4} + \frac{\underline{h}_4^\circ \partial_k \underline{h}_4}{2(\underline{h}_4)^2}, \\
{}_{[3]}R_{4k} &= w_k\left(\frac{h_4^*}{2h_4}\frac{\underline{h}_4^\circ}{2\underline{h}_4}\right) + n_k(\frac{\underline{h}_4^\circ}{2\underline{h}_4})^2 - \frac{\partial_k h_4}{2h_4}\frac{\underline{h}_4^\circ}{2\underline{h}_4} - \frac{\underline{h}_4^\circ}{2\underline{h}_4}\frac{\partial_k \underline{h}_4}{2\underline{h}_4}.
\end{aligned}
$$

We summarize the above three terms as follows

$$
\begin{aligned}
\widehat{R}_{4k} &= -n_k^{**}\frac{h_4}{2h_3}\underline{h}_4 + n_k^*\left(-\frac{h_4^*}{2h_3} + \frac{h_4^* h_3^*}{2(h_3)^*} - \frac{h_4^* h_3^*}{4(h_3)^*} - \frac{h_4^*}{4h_3}\right)\underline{h}_4 \\
&+ n_k\left(-\frac{\underline{h}_4^{\circ\circ}}{2\underline{h}_4} + \frac{(\underline{h}_4^\circ)^2}{2(\underline{h}_4)^2} + \left(\frac{\underline{h}_4^\circ}{2\underline{h}_4}\right)^\circ - \frac{\underline{h}_4^\circ}{2\underline{h}_4}\frac{\underline{h}_4^\circ}{2\underline{h}_4} + (\frac{\underline{h}_4^\circ}{2\underline{h}_4})^2\right) + \frac{\partial_k \underline{h}_4^\circ}{2\underline{h}_4} - \frac{\underline{h}_4^\circ \partial_k \underline{h}_4}{2(\underline{h}_4)^2} \\
&+ \frac{\partial_k h_4}{2h_4}\frac{\underline{h}_4^\circ}{2\underline{h}_4} + \frac{\partial_k \underline{h}_4}{2\underline{h}_4}\frac{\underline{h}_4^\circ}{2\underline{h}_4} - \frac{\partial_k \underline{h}_4^\circ}{2\underline{h}_4} + \frac{\underline{h}_4^\circ \partial_k \underline{h}_4}{2(\underline{h}_4)^2} - \frac{\partial_k h_4}{2h_4}\frac{\underline{h}_4^\circ}{2\underline{h}_4} - \frac{\underline{h}_4^\circ}{2\underline{h}_4}\frac{\partial_k \underline{h}_4}{2\underline{h}_4},
\end{aligned}
$$

and thus we have proved equations (44).

For $\widehat{R}^i{}_{jka} = \frac{\partial \widehat{L}^i_{jk}}{\partial y^k} - \left(\frac{\partial \widehat{C}^i_{ja}}{\partial x^k} + \widehat{L}^i_{lk}\widehat{C}^l_{ja} - \widehat{L}^l_{jk}\widehat{C}^i_{la} - \widehat{L}^c_{ak}\widehat{C}^i_{jc}\right) + \widehat{C}^i_{jb}\widehat{T}^b_{ka}$ from (15), we obtain zero values because $\widehat{C}^i_{jb} = 0$ and $\widehat{L}^i_{jk}$ do not depend on y^k. So, $\widehat{R}_{ja} = \widehat{R}^i{}_{jia} = 0$.

Taking $\widehat{R}^a{}_{bcd} = \frac{\partial \widehat{C}^a_{.bc}}{\partial y^d} - \frac{\partial \widehat{C}^a_{.bd}}{\partial y^c} + \widehat{C}^e_{.bc}\widehat{C}^a_{.ed} - \widehat{C}^e_{.bd}\widehat{C}^a_{.ec}$ from (15) and contracting the indices in order to obtain the Ricci coefficients, $\widehat{R}_{bc} = \frac{\partial \widehat{C}^d_{bc}}{\partial y^d} - \frac{\partial \widehat{C}^d_{bd}}{\partial y^c} + \widehat{C}^e_{bc}\widehat{C}_e - \widehat{C}^e_{bd}\widehat{C}^d_{ec}$, we compute $\widehat{R}_{bc} = (\widehat{C}^3_{bc})^* + (\widehat{C}^4_{bc})^\circ - \partial_c \widehat{C}_b + \widehat{C}^3_{bc}\widehat{C}_3 + \widehat{C}^4_{bc}\widehat{C}_4 - \widehat{C}^3_{b3}\widehat{C}^3_{3c} - \widehat{C}^3_{b4}\widehat{C}^4_{3c} - \widehat{C}^4_{b3}\widehat{C}^3_{4c} - \widehat{C}^4_{b4}\widehat{C}^4_{4c}$. There are nontrivial values,

$$
\begin{aligned}
\widehat{R}_{33} &= \left(\widehat{C}^3_{33}\right)^* + \left(\widehat{C}^4_{33}\right)^\circ - \widehat{C}_3^* + \widehat{C}^3_{33}\widehat{C}_3 + \widehat{C}^4_{33}\widehat{C}_4 - \widehat{C}^3_{33}\widehat{C}^3_{33} - 2\widehat{C}^3_{34}\widehat{C}^4_{33} - \widehat{C}^4_{34}\widehat{C}^4_{43} = \left(\frac{h_3^*}{2h_3}\right)^* \\
&\quad -\left(\frac{h_3^*}{2h_3} + \frac{h_4^*}{2h_4}\right)^* + \frac{h_3^*}{2h_3}\left(\frac{h_3^*}{2h_3} + \frac{h_4^*}{2h_4}\right) - \left(\frac{h_3^*}{2h_3}\right)^2 - \left(\frac{h_4^*}{2h_4}\right)^2 = -\frac{1}{2}\frac{h_4^{**}}{h_4} + \frac{1}{4}\frac{(h_4^*)^2}{(h_4)^2} + \frac{1}{4}\frac{h_3^*}{h_3}\frac{h_4^*}{h_4}, \\
\widehat{R}_{44} &= \left(\widehat{C}^3_{44}\right)^* + \left(\widehat{C}^4_{44}\right)^\circ - \partial_4 \widehat{C}_4 + \widehat{C}^3_{44}\widehat{C}_3 + \widehat{C}^4_{44}\widehat{C}_4 - \widehat{C}^3_{43}\widehat{C}^3_{34} - 2\widehat{C}^3_{44}\widehat{C}^4_{34} - \widehat{C}^4_{44}\widehat{C}^4_{44} \\
&= -\frac{1}{2}\left(\frac{h_4^*}{h_3}\right)^*\frac{\underline{h}_4}{\underline{h}_3} - \frac{h_4^*\underline{h}_4}{2h_3\underline{h}_3}\left(\frac{h_3^*}{2h_3} + \frac{h_4^*}{2h_4} - \frac{h_4^*}{h_4}\right) = -\frac{1}{2}\frac{h_4^{**}}{h_3}\underline{h}_4 + \frac{1}{4}\frac{h_3^* h_4^*}{(h_3)^2}\underline{h}_4 + \frac{1}{4}\frac{h_4^*}{h_3}\frac{h_4^*}{h_4}\underline{h}_4.
\end{aligned}
$$

We get the nontrivial v–coefficients of the Ricci d–tensor,

$$
\widehat{R}^3{}_3 = \frac{1}{h_3\underline{h}_3}\widehat{R}_{33} = \frac{1}{2h_3h_4}[-h_4^{**} + \frac{(h_4^*)^2}{2h_4} + \frac{h_3^*h_4^*}{2h_3}]\frac{1}{\underline{h}_3}, \ \widehat{R}^4{}_4 = \frac{1}{h_4\underline{h}_4}\widehat{R}_{44} = \frac{1}{2h_3h_4}[-h_4^{**} + \frac{(h_4^*)^2}{2h_4} + \frac{h_3^*h_4^*}{2h_3}]\frac{1}{\underline{h}_3},
$$

i.e., equation (42).

A.4. Zero torsion conditions

Let us analyze how to solve the equation

$$\widehat{T}^4_{4k} = \widehat{L}^4_{4k} - e_4(N^4_k) = \frac{\partial_k(h_4\underline{h}_4)}{2h_4\underline{h}_4} - w_k\frac{h_4^*}{2h_4} - n_k\frac{\underline{h}_4^\circ}{2\underline{h}_4} = 0,$$

which follows from formulas (A.5) for a vanishing torsion for $\widehat{\mathbf{D}}$. Taking any $\underline{h}_4$ for which

$$n_k\underline{h}_4^\circ = \partial_k\underline{h}_4, \tag{A.6}$$

the condition $n_k\frac{h_4^*}{2h_4}\frac{\underline{h}_4^\circ}{2\underline{h}_4} - \frac{h_4^*}{2h_4}\frac{\partial_k\underline{h}_4}{2\underline{h}_4} = 0$ is satisfied. For instance, parameterizing $\underline{h}_4 = {}^h\underline{h}_4(x^k)\underline{h}(y^4)$, the equation (A.6) is solved by any $\underline{h}(y^4) = e^{\varkappa y^4}$ and $n_k = \varkappa\partial_k[\ {}^h\underline{h}_4(x^k)]$, for $\varkappa = cont$.

We conclude that for any n_k and $\underline{h}_4$ related by conditions (A.6) the zero torsion conditions (A.5) are the same as for $\underline{h}_4 = const$. Using a similar proof from [21, 22], it is possible to verify by straightforward computations that $\widehat{T}^\alpha_{\beta\gamma} = 0$ if the equations (32) are solved.

[1] R. Bartnik and J. McKinnon 1988 Phys. Rev. Lett. **61** 141
[2] Y. Brihaye, A. Chakbarati and D. H. Tchrakian 2003 Class. Quant. Grav. **20** 2765
[3] Y. Brihaye, B. Hartmann and E. Radu 2006 Phys. Rev. Lett. **96** 071101
[4] Y. Brihaye, B. Hartmann, E. Radu and C. Stelea 2007 Nucl. Phys. **B 763** 115
[5] Y. Choquet–Bruhat and R. Geroch, Commun. Math. Phys. 1969 **14** 329
[6] Y. Choquet–Burhat and J. York 1989 The Cauchy problem, in: General Relativity, ed. A. Held, (Plenum Press, NY) p 99
[7] D. Christodoulou 2009 The formation of black holes in general relativity (European Mathematical Society Publishing House, Zürich, Switzerland)
[8] P. Chruściel and H. Friedrich (editors) 2004 The Einstein Equations and the Large Scale Behaviour of Gravitational Fields. 50 Years of the Cauchy Problem in General Relativity (Birkhäuser)
[9] P. Chruściel, G. Galloway and D. Pollack 2010 Bulletin (New Series) of the American Mathematical Society, **47** 567
[10] L. C. Evans 1998 Partial Differential Equations, Graduate Studies in Mathematics, vol. 19 (American Mathematical Society, Providence, RI)
[11] T. Harko, F. S. N. Lobo, Shin'chi Nojiri, S. D. Odintsov 2012 Phys. Rev. **D 84** 024020
[12] G. 't Hooft 1974 Nucl. Phys. **B79** 276
[13] B. B. Kadomtsev and V. I. Petviashvili 1970 Doklady Akademii Nauk SSS **192** 753 [in Russian]; Sov. Phys. Dokl. 1970 **15** 539 [English translation]
[14] S. Klainerman and F. Nicolo 2003 The Evolution Problem in General Relativity, Progress in Mathematical Physics **25** (Birkhouser, Boston Inc, Boston, MA)
[15] S. Klainerman and I. Rodnianski 2005 Ann. Math. (2) **161** 1195
[16] J. E. Lidsey, D. Wands and E. J. Copeland 2000 Phys. Rept. **337** 343
[17] J. M. Overduin and P. S. Wesson 1997 Phys. Rept. **283** 303
[18] I. Rodnianski 2006 in: Proceedings of the Interantional Congress of Mathematicians, Madrid, Spain, 2006, Vol. III, p 421
[19] H. F. Smith and D. Tataru 2005 Ann. Math. (2) **162** 291
[20] S. Vacaru 2007 Int. J. Geom. Methods. Mod. Phys. **4** 1285
[21] S. Vacaru 2011 Int. J. Geom. Meth. Mod. Phys. **8** 9
[22] S. Vacaru 2010 Int. J. Theor. Phys. **49** 884
[23] S. Vacaru 2010 Acta Applicandae Mathematicae **110** 73
[24] S. Vacaru and D. Singleton 2002 J. Math. Phys. **43** 2486

Journal of Physics: Conference Series **453** (2013) 012022 doi:10.1088/1742-6596/453/1/012022

Hořava–Lifshitz gravity with detailed balance

Daniele Vernieri and Thomas P. Sotiriou

SISSA, Via Bonomea 265, 34136, Trieste, Italy and INFN, Sezione di Trieste, Italy

E-mail: daniele.vernieri@sissa.it, sotiriou@sissa.it

Abstract. Hořava–Lifshitz gravity with "detailed balance" but without the projectability assumption is discussed. It is shown that detailed balance is quite efficient in limiting the proliferation of couplings in Hořava–Lifshitz gravity, and that its implementation without the projectability assumption leads to a theory with sensible dynamics. However, the (bare) cosmological constant is restricted to be large and negative.

1. Introduction

Hořava–Lifshitz gravity has been proposed as a power-counting renormalizable gravity theory, and a potentially ultra-violet (UV) complete theory of quantum gravity [1]. The improved UV behaviour is achieved by giving up Lorentz symmetry and adding terms to the action that are higher order in spatial derivatives, which suitably modifies the propagators. The theory is most straightforwardly defined in a space-time decomposition (ADM splitting) where a preferred foliation has been imposed and N is the lapse, N_i is the shift and g_{ij} the 3-metric induced on the spacelike hypersurfaces. The action of the theory is then

$$S = S_K - S_V\,, \tag{1}$$

where the kinetic term S_K is given by

$$S_k = \frac{2}{k^2}\int dt d^3x \sqrt{g} N K_{ij} G^{ijkl} K_{kl}, \tag{2}$$

K_{ij} is the extrinsic curvature of the spacelike hypersurfaces, k a coupling constant with suitable dimensions and

$$G^{ijkl} = \frac{1}{2}\left(g^{ik}g^{jl} + g^{il}g^{jk}\right) - \lambda g^{ij}g^{kl}\,, \tag{3}$$

is the generalized DeWitt metric. The potential term can be written as

$$S_V = \frac{k^2}{8}\int dt d^3x \sqrt{g} N\, V[g_{ij}, N]. \tag{4}$$

The action is required to be invariant under the subclass of diffeomorphisms that leave the foliation intact, *i.e.* $t \to \tilde{t}(t)$ and $x^i \to \tilde{x}^i(t, x^i)$, so in principle V should include all terms compatible with this symmetry. In addition, V should contain terms which are at least 6th order in spatial derivatives for the theory to have the desirable UV properties [1, 2, 3].

An unappealing feature of the full theory is that, without any further restrictions or symmetries, V will contain a very large number of terms (and independent couplings). Two restrictions have been considered as the means to limit the proliferation of couplings in the original proposal: *projectability* and *detailed balance*. Projectability sums up to the requirement that $N = N(t)$, whereas detailed balance requires V to be derived from a "superpotential" (see below).

An important characteristic of the theory is that it propagates not only a spin-2, but also a spin-0 mode. The assumption of projectability, with or without detailed balance [4] leads to pathologies in the dynamics of the scalar mode, in particular instabilities and strong coupling at low energies [5, 6, 7, 8, 9, 10]. For this reason, here we discuss a version of the theory which satisfies the detailed balance condition without assuming projectability.

This proceedings contribution is based on Ref. [11], where a more detailed discussion can be found. See also Ref. [12] for a brief review on the various versions of Hořava–Lifshitz gravity.

2. Detailed balance without projectability

Detailed balance prescribes that the potential V is derived from a superpotential W as follows

$$V = \frac{1}{\sqrt{g}}\frac{\delta W}{\delta g_{ij}} G_{ijkl} \frac{1}{\sqrt{g}}\frac{\delta W}{\delta g_{kl}}\,, \tag{5}$$

where G_{ijkl} is the inverse of the DeWitt metric G^{ijkl}. Once W has been chosen the action is fully determined. W can be a functional of the metric g_{ij} but also of the lapse N, and in particular of the combination $a_i = \partial_i \ln N$, if invariance under foliation preserving diffeomorphisms is imposed [13]. Furthermore, in order to have a power-counting renormalizable theory one needs at least sixth order spatial derivatives in the action [1, 2, 3], thus the superpotential W must contain at least third order spatial derivatives. Then, the most general W containing all of the possible terms up to third order spatial derivatives is

$$W = \frac{M_{\rm pl}^2}{2M_6^2}\int \omega_3(\Gamma) + \frac{M_{\rm pl}^2}{M_4}\int d^3x\sqrt{g}\left[R - 2\xi(1-3\lambda)M_4^2\right] + \beta\int d^3x\sqrt{g}\,a_i a^i, \tag{6}$$

where $\omega_3(\Gamma)$ is the gravitational Chern-Simons term, $M_{\rm pl}$, M_6, M_4 and β have dimensions of a mass, whereas ξ is a dimensionless coupling. The corresponding action is

$$\begin{aligned} S_H \;=\; & \frac{M_{\rm pl}^2}{2}\int dt d^3x\sqrt{g}N\Bigg\{K_{ij}K^{ij} - \lambda K^2 + \xi R - 2\Lambda + \eta\, a^i a_i - \frac{1}{M_4^2}R_{ij}R^{ij} + \frac{1-4\lambda}{4(1-3\lambda)}\frac{1}{M_4^2}R^2 \\ & + \frac{2\eta}{\xi M_4^2}\left[\frac{1-4\lambda}{4(1-3\lambda)}Ra^i a_i - R_{ij}a^i a^j\right] - \frac{\eta^2}{4\xi^2 M_4^2}\frac{3-8\lambda}{1-3\lambda}(a^i a_i)^2 + \frac{2}{M_6^2 M_4}\epsilon^{ijk}R_{il}\nabla_j R^l_k \\ & + \frac{2\eta}{\xi M_6^2 M_4}C^{ij}a_i a_j - \frac{1}{M_6^4}C_{ij}C^{ij}\Bigg\}, \end{aligned} \tag{7}$$

where C_{ij} is the Cotton–York tensor and

$$\eta = \frac{\beta\,\xi M_4}{M_{\rm pl}^2}\,, \qquad \Lambda = \frac{3}{2}\xi^2(1-3\lambda)M_4^2. \tag{8}$$

For $\xi, \lambda \sim 1$, which are the values these parameters have in general relativity, the bare cosmological constant, Λ, is negative [4]. Moreover, the magnitude of Λ is directly related to M_4 [14, 11] which is the mass scale at which Lorentz-violating effects become manifest as modifications to the dispersion relation. The mildest constraint on the value of M_4, which comes by the fact that the gravitation interaction has been tested down to mm scales, is $M_4 \geq 1 \div 10\text{meV}$.[1] Thus, the bare cosmological constant is at best 60 orders of magnitude bigger than the observed value.

The overall value of the cosmological constant would be the sum of the bare cosmological constant and the contribution from the vacuum energy of matter fields. Therefore, the size and not the sign of the bare cosmological constant is the main concern here. The contribution of the vacuum energy is not known with any certainty, but one can only hope for a miraculous cancelation if the overall value is to match the observed value. There is no known reason to expect such a cancelation (see, however, Ref. [14]).

3. Perturbations and power-counting renormalizability

In order to highlight that the magnitude of the cosmological constant is the only real concern and that the action constructed here describes a dynamically sensible theory, we will proceed as follows: we will set $\Lambda = 0$ *by fiat* (to simplify the analysis) and consider perturbations around a flat background. We focus in particular to scalar perturbations, as the main concern is the dynamics of the scalar mode. Thus we have

$$N = 1 + \alpha\,, \qquad N_i = \partial_i y\,, \qquad g_{ij} = e^{2\zeta}\delta_{ij}\,, \tag{9}$$

and at quadratic order the action becomes

$$S^{(2)} = \frac{M_{\rm pl}^2}{2}\int dt d^3x \left\{\frac{2(1-3\lambda)}{1-\lambda}\dot{\zeta}^2 + 2\xi\left(\frac{2\xi}{\eta}-1\right)\zeta\partial^2\zeta - \frac{2(1-\lambda)}{1-3\lambda}\frac{1}{M_4^2}(\partial^2\zeta)^2\right\}, \tag{10}$$

where we have integrating out the nondynamical degrees of freedom α and y. The corresponding dispersion relation for the scalar is

$$\omega^2 = \xi\left(\frac{2\xi}{\eta}-1\right)\frac{1-\lambda}{1-3\lambda}p^2 + \frac{1}{M_4^2}\left(\frac{1-\lambda}{1-3\lambda}\right)^2 p^4. \tag{11}$$

Then we can conclude that there is a suitable choice of the parameters such that the scalar has positive energy (the right sign is determined by the sign of the kinetic term of the spin-2 graviton), and at the same time is classically stable:

$$2\xi > \eta > 0\,. \tag{12}$$

The coefficient of the p^4 term in eq. (11) is manifestly positive, so there are no short-wavelength instabilities. However, the dispersion relation does not contain any sixth order terms. This is because the only sixth order term in action (7) does not contribute to the dynamics of the scalar more. The absence of sixth order terms in the dispersion relation poses a threat for renormalizability. However, a straightforward resolution is to add to the superpotential W fourth order terms, so as to generate sixth (and eight order terms) in the action, which would restore the appealing renormalizability properties of the theory. The minimal consistent prescription would be to add to W all possible fourth order terms that respect invariance under foliation preserving diffeomorphisms, but also impose parity invariance. Then the terms added would be

$$R^2\,, \quad R^{\mu\nu}R_{\mu\nu}\,, \quad R\nabla^i a_i\,, \quad R^{ij}a_ia_j\,, \quad Ra_ia^i\,, \quad (a_ia^i)^2\,, \quad (\nabla^i a_i)^2\,, \quad a_ia_j\nabla^i a^j\,. \tag{13}$$

One would be left with 12 free couplings, which is an appreciable improvement with respect to the theory without detailed balance.

[1] Stronger constraints can be obtained for M_4 but depend on the details of how Lorentz violations percolate into the matter sector, see *e.g.* Ref. [15].

Journal of Physics: Conference Series **453** (2013) 012022 doi:10.1088/1742-6596/453/1/012022

4. Conclusions

We have put together an action for Hořava–Lifshitz gravity that satisfies detailed balance without assuming projectability. This version appears to lead to sensible dynamics for the scalar mode and be free of other pathologies that plague the projectable version of Hořava–Lifshitz gravity. Additionally, it has the appealing renormalizability properties of the most general non-projectable theory, with about an order of magnitude less independent couplings.

However, imposing detailed balance leads to a large and negative bare cosmological constant. Barring a miraculous cancelation with the (yet to be understood) vacuum energy contribution to the cosmological constant, this is obviously hard to reconcile with low energy gravitational phenomenology. An additional caveat in the use of detailed balance is the fact that it is not clear if or why it should be robust against radiative corrections. The implementation of detailed balance presented here leads to a consistent theory which is free of pathologies and, therefore, allows to consider this question rigorously.

References

[1] Hořava P 2009 *Phys. Rev.* D **79** 084008
[2] Visser M 2009 *Phys. Rev.* D **80** 025011
[3] Visser M 2009 arXiv:0912.4757 [hep-th]
[4] Sotiriou T P, Visser M and Weinfurtner S 2009 *Phys. Rev. Lett.* **102** 251601
[5] Sotiriou T P, Visser M and Weinfurtner S 2009 *JHEP* **0910** 033
[6] Charmousis C, Niz G, Padilla A and Saffin P M 2009 *JHEP* **0908** 070
[7] Blas D, Pujolas O and Sibiryakov S 2009 *JHEP* **0910** 029
[8] Wang A and Maartens R 2010 *Phys. Rev.* D **81** 024009
[9] Afshordi N 2009 *Phys. Rev.* D **80** 081502
[10] Koyama K and Arroja F 2010 *JHEP* **1003** 061
[11] Vernieri D and Sotiriou T P 2012 *Phys. Rev.* D **85** 064003
[12] Sotiriou T P 2011 *J. Phys. Conf. Ser.* **283** 012034
[13] Blas D, Pujolas O and Sibiryakov S 2010 *Phys. Rev. Lett.* **104** 181302
[14] Appignani C, Casadio R and Shankaranarayanan S 2010 *JCAP* **1004** 006
[15] Liberati S, Maccione L and Sotiriou T P 2012 *Phys. Rev. Lett.* **109** 151602

Journal of Physics: Conference Series **453** (2013) 012023 doi:10.1088/1742-6596/453/1/012023

Causal Sets Dynamics: Review & Outlook

Petros Wallden

1. University of Athens, Physics Department, Nuclear & Particle Physics Section, Panepistimiopolis 157-71, Ilissia Athens, Greece
2. Technological Educational Institute (TEI) of Chalkida, General Science Department, Psahna-34400, Greece

E-mail: petros.wallden@gmail.com, pwallden@phys.uoa.gr

Abstract.
Causal sets is an approach to quantum gravity, where spacetime is replaced by a causal set. It is fundamentally discrete, and the causal relations between spacetime elements is the only structure that remains. A complete theory should have (i) kinematics (ii) dynamics and (iii) phenomenology. In this contribution we will explore the dynamical part of the theory, focusing on recent developments. We will analyse (a) classical dynamics *of* the causal set, (b) quantum dynamics of matter and fields on a classical causal set and finally (c) quantum dynamics of the causal set.

1. Motivation

Constructing a quantum theory of gravity, consistent, widely accepted and confirmed by experiments is probably the most important open problem in current theoretical physics. The mathematical and conceptual difficulties that arise, suggest we may need to abandon some of the a-priori assumptions we make about nature. This is essentially done in all approaches to quantum gravity. In this contribution, we examine the causal sets approach (first appeared in [1] while in [2] some reviews can be found). It is the nature of spacetime as a differential manifold with a Lorentzian metric on it that is questioned in this approach. It is replaced by a causal set that is (i) a discrete entity with (ii) only remaining structure the causal relations between spacetime points. Both assumptions, discreteness and causality are based on indications from current physics and naturality arguments.

Approaches to quantum gravity, that start with a continuous spacetime, observe a possibly effective, discreteness. Examples of the above, are the discreteness of the volume operator in Loop Quantum Gravity and the dualities between small and large scales in string theory. Discreteness of spacetime, could also remove the infinities we face in quantum field theory, which are problematic even though we have "learned living" with them. Furthermore, singularities in general relativity could be avoided. Indications for discreteness of spacetime, exist in black hole thermodynamics, where the finiteness of the entropy that is proportional with the area of the horizon can be understood easier if that corresponded to the number of fundamental area units that cover the horizon. Finally, certain modifications of gravity that account for the smallness of the cosmological constant also point towards discreteness of spacetime. All these arguments suggest, that a natural starting point for a quantum gravity theory, would be a discrete spacetime.

The second feature that is used in causal sets in order to replace the standard view of spacetime, is the *causality*. The concept that two spacetime points are causally related, has the direct explanation, that objects/fields at the one (the one being "earlier") can affect objects/fields at the other. It is a natural and intuitive concept. Even though we are used to the standard view of spacetime, arguably the topology and differential structure and the Lorentzian signature metric, are far more abstract concepts than the simple notion of causality. Moreover, as we will see below, the causal relation can encode most of the information of the Lorentzian signature metric on a manifold.

In section 2 we will define mathematically a causal set and explain in which sense it can replace spacetime, give some basic definitions that will be required later on and divide the analysis of a theory on 3 parts, the kinematics, dynamics and phenomenology. Section 3 is the main part, dealing with the dynamics of causal sets. In section 3.1 we examine the classical dynamics of a causal set, in section 3.2 the developments that have occurred in the quantum dynamics of matter and fields on a classical/fixed causal set and in section 3.3 the quantum dynamics of the causal sets themselves. Finally in section 4 we will summarise and conclude.

2. Definition and basics

Causal set is the mathematical entity that replaces spacetime in this approach to quantum gravity. Mathematically is a set C with the following features.

(i) A partial order relation $\prec$ which is (a) irreflexive ($x \not\prec x$) and (b) transitive ($x \prec y \prec z \Rightarrow x \prec z$)

(ii) The partial order corresponds to the causal relation between elements of C, so if $x \prec y$ it means that x is in the past of y.

(iii) Locally finite: $[x, y] \equiv |\{y \text{ such that } x \prec y \prec z\}| < \infty \ \forall \ \ x, z \in C$. And $|A|$ indicates cardinality of the set A. It is this condition that imposes the discreteness of spacetime, since it requires that between every pair of elements of the causal set, to be only a finite number of other elements.

Here we should point out that the standard spacetime, if we use the metric to define causal relations, satisfies the first two conditions we required above and it is only the locally finiteness condition that is not satisfied.

Following the definition of a causal set, an explanation is needed on why could such an entity replace the continuum spacetime. It is based on works of Hawking [3] and later of Malament [4] and in particular on this theorem by Malament:

> The metric of a globally hyperbolic spacetime can be reconstructed uniquely from its causal relations up to a conformal factor.

This theorem states that most of the information of the metric is encoded in the causal relations of the spacetime points. Having the conformal metric, means that the metric is completely fixed, once we determine the spacetime volume of each neighborhood. However, the discreteness of spacetime provides a very natural way to fix the volume. Since each spacetime interval, by definition, has a finite number of elements, by assuming that each element corresponds to some fundamental discrete volume (Plank volume), the volume of each spacetime interval is uniquely determined. In other words we make the correspondence "Volume = number of elements × Plank volume". In a schematic way we have that "Order + Number = Geometry ". This leads to the central conjecture of causal sets (also called the Hauptvermutung):

> **Central Conjecture**: Two distinct, non-isometric spacetimes cannot arise from a single causal set.

The conjecture is still not proven, however it has very solid theoretical and numerical evidence for its validity[1]. Having a causal set as the real structure existing in nature, we may ask under what circumstances a spacetime (Lorentzian manifold) can be said to approximate a given causal set. To make the above precise we define the concept of a *faithful embedding*:

A faithful embedding is a map ϕ from a causal set $\mathcal{P}$ to a spacetime $\mathcal{M}$ that:

(i) preserves the causal relation ($i.e., x \prec y \iff \phi(x) \prec \phi(y)$)

(ii) is "volume preserving", meaning that the number of elements mapped to every spacetime region is Poisson distributed, with mean the volume of the spacetime region in fundamental units, and

(iii) $\mathcal{M}$ does not possess curvature at scales smaller than that defined by the "intermolecular spacing" of the embedding (discreteness scale).

The central conjecture can be restated as "a causal set cannot be faithfully embedded in two non isometric spacetimes".

While physically, it is the spacetime approximating the causal set, we can ask the inverse question and in particular to ask to find a causal set that approximates a given spacetime. This will happen, when the number of elements of the causal set is large and we speak of continuum approximation, which is the analogue of a continuum limit[2].

An interesting thing to point out, is that following the definition of a faithful embedding, we can see that a regular lattice does not correspond to Minkowski spacetime since there is no faithful embedding between them. The condition (ii) states that *every* spacetime interval should have a number of elements of the causal set mapped that is proportional to the volume (on average). Considering a boosted frame one can easily notice that there are large volumes with no elements at all. It is well known that there is a tension between discreteness and Lorentz invariance. However, we *can* have a causal set such that it faithfully embeds in Minkowksi spacetime, by considering a random lattice. One can generate such a lattice, by sprinkling elements in Minkowski spacetime, randomly with probability $P(n) = \frac{(\rho V)^n \exp^{-\rho V}}{n!}$ and variance $\sqrt{V}$ which is called a Poisson sprinkling.

Causal sets constructed in this manner, respect Lorentz invariance at the kinematic level, and evade the above mentioned tension between discreteness and Lorentz invariance. Other discrete approaches, do not satisfy this condition on kinematic level and expect Lorentz-invariance to emerge at a later stage (e.g. through superposition of Lorentz-violating objects). Further analysis of the above can be found in [11].

We should note that according to the definition, a given causal set is not necessarily faithfully embedded in a manifold. In other words, a random causal set does not always correspond to some continuous spacetime. Moreover, if we take a typical causal set consisting of a large number of elements n, it does not look like a manifold, but rather as a Kleitman-Rothschild

[1] Most of the work on the kinematics of causal sets, regarding the dimension (e.g. [5, 6]), topology ([7]), timelike and spacelike distances ([8, 9]), provide support to the conjecture.

[2] It is analogue, since the actual limit of taking some scale going to exactly zero, is never taken as for example in CDT [10].

(KR) order [12]. A typical causal set is selected, by pure counting, if we assume that any possible order (causal set) has equal weight. The KR orders have only three layers (i.e. longest chain is a 3-element chain) and 1/4 of elements are in the first layer, 1/2 in the second and the remaining 1/4 in the third. These causal sets are clearly not manifold-like (they have only "three moments of time"). It remains a task for the dynamics of the theory, to select a causal set that can be faithfully embedded on a continuous spacetime rather than one that is of the KR type.

To this end, we will give some definitions for causal sets that will be needed later in the text. We call *links* a pair of elements $x \prec y$ such that $\nexists\ z$ such that $x \prec z \prec y$. In other words two elements are linked, if there is no element "between" them. We call *chain* C, a collection of elements that $\forall \quad x, y \in C$, either $x \prec y$ or $y \prec x$. In other words, all the elements in the cain are ordered, and it is the causal set analogue of a timelike curve. Closely related is the definition of a *path* which is a chain with the extra condition that each pair of consecutive elements are links. An *anti-chain* is a collection of elements A, that $\forall \quad x, y \in A$ neither $x \prec y$ nor $y \prec x$. I.e. it is a collection of unrelated elements. To consider a *maximal* anti-chain, means that it is an anti-chain such that all the remaining elements of the causal set are to the past or to the future of at least one element of the anti-chain, i.e. we cannot extend the anti-chain and still remain anti-chain. The maximal anti-chain is the causal set analogue of spacelike surface.

When constructing a theory, there are three elements that one needs to address. First is the kinematics of the theory, and for causal sets, one can see how the concepts of dimension [5, 6], topology [7], distances (temporal and spatial) [8, 9], and geometry can arise. Second is the dynamics, that is the topic of this contribution. Finally, we have the phenomenology. It deals with the observable consequences that the novel elements of the theory lead us. For causal sets people have explored the cosmological constant problem that attains a potential solution [13, 14], entropy bounds and black hole entropy and thermodynamics [15, 16] and deviations/diffusion of motion due to the fundamental discreteness [17].

3. Dynamics

While from the kinematics of a theory one can gain some insight in the understanding of nature, the basic physical content of a theory is encoded in the dynamics. We will divide our analysis into three parts. Firstly we will examine the possible classical dynamics that a causal set can have. This should be done in a way that is intrinsic to the causal set itself and make use solely of the causal relations. The analogue in continuum physics, would be the theory of general relativity which is the classical theory for spacetime. Secondly, we will explore, the dynamics of quantum matter and fields on a given "classical" causal set. This is the analogue of quantum field theory on a fixed curved spacetime. Finally we will analyse, the possible quantum dynamics of the causal set itself, which is the final aim in order to construct a quantum theory for gravity.

There are two issues that one needs to consider when, in the attempt to quantise gravity, takes as starting point an alternative (possibly discrete) structure for spacetime. The first issue is the so-called "entropy" problem. It is common in all discrete approaches. By pure counting, a typical discrete structure (in our case a causal set) does not correspond in the appropriate limit to a continuous spacetime. As we have seen the typical causal set is a KR order. In constructing the appropriate dynamics, one should overcome this difficulty and obtain a dynamical way of selecting causal sets that do correspond to continuous spacetimes.

The second issue, is which direction one would take for the (classical or quantum) dynamics. On can have a *bottom-up* point of view, where we start from the fundamental relations of causal sets and by imposing conditions that are natural to the order itself we obtain the dynamics. This is the most ambitious but also principled way. Alternatively, we can have a *top-down* approach, where one gets motivation from continuous spacetime, rephrase the concepts in terms of the causal sets relations and uses them in order to obtain the dynamics. While, this is not conceptually preferable, if assume that nature *is* a causal set in reality, technically is easier to proceed and it may lead to useful conclusions.

Finally, having formally quantise causal sets is not the end of the story. To make contact with experiments, one needs a proper way to interpret the theory. There are two aspects that need consideration. First one needs to formally construct questions that are invariant from the arbitrary choices that were made during the quantisation process and do not correspond to physical degrees of freedom. This is guaranteed by dealing with re-labelling invariant questions, which is the discrete analogue of diffeomorphisms invariance. The second aspect has to do with the interpretation of quantum theory for the case that we have a single closed no-repeatable system. The problem is twofold, first we need an interpretation that does not require an external observer and secondly that gives predictions for single systems rather than ensembles of identically prepared systems. These thoughts, along with the inability of having a canonical approach due to the fundamental spacetime nature of causal sets, leads to interpretations based on histories, such as the co-event formulation [18] or the decoherent histories approach [19].

3.1. Classical Dynamics for Causal Sets

The classical dynamics of causal sets that have been analysed so far, belong to the "bottom-up" approach. We start with the basic concepts of causal sets, and by imposing certain physical conditions we generate the most general dynamics that can be obtained. The dynamics that we consider, have certain free-parameters that along with assumptions on the initial condition, determine the type of "evolution" of a causal set. In this general set up, the dynamics are *not* deterministic, but rather stochastic. For a more detailed description the reader is directed to the original references [20].

The arena that the dynamics of causal sets take place, is the set of all causal sets. This set has a natural partial order on it, which is that two elements (i.e. causal sets), are related, if the one is subset of the other, when viewed as partial orders. This partial order of partial orders, is called *poscau* (see Fig. 3.1). It is the analogue of superspace for geometrodynamics, or more precisely in our case, the space of all spacetimes.

The dynamics we will consider, generate the causal set in the following manner. We start with one element, and then grow the causal set by adding one more element at every step/level. In each level, the newborn element has to be determined whether it is at the past or future or unrelated to every other, already existing element of the causal set. This corresponds to a move from one point of poscau to another, since the old causal set is obviously subset of the new. The most general dynamics, would allow with some probability any of those transitions. This process of growing the causal set with adding a new element, with some probability of where to add it, is called *Classical Growth Dynamics* (CGD). This process has an "internal/parameter time" which is the number of born elements, which however, should not be confused with the time observed by observers in the causal set itself. These stochastic dynamics are actually generalisations of random walks.

However we are not totally free to choose any parameter for the probabilities of the newborn elements. There are the following physical conditions that our probabilities are required to satisfy, that turn out to be very restrictive.

(i) **Internal temporality**. Each new element is born to the future or spacelike to all existing

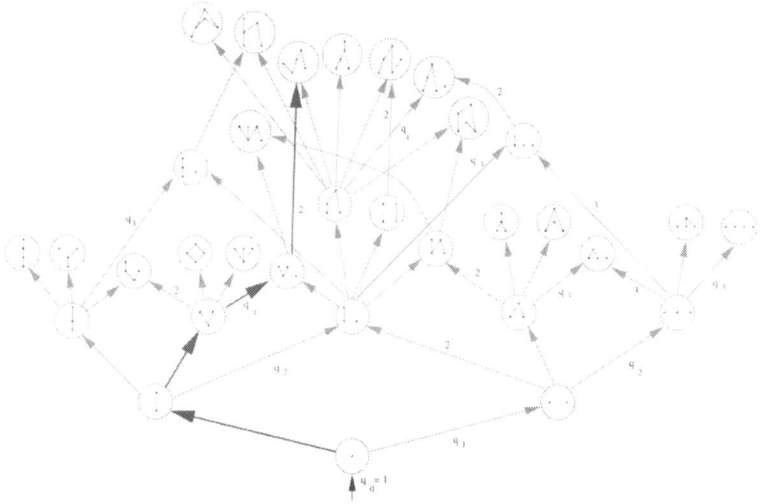

The poset of finite causal sets

elements (never at the past). A natural requirement, that guarantees that the "present time" causal set (in the internal time defined above), is not affected by adding later more elements.

(ii) **Discrete general covariance**. The probability of arriving at a particular causal set does not depend on the order the elements are born. In other words the probability does not depend on the path chosen on the Figure, but only on the starting and ending point on the poscau. This condition guarantees label-invariance of the theory and it is the discrete analogue of diffeomorphism invariance at general relativity.

(iii) **"Bell's causality"**. The probability for the newborn element depends only on the elements of the existing causal set that belong to the past of the new element and *not* from the total causal set. This condition guarantees some sense of locality for the evolution of the causal set. Quantum paradoxes, such as the Bell inequalities, that arise due to non-locality gave the name for this condition. It is therefore believed, that for quantum dynamics of the causal set this condition might need to either be dropped or relaxed to accommodate those paradoxes. However, non-local effects on a causal set may arise even if the dynamics of the causal set itself are local in the way defined above and therefore altering this condition is not theoretically necessary.

(iv) **The Markov Sum rule**. Starting from a given causal set with n elements (i.e. a point at the poscau) and adding a single element, give rise to many possible different causal sets with $n+1$ elements. The sum of the probabilities for all these $n+1$ elements causal sets, need to sum up to one.

The above 4 conditions, are very well physically motivated and seem very general. However, they restrict the possible parameters and there is only *one* free parameter for each level of the growth process. By each level, we mean, the total number of elements of the causal set (see details in [20]). The relation these parameters have between them give rise to different dynamics. For example, setting them equal, give rise to the well known and explored example of transitive percolation (see references in [20]). The choices for dynamics are still many (remember that a typical realistic causal set has huge number of elements), and the resulting causal sets differ. Since we have started with concepts intrinsic to causal sets we have followed a strictly bottom-up approach, and this is how far we can reach without resorting to concepts arising from continuous spacetimes. Despite the fact that certain choices of the parameters give rise to causal sets with notable similarities with cosmological models [21] (and certainly nothing like the KR typical causal sets), it is believed that the resulting causal sets are not in general manifold-like. It is thus believed that in order to explain the continuum-like behaviour, we may need to resort to quantum dynamics, as we will see in section 3.3.

3.2. Quantum Matter on a fixed Causal Set

The second type of dynamics we consider, is the dynamics of matter and fields on a single fixed causal set. As already stated, this is the analogue of fixing the background to a curved, in general, background. One needs to first recover the standard results from continuous spacetimes, and then explore the possibility for deviations from them due to the underlying fundamental discreteness. The effects are typically small and may depend on free parameters and possibly on the full quantum dynamics. In any case, this analysis may give us hints and directions both for developing quantum dynamics and for testing (with experiments eventually) the theory.

Matter on a causal set may arise in two ways. Either from the relations between elements of the causal set or by adding matter or fields on a given causal set. The first one is more intrinsic to the causal set but will not be explored here, since not much progress has been made. Matter degrees of freedom could appear e.g. in a Kaluza-Klein way, from variations of the causal structure at different parts of the total causal set. The second one will be explored in the following.

First we will explore the case of a causal set that faithfully embeds into Minkowski spacetime. The way this causal set arises is the issue of quantum dynamics of the causal set and indications for that will be reported in the next subsection. Secondly, using certain results from the flat case, we will consider causal sets that faithfully embed into curved spacetimes.

One thing to do is to consider point particles (massive for now), moving along a chain of the causal set that faithfully embeds into Minkowski spacetime. Several models were considered [17, 22] that shared the common feature, that in the continuum approximation the particle followed roughly a timelike geodesic, slightly deviating (swerving) from this motion. The resulting effect, is that of a drift motion, and the evolution equation turns out to be a diffusion equation that depends on a single parameter the "diffusion strength" k. We will present here briefly the first model as in [22]. The particle follows trajectories that are chains of the causal set $\{e_1, e_2, \cdots\}$. There is a time t_f, below which the causal set may behave non-locally. It is called "forgetting time" and above this it behaves normally. If the particle has position e_n with four momentum p_n the next element is chosen such that:

- e_{n+1} is at the causal future of e_n and at proper time t_f.
- The momentum change $|p_n - p_{n+1}|$ is minimized. Note, that p_{n+1} is on the mass shell and proportional to the vector between e_n and e_{n+1}.

The effect of this, is a motion which is as straight as possible, with some deviations, due to random fluctuations that the particle's momentum has. Taking the continuum approximation leads to the following diffusion equation (see [22]).

$$\frac{\partial \rho}{\partial \tau} = k\nabla^2\rho - \frac{1}{m}p^\mu\partial_\mu\rho \qquad (1)$$

The next stage is to examine the consequences of discreteness for massless particles. Technically, there is an extra difficulty, which arises due to the fact that the trajectories are no longer chains, since chains correspond by definition to timelike curves. We are lead to an equation with two parameters (diffusion and drift) k_1, k_2. The reader is referred to [17].

The next thing to consider are quantum fields, and in order to do so, we have to look on how to obtain the d'Alembertian and the propagators. In [23], the d'Alembertian was introduced. Here we follow [24], to obtain the retarded propagator for a particle on a causal set.

To get the propagator $K(x, y)$ from an element x to y in its future, one needs to sum over all the chains (or paths) on a causal set from element x to element y, weighted with a particular amplitude. This is in direct analogy with quantum mechanics where the quantum amplitudes to go from one point to another are obtained by summing over all the possible trajectories. It is required to choose what is the analogue for the causal set, of the trajectories that we will sum over. It can be either the chains or (more restricted) the paths. The choices made should lead to the desired behaviour when considering causal sets faithfully embedded in Minkowski spacetime, i.e. give rise to the propagator of the Klein-Gordon equation (with the choices made in [24] we will get the retarded propagator). The amplitude depends on two parameters a which is the probability that the particle "hops" once along the trajectory from one element to another, and b which is the probability that the particle stops at an element of the trajectory. For a chain of length n (where we have n hopes and $n-1$ intermediate stops) the amplitude would be $a^n b^{n-1}$. For causal set in 1+1 Minkowski and 3+1 Minkowski we can choose the values of a and b in such a way to recover the retarded Klein-Gordon propagator. The values depend only on the mass of the particle and the volume corrseponding to each causal set element, i.e. the density. For 1+1 dimension these values are:

$$a = \frac{1}{2}, \quad b = -\frac{m^2}{\rho} \tag{2}$$

while for 3+1 dimensions the values are:

$$a = \frac{\sqrt{\rho}}{2\pi\sqrt{6}}, \quad b = -\frac{m^2}{\rho}. \tag{3}$$

Further details can be found in [24]. Continuing in the same line of research, in [25] scalar quantum field on a causal set was considered, and the Feynman propagator was computed. Of crucial importance for that work, was the use of Pauli-Jordan function [26] $\Delta(x) := G_R(x) - G_A(x)$ and its analogue for a causal set. Further generalisations involving curved spacetimes could be possible.

A different direction was taken in [23] and [27]. A slowing varying (at some frame) field $\phi(x)$ was considered and they used:

$$\Box_d\phi(u, v) = \frac{1}{a^2}(\phi(u, v) - \phi(u-a, v) - \phi(u, v-a) + \phi(u-a, v-a)) \tag{4}$$

in order to define the following B operator:

$$\begin{aligned} B\phi(x) &= \frac{4}{a^2}(-\frac{1}{2}\phi(x) + \sum_{y\in N_1(x)}\phi(y) \\ &\quad -2\sum_{y\in N_2(x)}\phi(y) + \sum_{y\in N_3(x)}\phi(y)) \end{aligned} \tag{5}$$

$N_i(x)$ is the set of elements y, such that the number of elements z that satisfy $y \prec z \prec x$ is exactly i (i.e. i inclusive intervals). It has be shown (and confirmed by simulations) that this operator, on average, gives the value of the d'Alembertian for flat space in the suitable limit. Further care is needed to guarantee the variations are also controlled [23].

This work is the starting point on extending the work for causal sets that faithfully embed in curved spacetime. Using expressions for the Ricci scalar in terms of volume and proper distance of small causal intervals [6, 28] the expression for the operator B (d'Alembertian) in flat space changes by a term:

$$B\phi(x) = (\Box - \frac{1}{2}R)\phi(x) \tag{6}$$

However, if we apply this expression to a constant field $\phi(x) =$ constant it gives an expression of the Ricci scalar for the causal set. This leads to an important step for causal sets, which is the derivation of an analogue of the Einstein-Hilbert action in curved spacetime called the *Benincasa-Dowker* (BD) Action (see [27]). The expressions for 2 and 4 dimensions are the following:

$$\frac{1}{\hbar}S^{(2)} = N - 2N_1 + 4N_2 - 2N_3 \tag{7}$$

$$\frac{1}{\hbar}S^{(4)} = N - N_1 + 9N_2 - 16N_3 + 8N_4 \tag{8}$$

N_i are the i-inclusive intervals as defined above. Generalisation for arbitrary dimensions also exist. The importance of this development, is that one can use this BD action, for constructing quantum dynamics.

Finally, recently an alternative expression for the Ricci scalar was derived in [29], having used the concept of neighborhood. Rather than using intervals as in the BD action, the authors used chains (see eq. 47 of [29]). While the expressions for arbitrary dimensions were clearer and could be used as an improved dimension estimator, this action, cannot be used directly for quantum dynamics, since it is local and requires the concept of a neighborhood. It is difficult to define such concept from the causal relations themselves without resorting to the background structure.

3.3. Quantum Causal Sets

The first important observation, is that causal sets have a spacetime nature, from their definition. A spacelike surface is simply a collection of unrelated elements, and other candidates for this concept, such as "thickened" anti-chains (see [7]) are not intrinsic to the causal set. It follows that a canonical quantisation of causal sets, even if possible, would contradict the spirit of the approach. This leads, to path integral quantisations and in general in "histories" formulations.

The mathematical aim is the assignment of a quantum amplitude to each causal set. Alternatively, one can define a quantum measure (see [30]) on the space of all causal sets (poscau). Interpreting the quantum measure (or the amplitudes) is complicated as already mentioned at the end of section 2 and a histories approach ([18, 19]) should be preferred.

Technically, there are two ways to proceed. The first one, is to attempt to generate the quantum measure, through a growth process, similarly with the CGD and is a bottom-up approach. It is a generalisation of quantum random walks. The second is to directly assign a weight on causal sets and this is usually done in a manner justified from continuum physics and is thus a top-down approach.

Quantum Growth Dynamics: There are two notable differences, in the quantum case. First of all, instead of a classical measure, we have either a quantum measure, or quantum amplitude.

In those cases, it is technically more involved to extend the measure to the full histories space. However, progress in this direction has been made for measurable sets of histories (see [31]). The second issue, is that the requirement of Bell's causality, may not be a suitable choice for quantum causal sets. It is well known in quantum theory, that the Bell's inequalities are violated and thus we require a novel locality condition that allows for these violations. For the moment, there is no widely accepted suitable generalisation of this condition.

Path Integral Approach: A first and very interesting result appeared in [32], from consideration of the so-called 2-dimensional orders. To this end, we should stress what is meant by a 2-dimensional order. It is a technical term for partial orders, and in general has nothing to do with spacetime dimension. The definition is the following:

Assume that we have a set $\mathcal{P} = \{e_1, e_2, \cdots\}$ along with two linear orderings of this set. Then we take the intersection of this linear orderings, which means that an element $e_1 \prec e_2$ iff e_1 is before e_2 in both linear orderings. The resulting structure is a partial order that is called 2-dimensional. Conversely,if we have a partial order that can be generated from the intersection of 2 linear orderings it is called 2-dim order. By analogy we can define n-dimensional partial order if it is the intersection of n linear orderings. For the special case of 2-dimensional partial orders, it turns out that there is a correspondence with 2-dimensional spacetimes. However, this analogy breaks down for higher dimensions.

Brightwell et al in [32] considered 2-dim orders and attempted to find what the typical 2-dim order would look like. Note, that as we have already mentioned a typical causal set is not manifold-like but a KR order, which is the source of the "entropy" problem. In Causal Dynamical Triangulations approach to quantum gravity, the crucial observation they made, was that instead of summing over all possible triangulations, they restricted attention only to the triangulations that obeyed their "causality" condition. In similar spirit, in [32] they restricted their potential causal sets, to 2-dim partial orders. The surprising result, was that within the 2-dim orders the major contribution comes from causal sets that are manifold-like and moreover that are faithfully embeddable in 2-dim Minkowski. We can view this as an example of how manifold-like behaviour (and more precisely flat) emerges from general considerations on causal sets.

Finally, one can use expressions for the causal set analogue of the Einstein-Hilbert action, to assign quantum amplitudes and examine the consequences. The more suitable choice, is the BD action eqs. (7, 8). However, other attempts may also be worth exploring, such as the expressions for the Ricci scalar found in [29] or in [33].

4. Summary and Conclusion

We introduced the causal sets which is a discrete replacement for standard continuous spacetime. We then focused on the dynamics of the theory, which divides to three parts. (1) The classical dynamics of causal sets, (2) the quantum dynamics of matter and fields on a classical causal set and finally (3) the quantum dynamics of the causal set. The following major issues concerning all three parts, were stressed. The "entropy" problem, which is that a typical causal set is not manifold-like. The direction chosen for the dynamics i.e. whether we take a bottom-up approach starting from concepts intrinsic to causal sets or the top-down approach where motivation for our choices comes from standard continuous physics. The interpretation of quantum causal sets, has the following two aspects. Constructing label-invariant questions, and constructing an interpretation of quantum theory suitable for quantum causal sets. The issues taken into consideration are that (a) we need a path integral approach (rather than canonical), (b) an observer independent formulation (since nothing is outside the total causal set) and (c) an interpretation that is valid for single, non repeatable system. The suggested interpretations were the co-event formulation and decoherent histories approach. Applications of those formulations to quantum causal sets, is definitely something that needs to be pursued.

The classical dynamics presented, were bottom-up and in general stochastic [20]. Despite the fact that we constructed the most general dynamics, four physical conditions, restricted vastly the possible dynamics. The typical causal sets for any given choice of the parameters differs, and while it avoids the KR causal sets it appears that does not in general give manifold-like causal sets.

Particles moving on a given fixed causal set, move as expected with some small deviation that appears as a drift. With reasonable choice of parameters, however, this deviation is not observable. There are several attempts to define the d'Alembertian and using this to obtain the propagator. This was done initially on a causal set that faithfully embeds in flat spacetime. In [24, 25] the retarded Klein-Gordon propagator and eventually the Feynman propagator were computed. The greatest development is, however, the Benincasa-Dowker [27] action that applies to general causal sets that faithfully embed in curved spacetimes.

Finally, we explored the developments on the quantum dynamics of causal sets. There are two directions, the one being the generalisation of the CGD, and the other is the direct assignment of amplitudes to causal sets. The difficulties for the first approach were both technical (how to extend the quantum measure) and conceptual (what replaces the Bell's causality condition). For the second approach, two major developments have been made. The first was in [32] where for some sub-class of causal sets, namely the 2-dim orders, the typical causal set not only faithfully embeds in a manifold, but it is 2-dim Minkowski. This is very interesting, since it shows how quantum dynamics may evade the "entropy" problem. The second development, is the existence of an analogue of the Einstein-Hilbert action in arbitrary dimensions. One can use this, namely the BD action, to attribute weights to different causal sets and analyse the consequences. This direction is currently the priority, and accomplishing this, even for some special class of causal sets, would give a well defined quantum causal sets theory.

Acknowledgments
The author acknowledges all the people that he collaborated and discussed about causal sets and in particular Rafael Sorkin, Fay Dowker, Sumati Surya, David Rideout and Yousef Ghazi-Tabatabai. This work is partly supported by COST Action MP1006 "Fundamental Problems in Quantum Physics".

References
[1] L. Bombelli, J.H. Lee, D. Meyer, and R. Sorkin, *Phys. Rev. Lett.* **59** (1987) 521.
[2] F. Dowker (2005). Causal sets and the deep structure of spacetime. in "100 Years of Relativity, Space-Time Structure: Einstein and Beyond", p445, World Scientific; J. Henson, preprint arXiv:1003.5890 (gr-qc), (2010); S. Surya, preprint arXiv:1103.6272 (gr-qc), (2011).
[3] S. Hawking, A. King and P. McCarthy, *J. Math. Phys.* **17** (1976) 174.
[4] David. Malament, J. Math. Phys. **18**, 1399 (1977).
[5] D.A. Meyer, *Order* **10** 227–237 (1993).
[6] J. Myrheim, CERN preprint TH-2538 (1978).
[7] S. Major, D. Rideout, and S. Surya, *J. Math. Phys.* **48**, 032501, (2007); Sumati Surya, *Theoretical Computer Science* **405**, 1–2, pp. 188–197 (2008).
[8] G. Brightwell and R. Gregory, Phys. Rev. Lett. 66: 260-263 (1991).
[9] D. Rideout and P. Wallden, *Class. Quant. Grav.* **26** (2009) 155013.
[10] J. Ambjorn, J. Jurkiewicz, R. Loll, To appear as part of a Springer Lecture Notes in Physics publication: "Quantum Gravity - New Paths Towards Unification", [arXiv:gr-qc/0906.3947].

[11] L. Bombelli, J. Henson and R. D. Sorkin, *Mod. Phys. Lett.* **A24** (2009), 2579; J. Henson *J. Phys. Conf. Ser.* **174** (2009) 012020.
[12] D. Kleitman and B. L. Rothschild, *Trans. Am. Math. Soc.* **205**, (1975) 205.
[13] Rafael Sorkin, in Relativity and Gravitation: Classical and Quantum, pp. 150-173. World Scientific, Singapore, 1991.
[14] M. Ahmed, S. Dodelson, P. Greene and R. Sorkin, *Phys. Rev. D* **69** (2004) 103523; Ahmed, M. and Sorkin, R. (2012) preprint arXiv:1210.2589 (gr-qc), (2012).
[15] L. Bombelli, R. Koul, J. Lee and R. Sorkin, *Phys. Rev. D* **34** (1986) 373; M. Srednicki, *Phys. Rev. Lett.* **71** (1993) 666.
[16] D. Dou and R. Sorkin, *Found. Phys.* **33** (2003) 279; D. Rideout and S. Zohren, *Class. Quant. Grav.* **23** (2006) 6195.
[17] F. Dowker, L. Philpott and R. Sorkin, *Phys. Rev. D* **79** (2009) 124047.
[18] R. D. Sorkin, J. Phys. A **40**, 3207 (2007); R. D. Sorkin, J. Phys. Conf. Ser. **67**, 012018 (2007); F. Dowker and Y. Ghazi-Tabatabai, J. Phys. A **41**, 105301 (2008); Y. Ghazi-Tabatabai and P. Wallden, J. Phys. A: Math. Theor. **42** 235303 (2009); P. Wallden, preprint arXiv:1211.2084 (quant-ph), (2012).
[19] R.B. Griffiths, J. Stat. Phys., **36** 219, (1984); R. Omnès. J. Stat. Phys. **53**, 893 (1988); M. Gell-Mann and J. Hartle, in *Complexity, Entropy and the Physics of Information, SFI Studies in the Science of Complexity, Vol. VIII,* edited by W. Zurek, (Addison-Wesley, Reading, 1990); M. Gell-Mann and J. Hartle. Phys. Rev. D **47**, 3345 (1993).
[20] D. Rideout and R. Sorkin, *Phys.Rev.* **D61** (2000) 024002 ; D. Rideout and R. Sorkin, *Phys.Rev.* **D63** (2001) 104011.
[21] M. Ahmed and D. Rideout, preprint [arXiv:gr-qc/0909.4771].
[22] F. Dowker, J. Henson and R. Sorkin, *Mod. Phys. Lett.* **A19** (2004) 1829.
[23] R. D. Sorkin, in *Approaches to Quantum Gravity*, D. Oriti (ed.), Cambridge University Press (2009).
[24] Steven Johnston, *Class. Quant. Grav.* **25** (2008) 202001.
[25] Steven Johnston, *Phys. Rev. Lett.* **103** (2009) 180401.
[26] N. N. Bogoliubov and D. V. Shirkov, Introduction To The Theory Of Quantized Fields Interscience Monographs and Texts in Physics and Astronomy (Wiley Interscience, London, 1959), Vol 3.
[27] D. Benincasa and F. Dowker, Phys. Rev. Lett. **104**, (2010) 181301.
[28] G. W. Gibbons and S. N. Solodukhin, Phys. Lett. B 649: 317 (2007).
[29] M. Roy, D. Sinha and S. Surya, preprint arxiv:1212.0631 (gr-qc), (2012).
[30] R. D. Sorkin, Mod. Phys. Lett. A **9**, 3119 (1994); R. D. Sorkin, *Quantum Classical Correspondence: Proceedings of the 4th Drexel Symposium on Quantum Nonintegrability,* Philadelphia, September 8-11, 1994, pages 229-251, (International Press, Cambridge Mass. 1997) D.H. Feng and B-L Hu, editors. [arXiv:gr-qc/9507057].
[31] F. Dowker, S. Johnston and S. Surya, *J. Phys A* **43**, (2010) 505305.
[32] G. Brightwell, J. Henson and S. Surya, *Class. Quant. Grav.* **25** (2008) 105025.
[33] Roman Sverdlov and Luca Bombelli, *Class. Quant. Grav.* **26** (2009) 075011.

Radiation from horizons, chirality and the principle of effective theory

Adamantia Zampeli
Department of Physics and Astronomy,
University of Lethbridge,
Lethbridge, Canada T1K 3M4

E-mail: a.zampeli@uleth.ca

Abstract. We review a modification of the gravitational anomalies method for the calculation of Hawking-like radiation from horizons of n-dimensional static or stationary spacetimes with horizon(s). This modification uses (i) the chirality near the horizon and (ii) the principle of effective theory of gravity, in order to resolve some of its defects and apply it to spacetimes with constant curvature. The new approach can be smoothly applied in the same class of spacetime as the original gravitational anomalies method.

1. Introduction

The discovery of Hawking radiation from black holes [1] is a particularly important result for theoretical physics. It implies that in the semiclassical picture of a background gravitational field on which the quantum fields propagate, matter and energy can escape from the black holes, which, when considered as objects of the classical gravitational theory, do not permit even to light to escape. Since then, it is proved that Hawking-like radiation can be produced in a series of Lorentz spacetimes with horizons, such as the Hawking-Gibbons radiation which arises at the cosmological horizon of the de Sitter spacetime [2] and the Unruh radiation arising from the horizon for an accelerating observer in the flat spacetime (Rindler spacetime) [3]. Other methods apart from the original one have also been proposed for the calculation of the radiation flux from a horizon, a conceptually clear and easily visualisable method being the tunneling [4], which pictures the radiation as the escape of virtual particles from the horizon, and the trace anomaly method [5] which is applicable only in two dimensions. The latter gave insight on the derivation of the Hawking flux from quantum gravitational anomalies near the region of the horizon and led to the gravitational anomalies method [6]. In this method, the flux is calculated from the local theory near the horizon, which is chiral and two-dimensional. The structure of the spacetime plays no other role except to impose appropriate boundary conditions so as to derive the flux. Although the gravitational anomalies method has been successfully applied in a series of spacetimes of any dimension, it fails to give a correct value for the cases of the de Sitter and the Rindler spacetimes because the quantum anomalies vanish [7]. In [8], a resolution was given by pointing out the fact that it is the chirality in the near horizon region that plays the fundamental role in the method and quantum anomalies is just a consequence of the chirality. Additionally, as postulated in the principle of effective theory [9], different observers[1], who

[1] Here, by observer we mean a congruence of timelike curves.

have access to different regions of the spacetime manifold will have access to different amount of information. Then, it is straightforward to define a new effective stress-energy tensor which gives the correct radiation flux for all previous studied cases and in addition for the two pathological cases (Rindler and de Sitter spacetime).

In the following sections, we briefly summarize the gravitational anomalies method in order to introduce the reader to the technical framework. Then, we explain how the principle of effective theory combined with chirality leads to the new approach and apply it to the prototype spacetimes.

2. Gravitational anomalies method

We start by describing the gravitational anomalies method in the four-dimensional Schwarzschild black hole. The metric in static (Schwarzschild) coordinates has the form

$$ds^2 = -f(r)dt^2 + \frac{dr^2}{f(r)} + r^2 d\Omega^2 \tag{1}$$

where $f(r) = 1 - 2M/r$, M the mass of the black hole. The horizon is at r_H where $f(r_H) = 0$ with the surface gravity given by $\kappa = \frac{f'(r_H)}{2}$. Near the horizon, the theory is reduced to a two dimensional field theory. Thus, a four dimensional quantum field acts as a two dimensional massless free field and the ingoing and outgoing modes of the field are plane waves. Since the horizon surface acts like a one-membrane surface, the ingoing modes of the field are excluded from the external region of the spacetime. Thus, the theory near the horizon becomes chiral and, as a result, quantum chiral anomalies appear in the conservation equations. These anomalies are of the form[2] [10]

$$\nabla_\mu T^\mu_{\nu(H)}(r) = \mathcal{A}_\nu = \partial_\mu N^\mu_\nu(r) \,, \qquad N^\mu_\nu(r) = \frac{1}{96\pi}\epsilon^{\beta\mu}\partial_\alpha \Gamma^\alpha_{\nu\beta} \,. \tag{2}$$

where the symbol $\epsilon^{\beta\mu}$ is the totally antisymmetric tensor in two dimensions and the N^μ_ν denotes the contribution from the field modes near the horizon. On the other hand, far from the horizon, the theory is four dimensional and the energy-momentum tensor satisfies the equation

$$\nabla_\mu T^\mu_{\nu(4)}(r) = 0. \tag{3}$$

We have thus resulted with a four dimensional theory far from the horizon and an effective two dimensional local theory in the region of the horizon. To combine the theories in the two separate regions, we ignore the transverse dimensions in the region far from the horizon, since they play no role on our discussion. Then, the energy-momentum tensor is derived by integrating over the transverse dimensions its T^r_t component given by

$$T^r_{t(o)} \equiv \int d\Omega^2 r^2 T^r_{t(4)} \tag{4}$$

and the differential equation is

$$\partial_r T^r_{t(o)} = 0. \tag{5}$$

The next step is to integrate this differential equation and eq. (2) in order to obtain the flux from the horizon. By integrating eq. (2) from r_H to r, the $\nu = t$ component of (2) yields

$$T^r_{t(H)}(r) - N^r_t(r) = T^r_{t(H)}(r_H) - N^r_t(r_H) \tag{6}$$

[2] For simplicity, we denote as T^μ_ν the expectation value of the energy-momentum tensor with respect to the time-asymmetric vacuum state analogue to the Unruh vacuum state in the Schwarzschild spacetime $< U|T^\mu_\nu|U >$.

where $N_t^r = \frac{1}{192\pi}(ff'' + f'^2)$, while integration of (5) from r to ∞ gives $T^r_{t(o)}(r) = c_0 = \Phi$ with c_0 being the constant flux at infinity. The radiation flux is obtained from the condition that the value of the energy-momentum tensor far from the horizon should be equal to the limit of the near-horizon energy-momentum tensor at infinity

$$T^r_{t(o)}(r) = T^r_{t(H)}(r \to \infty). \tag{7}$$

The physical meaning of this condition is the assumption that no other sources of energy are present in the spacetime. Thus we have to calculate the limit of (6) at infinity. For black hole spacetimes which are asymptotically flat, the component $N_t^r(r \to \infty)$ vanishes. Therefore, the flux can be written

$$T^r_{t(o)}(r) = \Phi = T^r_{t(H)}(r \to \infty) = T^r_{t(H)}(r_H) - \frac{1}{192\pi} f'^2(r_H)\ . \tag{8}$$

The only missing information is the value of the energy-momentum tensor on the horizon. This can be calculated by the covariant condition $\tilde{T}^r_t(r_H) = 0$, where $\tilde{T}^r_t$ is the covariant tensor defined by [11]

$$\tilde{T}^r_t(r) = T^r_t(r) + \frac{1}{192\pi}(ff'' - 2f'^2) \tag{9}$$

while the T^r_t tensor we used in our relations till now is called consistent. The covariant condition is imposed in order to ensure that the physical quantities on the horizon do not diverge. Using (9) and the fact that $f(r_H) = 0$ one obtains $T^r_t(r_H) = 2\,\frac{f'^2(r_H)}{192\pi} = 2N^r_t(r_H)$. By imposing the covariant condition [12], the flux of Hawking radiation reads

$$\Phi = N^r_t(r_H) = \frac{f'^2(r_H)}{192\pi} = \frac{\kappa^2}{48\pi} = \frac{\pi}{12}T_H^2\ , \tag{10}$$

where κ is the surface gravity of the black hole horizon and $T_H = \frac{\kappa}{2\pi}$ is the Hawking temperature.

Up to now, we have shown that the flux can be computed when one employs the conservation and anomaly equations of the consistent energy-momentum tensor. One can derive exactly the same results if the conservation and anomaly equations of the covariant energy-momentum tensor are considered. In particular, the anomaly equation of the covariant energy-momentum tensor reads [12]

$$\nabla_\mu \tilde{T}^\mu_\nu(r) = \tilde{\mathcal{A}}_\nu = \frac{1}{96\pi}\epsilon^\mu_\nu \nabla_\mu R\ , \tag{11}$$

where R is the Ricci scalar, and $\epsilon_{\mu\nu}$ is the totally antisymmetric tensor in two dimensions. The $\nu = t$ component of the anomalies is given by

$$\tilde{\mathcal{A}}_t = \frac{1}{\sqrt{-g}}\partial_r \tilde{N}^r_t \qquad \tilde{N}^r_t(r) = \frac{1}{96\pi}(ff'' - \frac{f'^2}{2})\ . \tag{12}$$

Following the same procedure and imposing the same boundary conditions as before (i.e. $\tilde{T}^r_t(r_H) = 0$), one finds that the flux is

$$\Phi = -\tilde{N}^r_t(r_H) = \frac{1}{96\pi}\frac{f'^2(r_H)}{2} = \frac{\pi}{12}T_H^2\ . \tag{13}$$

This flux is identical to the one given by the consistent equations (10). This method is successful in a large class of static and stationary n-dimensional spacetimes with horizons with metric given by the general form

$$ds^2 = -f(r)dt^2 + h^{-1}dr^2 + P(r)d\sigma_\perp^{n-2}. \tag{14}$$

The horizon is at r_H where $f(r_H) = h(r_H) = 0$. $P(r)$ is a smooth function and $d\sigma_\perp^{n-2}$ is the metric of the transverse dimensions. The surface gravity of the horizon is given by $\kappa = \frac{1}{2}\sqrt{f'(r_H)h'(r_H)}$. The dimensional reduction should be performed again in a way that the conformal factor preserves the roots of $f(r)$ and $h(r)$ in order to have physically acceptable solutions [13]. Moreover, the spacetime need not be asymptotically flat, as is the case of the black hole spacetime we studied in this section, but further modifications are needed in order to obtain the correct flux, as was shown in [14], because the $N_t^r(r)$ does not always vanish at infinity. In this category also falls the case of the spacetimes with multiple horizons such as the Schwarzschild-de Sitter spacetime. In spite of the success of this method to a broad class of metrics, there are several problems yet unresolved which raise questions whether the method can be smoothened and become more physically intuitive. First, it would appear more natural to express the law of energy-momentum conservation in a unified form instead of having two equations, one normal and one anomalous. Second, it is important to be explored which form of the equations, consistent or covariant is most physically appropriate for the calculation of the radiation flux. Important explorations are also on call for the cases of time-dependent spacetimes. Although there is work done in this direction (see e.g.[15]), further research is needed. Finally, a major defect of the method is the fact that it cannot be applied in spacetimes with constant curvature. In the next section, we review the suggestion in [8] for the resolution of this problem.

3. A new approach

The gravitational anomalies appear in the area near the horizon because the horizon acts as a one way membrane and therefore the ingoing modes of any quantum field are excluded from the external part of the manifold. This makes the theory chiral and consequently the conservation equations become anomalous. However, in the case of certain spacetimes, the derivative of the anomalous flux vanishes and we get the unexpected result that there is no radiation flux from the horizon, despite the fact that the anomalous flux N_t^r does not vanish. This points to the fact that the anomalies are not the fundamental quantity but rather it is chirality near the region of the horizon that one should take into account in order to calculate the radiation flux.

The second observation leads to the principle of effective theory applied in gravity. This principle states that *physical theories in a given coordinate system must be formulated entirely in terms of the variables that an observer using that coordinate system can access* [9]. This can be seen as a natural interpretation of general covariance [16]. Therefore, the conservation law for the energy-momentum tensor defined only in the external region of the horizon is

$$\nabla_\mu \mathcal{T}_\nu^\mu = 0, \quad \nabla_\mu \tilde{\mathcal{T}}_\nu^\mu = 0 \tag{15}$$

where now the $\mathcal{T}_\nu^\mu \equiv T_\nu^\mu - N_\nu^\mu$ is the consistent *effective* energy-momentum tensor, T_ν^μ is the consistent energy-momentum tensor it was considered in the anomalies method, while N_ν^μ is the contribution of the modes near the horizon as discussed already. The second equation is the analogue for the covariant form of the equations. The $r-t$ component of the conservation equation is

$$\partial_r(T_t^r \mp N_t^r) = 0 \quad \text{and} \quad \partial_r(\tilde{T}_t^r \mp \tilde{N}_t^r) = 0 \tag{16}$$

where we have written both the consistent and covariant form of the equations. We have also included $\mp$ signs to introduce the generalization of the method in spacetimes in which the observer has access in the internal of the horizon part of the manifold as in the case that the horizon is cosmological and an observer exists in the inner to the horizon part of the manifold. Then the ingoing modes are to be taken into account and the outgoing modes are excluded. The minus represents the consideration of the outgoing modes as we considered till this point while the plus sign the consideration of the ingoing. We will follow this notation in the following where necessary. These effective energy-momentum tensors reproduce all the correct results for the Hawking radiation fluxes of asymptotically flat spacetimes, but in addition they work for the non-asymptotically flat, constant curvature spacetimes of de Sitter and Rindler. In the next section, we will calculate the flux for these two spacetimes. By integrating eq. (15), we find

$$T^r_t(r \to r_0) \mp N^r_t(r \to r_0) = T^r_t(r_H) \mp N^r_t(r_H) = \Phi = \pm N^r_t(r_H) \ . \quad (17)$$

The integration constant, Φ, is the flux defined previously in (10). We have evaluated the integrated quantity at two points $r = r_0$ and $r = r_H$. Previously $r_0 \to \infty$ since we were dealing with asymptotically flat spacetimes. We have generalized (6) to (17) since de Sitter and Rindler spacetimes are not asymptotically flat so we will need to take a limit different than $r_0 \to \infty$. In a similar manner we can integrate the second relationship in (16) to obtain the covariant expression

$$\tilde{T}^r_t(r \to r_0) \mp \tilde{N}^r_t(r \to r_0) = \tilde{T}^r_t(r_H) \mp \tilde{N}^r_t(r_H) = \Phi = \pm N^r_t(r_H) \ . \quad (18)$$

As previously the integration constant, Φ, is the flux defined in (13) and we have used the fact that $N^r_t(r_H) = -\tilde{N}^r_t(r_H)$. The proper limit r_0 will be specified from the geometry of the spacetime.

4. Constant curvature spacetimes

4.1. De Sitter spacetime

The metric of de Sitter spacetime in static coordinates is written as

$$ds^2 = -\left(1 - \frac{r^2}{a^2}\right) dt^2 + \left(1 - \frac{r^2}{a^2}\right)^{-1} dr^2 \quad (19)$$

with the cosmological horizon located at $r_H = a$. The method of gravitational anomalies can be applied to de Sitter spacetime only by using the consistent expressions of the equations and fails if one applies the covariant expressions of the equations [7]. The reason for this failure is that the covariant anomaly vanishes since the Ricci scalar is a constant, i.e. $R = \frac{2}{a^2}$. However, the elements N^r_t and $\tilde{N}^r_t$ do not vanish at the proper limit, $r_0 = 0$, namely $N^r_t(r \to 0) = \frac{1}{192\pi}\left(-\frac{2}{a^2}\right)$ and $\tilde{N}^r_t(r \to 0) = \frac{1}{96\pi}\left(-\frac{2}{a^2}\right)$. The proper limit is $r_0 = 0$ since the flux is flowing from the cosmological horizon inwards. The accessible part of the manifold inside the cosmological horizon and the flux of Hawking radiation has to be calculated at the point $r_0 = 0$. Using the equations for the effective consistent energy-momentum tensor using the lower signs for these two equations since the flux is coming inward from the cosmological horizon at r_H, we obtain

$$\Phi = T^r_t(r \to 0) = T^r_t(r_H) + N^r_t(r_H) = -N^r_t(r_H) \ . \quad (20)$$

Similarly using the equations for the effective covariant energy-momentum tensor (17) and (18) (again using the lower sign in these two equations) we obtain

$$\Phi = \tilde{T}^r_t(r \to 0) = \tilde{T}^r_t(r_H) + \tilde{N}^r_t(r_H) = -N^r_t(r_H) \quad (21)$$

It is evident that employing either the effective consistent energy-momentum tensor or the effective covariant energy-momentum tensor, the flux of the Hawking radiation in de Sitter spacetime is equal to

$$\Phi = -N^r_t(r_H) = +\tilde{N}^r_t(r_H) = -\frac{f'^2(r_H)}{192\pi} = -\frac{\pi}{12}T^2_{GH} \ , \quad (22)$$

where T_{GH} is the Gibbons-Hawking temperature of the cosmological horizon [2], i.e. $T_{GH} = 1/2\pi a$.

4.2. Rindler spacetime

The metric of the Rindler spacetime in static coordinates is

$$ds^2 = -(1+2ar)dt^2 + (1+2ar)^{-1}dr^2 , \tag{23}$$

where a is the acceleration of an observer, a horizon at $r_H = -1/2a$ blocks the observer from accessing the whole manifold. With respect to this observer, there exists a radiation flux, called Unruh flux [17]. Thus, it is expected that this flux can be calculated with the gravitational anomalies method. However, in [7] it was shown that both forms of the anomaly method, consistent and covariant, fail to give the correct value of the Unruh flux, since the curvature of the spacetime in (23) is zero, $R = 0$. However, the components N_t^r and $\tilde{N}_t^r$ of the anomaly part of the equations, were non-zero and constant at the limit point $r_0 = 0$, where the static observer resides in that coordinate system. We need the quantities $N_t^r = a^2/48\pi$ and $\tilde{N}_t^r = -a^2/48\pi$ to calculate the flux. The limit point is at $r_0 = 0$ rather than infinity. However the horizon is located at $r_H = -1/2a$ so that the flux is "outward" from negative r toward positive r. Thus one should use the upper signs of the equations since the flux is "outward". The flux then reads

$$\Phi = \mathcal{T}_t^r(r \to 0) = T_t^r(r_H) - N_t^r(r_H) = N_t^r(r_H) \tag{24}$$

and using the covariant expressions

$$\Phi = \tilde{\mathcal{T}}_t^r(r \to 0) = \tilde{T}_t^r(r_H) - \tilde{N}_t^r(r_H) = -\tilde{N}_t^r(r_H) = N_t^r(r_H) . \tag{25}$$

In both of the above equations $r_0 = 0$ and $r_H = -1/2a$. It is again evident that employing either the equations of the effective consistent energy-momentum tensor, or the equations for the effective covariant energy-momentum tensor, the flux of the Unruh radiation in Rindler spacetime is equal to

$$\Phi = N_t^r(r_H) = -\tilde{N}_t^r(r_H) = \frac{f'^2(r_H)}{192\pi} = \frac{\pi}{12}T_H^2 \tag{26}$$

where $T_H = a/2\pi$ is the Unruh temperature seen by the accelerated observer.

5. Conclusions

In this article, we reviewed the method of gravitational anomalies and discussed a modification introduced in [8] for the calculation of the flux radiation and temperature of horizons. This modification resolved the problem the gravitational anomalies method had when applied in spacetimes with constant curvature. The definition of the effective stress-energy tensor allowed the equations to take a general form independent of the spacetime asymptotic behaviour. However, further investigation is needed in order to understand and clarify other aspects of this method such as why both the consistent and the covariant form of the equations give equivalent results and whether this modification can be applied in time-dependent horizons.

Acknowledgments

The author would like to thank D. Singleton and E. Vagenas for comments and corrections on a previous draft. This work is supported by the Government of Canada through an NSERC grant.

[1] Hawking S W 1975 *Commun. Math. Phys.* **43** 199–220
[2] Gibbons G and Hawking S 1977 *Phys.Rev.* **D15** 2738–2751
[3] Unruh W G 1981 *Phys. Rev. Lett.* **46** 1351–1353
[4] Parikh M K and Wilczek F 2000 *Phys. Rev. Lett.* **85** 5042–5045 (*Preprint* hep-th/9907001)
[5] Christensen S M and Fulling S A 1977 *Phys. Rev.* **D15** 2088–2104

Journal of Physics: Conference Series **453** (2013) 012024 doi:10.1088/1742-6596/453/1/012024

[6] Robinson S P and Wilczek F 2005 *Phys. Rev. Lett.* **95** 011303 (*Preprint* gr-qc/0502074)
[7] Akhmedova V, Pilling T, de Gill A and Singleton D 2009 *Phys. Lett.* **B673** 227–231 (*Preprint* 0808.3413)
[8] Zampeli A, Singleton D and Vagenas E C 2012 *JHEP* **1206** 097 (*Preprint* 1206.0879)
[9] Padmanabhan T and Patel A 2003 (*Preprint* gr-qc/0309053)
[10] Iso S, Umetsu H and Wilczek F 2006 *Phys. Rev. Lett.* **96** 151302 (*Preprint* hep-th/0602146)
[11] Bertlmann R and Kohlprath E 2001 *Annals Phys.* **288** 137–163 (*Preprint* hep-th/0011067)
[12] Banerjee R and Kulkarni S 2008 *Phys. Rev.* **D77** 024018 (*Preprint* 0707.2449)
[13] Wu S Q, Peng J J and Zhao Z Y 2008 *Class. Quant. Grav.* **25** 135001 (*Preprint* 0803.1338)
[14] Papantonopoulos E and Skamagoulis P 2009 *Phys. Rev.* **D79** 084022 (*Preprint* 0812.1759)
[15] Vagenas E C and Das S 2006 *JHEP* **10** 025 (*Preprint* hep-th/0606077)
[16] Padmanabhan T 2005 *Phys. Rept.* **406** 49–125 (*Preprint* gr-qc/0311036)
[17] Unruh W G 1976 *Phys. Rev.* **D14** 870

Quasinormal modes and thermodynamical aspects of the 3D Lifshitz black hole

B Cuadros-Melgar[1], **J de Oliveira**[2] **and C E Pellicer**[3]

[1]Physics Department, University of Buenos Aires, FCEN-UBA and IFIBA-CONICET, Ciudad Universitaria, Pabellón 1, 1428, Buenos Aires, Argentina
[2]Instituto de Física, Universidade Federal do Mato Grosso, 78060-900, Cuiabá, Brazil
[3]Instituto de Física, Universidade de São Paulo, C.P.66318, CEP 05315-970, São Paulo, Brazil

E-mail: berthaki@gmail.com, jeferson@fisica.ufmt.br, carlosep@fma.if.usp.br

Abstract. We consider scalar and spinorial perturbations on a background described by a $z = 3$ three-dimensional Lifshitz black hole. We obtained the corresponding quasinormal modes which perfectly agree with the analytical result for the quasinormal frequency in the scalar case. The numerical results for the spinorial perturbations reinforce our conclusion on the stability of the model under these perturbations. We also calculate the area spectrum, which prove to be equally spaced, as an application of our results.

1. Introduction

New Massive Gravity (NMG) is a novel parity-preserving, unitary [1], power-counting super-renormalizable [2], three-dimensional model describing the propagation of a massive positive-energy graviton with two polarization states of helicity ± 2 in a Minkowski vacuum, whose linearized version is equivalent to the Pauli-Fierz theory for a massive spin-2 field in three dimensions. The action of NMG consists of a "wrong sign" Einstein-Hilbert term plus a quadratic curvature term given by a precise combination of the Ricci tensor and the curvature scalar, which introduces a mass parameter [3]. As with other massive gravity theories, NMG also admits black hole-type solutions with several asymptotics and additional parameters [4, 5]. Even though this last feature could challenge the usual Einstein-Hilbert gravity, it is seen that the definition of mass in this new type of black holes is a conserved charge computed from a combination of the black hole parameters, which satisfies the first law of thermodynamics.

In this contribution we present the results of a study on the stability of the three-dimensional Lifshitz black hole found in the context of the so-called new massive gravity (NMG) [6]. These solutions exhibit the anisotropic scale invariance, $t \to \lambda^z t$, $\vec{x} \to \lambda \vec{x}$, where z is the dynamical critical exponent. Specifically, we deal with the solutions found for the particular case of $z = 3$ and a precise value of the mass parameter.

We aimed to determine the stability of this kind of black holes by calculating the quasinormal modes (QNM) and quasinormal frequencies (QNF) of scalar and spinorial matter fields in the probe limit, *i.e.* there are no backreaction effects upon the asymptotically Lifshitz black hole metric. QNM do not depend of the particular initial perturbation that excited the black holes, while QNF depend only on the parameters of the black hole. For this reason QNF are considered the "footprints" of this structure. The study of QNM has motivated the development of numerical and analytical techniques for their computation ([7, 8, 9]). QNM are important in astrophysical grounds as a way of possible detection of black holes through the observation of its gravitational wave spectrum. They are also relevant in the context of gauge/gravity correspondence [10, 11], where the Hawking temperature of the black hole is related to the temperature of a thermal field theory defined at the boundary, and QNF can be related to the dual field theory relaxation time. In this work we focus in the application of QNM as probes of stability. Moreover, the QNF we obtain are useful to calculate the area spectrum of Lifshitz black holes. Actually, according to Bekenstein the horizon area of a black hole must be quantized [12], so that the area spectrum has the form $A_n = \gamma n\hbar$, with γ a dimensionless constant to be determined. In this way, we find this spectrum by using an approach suggested by Maggiore [13], who claims that QNM should be described as damped harmonic oscillators, and the module of the entire QNF is the proper physical frequency that should be used in the Bohr-Sommerfeld quantization of the adiabatic invariant $I = \int dE/\omega(E)$.

In the following sections we describe the model we dealt with, the equations for the scalar and spinorial perturbations, the numerical analysis, and the calculation of the area spectrum. In the last section we discuss our results and conclude.

2. The black hole solution

The black hole we study along this paper is a solution of the NMG $(2+1)$-dimensional action,

$$S = \frac{1}{16\pi G}\int d^3x\sqrt{-g}\left[R - 2\lambda - \frac{1}{m^2}\left(R_{\mu\nu}R^{\mu\nu} - \frac{3}{8}R^2\right)\right], \tag{1}$$

where m is the so-called "relative" mass parameter, and λ is the three-dimensional cosmological constant. This theory has a black hole solution with Lifshitz scaling $z = 3$ given by

$$ds^2 = -a(r)\frac{\Delta}{r^2}dt^2 + \frac{r^2}{\Delta}dr^2 + r^2 d\phi^2, \tag{2}$$

where

$$a(r) = \frac{r^4}{l^4}, \tag{3}$$

and

$$\Delta = -Mr^2 + \frac{r^4}{l^2}, \tag{4}$$

with M an integration constant and $l^2 = -\frac{13}{2\lambda} = -\frac{1}{2m^2}$. Also, the NMG admits as a solution, the well-known Bañados-Teitelboim-Zanelli (BTZ) black hole with the dynamical critical exponent $z = 1$. The Kretschmann scalar for the metric (2),

$$R_{\mu\nu\lambda\sigma}R^{\mu\nu\lambda\sigma} = -\frac{4}{l^4 r^4}\left[8r_+^4 - 48r_+^2 r^2 + 91r^4\right], \tag{5}$$

diverges for $r \to 0$. Thus, the black hole solution has a genuine spacetime singularity at the origin. From the Penrose-Carter diagram in Fig.1 we see that the spacetime singularity at $r = 0$ is light-like covered by a regular event horizon at $r = r_+ = l\sqrt{M}$.

3. Scalar perturbation equation

In order to consider the response of the Lifshitz black hole background to a scalar perturbation we consider a scalar field obeying the Klein-Gordon equation,

$$\Box\Phi = \frac{1}{\sqrt{-g}}\partial_M\left(\sqrt{-g}g^{MN}\partial_N\right)\Phi = m^2\Phi, \tag{6}$$

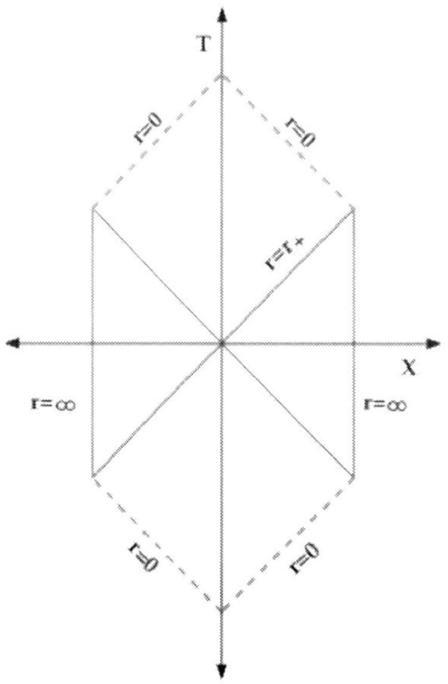

Figure 1. Penrose-Carter diagram for the Lifshitz black hole. The singularity at $r = 0$ is light-like and covered by a regular event horizon r_+.

where m is the mass of the field $\Phi(t, r, \phi)$. After decomposing the angular part of the field this equation takes the following form,

$$-\partial_t^2\Psi + \frac{r^4}{l^6}\left(1 - \frac{Ml^2}{r^2}\right)\left(\frac{5r^3}{l^2} - 3Mr\right)\partial_r\Psi + \frac{r^8}{l^8}\left(1 - \frac{Ml^2}{r^2}\right)^2\partial_r^2\Psi - \\ -\frac{r^4}{l^6}\left(m^2r^2 + \kappa^2\right)\left(1 - \frac{Ml^2}{r^2}\right)\Psi = 0. \tag{7}$$

Although this equation has an analytical solution, which we will show later, it is a useful test for our numerical code to check the numerical results. With this goal we need to rewrite (7) in a Shrödinger-like form,

$$-\partial_t^2 X + \partial_{r_*}^2 X = V(r)X\,, \tag{8}$$

where r_* is the so-called *tortoise* coordinate given by

$$r_* = l^4\left[-\frac{1}{M^{3/2}l^3}\,\text{arccoth}\left(\frac{r}{l\sqrt{M}}\right) + \frac{1}{Ml^2r}\right], \tag{9}$$

and the scalar effective potential can be written as

$$V(r) = \left(\frac{7}{4l^8} + \frac{m^2}{l^6}\right)r^6 - \left(\frac{5M}{2l^6} + \frac{Mm^2}{l^4} - \frac{\kappa^2}{l^6}\right)r^4 + \left(\frac{3M^2}{4l^4} - \frac{M\kappa^2}{l^4}\right)r^2. \tag{10}$$

Now let us return to Eq.(7) and discuss its analytical solution. First of all we set the time dependence of the scalar field as $\Psi(t, r) = R(r)e^{-i\omega t}$, and then redefine the radial coordinate as $r = l\sqrt{M/y}$. The resultant equation has the following solution in terms of the Heun confluent functions,

$$R(y) = C_1 y^{2+\alpha}(1-y^2)^{\beta/2}\,\mathrm{HeunC}\left(0,\alpha,\beta,-\frac{\beta^2}{4},\frac{\alpha^2}{4}+\frac{\kappa^2}{4M},y^2\right)+ \\ +C_2 y^{2-\alpha}(1-y^2)^{\beta/2}\,\mathrm{HeunC}\left(0,-\alpha,\beta,-\frac{\beta^2}{4},\frac{\alpha^2}{4}+\frac{\kappa^2}{4M},y^2\right), \quad (11)$$

where C_1 and C_2 are integration constants, while $\alpha = \sqrt{4+m^2l^2}$, and $\beta = -il\omega/M^{3/2}$. After imposing Dirichlet condition at infinity and the boundary condition of ingoing waves at the horizon we obtain the quasinormal frequency spectrum given by

$$\omega = 2i\frac{M^{3/2}}{l}\left[1+2N+\sqrt{4+m^2l^2}-\sqrt{7+\frac{3}{2}m^2l^2+\frac{\kappa^2}{2M}+(3+6N)\sqrt{4+m^2l^2}+6N(N+1)}\right], \quad (12)$$

where N is a positive integer. These same frequencies were also obtained in subsequent works[14, 15] by using equivalent analytical procedures. The imaginary part of the fundamental frequency ($N=0$) is negative provided that

$$\sqrt{7+\frac{3}{2}m^2l^2+\frac{\kappa^2}{2M}+3\sqrt{4+m^2l^2}} > 1+\sqrt{4+m^2l^2}\,. \quad (13)$$

While the asymptotic frequency ($N\to\infty$) is given by

$$\omega_\infty = -2(\sqrt{6}-2)\,i\,\frac{M^{3/2}}{l}N < 0\,. \quad (14)$$

Thus, since the imaginary part of the quasinormal frequencies is negative provided that the parameters respect the relation (13), we can conclude that the model is stable under scalar perturbations.

4. Spinorial perturbation equation

Let us consider a spinorial field perturbing the Lifshitz black hole background, which obeys the covariant Dirac equation,

$$i\gamma^{(a)}e_{(a)}{}^{\mu}\nabla_\mu\Psi - \mu_s\Psi = 0\,, \quad (15)$$

where Greek indices refer to spacetime coordinates (t,r,ϕ), and the Latin indices enclosed in parentheses describe the flat tangent space in which the triad basis $e_{(a)}{}^{\mu}$ is defined. The spinor covariant derivative ∇_μ is given by

$$\nabla_\mu = \partial_\mu + \frac{1}{8}\omega_\mu{}^{(a)(b)}\left[\gamma_{(a)},\gamma_{(b)}\right], \quad (16)$$

where $\omega_\mu{}^{(a)(b)}$ is the spin connection,

$$\omega_\mu{}^{(a)(b)} = e_\nu{}^{(a)}\partial_\mu e^{(b)\nu} + e_\nu{}^{(a)}\Gamma^\nu{}_{\mu\sigma}e^{\sigma(b)}\,, \quad (17)$$

$\Gamma^\nu{}_{\mu\sigma}$ are the metric connections, and the gamma matrices we employ along this work are $\gamma^{(0)} = i\sigma_2$, $\gamma^{(1)} = \sigma_1$, and $\gamma^{(2)} = \sigma_3$.

In order to solve the Dirac equation in a numerical way once again we should adapt it to a Shrödinger-like form. First, we redefine the two component spinor as,

$$\Psi = \begin{pmatrix} \Psi_1(t,r,\phi) \\ \Psi_2(t,r,\phi) \end{pmatrix} , \tag{18}$$

as

$$\begin{aligned} \Psi_1 &= i\left[a(r)\Delta\right]^{1/4} e^{-i\omega t+im\phi} e^{i\theta/2} R_+(r), \\ \Psi_2 &= \left[a(r)\Delta\right]^{1/4} e^{-i\omega t+im\phi} e^{-i\theta/2} R_-(r) , \end{aligned} \tag{19}$$

where

$$\theta = \arctan(\frac{\mu_s r}{\hat{m}}) , \qquad m = i\hat{m} . \tag{20}$$

And in addition we make $X_\pm = R_+ \pm R_-$ to finally have

$$\left(\partial^2_{\hat{r}_*} + \omega^2\right) X_\pm = V_\pm X_\pm , \tag{21}$$

where $V_\pm$ are the superpartner potentials,

$$V_\pm = W^2 \pm \frac{dW}{d\hat{r}_*} , \tag{22}$$

written in terms of the so-called superpotential W,

$$W = \frac{i\sqrt{a(r)\Delta}\,(\hat{m}^2 + {\mu_s}^2 r^2)^{3/2}}{r^2(\hat{m}^2 + \mu_s^2 r^2) + \frac{\mu_s \hat{m}\sqrt{a(r)\Delta}}{2\omega}} . \tag{23}$$

In the case of a massless spinor ($\mu_s = 0$) the superpartner potentials reduce to

$$V_\pm = \left(-\frac{m^2 M}{l^4} \mp \frac{mM}{l^5}\sqrt{r^2 - Ml^2}\right) r^2 + \left(\frac{m^2}{l^6} \pm \frac{2m}{l^7}\sqrt{r^2 - Ml^2}\right) r^4 . \tag{24}$$

5. Numerical results

Although in the scalar case we already found the corresponding QNF, the numerical analysis is an interesting tool to verify the applicability of certain numerical methods in asymptotically Lifshitz spacetimes. Specifically, here we use the Finite Difference and the Horowitz-Hubeny [16] methods. Let us begin with the scalar perturbation. By rewriting Eq.(10) in terms of a new variable $z = r^2$, we obtain

$$V(r) = \frac{z}{l^8}\left[\left(\frac{7}{4} + m^2 l^2\right) z^2 - \left(\frac{5}{2} + m^2 l^2 - \frac{\kappa^2 l^2}{z_h}\right) z_h z + \left(\frac{3}{4} - \frac{\kappa^2 l^2}{z_h}\right) z_h^2\right] , \tag{25}$$

where $z_h = r_h^2$. The parable in brackets tends to infinity as long as $\left(\frac{7}{4} + m^2 l^2\right) > 0$, which is consistent with the Breitenlohner-Freedman-type bound for the present case. The roots of this polynomial potential are given by

$$\begin{aligned} z_0 &= 0 , \\ z_+ &= z_h , \end{aligned} \tag{26}$$

$$z_- = z_h \left[\frac{\frac{3}{4} - \frac{\kappa^2 l^2}{z_h}}{\frac{7}{4} + m^2 l^2}\right] . \tag{27}$$

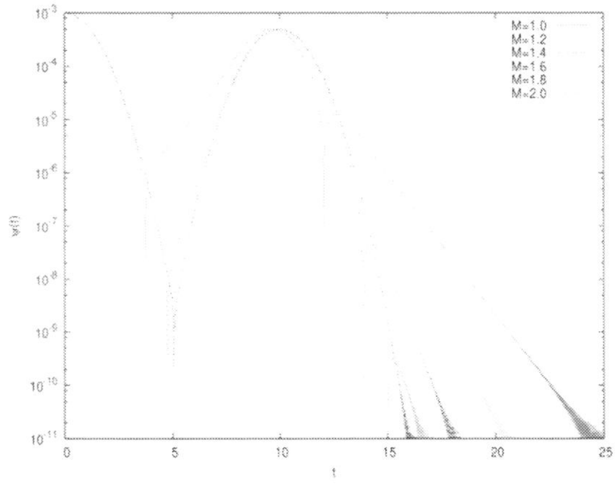

Figure 2. Decay of scalar field with mass $m = 1$ and $l = 1$ for different values of black hole mass M.

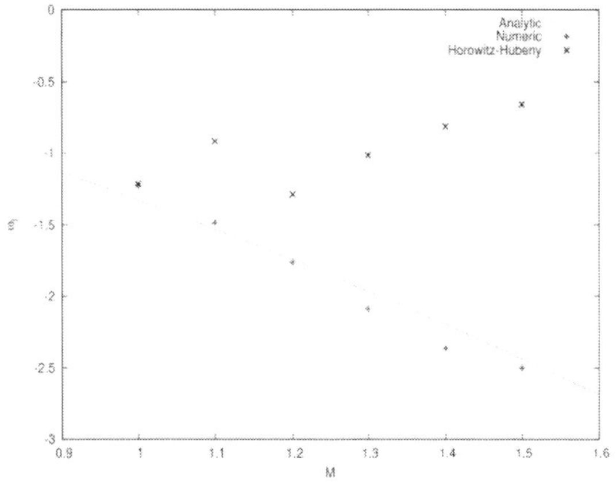

Figure 3. Imaginary part of scalar quasinormal frequencies. We display the results obtained using different approaches.

If $m^2l^2 > -1$, we see that $z_- < z_+$. Thus, going back to the original variable r, the roots of the potential are $r = 0$ with double multiplicity, $r = \sqrt{z_-}$ and $r = r_h$ ($r = -\sqrt{z_-}$ and $r = -r_h$ are excluded as $r > 0$). Then, since r_h is the biggest root and $V(r)$ tends to ∞ when r tends to ∞, the potential is positive-definite in the region (r_h, ∞). Therefore, the quasinormal modes for $m^2l^2 > -1$ are necessarily stable [16].

In Figs.2-3 we show the numerical results for the decay of the scalar field and a comparison of the scalar QNF obtained by three different approaches, analytical calculation, finite difference, and Horowitz-Hubeny methods. Our results reinforce the conclusion already found analytically, the $z = 3$ Lifshitz black hole is stable under scalar perturbations. Moreover, according to Fig.3, the numerical results have a very good agreement with the analytical calculation. However, the Horowitz-Hubeny method gives unreliable results. We discarded explanations related to the not-convergence of the frequencies or the size of the black holes ocurring in other cases [17, 18], in spite of that, there is no clear explanation for the limitation of the method. In our case the asymptotic behavior of the black hole under study might play an important role in the convergence of the method. Nonetheless, a general criteria for the convergence of the Horowitz-Hubeny method remains an open question.

Now turning to the case of spinorial perturbations, we can analogously analyze the superpartner potentials (24) and its derivatives, which we rewrite as

$$V_{\pm} = \frac{1}{l^8}\left[(ml)^2 r^2 (r^2 - r_+^2) \pm (ml) r^2 (2r^2 - r_+^2)\sqrt{r^2 - r_+^2}\right], \tag{28}$$

$$V'_{\pm} = \frac{1}{l^8}\left\{(ml)^2 r (2r^2 - r_+^2) \pm (ml)\left[2r(r^2 - r_+^2)\sqrt{r^2 - r_+^2} + r^3\frac{2r^2 - r_+^2}{\sqrt{r^2 - r_+^2}}\right]\right\}. \tag{29}$$

We see that outside the event horizon V_+ is positive-definite if $ml > 0$, and $\lim_{r\to\infty} V_+(r) = -\infty$ if $ml < 0$. Whereas V_- is positive-definite if $ml < 0$, and $\lim_{r\to\infty} V_-(r) = -\infty$ if $ml > 0$. Furthermore, we notice that if $ml = 0$, we have a free-particle case. The decaying behavior of the massless spinor is given in Figs.4-5. Thus, we conclude that the $z = 3$ Lifshitz black hole is stable under massless spinorial perturbations.

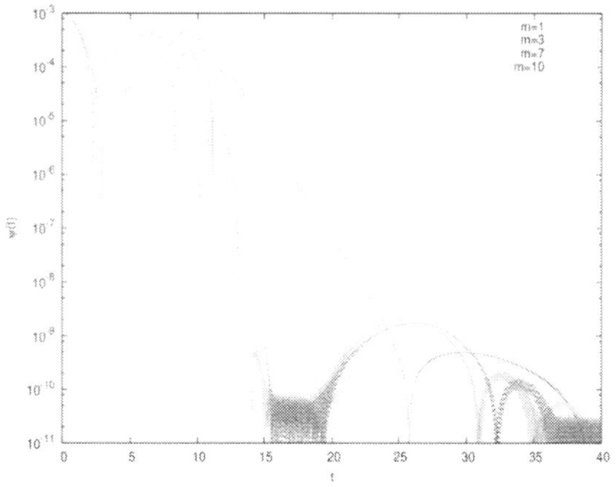

Figure 4. Decay of massless spinor with $l = 1$ and black hole mass $M = 1.0$ for different values of the azimuthal parameter m.

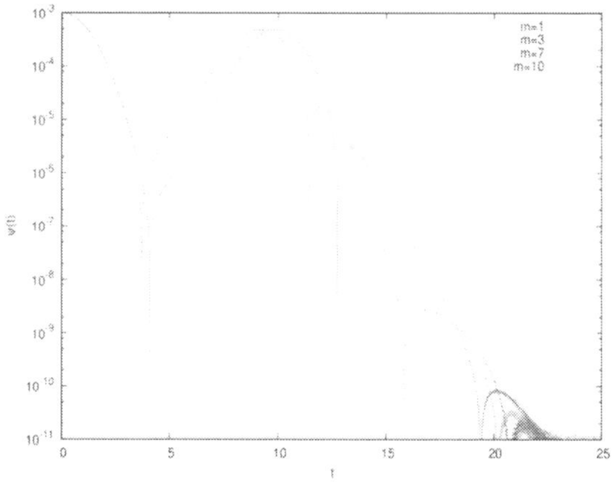

Figure 5. Decay of massless spinor with $l = 1$ and black hole mass $M = 1.5$ for different values of the azimuthal parameter m.

6. Applying QNM to thermodynamics

In this section we use our analytic results for the QNF to find the area spectrum of the Lifshitz black hole. In order to apply Maggiore's method we need the asymptotic value of the QNF given in (14), so that the proper physical frequency of the damped harmonic oscillator equivalent to the black hole QNM turns out to be

$$\omega_p = 2(\sqrt{6}-2)\frac{M^{3/2}}{l}N\,. \tag{30}$$

Now we can calculate the adiabatic invariant

$$I = \int \frac{d\mathcal{M}}{\Delta\omega} = \int \frac{M}{\Delta\omega}dM\,, \tag{31}$$

where $\mathcal{M} = M^2/2$ is the ADM mass of the Lifshitz black hole, and $\Delta\omega$ is the change of proper frequency between two neighboring modes, which can be obtained from (30). Thus,

$$I = \frac{lM^{1/2}}{(\sqrt{6}-2)}\,, \tag{32}$$

which is quantized via Bohr-Sommerfeld quantization in the semiclassical limit. Then, expressing this equation in terms of the horizon area $A = 2\pi r_+$, with $r_+ = l\sqrt{M}$, we finally obtain the spectrum we look for,

$$A = 2\pi(\sqrt{6}-2)n\hbar\,, \tag{33}$$

with n an integer number. As we can see, the horizon area of the Lifshitz black hole is quantized and equally spaced, a result not expected from a theory containing higher order curvature corrections. In addition, since this black hole has an entropy proportional to its horizon area [19, 20], our result (33) also says that the entropy is quantized with an evenly spaced spectrum. Subsequent studies on area spectrum of other kind of Lifshitz black holes [14] seem to confirm this last conclusion.

7. Conclusions

We have studied the stability of the three-dimensional Lifshitz black hole under scalar and spinorial perturbations in the probe limit through the computation of quasinormal modes. In addition, we have found the event horizon area quantization as an application of the results for quasinormal modes using Maggiore's prescription.

Regarding the scalar perturbations, we found that the QNF are purely imaginary and negative pointing out a stable equilibrium configuration for the Lifshitz black hole. Our numerical results perfectly agree with the analytical calculation in the case of the finite difference method, which allows to obtain the temporal behavior of the fields showing all the stages of the decay. However, in the case of the Horowitz-Hubeny method the frequencies values fail to be obtained. An important observation is that apart from the numerical factor, the asymptotic scalar frequency is the same as that calculated in the hydrodynamic limit in the context of Gauge/Gravity duality [21].

Concerning the spinorial perturbation, our numerical results also show probe massless spinors decaying, and therefore, the $z = 3$ Lifshitz black hole is stable under spinorial perturbations too.

Additionally, using the asymptotical form of the analytical QNF we calculated the area spectrum of the Lifshitz black hole using Maggiore's method. Our results show that the horizon area and the entropy are quantized and their spectra are equally spaced.

At last, we should stress that the definite answer on the stability of the Lifshitz black hole should come from gravitational perturbations, mainly from the tensor part of the metric perturbation, since NMG admits the propagation of gravitational waves contrary to Einstein gravity in three dimensions, that has no propagating degrees of freedom.

Acknowledgments

This work was supported by Fundação de Amparo à Pesquisa do Estado de São Paulo **(FAPESP-Brazil)** and Consejo Nacional de Investigaciones Científicas y Técnicas **(CONICET-Argentina)**.

References

[1] Nakasone M and Oda I, *Prog. Theor. Phys.* **121**, 1389 (2009). Deser S, *Phys. Rev. Lett.* **103**, 101302 (2009).
[2] Oda I, *JHEP* **0905**, 064 (2009).
[3] Bergshoeff E, Hohm O and Townsend P, *Phys. Rev. Lett.* **102**, 201301 (2009).
[4] Bergshoeff E, Hohm O and Townsend P, *Phys. Rev. D* **79**, 124042 (2009).
[5] Clement G, *Class. Quant. Grav.* **26**, 105015 (2009); Oliva J, Tempo D and Troncoso R, *JHEP* **0907**, 011 (2009); Giribet G, Oliva J, Tempo D and Troncoso R, *Phys. Rev. D* **80**, 124046 (2009); Ghodsi A and Moghadassi M, *Phys. Lett. B* **695**, 359 (2011).
[6] Ayón-Beato E, Garbarz A, Giribet G and Hassaïne M, *Phys. Rev. D* **80**, 104029 (2009).
[7] Kokkotas K D and Schmidt B G, *Liv. Rev. Rel.* **2**, 2 (1999).
[8] Nollert H P, *Class. Quant. Grav.* **16**, R159-R216 (1999).
[9] Berti E, Cardoso V and Starinets A O, *Class. Quant. Grav.* **26**, 163001 (2009).
[10] Kachru S, Liu X and Mulligan M, *Phys. Rev. D* **78**, 106005 (2008).
[11] Azeyanagi T, Li W and Takayanagi T, *JHEP* **0906**, 084 (2009).
[12] Bekenstein J D, *Lett. Nuovo Cim.* **11**, 467 (1974).
[13] Maggiore M, *Phys. Rev. Lett.* **100**, 141301 (2008).
[14] Myung Y S and Moon T, *Phys. Rev. D* **86**, 064017 (2012);
[15] Lepe S, Lorca J, Pena F and Vasquez Y, 2012, *Preprint* arXiv:1205.4460.
[16] Horowitz G T and Hubeny V E, *Phys. Rev. D* **62**, 024027 (2000).
[17] Abdalla E, Pellicer C E, De Oliveira J and Pavan A B, *Phys. Rev. D* **82**, 124033 (2010).
[18] Morgan J, Cardoso V, Miranda A S, Molina C and Zanchin V T, *JHEP* **0909**, 117 (2009).
[19] Myung Y S, Kim Y W and Park Y J, *Eur. Phys. J. C* **70**, 335-340 (2010).
[20] Cai R G, Liu Y and Sun Y W, *JHEP* **0910**, 080 (2009); Hohm O and Tonni E, *JHEP* **1004**, 093 (2010); Gonzalez H, Tempo D and Troncoso R, *JHEP* **11**, 066 (2011).
[21] Abdalla E, De Oliveira J, Lima-Santos A and Pavan A B, *Phys. Lett. B* **709**, 276-279 (2012).

AUTHOR INDEX

Apostolatos, T A ... 1
Basilakos, S ... 110
Chatziioannou, K ... 1
Christodoulakis, T ... 6
Contopoulos, G ... 30
Contopoulos, I ... 15
Cuadros-Melgar, B ... 286
De Oliveira, J ... 286
Dimakis, N ... 6
Doneva, D D ... 195
Doulis, G ... 6
Gaertig, E ... 102
Grammenos, Th ... 6
Iida, K ... 189
Iosif, P ... 25
Kishi, R ... 47
Kokkotas, K D ... 102, 195
Lukes-Gerakopoulos, G ... 30
Melas, E ... 6
Miritzis, J ... 40
Murakami, H ... 47
Murata, J ... 47, 203
Nakaya, Y ... 203
Nakazato, K ... 189
Ninomiya, K ... 47, 203
Nishio, H ... 47, 203
Nungesser, E ... 57
Ogawa, N ... 47
Onishi, J ... 203
Openshaw, R ... 203
Otoya, V J V ... 124
Oyamatsu, K ... 189
Paliathanasis, A ... 68
Papadopoulos, D B ... 15
Papakostas, T ... 84
Pappas, G ... 1
Passamonti, A ... 102
Pearson, M ... 203
Pellicer, C E ... 286
Pliakis, D ... 84
Pouri, A ... 110
Skákala, J ... 118
Sobreiro, R F ... 124
Solà, J ... 141
Sotani, H ... 189
Sotiriou, T P ... 263
Spanou, A ... 6
Stefanov, I Z ... 195
Stergioulas, N ... 25
Taketani, A ... 47
Tanaka, S ... 203
Tartaglia, A ... 210
Terzis, P A ... 6
Todorov, M D ... 195
Tomaz, A A ... 124
Totsuka, Y ... 203
Tsamparlis, M ... 219
Vacaru, S I ... 234
Vernieri, D ... 263
Wallden, P ... 267
Yazadjiev, S S ... 195
Zampeli, A ... 279